Plumbing 201

PHCC Educational Foundation Plumbing Apprentice & Journeyman Training Committee

SIXTH EDITION

Cengage

Australia • Brazil • Canada • Mexico • Singapore • United Kingdom • United States

Cengage

Plumbing 201, **Sixth Edition**
Plumbing-Heating-Cooling-Contractors—National Association Educational Foundation

SVP, GM Skills & Global Product Management: Dawn Gerrain

Product Director: Matthew Seeley

Product Manager: Vanessa Myers

Senior Director, Development: Marah Bellegarde

Senior Product Development Manager: Larry Main

Senior Content Developer: Meaghan Tomaso

Product Assistant: Andrew Ouimet

Vice President, Marketing Services: Jennifer Ann Baker

Marketing Manager: Scott Chrysler

Senior Production Director: Wendy Troeger

Production Director: Andrew Crouth

Senior Content Project Manager: Glenn Castle

Senior Art Director: Benjamin Gleeksman

Cover Image(s): © nikkytok/veer.com
© totalpics/veer.com

WCN: 01-100-496

Library of Congress Control Number: 2015939529

ISBN: 978-1-305-40164-8

Cengage
5191 Natorp Boulevard
Mason, OH 45040
USA

Cengage is a leading provider of customized learning solutions. Our employees reside in nearly 40 different countries and serve digital learners in 165 countries around the world. Find your local representative at: **www.cengage.com.**

To learn more about Cengage platforms and services, register or access your online learning solution, or purchase materials for your course, visit **www.cengage.com.**

Notice to the Reader

Publisher does not warrant or guarantee any of the products described herein or perform any independent analysis in connection with any of the product information contained herein. Publisher does not assume, and expressly disclaims, any obligation to obtain and include information other than that provided to it by the manufacturer. The reader is expressly warned to consider and adopt all safety precautions that might be indicated by the activities described herein and to avoid all potential hazards. By following the instructions contained herein, the reader willingly assumes all risks in connection with such instructions. The publisher makes no representations or warranties of any kind, including but not limited to, the warranties of fitness for particular purpose or merchantability, nor are any such representations implied with respect to the material set forth herein, and the publisher takes no responsibility with respect to such material. The publisher shall not be liable for any special, consequential, or exemplary damages resulting, in whole or part, from the readers' use of, or reliance upon, this material.

Printed in the United States of America
Print Number: 27 Print Year: 2026

Table of Contents

About PHCC and the PHCC Educational Foundation xxi

Acknowledgments xxiii

About the Author xxv

Preface xxvii

1 COMMUNICATION SKILLS AND PROFESSIONALISM 2

CUSTOMER EXPECTATIONS 3

PRODUCTIVITY 5

ORGANIZATION 5

SORT 5

STRAIGHTEN 6

SHINE 6

STANDARDIZE 6

SUSTAIN 6

REVIEW QUESTIONS 7

2 PIPE JOINING TECHNIQUES 8

HISTORY OF PLASTIC MATERIALS 9

CHEMICAL, MECHANICAL, AND ELECTRICAL BACKGROUND 9

Flammability and Electrical Properties 10

PLASTIC PIPE AND SIZING TERMS 10

ADVANTAGES OF PLASTIC PIPING AS COMPARED TO METALLIC PIPING 13

AVAILABLE THERMOPLASTIC PIPING MATERIALS 15

Rigid and Flexible Pipes 15
Nonpressure Applications 15
Pressure Applications 16
JOINING PLASTIC PIPE 17
Solvent Cement Joining 17
Elastomeric Sealing Gasket 18
Mechanical Fitting Joints 19
Flanges 19
Heat Fusion Joining 20
Threaded Joints 20
PROPERTIES OF PLASTIC PIPING 20
Physical Characteristics 20
PIPING STANDARDS: DIMENSIONS AND TERMINOLOGY 21
IPS—Schedules 40, 80, and 120 21
DR, SDR, and SIDR 22
Pressure Piping 23
Nonpressure Piping 24
Temperature Effects 24
Expansion and Contraction 24
Chemical Resistance 25
POTABLE WATER PIPING 25
PLASTIC DWV AND SEWER PIPING 25
PROCESS PIPING 26
PRESSURIZED GASES 26
Compressed Air Piping Systems 26
ACRYLONITRILE-BUTADIENE-STYRENE (ABS) PLASTIC PIPING 26
Ultraviolet Radiation 28
Uses 28
Compressed Gas Use to Be Avoided 28
Code Status 29
Installed Locations 29
Installation Methods 29
Testing 30
Expansion and Contraction 30
Joining 30
Marking 32

COPPER 33

- Types of Copper Tube and Use 33
- Installation Methods 34
- Support 34
- Testing 34
- Expansion and Contraction 34
- Joining 34
- Soldering and Brazing 35

HUBLESS CAST IRON SOIL PIPE 40

- Support 40
- Testing 40
- Expansion and Contraction 41
- Cutting and Joining 41

HUBBED CAST IRON SOIL PIPE 41

- Cutting 41
- Caulked Joints 42
- Compression Joints 43

CHLORINATED POLYVINYL CHLORIDE (CPVC) PLASTIC PIPING 44

- Storage and Handling 45
- Uses 45
- Code Status 45
- Availability 46
- Installed Locations 46
- Installation Methods 46
- Thermal Expansion and Contraction 48
- Joining Methods 48
- Marking 50
- When Installing CPVC Piping Systems 51

POLYBUTYLENE (PB) PLASTIC PIPING 52

POLYETHYLENE (PE) PLASTIC PIPING 53

- Ultraviolet Radiation 54
- Uses 55
- Code Status 55
- Availability 55
- Installed Locations 56
- Installation Methods 56
- Expansion and Contraction 57

Joining Methods 57
Marking 60
CROSS-LINKED POLYETHYLENE (PEX) PLASTIC PIPING 60
Uses 61
Code Status 61
Storage and Handling 61
Installation Locations 61
Installation Methods 62
Tubing Support 62
Connection to Fixtures 63
Sleeving 63
Bending 63
Expansion and Contraction 63
Manifold Plumbing Systems 64
Joining Methods 64
POLYVINYL CHLORIDE (PVC) PLASTIC PIPING 64
Storage and Handling 66
Temperature Limitations 66
Ultraviolet Radiation 66
Uses 67
Compressed Gas Use 67
Code Status 67
Availability 68
Installed Locations 68
Installation Methods 68
Testing Piping Installations 70
Expansion and Contraction 70
Joining Methods 71
Marking 74
When Installing PVC Piping Systems 74
STEEL PIPE 76
Code Status 76
Support 76
Testing 76
Expansion and Contraction 76
Joining 76
CORRUGATED STAINLESS STEEL TUBING 78
Gas Piping 78
REVIEW QUESTIONS 79

3 SPECIAL APPLICATIONS OF PLASTIC PIPE AND FITTINGS 80

PLASTIC PIPING IN FIRE-RESISTIVE CONSTRUCTION 81
- Plastics and Combustion 81
- Penetration of Rated Assemblies 83

MASTER COMPARISON OF THERMOPLASTIC PIPING MATERIALS 84
- Plastic Pipe Selection 84
- Typical Applications 85
- Industrial Special Applications 85
- Heating and Cooling Applications 85
- Sizing Plastic Piping for Plumbing Systems 86

DETERMINATION OF LOADS IN PLUMBING SYSTEMS 86
- Water Supply Fixture Units (WSFU) 87
- Drainage Fixture Units (DFU) 87
- Hunter's Curve 87
- Maximum Probable Demand 88
- Water Pipe Sizing 88
- Drain Pipe Sizing 88
- Vent Pipe Sizing 88
- Storm Drainage 90

VENT TERMINALS 91
- Air Admittance Valves 91

PLASTIC PIPING IN NONPOTABLE WATER SYSTEMS 93
- Radiant Heat 94
- Snow Melting Systems 95
- Geothermal Ground Source Systems 95
- Ice Rinks 96
- Freezer Warehouses 96
- Water Source Heat Pumps 96
- Gas Piping 96
- Under Slab Warm Air Duct 97
- Chilled Water Cooling Systems 97

CORROSIVE FUME HANDLING SYSTEMS 97

INDUSTRIAL APPLICATIONS 98

CHEMICAL RESISTANCE 98
- Plumbing Piping Systems 98
- Other Piping Systems 99

REVIEW QUESTIONS 99

4 INTRODUCTION TO WELDING 100

HISTORY OF WELDING 101
PRACTICE OF WELDING 101
WELDING STANDARDS 102
GAS WELDING EQUIPMENT AND SAFETY 102
- Welding Gases 102
- Safe Handling of Welding Cylinders 104

GAS WELDING EQUIPMENT 104
- Pressure Regulator 104
- Gauges 105
- Hoses 105
- Torches 106

FILLERS AND FLUXES FOR SOLDERING AND BRAZING 109
WELDING RODS 110
SAFETY AND SAFETY EQUIPMENT 110
- Fire and Burn Hazards 110
- Fire Watch 112
- Asphyxiation 113
- Eye and Face Protection 113
- Safety Procedures 114

REVIEW QUESTIONS 115

5 SOLDERING, BRAZING, CUTTING, AND GAS WELDING 116

DETERMINING THE APPROPRIATE JOINING METHOD 117
- Soldering Defined 117
- Brazing Defined 117

SOLDERING 118
- Approved Solders and Fluxes 118

BRAZING 118
- Brazing Technique 119

WELDING 119
FLAME CUTTING 119
- Cutting Procedure 120

PLASMA ARC CUTTING 122
OXY-ACETYLENE FLAT WELDING 122
PERSONAL PROTECTION 123
REVIEW QUESTIONS 123

6 PIPING MATERIALS, SOURCES, AND DISTRIBUTION FOR POTABLE WATER 124

STANDARDS 125
INSTALLATION CONSIDERATIONS 126
MATERIAL CHARACTERISTICS 128
- Asbestos-Cement Pipe 128
- Plastic Pipe 128
- Glass Pipe 131
- Metallic Pipe 132
- Summary 133

SOURCES AND TREATMENT OF WATER 133
- Individual or Private Systems 134
- Public/Private Systems 135

WATER TREATMENT 135
- Suspended Solids 135
- pH Value 135
- Carbon Dioxide (CO_2) Content 136
- Dissolved Air Content 136
- Hardness 136
- Bacterial Content and Effects of Water Temperature 136
- Flushing and Sterilization 136

PUBLIC WATER DISTRIBUTION SYSTEMS 138
- Distribution 138
- Distribution Layout 138
- Components and Operation of a Public Water System 139
- Water Main Installations 139
- Pressurization Methods 140
- System Water Pressure Range 141
- Building Services 142
- Pressure-Reducing Valves 144
- Strainers 146
- Hose Bibb 147

PRIVATE WATER DISTRIBUTION SYSTEMS 147
NONPOTABLE WATER DISTRIBUTION SYSTEMS 148
SUMMARY 148
REVIEW QUESTIONS 149

7 WATER PIPE SIZING: MAIN-AND-BRANCH AND INDIVIDUAL RUN SYSTEMS 150

THE GOAL OF BUILDING PIPING DESIGN 151

WATER PRESSURE 151

Maintaining Water Pressure in the Community 153

Considerations of Flow in Public Water Distribution Systems 154

Considerations of Flow in Building Systems 154

Pressure, Velocity, Friction, and Flow 155

DISTRIBUTION PIPING SIZED BY AVAILABLE PRESSURE 157

WATER SUPPLY FIXTURE UNITS 159

Sizing Based on Velocity Limitations and Pressure Drop 167

INDIVIDUAL RUN SYSTEMS 170

Typical Materials 172

Sizing 173

Operating Advantages 174

System Disadvantages 174

Cost Considerations 174

Other Uses 175

PRESSURE BOOSTING SYSTEMS 176

Hydropneumatic Tank 176

Gravity Tank 176

Booster Pump 176

REVIEW QUESTIONS 179

8 WATER HEATERS AND HOT WATER DISTRIBUTION 180

DEFINITIONS 181

WATER HEATER TYPES 181

Direct-Fired 182

Indirect-Fired 184

Scaling 185

Circulation of Hot Water 185

Water Heater Controls and Markings 185

MIXED WATER TEMPERATURE CONTROL 185

WATER HEATER REPLACEMENT 187

SAFETY PRECAUTIONS 189

GAS-FIRED WATER HEATERS: PRIOR TO JANUARY 1, 2004 190

Safe Lighting Interlock and Flame-Failure Protection 190
Thermostat 190
Energy Cutoff 190
Thermocouple 190
Gas Control Replacement 190
GAS-FIRED WATER HEATERS MANUFACTURED AFTER JANUARY 1, 2004 193
ELECTRIC ELEMENTS 194
Heating Element Failures 195
Electric Heater Thermostats 195
Relief Valves 196
OIL BURNERS 196
NAMEPLATE INFORMATION 197
INSTANTANEOUS WATER HEATERS 197
Direct-Fired 197
Indirect-Fired 199
Controls and Accessories 199
RESIDENTIAL SIZING 201
Storage Tank Type 201
Tankless Type 202
Increasing Capacity 203
Circulating Systems 203
General Considerations 204
Tank-Type Water Heaters 204
HYDRONIC HEATING SYSTEMS 205
Overview 205
Modes of Heat Transfer 206
System Sizing 207
Heating Terminals 208
Components 210
REVIEW QUESTIONS 216

9 CROSS-CONNECTION PROTECTION AND PIPE IDENTIFICATION 218

BACKFLOW PREVENTION 219
Gravity Backflow 220
Back-Pressure Backflow 220
Back-Siphonage Backflow 220

DEGREE OF HAZARD 221
Toxic Substance 221
Nontoxic Substance 221
Polluted 221
Contaminated 221
CROSS-CONNECTION PROTECTION 221
Air Gap 221
Reduced Pressure Zone Backflow Preventer 223
Double Check Valve Assembly 223
Pressure-Type Vacuum Breaker 224
Atmospheric-Type Vacuum Breaker 225
PIPE IDENTIFICATION 227
REVIEW QUESTIONS 227

10 SEWERS & SEWAGE DISPOSAL METHODS 228
SEWAGE DISPOSAL 229
PUBLIC SYSTEMS 230
PRIVATE SYSTEMS 231
Septic Tank Detail 232
Distribution Box 233
Disposal Fields 234
Dosing Siphon and Dosing Tank 238
PRIVATE SYSTEM MAINTENANCE 239
BUILDING SEWER 240
MATERIALS 240
INSTALLATION METHODS 243
General Rules 243
Cleanouts 245
Thrust Blocks 247
REVIEW QUESTIONS 249

11 DRAINAGE FIXTURE UNITS 250
DRAINAGE FIXTURE UNITS 251
Offsets 256
Summary of Sizing Methods 257
REVIEW QUESTIONS 262

12 **BUILDING DRAINS 264**
DEFINITION OF BUILDING DRAIN 265
ELEVATION 265
MATERIALS 266
FITTINGS 266
BACKFLOW PROTECTION 267
SIZING 268
COMBINED DRAINS 268
FLOW RATE 269
GREASY WASTES 271
Grease Traps 271
FLOOR DRAINS 272
PLANNING THE BUILDING DRAIN INSTALLATION 273
FINAL CONSIDERATIONS 274
HYDRAULIC JUMP 275
CLEANOUTS 277
INSTALLATION OF BUILDING DRAINS 277
Support 277
PROTECTION 279
REVIEW QUESTIONS 279

13 **STACKS 280**
STACKS 281
MATERIALS 281
LOADING AND CAPACITY TABLE 282
TERMINAL VELOCITY 283
VENTING 283
STACK SUPPORT 283
SLEEVES AND THIMBLES 284
FITTINGS AND SUPPORT 285
CLEANOUTS 285
FULL-SIZE STACKS 285
Typical Branches 286
Branch Loads 286
STACK FITTINGS 287
STACK ROUTE 288

ROOF PENETRATIONS 291
TESTING STACKS 293
REVIEW QUESTIONS 293

14 APPLIED PLUMBING MATHEMATICS 294

FRACTIONS 295
  Improper Fractions 295
  Adding or Subtracting Fractions 296
  Multiplying Fractions 296
  Dividing Fractions 296
  Mixed Numbers 297
DECIMALS 297
  Decimal to Fraction 297
  Fraction to Decimal 297
SYSTEM OF MEASUREMENT 297
LINEAR MEASURE 298
  Review 304
SQUARES AND SQUARE ROOTS 304
CUBES 307
CUBE ROOTS 308
REVIEW QUESTIONS 309

15 APPLIED GEOMETRIC CONCEPTS 310

OFFSETS 311
OFFSETS WITH OTHER ANGLES 313
CIRCLE 317
SQUARE 318
RECTANGLE 319
TRIANGLE 320
AREA CALCULATIONS 321
  Area of a Circle 321
  Area of a Square 322
  Area of a Rectangle 323
  Area of a Triangle 324
VOLUME CALCULATIONS 324
  Volume of a Cylinder 324
  Volume of a Rectangular Prism 325
  Volume of a Cube 326

Volume of a Triangular Prism 326
Degree of Precision with Excavation Problems 327
LEAD AND OAKUM 327
REVIEW QUESTIONS 330

16 PROPERTIES OF WATER AND AIR 332
WATER 333
Solid Form 333
Liquid Form 334
Gas Form 334
Water as a Solvent 335
Potable Supplies 335
Compressibility 335
Mass and Force 335
Water Pressure 336
Pressure Independent of Shape of Container 339
HYDRAULICS 340
AIR 341
Pneumatics 342
Drain Lines 342
AIR IN HEATING LINES AND WATER LINES 343
Compressed Air 343
SAFETY ISSUES WHEN TESTING WATER PIPING SYSTEMS 345
MANOMETERS 346
BOURDON TUBE 347
Gauge Selection and Use 348
INSTALLATION TESTS 349
Rough-in Pressure Test 349
NOISES IN WATER SYSTEMS 351
Vibration 351
Turbulence 352
Water Hammer 352
Locating Shock Arrestors 355
Sizing Water Hammer Arrestors 356
MECHANICAL SHOCK ARRESTOR SIZING 358
REVIEW QUESTIONS 358

17 VENTING PLUMBING DRAINAGE SYSTEMS 360

PLUMBING VENT SYSTEMS 361
VENT PIPING MATERIALS 364
- Above Ground 364
- In Ground 364
- Chemical Piping 364
- Installation Practices 364

VENT SIZING 366
EXCEPTIONS AND SPECIAL CASES 371
TOTAL AREA OF VENT PIPES 377
- Economical Sizing 377

VENT HANGERS 377
- Cast Iron Soil Pipe 379
- Steel Threaded Pipe 380
- Copper Tube 380
- Plastic 380
- Stainless Steel 380
- General Considerations 380

VENT TYPES 381
- Back Vent 381
- Branch Vent 381
- Circuit Vent 381
- Combination Waste and Vent 382
- Common Vent or Dual Vent 383
- Continuous Vent 383
- Crown Vent (Prohibited) 383
- Individual Vent 383
- Local Vent 384
- Loop Vent 384
- Main Vent 384
- Relief Vent 384
- Side Vent 384
- Soil Vent or Stack Vent 385
- Vent Stack 386
- Wet Vent 386
- Yoke Vent 387

TRAP ARM SLOPE 387
SEPARATE VENTS REQUIRED 387

SUDS 388
ENGINEERED SYSTEMS 388
Roof Terminals 389
REVIEW QUESTIONS 390

18 SUMP PUMPS, SEWAGE PUMPS, AND SEWAGE EJECTORS 392

BUILDING SUBDRAIN SYSTEMS 393
SUMP PUMPS 393
Sump Pump Configurations 395
SEWAGE PUMPS 398
SEWAGE EJECTORS 399
DIFFERENCES AND DISTINCTIONS 400
Sump Pump Installation 400
Connections and Controls 400
Sewage Pump Installation 401
Sewage Ejector Installation 401
REVIEW QUESTIONS 401

19 PLUMBING TRAPS 402

PLUMBING TRAPS 403
Trap Requirements 404
Permitted and Prohibited Traps 405
Building Traps 407
P-Trap 407
LOSS OF TRAP SEAL 408
Siphonage 408
Aspiration 410
Momentum 410
Oscillation 410
Back-Pressure 410
Evaporation 411
Capillary Action 411
Cracked Pipes 411
TRAP PRIMER 411
TRAP INSTALLATION 412
Connections Permitted to Traps 412
REVIEW QUESTIONS 413

20 FIXTURE SUPPORTS 414

FASTENERS 415
- Strength Requirements 415
- Ease of Assembly and Disassembly 416

FIXTURES 416
- Water Closet 416
- Urinal 418
- Lavatory and Service Sinks 419
- Shower Pan 420
- Handrails and Grab Bars 422

REVIEW QUESTIONS 423

21 HOISTING AND RIGGING 424

CRANES 425

HOISTS 425

BLOCK AND TACKLE 426
- Center of Gravity 426

RIGGING CONCEPTS 428

RIGGING METHOD 429
- Slings 429

KNOTS AND HITCHES 431
- Knots 431
- Hitches 432
- Grasping Loads with Hitches 433

REVIEW QUESTIONS 437

22 SAFETY CONCEPTS 438

PREVENTING INJURY AND LOSS 439

SAFETY 439

SAFETY GUIDELINES FOR HOISTS AND CRANES 440

FORKLIFT SAFETY 441

SAFE RIGGING PRACTICES 442

LADDERS 442
- Straight Ladders 442
- Extension Ladders 443
- Step Ladders 443

OSHA GUIDELINES FOR LADDERS 444

Loads 444
Angle 444
Rungs 444
Slipping 445
Other Requirements 445
Proper Selection of Ladders 446
Inspection Before Each Use 446
Ladder Accidents 446
Proper Ladder Setup and Use 447
Proper Ladder Care and Storage 447
SCAFFOLDS 447
Pre-Manufactured (Bridgework) 447
Accessories 448
Guidelines for Safe Use of Scaffolds 448
Scaffold Hazards 449
Scissor Lifts and Power Scaffolds 449
Summary 450
TRENCHING AND SHORING 450
Cave-Ins 451
Materials and Tools Falling into the Ditch 455
Excavated Debris Falling into the Trench 455
Machinery Falling into the Excavation 455
Debris Falling from Working Machinery 455
Machinery Hitting Power Lines and Similar Utilities 455
Passersby Falling into an Open Trench 456
PREPARING FOR EXCAVATION WORK 456
CONFINED SPACES 458
REVIEW QUESTIONS 459

23 ROUGH-IN SHEETS, SINGLE LINE DRAWINGS, AND ISOMETRIC DRAWINGS 460

SCALE RULER REVIEW 461
ROUGH-IN SHEETS 461
INSTALLATION TOLERANCES 465
CREATING DRAWINGS AND DIAGRAMS 472
Single Line Drawings 472
ISOMETRIC VIEWS 476
COMMERCIAL PROJECTS 479

INDUSTRIAL PIPING LAYOUTS 495
INSTITUTIONAL INSTALLATIONS 503
REVIEW QUESTIONS 508

24 DETAILS, SECTIONS, AND EXPLODED VIEW DRAWINGS 510

DRAWING PROJECTION REVIEW 511
SECTION VIEWS 514
- Section Types 517
- Cutting Plane 518
- Cross-Hatching 518

EXPLODED VIEWS 521
REVIEW QUESTIONS 521

APPENDIX A—OXY-ACETYLENE WELDING, SHIELDED METAL-ARC WELDING, AND WELD TESTING 522

APPENDIX B—ANSWERS TO REVIEW QUESTIONS 557

APPENDIX C—REFERENCES 565

APPENDIX D—CONTRIBUTORS 567

Index 569

About PHCC and the PHCC Educational Foundation

PHCC EDUCATIONAL FOUNDATION

The Plumbing-Heating-Cooling Contractors—National Association Educational Foundation was incorporated in 1986 as a 501(c)(3) tax-exempt, charitable organization. The purpose of the Foundation is to help shape the industry's future by developing and delivering educational programs that positively impact every aspect of the plumbing-heating-cooling contractor's business. Major programs include apprentice, journeyman, and business management training.

The Foundation relies on contributions from many sources. Gifts to the Foundation Endowment Fund are noninvadable, ensuring the Foundation's financial integrity for generations to come. Donors include contractors, manufacturers, suppliers, and other industry leaders who are committed to preparing plumbing-heating-cooling contractors and their employees to meet the challenges of a constantly changing marketplace.

The PHCC Educational Foundation mission statement sets forth this vision:

> *The Plumbing-Heating-Cooling Contractors–National Association Educational Foundation engages in workforce development, management education, and technical training activities for the growth and career success of those employed in the plumbing- heating-cooling industry.*

PHCC—NATIONAL ASSOCIATION

The Plumbing-Heating-Cooling Contractors—National Association (PHCC) has been an advocate for plumbing, heating, and cooling contractors since 1883. As the oldest trade organization in the construction industry, approximately 4,000 member companies nationwide put their faith in the Association's efforts to lobby local, state, and federal government; provide forums for networking and educational programs; and deliver the highest quality of products and services.

PHCC's mission statement is the guiding principle of the Association:

The Plumbing-Heating-Cooling Contractors—National Association is dedicated to the promotion, advancement, education, and training of the industry for the protection of the environment, and the health, safety, and comfort of society.

A complete account of PHCC's history is included in the PHCC Educational Foundation publication *A Heritage Unique*.

Acknowledgments

The following subject matter experts provided their time and expertise to the writing of this book:

Charles L. Chalk, *Maryland PHCC, Ellicott City, Md.*

Laurie C. Crigler, *L & D Associates, Inc., Aroda, Va.*

Dave Jones, *Dave Jones, Inc., Madison, Wisc.*

Michael J. Kastner, Jr., *Kastner Plumbing & Heating, West Friendship, Md.*

Richard R. Kerzetski, *Universal Plumbing & Heating Co., Las Vegas, Nev.*

Robert Muller, *John J. Muller Plumbing & Heating, Matawan, N.J.*

Vic Poma, *Universal Plumbing & Heating Co., Las Vegas, Nev.*

Larry Rothman, *Roto Rooter Services Company, Cincinnati, Ohio*

About the Author

Edward T. Moore is the author of the *Residential Construction Academy Plumbing* video set and *Plumbing 101*, *Plumbing 201,* and *Plumbing 301*. He holds a Bachelor of Science degree in Mechanical Engineering Technology and a master's degree in Technology Systems with a concentration in Manufacturing Engineering. Ed is currently the Department Manager and an Instructor for the Building Construction Trades and Air Conditioning Program at York Technical College in Rock Hill, S.C. He has also served as an instructor for the Industrial Maintenance and Welding programs as well. A licensed Master Plumber for South Carolina, Ed is also the owner of Moore Plumbing and Cabinetry. He has earned the NATE certification in Air Conditioning and Heat Pumps. Ed was previously certified as a HERS (Home Energy Rating Services) instructor through RESNET.

Ed currently resides in Clover, S.C., with his wife and three children.

Preface

This is the second in a series of four plumbing textbooks suitable for training in an industry or academic setting. The content has been selected and organized to suit students and trainees who have had at least 1 year of formal or on-the-job training. The subject matter progresses on the assumption that students are applying what they learn in a school shop or field environment.

In addition to the new full-color format, the Sixth Edition has been updated to reflect the latest methods, components, codes, and standards. Many illustrations have been added to help readers grasp concepts more completely. The examples have been reworked to show a more logical, step-by-step process. The section on welding has been vastly improved and updated. Finally, there are new field applications and revised safety information.

The curriculum that supports this book and the content itself have been developed and approved by the PHCC Educational Foundation Plumbing Apprentice & Journeyman Training Committee. The committee has endeavored to respond to industry demands for training material best suited to a field-oriented program. It is the combination of classroom and field experience that the committee wishes to promote throughout the industry.

Chapter 1

LEARNING OBJECTIVES

The student will:

- Explain the key factors concerning interaction with the customer and coworkers.
- Discuss the traits of a good listener.
- Explain the concept of 5 S.

Communication Skills and Professionalism

CUSTOMER EXPECTATIONS

In Chapter 1 of *Plumbing 101*, eleven employer/customer expectations were presented so that soft skills could be developed as well as technical skills. Below are some examples of how those same expectations apply to dealing with customers. In many cases, the plumbing technician is the first person that customers meet and will be the basis that they will use to form their opinion of the company. As you go through each of these items, think of how you would feel if you were the customer and remember: you get only one chance to make a first impression.

1. **Arrive to the job on time.** Just as you are expected to show up to work on time, you should show up to the customer on time. In many cases, customers may have had to leave work early or take a day off so that they could meet with you. In the case of a service or repair call, customers could be upset with the problems they are having and not showing up on time will only make matters worse. If you are going to be late, you should call ahead and let the customer know you are going to be late and ask if the customer would like the service to be rescheduled. This shows customers that you value their time. This also forces you as a technician to manage your time and plan ahead.
2. **Drive the company vehicle more carefully than your own.** When you are a service technician, the work truck is your office, your warehouse, and sometimes your break room. You should keep it clean and well organized. Also, when you arrive to the jobsite, you should always consider where you park the vehicle. In the case of a residential jobsite, it would not be polite to take up space in the driveway or to block in the customer. Once you meet with the customer, you should ask if they prefer you to park the vehicle somewhere in particular and if you must drive into the yard, consult with them as to why and what possible route they would prefer. In the case of a commercial jobsite, you should talk to the project or site manager and see if there are some designated parking spaces. When there are high traffic areas and heavy equipment, it is easy to hold up production or cause an accident by parking improperly. In highly congested areas, parking on the street may result in a ticket, so don't assume it is OK to park in public areas.

3. **Be neat and clean in appearance.** Chances are the service technician will be the first person from the company that the customer will see, and appearances matter. Companies either supply a company uniform or have a dress code for two reasons: safety and professional appearance. When the technician approaches customers, they should easily recognize that the service person is from the company; they might be apprehensive to open their doors otherwise. Some companies have strict grooming standards that reflect their preference toward length of hair, tattoos, or piercings, but it is expected by all companies that you have good personal hygiene. You should also have clean hands when you first greet customers.
4. **Be honest.** As more and more companies are putting service technicians on commission, there is an increasing mistrust from customers. When repairs are needed, present customers with all of the options and let them decide. For example, while your first inclination to fix a leaking ballcock on a 10-year-old water closet is replacing the entire water closet, you should discuss with the customer the options for making the repair. Give the customer the pricing information and point out the pros and cons of each option. This is an excellent time to offer any special services that your company provides, such as home warranties, volume discount on multiple repairs, or maintenance service programs.
5. **Listen.** Learn to become an active listener. When you first meet with customers, don't assume you know all the details. Ask them to explain the problem and what they would like to have done. Take notes and listen patiently; they may be frustrated and need to vent some of their frustration. Wait until customers reach a stopping point before asking any questions and then repeat back what you believe to be the problem. Try not to take any of the comments personally; simply assure customers that you are there to assist them and that you want to help solve the problem.

   When trying to troubleshoot a system problem, it is important to find out details such as the following: how often does the problem occur, under what conditions does the problem occur, and whether anything has been modified recently. It is not uncommon to find that when one part of the system is repaired, another part is affected. For example, aerators are commonly clogged with debris from old pipe corrosion or dirt from earlier pipe repairs. When talking to customers, it is important to use proper terminology, but be prepared to explain in "layman's terms" what you mean. For instance, while *water closet* may be the technical term, *toilet* may be a term with which they are more familiar. If possible, show customers the item or walk them through the steps that you will be performing to do the repairs. This may be a good time to offer them some upgrades to their existing system, such as additional hose bibbs, new fixtures, or a complete remodel or re-piping.
6. **Treat the customer's property better than you treat your own, and keep your work area safe and clean.** Before you begin any work, survey the work site and make sure that none of the surface finishes or any of the customer's belongings could become damaged when performing the work. If there is any damage to the flooring or other parts of the structure before you begin your

work, you may want to point it out to the customer or document it through photographs before beginning any work. Before the work begins is a good time to think about safety. Plan where you intend to store supplies and debris as well as how you plan to transport materials. On large jobsites or new construction, there may be other trades working in the area, so be aware of how you may be interfering with each other. Always make sure you clean up any mess before you leave.

PRODUCTIVITY

You may have heard the expression that time is money; in the construction trades nothing could be more true. In today's market, companies need to work smarter and more efficiently to stay competitive. As a plumbing technician, you should always look at ways to streamline the work. This does not mean to cut corners or to not correct mistakes; this means to plan your work and remove any wasted effort. The best way to achieve this is by preplanning the work.

Before the work begins, prepare a list of steps that need to occur to do the work. Think of how the work should progress and take note of any materials or safety precautions that are needed. Taking a few minutes to prepare can help you arrange the materials and tools in the order they are needed. This prevents the need to move materials multiple times and keeps unnecessary objects out of the work area. Take note of jobsite conditions while performing this planning because it may necessitate the need to work around other trades or store some materials off-site. Many commercial jobsites have limited space or problems with theft, so planning is paramount.

ORGANIZATION

For years, the manufacturing industry has used a method known as "5 S" as a means to increase productivity and efficiently utilize work space. When Japan was rebuilding its manufacturing base after World War II, the need for a clean, clutter-free work space was recognized and the following steps to achieve this goal were adopted. This same philosophy can easily be applied to the construction trades. The name 5 S comes from the original Japanese terms that were used. It is important to note that the steps should occur in the order presented and that it is an ongoing process. As the work changes, there will always be a need to arrange and reorganize your area. It is also important to note that the company vehicle should be organized in the same manner. Just think of all the freed up space and better gas mileage a well-organized vehicle would have.

SORT

The Japanese used the word *seiri*, which means to remove any unnecessary items and dispose of them properly. Unnecessary items get in the way, become trip hazards, and consume time having to sort through them to find the necessary item. Look at each item and ask, "Do I need this for this job?" If the answer is "no" then remove it. This applies to tools as well as supplies. Sometimes, it is more efficient to have two smaller tool bags as opposed to a single big one. Focus on the work truck as well as the jobsite.

STRAIGHTEN

The Japanese used the word *seiton*, which means to straighten or arrange all items so they are in the order they are needed to the work and easily on hand. A tremendous amount of time can be lost looking for tools or parts. Think about how the work will flow and then organize the tools and supplies in that order. Use this concept when bringing supplies to the job—not all supplies are needed at the same time so plan ahead so you don't have to move items more than once. A lot of time and energy is wasted moving material. This step is sometimes called "set in order."

SHINE

The Japanese used the word *seiso*, which means to shine or clean up. Keep your work area and tools in good working order and clean. Debris from packing boxes and scrap material can become a trip hazard and waste time if left scattered on the jobsite. The same is true with broken tools; if a tool is in need of repair, bring it to your supervisor's attention and get a replacement.

STANDARDIZE

The Japanese used the word *seiketsu*, which means to standardize. This is sometimes interpreted to mean only to buy the same brand parts and tools, but it can also mean to use the same methodology to organize your work area. Having the same brand equipment does allow more flexibility such as swapping out batteries on drills and other power tools, but it also removes the need for having to become familiar with different operating procedures. Try to keep your truck or workspace arranged in the same order; it will make it easier to move from jobsite to jobsite or between vehicles. Decide where something should be and keep it in its proper place.

SUSTAIN

The Japanese used the word *shitsuke*, which means to sustain. The steps just discussed should not be a once-a-year task; it should be ongoing. As the jobsite and job functions change the process needs to be repeated. You need to constantly review the process and see if items need to be added or removed. Is there a better way to organize? Work with your teammates and come to a consensus on what is best for your work environment. This may sound very time consuming but it is a great activity for periods of time when the work is slow.

REVIEW QUESTIONS

True or False

1. __________ Calling customers ahead of time to confirm appointments or to let them know you are running late does not cost the company money.
2. __________ Asking customers questions is a sign that the service technician is still learning the trade.
3. __________ Safety should be considered in planning the work ahead.
4. __________ The steps to 5 S are sort, shine, standardize, straighten, and sustain.
5. __________ Time spent organizing a messy jobsite can increase job productivity.

Chapter 2

LEARNING OBJECTIVES

The student will:

- Compare the advantages and disadvantages of thermoplastic and metallic piping materials used in plumbing systems.
- Discuss joining techniques as they apply to various piping types.

Pipe Joining Techniques

The Plastic Piping Educational Foundation commissioned the book that forms the basis of much of this chapter and Chapter 3. Portions of the book have been reprinted with permission.

Plastics can be divided into two general types:

- Thermoplastics, which can be softened by the application of heat.
- Thermosets, which cannot be softened by the application of heat.

HISTORY OF PLASTIC MATERIALS

The plastic industry traces its beginning back to the scarcity of ivory in the manufacturing of billiard balls. John Wesley Hyatt, a chemist, mixed pyroxylin (a derivative of cotton) and nitric acid with camphor to form a product he called celluloid, a thermoset plastic material.

Like so many other technological discoveries, there was little advancement until war created a need for alternative materials. Although some plastics were introduced into piping systems in the 1930s, many of today's plastics—polyvinyl chloride (PVC), polyethylene (PE), and reinforced plastics—were not developed until just before World War II.

In the United States, plastic piping systems obtained wide acceptance in the late 1950s and early 1960s. Since then, plastic pipe usage has increased at an astounding rate.

Water mains, hot and cold water distribution, drain, waste, and vent (DWV), sewer, gas distribution, irrigation, conduit, fire sprinkler, and process piping are the major markets for plastic piping systems throughout the world. Underground piping makes up the largest part of the market.

CHEMICAL, MECHANICAL, AND ELECTRICAL BACKGROUND

A carbon atom's ability to bond with many different elements accounts for the great number of plastic materials available in today's market. Carbon is the common element in polymer chains, but other elements are necessary to obtain the chemical structures of plastics. Hydrogen, oxygen, nitrogen, chlorine, fluorine, and occasionally other elements, such as sulfur and silicon, are used to create the wide range of properties found in plastic piping materials.

The nature of molecules and polymer chains make plastics highly resistant to chemical attack by most products encountered in the home, office, and factory. The organic chemicals in plastic pipe, however, do not withstand the energy in ultraviolet (UV) rays, which is why most manufacturers incorporate a shielding pigment in the plastic formula or warn that the pipe must be protected against UV exposure.

The following compares the mechanical properties of plastic pipe to the metallic pipe also used in the industry:

- Plastic pipe has a higher resistance to corrosion than metallic pipe.
- Plastic pipe has lower strength compared to metallic pipe but still high enough to be suitable for many piping applications.
- Plastic pipe is about ten times more flexible than steel.
- Plastic pipe is softer (easier to scratch or cut as compared to metal piping).
- Plastic pipe is lighter than metals and easier to handle.
- At low temperatures (below 35°F), plastic pipe has lower resistance to impacts than metallic pipe.
- Plastic pipe loses strength and hardness much faster as temperature increases, as compared to metallic pipe.
- Plastic piping's thermal expansion is about eight times higher than that of steel.

Flammability and Electrical Properties

Most, but not all, plastics will burn in the presence of a flame. Some of these will not continue burning if the ignition source is removed. Burning plastics that have been chlorinated, such as chlorinated polyvinyl chloride (CPVC) or polyvinyl chloride (PVC) can release dioxins. Dioxins are strong carcinogens and are extremely hazardous to human health. See Chapter 3 for a more detailed discussion. Unlike metallic pipe, most plastic materials are inherently safe if used in and around electric circuits because of the low electrical conductivity of plastics. However, plastics cannot be used as electrical grounds since they are nonconductors. Since older metallic piping systems were used as grounding circuits in the past, it is important to make sure a suitable replacement ground is installed before replacing it with plastic piping. Several plastics are even used as insulators.

PLASTIC PIPE AND SIZING TERMS

As the new material in the piping industry, plastics adopted many of the metal piping sizing systems and developed some of their own. Pipe sizing systems can be divided into two categories:

- OD—Outside Diameter Controlled
- ID—Inside Diameter Controlled

The joining system to be used depends on whether the pipe is OD or ID controlled. In the OD-controlled system, normally couplings or sockets join to the outer surface of the pipe. The ID-controlled pipe or tube uses insert fittings. For OD-controlled pipe, an outside diameter is established for each nominal pipe size and all wall thickness changes needed for different pressure ratings affect the inside diameter.

IPS (iron pipe size) pipes with schedule walls are OD controlled and have been used in plumbing for many years. For this reason, many of the first plastic pipes were made to the metal pipe IPS Schedule 40 and 80 dimensions (see ASTM D 1527 for ABS, D 1785 for PVC, D 2104 for PE, and F 441 for CPVC).

Another OD-controlled system is the CTS (Copper Tube Size) system in which the outside diameter is the nominal size in inches. Some plastic tubing standards (D 2666 and D 2737) have the term *tubing* in their title. Others (D 2846, D 3309, and F 877) have Hot and Cold Water Distribution Systems in their title.

If you inspect Table 2-1 for any of the plastic pipe IPS Schedule wall standards, you will see that no matter what wall schedule is selected (Sch 40 or 80), each pipe size will have a different pressure rating. It is also clear that as the pipe diameter increases, the pressure rating for that schedule pipe will decrease. To help solve this problem, the plastic piping industry developed the SDR-PR (Standard Dimension Ratio–Pressure Rating) system. In the SDR system, the pipe wall thickness increases as the outside diameter increases. This allows pipes with the same SDR rating to have the same pressure rating, regardless of size. For examples of this, see the pressure rating tables of ASTM D 2241 for PVC, D 2282 for ABS, D 3035 for PE, and F 442 for

TABLE 2-1 ABS Plastic Pipe Pressure Ratings at 73°F (psi)

Pipe	ABS 1316 (D 1527)		PVC 1120 (D 1785)		PE 3408 (D 2447)	
Size (in.)	Sch 40	Sch 80	Sch 40	Sch 80	Sch 40	Sch 80
$\frac{1}{2}$	480	680	600	850	240	340
$\frac{3}{4}$	390	550	480	690	195	275
1	360	500	450	630	180	250
$1\frac{1}{4}$	290	420	370	520	145	210
$1\frac{1}{2}$	260	380	330	470	130	190
2	220	320	280	400	110	160
$2\frac{1}{2}$	240	340	300	420	120	170
3	210	300	260	370	105	150
$3\frac{1}{2}$	190	280	240	350	95	140
4	180	260	220	320	90	130
5	160	230	190	290	80	115
6	140	220	180	280	70	110
8	120	200	160	250	60	100
10	110	190	140	230	55	95
12	110	180	130	230	55	90

CPVC. These are all IPS-OD pipes with SDR-PR wall thicknesses rather than Schedule wall thicknesses. Table 2-2 shows some examples of ASTM SDR pressure ratings and some AWWA DR pressure classes.

While some nonpressure piping is IPS-OD Sch 40 product (see D 2661 for ABS DWV and D 2665 for PVC DWV), there is provision for the PVC Sch 40 pipe to be dual-marked with D 1785 and D 2665. This enables the supply house to carry product that can be used for both pressure and DWV systems. Look for the marking on the pipe. Do not assume that all Sch 40 pipe qualifies for use as pressure pipe. There is Sch 40 (F 628 ABS and F 891 PVC) cellular core pipe that does not carry any pressure rating. Figure 2-1 shows both pressure-rated and nonpressure-rated schedule 40 PVC.

Sewer pipe sizes are also OD controlled, but they have smaller ODs than the IPS pipes. Sewer pipe is made from several materials (ABS, PE, and PVC) over a range of sizes. ASTM D 3034 even includes three different DR wall thicknesses.

In addition to the ASTM standards, there are American Water Works Association (AWWA) standards for both PVC and PE pipe. AWWA C 900 pipe (4 inches through 12 inches) is OD controlled but utilizes the cast iron/ductile iron pressure pipe ODs and assigns different safety factor/pressure ratings, even though they refer to the DR (dimension ratio) wall thickness system. Note the differences in Table 2-2 for PVC pipe made from PVC materials having a 200 psi HDS (hydrostatic design stress).

This comparison shows that even though AWWA DR 14 pipe has heavier walls than SDR 17 pipe, it has a lower pressure class. Therefore, we see that pressure rating and pressure class are not equivalent terms. Both systems use the SDR/DR wall thickness

TABLE 2-2 Pressure Ratings Versus Pressure Class

ASTM SDR and Pressure Ratings		AWWA DR and Pressure Classes	
SDR	Pressure (psi)	DR	Pressure (psi)
26	160	25	100
21	200	18	150
17	250	14	200

FIGURE 2-1 Foam core and solid wall piping look the same, but foam core is lighter and cannot be pressurized.

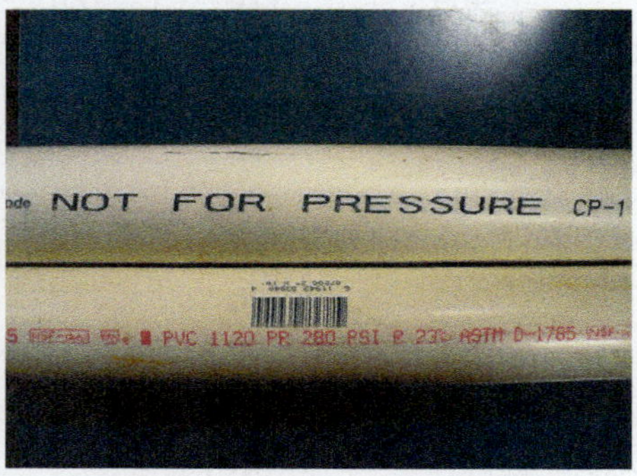

Photo by Ed Moore

approach so that all sizes of DR 18 pipe made from PVC carry a 150 psi pressure class marking.

Connections between ID-controlled pipe and OD-controlled pipe must be done with the proper adapter fittings to ensure that joint leaks do not occur. There are only a few ASTM standards for ID-controlled plastic piping, fully identified in their titles (see the following for examples).

D 2239—Polyethylene (PE) Plastic Pipe (SIDR-PR) Based on Controlled Inside Diameter

D 2662—Polybutylene (PB) Plastic Pipe (SIDR-PR) Based on Controlled Inside Diameter

This chapter is an introductory overview of the most widely used plastic piping products, standards, sizing systems, and terms used to identify any particular piping product.

ADVANTAGES OF PLASTIC PIPING AS COMPARED TO METALLIC PIPING

1. **Corrosion resistance.** Plastics are not conductive and are therefore immune to galvanic or electrolytic erosion. Because plastics are corrosion resistant, plastic pipe can be buried in acidic, alkaline, wet, or dry soils, and protective coatings are not required.
2. **Chemical resistance.** The variety of materials available allows plastic piping to handle a wide range of chemical solutions successfully.
3. **Low thermal conductivity.** All plastic piping materials have low thermal conductance properties. This feature maintains more uniform temperatures when transporting fluids in plastic versus metal piping. Low thermal conductivity of the wall of plastic piping may eliminate or greatly reduce the need for pipe insulation to control condensation on the outside of the pipe.
4. **Flexibility.** In general, thermoplastic piping is relatively flexible as compared to metal piping. This facilitates use of efficient installation techniques. Some of the more flexible materials allow for coiling, which permits long pipe runs with a minimum number of joints. The more rigid materials are typically made into 10- or 20-foot pipe lengths. Pipe size is also a factor in coiling and bending both rigid and flexible materials. Water mains and sewers made of plastic piping can be deflected to match the curve alignment of streets and rights-of-way. To avoid putting excessive strain on pipe fittings or joints when plastic is to be bent or deflected, consult the manufacturer's instructions.
5. **Low friction loss.** Because the interior surface of plastic piping is generally very smooth, less power may be required to transmit fluids in plastic piping compared with other piping systems. Furthermore, the excellent corrosion resistance of plastics means that the low friction loss characteristic will not change over time.
6. **Long-term performance.** Due to the relative chemical inertness and the minimal effects of internal and external corrosion, there is very little change in the physical characteristics of plastic piping over dozens of years. Examinations of pipe samples taken from some systems have shown no measurable degradation after 25 years of service.

7. **Light weight.** Most plastic piping systems weigh approximately one-sixth of the weight of steel piping. This feature means lower costs in many ways: lower freight charges, less manpower, simpler hoisting and rigging equipment, and so on. This characteristic has allowed some unique cost-saving installation procedures in several applications.
8. **Variety of joining methods.** Plastic piping can be joined by numerous methods. For each material, there are several appropriate methods. Some of the most common are solvent cementing, heat fusion, threaded joints, flanges, O-rings, rolled grooves, and mechanical compression joints.
9. **Nontoxic.** Plastic piping systems have been approved for potable water applications. As evidence of this, all plastic potable water piping materials and products are tested and listed for compliance to ANSI/NSF Standards 14 and 61. All ASTM and AWWA standards for plastic pressure piping that could be used for potable water contain a provision whereby the regulatory authority or user can require product that has been tested and found to be in conformance with ANSI/NSF Standard 61—Drinking Water System Components—Health Effects. When plastic pipe or fittings are ANSI/NSF Standard 14 listed and have the NSF-pw (potable water) mark, they also meet the ANSI/NSF Standard 61 requirements. To assure installers, regulators, and users of plastic piping that these products are acceptable for potable water applications, the plastic piping industry has dealt with the issue as follows:

    - All plastic piping material manufacturers have recognized the need to provide materials for all pressure pipes and all pressure pipe fittings that meet the requirements of ANSI/NSF Standard 61.
    - All manufacturers of these pipes and fittings have pressure pipe and fittings listed as meeting the requirements of ANSI/NSF Standard 61. These products can be identified by the NSF-pw, NSF-61, UL Classified/Std 61 mark or a similar pw mark (by another recognized listing agency) printed on the pipe or molded into the fitting. The NSF-pw mark certifies to installers, users, and regulators that the product meets the requirements of ANSI/NSF Std 14 for performance and the ANSI/NSF Std 61 for health effects. Products marked with Std 61 have only been evaluated as meeting the requirements of NSF 61 for health effects.
10. **Biological resistance.** To date, there are no documented reports of any fungal, bacterial, or termite attacks on any plastic piping system. In fact, because of its inertness, plastic piping is the preferred material in deionized and other high-purity water applications.
11. **Abrasion resistance.** Plastic piping materials provide excellent service in handling slurries such as fly ash, bottom ash, and other abrasive solutions. The material toughness and the smooth inner-bore of plastic piping make it ideal for applications where abrasion-resistance is needed.
12. **Colored piping.** Plastic piping is available in a variety of colors. However, do not depend on the pipe color as a factor in determining proper application. Read the printing on the pipe. Table 2-3 lists colors that are generally associated with different applications.
13. **Maintenance.** A properly designed and installed plastic piping system requires very little maintenance because there is no rust, pitting, or scaling to

TABLE 2-3 Plastic Pipe Coloring for Different Applications

Type of Piping	Color
Gas Distribution	Yellow or black with yellow stripes; formerly bright orange or tan
Water Distribution	Black, light blue, white, clear, or gray
Sewers	Green, white, black, or gray
DWV	Black, white, tan, or gray
Hot and Cold Water Distribution	Tan, red, white, blue, silver, or clear
Cable Duct	Variety of colors
Fire Sprinklers	Orange
Industrial Process	Dark gray (PVC) or light gray (CPVC)
Reclaimed Water	Purple or brown (local jurisdiction may set requirements)

contend with. The interior and exterior piping surfaces are not subject to galvanic corrosion or electrolysis. In buried applications, the plastic piping is not generally affected by chemically aggressive soil. However, installation in soils contaminated with hydrocarbons (gas, oil, etc.) should be avoided.

AVAILABLE THERMOPLASTIC PIPING MATERIALS

The variety of applications for plastics piping is quite diverse. Generally, these installations and the prevailing standards can be broken down into two groups: nonpressure systems and pressure systems. A brief discussion of the types of materials and prevailing industry standards for each of these two groups will be presented in the paragraphs that follow.

Rigid and Flexible Pipes

ABS, CPVC, and PVC pipe are generally referred to as *rigid* and are typically available in 10-foot and 20-foot straight lengths.

PE, PE-RT, PEX, and PP materials are referred to as *flexible* and are available in coils of various lengths. In many cases, special coil lengths can be ordered for tubing that is to be placed in radiant heat floor slabs, snow-melting slabs, or geothermal ground loops. These special-order coils assure minimum waste and enable designs in which no joints are made within the slab.

There are exceptions to these general classifications, however. In addition to being sold in coils, PE and PEX pipe are also available in 20- and 40-foot straight lengths. CPVC and PVC pipes are also available in coils in smaller sizes.

Nonpressure Applications

ABS, PVC, PE, and PP plastic pipe materials are used for nonpressure applications. There are separate ASTM standards for each plastic pipe based on material, dimensioning system, application, and sizing. They are listed later in this chapter.

Pressure Applications

Plastic pressure piping is used for many industrial processes, in heating and cooling systems, fire protection installations, gas distribution, and for water supply and distribution. Plastic piping materials that have a pressure rating Hydrostatic Design Basis, in accordance with ASTM D 2837, are published by the Plastics Pipe Institute in TR-4.

Potable water applications include cold water services from wells or water mains up to the building as well as hot and cold water distribution piping within buildings. The piping for hot and cold water distribution systems is also tested to 150 psi at 210°F for at least 48 hours to ensure that the system can withstand the conditions that relate to the pressure and/or temperature of water heater relief valves when operating.

All plastic pressure pipe standards have 73°F pressure ratings and this is printed on the pipe. In addition, certain plastic pipe materials and some of the pipe standards are rated for higher temperatures. Some CPVC, PE-RT, and PEX piping standards have been developed specifically for hot and cold water distribution systems and most plumbing codes allow for use of these products in water distribution systems.

Pressure-rated plastic pipes that have only a 73°F rating are not approved by codes for use in the cold water portions of the hot and cold water distribution systems. The reason for this prohibition is the fact that hot and cold water piping is installed simultaneously. Inspectors need to be certain that a piping rated for use with cold water only is not used by mistake in the hot water portion.

Surge Pressure (Water Hammer)

In dealing with pressure piping, you will find the following terms:

- **Gauge pressure.** The line pressure in the system that will tend to be fairly constant when there is no flow or steady flow.
- **Surge pressure.** Positive or negative change in pressure, which occurs whenever change in flow occurs.
- **Water hammer.** An extreme form of surge pressure.

One of the most common examples of water hammer can be found when line pressure and flow rate are high and a valve in the water supply line closes quickly. Typical examples are solenoid valves on laundry machines or dishwashers. The best protection against water hammer is to size the supply line to solenoid valves (dishwashers and laundry machines) so that the velocity in the line to the valve is not excessive. A second method of protection is to install a water hammer arrestor as close as possible to the quick-operating valve. Although air chambers are sometimes used, they have been shown to lose their effectiveness quickly. This is because the air is in contact with the water and is absorbed over time.

An additional risk of water hammer occurs in the filling and pressurization of an empty system—either the initial fill of a new system or refilling a system that has been drained. This situation is more likely to present problems in large-diameter, extensive systems. The best protection is to slowly fill the system and slowly vent air from high points and from only one point at a time if there is more than one point that develops an air trap.

Surge pressures are of very short duration, and they cannot be measured with standard pressure gauges. System design should provide for line pressure plus a surge allowance. See Chapter 16 for additional information concerning water hammer.

JOINING PLASTIC PIPE

Solvent Cement Joining

ABS, CPVC, and PVC plastic pipes are primarily joined by solvent cementing, but mechanical joints are also available. PE, PEX, and PP pipe cannot be joined with solvent cements.

Solvent cement joining always involves a pipe or tube end and fitting socket or pipe bell. The inside of the socket is slightly tapered from a diameter slightly larger than the pipe OD, at the entry, to a dimension at the root of the socket that is a few thousandths of an inch smaller than the pipe OD. Thus, the pipe-to-socket match-up results in an interference fit more-or-less midway in the socket. See Figure 2-2.

Solvent cement is applied to the outside of the pipe end and the inside of the socket. The pipe is then pushed into the socket until it bottoms. On smaller pipe sizes, it is a good practice to twist the pipe approximately $\frac{1}{4}$ turn as it is inserted into the fitting. This helps spread the glue evenly. Because the fitting is tapered, you should hold the joint together for a few seconds so that it does not push apart. Some codes and some manufacturers require a primer to be applied before the solvent cement. It should be pointed out that if the manufacturer's specifications are not followed, any warranties on the product are voided. Always review local codes and the manufacturer's instructions.

FIGURE 2-2 Solvent joint connects have a tapered ID so that there is an interference fit with the pipe. The solvent cement softens the pipe so this is possible.

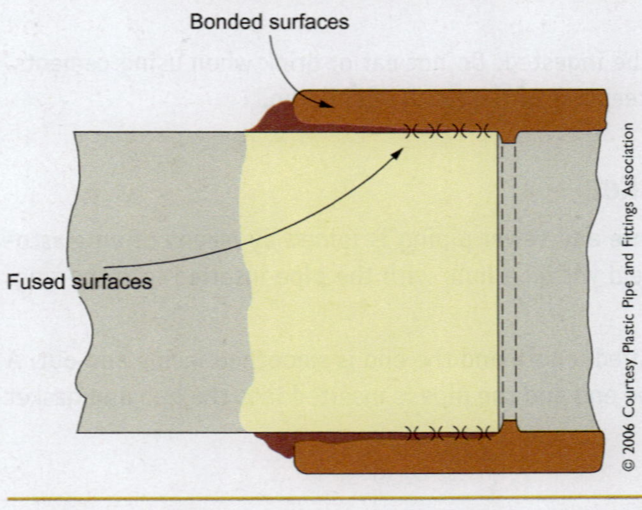

Pipe and fittings are bonded together by means of chemical fusion. Solvents contained in primer and cement soften and dissolve the surfaces to be joined. Once the pipe and fitting are assembled, a chemical weld occurs. This weld strengthens over time as the solvents evaporate.

See ASTM F 402 for safe handling of solvent cements, ASTM D 2855 for PVC instructions, ASTM F 493 for CPVC instructions, ASTM D 2235 for ABS instructions, or the cement manufacturer's instructions printed on the container label for further information.

IN THE FIELD
The fumes from chemical solvents can be very strong and harmful. Always work in a well-ventilated area.

Safe Handling Procedures

These cements, primers, and cleaners contain solvents that are classified as combustible, flammable, or extremely flammable. Keep these products away from all sources of ignition, such as sparks, heat, and open flames. Containers holding these products must be kept tightly closed except when in use.

Threshold limits for worker exposure during an 8-hour workday have been established for each of the solvents used in these products. Those limits are found on the Safety Data Sheets (SDS) for each product. It is very important to hold the air concentration of these solvents below the threshold limits. When using these products in an area of limited ventilation, a ventilating device such as a fan or air mover can be used to maintain a safe air concentration. Also, an air-purifying NIOSH-recognized respirator may be used. Any ventilating device must be selected and located so it cannot provide a source of ignition.

The solvents in these products should not come into contact with bare skin. Use the applicators provided to minimize skin contact. Use protective gloves whenever handling chemicals.

The solvents in these products will cause severe irritation if they come into contact with the eyes. Proper eye protection must be worn whenever working with chemicals. **Don't take chances!**

These cements must not be ingested. Do not eat or drink when using cements, primers, or cleaners. Always keep out of the reach of children.

Elastomeric Sealing Gasket

Most underground PVC pressure and sewer piping is joined by means of an elastomeric O-ring or seal that is held within a hub, with the pipe inserted into the ring. See Figure 2-3.

The pipe is cut to the desired length and the end is smoothed inside and out. A lubricant is applied to the pipe end and the pipe is inserted into the hub and gasket with a quick push.

FIGURE 2-3 Larger plastic sewer pipe relies on an O-ring to seal the joint. Care should be taken not to cut the O-ring when making the joint.

Photo by Ed Moore

Note that two different ASTM standards apply to elastomeric seal gravity joints and pressure joints (D 3212 and D 3139, respectively).

Mechanical Fitting Joints

PE and PEX tubing are often joined by mechanical means using fittings developed for that purpose. Several versions are available—each uses a metal or plastic insert stiffener inside the tube. Always check on the product label or in the manufacturer's literature or instructions to make sure that the mechanical fitting joining system is recommended for the type of service on which the fittings will be used (e.g., potable water, hydronic heating, gas service).

- **Crimp ring type.** A crimp ring surrounding the tube and insert is compressed by a special crimp tool after assembly. The crimp ring version is a one-time assembly—it must be destroyed to disassemble this joint. Some manufacturers' crimping tools must be calibrated periodically and the resultant crimp checked with the accompanying gauge.
- **Nut ferrule type.** A threaded nut is tightened onto a matching thread and compresses the tube or a ferrule over the insert as it is made tight. The threaded nut version can be taken apart and reassembled as necessary. Always use a plastic ferrule with plastic pipe.
- **Stab type.** The plastic pipe or tubing is cut, the end is chamfered, the stab depth is marked on the pipe or tubing, and then it is stabbed into the fitting.

Flanges

CPVC and PVC piping can be joined by bolting together flanges that are attached to the pipe end (usually by solvent cement joint). A thin, flat gasket is placed between the flange faces to make a leak-tight joint. Care must be taken to bolt the flanges together in the manner recommended by the manufacturer to develop a tight joint and avoid damaging the flanges. This method is extremely handy when installing valves that may need to be replaced or repaired.

Heat Fusion Joining

There are four types of heat fusion joints:

- Butt fusion
- Socket fusion
- Electrofusion
- Saddle fusion

Butt fusion joining is accomplished by heating the two ends of pipe or tubing to be joined to the required temperature in special heating devices and then quickly pushing the ends together with a controlled force. Special tools are used to obtain the required temperature and to control the mating force. This method is used on PE, PP, and PVDF pipe and produces a quality joint in these materials.

The socket fusion and saddle fusion processes are similar to butt fusion except that different heating tools are required. In the electrofusion process, the required temperature and heating time are controlled by passing current through an electrical resistance wire embedded in the socket.

Threaded Joints

Some plastic pipe fittings are available that consist of NPT thread on one end and plastic male (spigot) or female (socket) on the other. These fittings are used as adapters to join one piping material to another. Since the stresses in the female portion of the joint are tensile (trying to be pulled apart), use a metal female fitting with a plastic male. That way, the female portion is less likely to crack.

Schedule 80 CPVC and PVC plastic pipe can be threaded if special dies are used. Pipe threads must conform to ASTM F 1498. Threading pipe reduces its working pressure by 50 percent. The rated working pressure of systems assembled with threaded joints will be less than the working pressure with solvent cement joints. An advantage of this method is that disassembly is easily accomplished when necessary.

PTFE (i.e., DuPont Teflon®) tape is compatible and recommended for use with all plastic piping materials. Some paste-type sealants contain ingredients that can damage certain plastic piping system components. Use only paste thread sealants recommended for the specific plastic piping system being used.

PROPERTIES OF PLASTIC PIPING

Understanding plastic piping material characteristics helps to determine which plastic material is best for a given application.

Physical Characteristics

Rigid (Straight Pipe) Materials

ABS, CPVC, PVC, and PP materials are stiffer than the other plastic piping materials. These pipes retain their shape and are usually sold in straight, rigid lengths. They maintain their round cross section. It is possible to use solvent cement joints with socket fittings or elastomeric rings held in hubs that seal against the OD of the pipe.

These pipes, when installed in a horizontal position, can be supported with hangers at spacing of up to a few feet. Refer to your local code for specific requirements.

The straight, rigid lengths also assure a uniform slope for gravity drain lines so that sags and traps do not form in the line—sags and traps can lead to stoppages.

Nonrigid (Flexible Pipe) Materials

PE, PE-RT, and PEX piping materials, unlike those described above (ABS, CPVC, and PVC), are more flexible. They are available in coils of various lengths (up to 1,000 feet). The piping can be bent, thus minimizing fittings and joints, but it must be supported continuously or on close centers. The flexible tubing can be used for very small drains, serving clear water, such as condensate from cooling coils.

These materials cannot be solvent cemented. PE and PE-RT must be joined by heat fusion or with mechanical joints that seal against the pipe wall with external compressive force provided by draw bands, crimp rings, O-rings, or compression nuts. PEX can only be joined with mechanical fittings; it cannot be heat fused.

There are always exceptions. In small sizes, PVC and CPVC pipe and tube can be coiled.

PIPING STANDARDS: DIMENSIONS AND TERMINOLOGY

Plastic pipe and fittings are manufactured to conform to various widely recognized standard sets of dimensions.

There are several OD-controlled dimensioning systems (e.g., IPS, Sewer, and CTS), and each can have one or more wall thickness designations. They correlate, in general, with product applications and material. Most of the ASTM Standards for pressure pipe are based on the IPS (Iron Pipe Size) OD system or the CTS (Copper Tube Size) tube products. These pipes and tubes are made with both "schedule" and "SDR" wall thicknesses. Most sewer pipes are based on sewer pipe size ODs that are smaller than the IPS ODs; they use DR or pipe stiffness wall designations.

The CTS-OD system utilizes the same outside diameters that are used for copper water tubing. Most of these utilize the SDR wall thicknesses.

IPS—Schedules 40, 80, and 120

When standards for steel pipe were developed many years ago, certain wall thicknesses were given Schedule numbers (e.g., 40, 80, and 120). For a given nominal pipe size, the outside diameter (OD) of a pipe is the same for any schedule. Thus, this is an outside diameter–controlled system. The most common dimensions in this system are based on Sch 40, which is a moderately heavy wall material. For a still heavier wall, the dimension standard most often used is Sch 80. Schedule 120 (heavier still) is available in some plastic pipe materials.

When a single material is made into pipe of a given Schedule number (e.g., Sch 40), the rated working pressure of the pipe is affected by pipe diameter. Remember, the larger the pipe, the lower the pressure rating for the same Schedule number.

Note that a larger Schedule number means a heavier wall. At one time, the Sch 40 steel pipe was called "standard weight" and the Sch 80 steel pipe was called "extra heavy."

Table 2-4 shows dimensions for several sizes of pipe. Table 2-1 shows pressure ratings for Schedule 40 and 80 pipe made from ABS, PE, and PVC.

It is important to point out that the values in the tables are for unthreaded pipe. Threading pipe reduces pressure ratings and may not be allowed for some materials. See the individual material sections of this chapter for more information.

DR, SDR, and SIDR

The DR (Dimension Ratio) system produces a series of pipe for which the pressure rating is the same for all pipe sizes when they are made of the same plastic material. For any given pipe, the DR equals the pipe's outside diameter divided by the wall thickness:

$$\text{DR} = \frac{\text{pipe diameter}}{\text{wall thickness}}$$

In the SDR (Standard Dimension Ratio) system, a series of preferred numbers was selected in which each number is 25 percent greater than the previous one. Therefore, in the SDR pipes the pressure ratings also increase by 25 percent increments.

TABLE 2-4 Plastic Pipe Dimensional Data (Based on IPS Pipe Standard by AISI)

Size Nominal (in.)	Nominal OD (in.)	Min. Wall Thickness (in.)		
		Sch 40*	Sch 80**	Sch 120
	0.405	0.068	0.095	–
$\frac{1}{4}$	0.54	0.088	0.119	–
	0.675	0.91	0.126	–
$\frac{1}{2}$	0.84	0.109	0.147	0.17
$\frac{3}{4}$	1.05	0.113	0.154	0.17
1	1.315	0.133	0.179	0.2
$1\frac{1}{4}$	1.66	0.14	0.191	0.215
$1\frac{1}{2}$	1.9	0.145	0.2	0.225
2	2.375	0.154	0.218	0.25
$2\frac{1}{2}$	2.875	0.203	0.276	0.3
3	3.5	0.216	0.3	0.35
$3\frac{1}{2}$	4	0.226	0.318	0.35
4	4.5	0.237	0.337	0.437
5	5.563	0.258	0.375	0.5
6	6.625	0.28	0.432	0.562
8	8.625	0.322	0.5	0.718
10	10.75	0.365	0.593	0.843
12	12.75	0.406	0.687	1

*Originally called "standard weight" steel pipe

**Originally called "extra heavy" steel pipe

There are ASTM standards for all major plastic piping products based on the DR and SDR concept.

The Inside Diameter (ID) controlled pipes used with insert fittings also use the DR system by dividing ID by the DR to establish wall thickness. All standards for ID-controlled plastic pressure pipe have the term *SIDR* (Standard Inside Diameter Ratio) in the title.

Pressure Piping

Most plastic pressure piping is OD (outside diameter) controlled so that the fittings can be used with pipes having varying pressure ratings. The majority of the plastic piping standards are based on the SDR or DR system, but sometimes the terms (*DR* and *SDR*) are interchanged. For examples of SDR ASTM standards, see D 2241 (PVC), D 2282 (ABS), D 3035 (PE), and F 441 (CPVC). In these standards, you will see that SDRs 7, 9, 11, 13.5, 17, 21, 26, and 32.6 are used. Although much of the plastic pressure piping is based on IPS ODs, there are some ASTM standards based on CTS (copper tube size) ODs and some AWWA plastic pipe standards based on cast iron pressure pipe ODs. Table 2-5 shows some SDR PE pipe values taken from ASTM D 3035.

While there are only a few ASTM Standards for ID (Inside Diameter) controlled plastic piping (see PE and PEX sections), they must be recognized to avoid joining problems. The SIDR concept is explained earlier and details of dimensional differences are given in the ASTM standards. Table 2-5 also shows some SIDR PE pipe values taken from ASTM D 2239 and it shows how the SDR and SIDR numbers are related.

In Table 2-5, the SDR and SIDR pipes shown on the same line have the same pressure rating (e.g., SDR 9 and SIDR 7) when the same PE material is used. Here you can see how the pipe ODs change for the SIDR pipes and the IDs change for the SDR pipe.

It is important to remember that a smaller SDR (or SIDR) number means a heavier wall.

TABLE 2-5 Minimum Wall Thickness of 2-Inch PE Pipe Based on SDR/SIDR

IPS-OD SDR Outside Diameter = 2.375 in.		IPS-ID SIDR Inside Diameter = 2.067 in.	
SDR (D 3035)	Wall	SIDR (D 2239)	Wall
7	0.339	5.3	0.39
9	0.264	7	0.295
11	0.216	9	0.23
13.5	0.176	11.5	0.18
17	0.14	15	0.138
21	0.113	19	0.109
26	0.091	24	0.089
32.5	0.073	30.5	0.071

Nonpressure Piping

Some of the nonpressure piping standards show OD-controlled pipes (e.g., IPS Sch 40 for PVC and ABS DWV, Schedule 40 and Schedule 80 PE and PP per ASTM F 1412) and some have sewer pipe sizing (see D 2729, D 2751, and D 3034). The latter use both SDR and DR systems in their dimensioning. A review of the IPS and sewer pipe standards shows these OD differences:

4-inch IPS OD = 4.500 inches
4-inch sewer pipe OD = 4.215 inches

These OD differences continue over the whole range of pipe sizes. However, there are also several standards for ID-controlled nonpressure pipes. These include corrugated PE and PVC pipe, profile wall PE and PVC pipe, cellular core PE and PVC pipe, and even PVC/ABS truss pipe.

Temperature Effects

Materials with 73°F HDS Rating

ABS, PE, and PVC materials are all available with 73°F stress rating for use in pressure piping. PE piping is used extensively for cold water service lines outside the building. Its low-temperature flexibility makes it especially suited for use in applications where temperatures of 35°F and lower will occur.

ABS and PVC piping have been used for many years in residential DWV systems where intermittent temperature excursions up to 180°F can occur. The maximum temperature at which PE has an HDS rating is 140°F.

Materials with HDS Ratings for Higher Temperatures

Chlorinated polyvinyl chloride (CPVC), polyethylene-raised temperature (PE-RT), and cross-linked polyethylene (PEX) materials are available that are rated for long-term service at 180°F as well as for cold water applications. Hot and cold water distribution system piping made from these materials has a working pressure rating of 100 psi at 180°F. These systems are tested at 150 psi at 210°F for at least 48 hours to ensure integrity at conditions that may develop in the event the water heater controls malfunction. Thus, such materials are suitable for hot water distribution where water heaters are installed with relief valves set at 150 psi, 210°F.

All plumbing codes require the use of piping having the 100 psi at 180°F rating for both the hot and the cold water portions of the water distribution system.

Expansion and Contraction

While the coefficients of expansion for various plastic materials are not identical, in general plastic materials have about eight times the expansion coefficient of steel or four times the expansion coefficient of copper. While these facts must be recognized, proper allowance for the expansion characteristics can be readily accomplished.

To prevent expansion/contraction from adversely affecting a piping installation, the following techniques can be used:

- Underground piping with solvent cement joints or flexible pipe should be snaked in a ditch.
- For underground pressure and nonpressure lines, gaskets in bell joints can be used when they are available and suitable for the specific pipe material and the piping system.
- For long straight lines in buildings, use offsets or changes in direction.
- When none of the above options suffice, use expansion joints, per the pipe manufacturer's recommendations, and hangers that permit movement.

Chemical Resistance

Most plastics are derived from petroleum-based materials. As a result, they are generally not suitable as piping for petroleum liquids or for liquids where even small amounts of petroleum liquids are present. The main example of this last sort is a compressed air system where liquids may be present that are derived from oil-lubricated compressors. The vast majority of compressed air systems are of this type. The compressor lubricants, in very small amounts, gradually accumulate in the piping downstream of the compressor. Thus the compressed air piping becomes oil-coated on the inside surface after a very short time, which can cause premature failure of the plastic pipe.

An additional serious reason to avoid plastic pipe on compressed gas systems is discussed in the section that follows on pressurized gases.

Plastic materials are capable of containing many of the chemicals encountered in industry. For more details about resistance to specific reagents, mixtures, temperature, and pressure conditions, consult the pipe manufacturer.

POTABLE WATER PIPING

The following note is included in many of the ASTM standards that describe plastic pressure piping that may be used to convey potable water:

Potable Water Requirement: Products intended for contact with potable water shall be evaluated, tested, and certified for conformance with the ANSI/NSF Standard No. 61 or the health effects portion of NSF Standard No. 14 by an acceptable certifying organization when required by the regulatory authority having jurisdiction.

PLASTIC DWV AND SEWER PIPING

Plastic pipe is not affected by sanitary waste or drain cleaners. However, it has been found that creosote or some treatments that use petroleum-based carriers may affect plastic pipe.

PROCESS PIPING

Plastic pipe systems have given excellent service in water treatment plants, sewage treatment plants, and numerous industrial processes. Be aware that reagent concentration, stress level, operating temperature, and expected service life can all be factors in evaluating the plastic piping material. The most effective evaluation is often done under actual service conditions, but this must always be done with care for safety.

PRESSURIZED GASES

ABS, PVC, and CPVC pipe and fitting products cannot be used in piping systems intended to store and/or convey compressed air or other gases. Furthermore, these piping systems cannot be tested with compressed air or other gases unless the procedure being used has been clearly and specifically approved by the manufacturer(s) of the plastic products or system to be tested.

By virtue of their compressibility, compressed air and gases contain large amounts of stored energy, which present serious safety hazards should a piping system fail for any reason.

Compressed Air Piping Systems

Some manufacturers offer a piping material for compressed air applications. This type of pipe is tough enough that if the pipe is struck with sufficient force, a hole may form or the pipe may break, but it will not shatter. See the manufacturer's literature for complete details of pressure rating limits and use conditions.

The plumbing codes permit the use of buried polyethylene (PE) fuel gas piping for the distribution of natural gas, LP gas vapor, and other fuel gases. The same codes require that these piping systems be manufactured and marked ASTM D 2513 and be recommended by the manufacturer for fuel gas service. All such PE fuel gas piping must be installed underground except that PE gas piping may terminate above ground and outside of buildings when encased inside an anodeless riser designed and recommended for such uses. Check your local codes for any special requirements for such installations. Figure 2-4 shows a riser used to connect the PE gas line to the gas shutoff valve.

ACRYLONITRILE-BUTADIENE-STYRENE (ABS) PLASTIC PIPING

Acrylonitrile-butadiene-styrene (ABS) plastic materials are manufactured by extrusion in sizes and to the ASTM Standards shown in Table 2-6. Most ABS applications are for DWV uses, but ABS pressure pipe is available for certain industrial applications. The pipe is sold in 10-foot and 20-foot lengths in the United States and 12-foot lengths in Canada. Figure 2-5 shows an example of ABS piping.

ABS pipe and fittings for DWV use are made from ABS compounds meeting the requirements of ASTM D 3965 Cell Classification 4-2-2-2-2 for pipe and 3-2-2-2-2 for fittings.

FIGURE 2-4 This metallic riser connects the PE gas pipe buried in the ground with the gas shut-off valve. Notice there is a tracer wire for pipe detection.

Photo by Ed Moore

TABLE 2-6 ABS Pipe Standards

ABS	ASTM	Sizes (in.)	Temperature Range (°F)	Pressure Rating at 73°F (psi)	Dimension Standard
ABS DWV, solid core, pipe and Fitting	D 2661	$1\frac{1}{4}$–6	Up to 180°	N/A	Sch 40
ABS, DWV, cellular core, pipe	F 628	$1\frac{1}{4}$–6	Up to 180°	N/A	Sch 40
ABS plastic pipe (SDR-PR)	D 2282*	1–12	73°	80–250	SDR 13.5, 17, 21, 26
ABS plastic pipe fittings, Sch 40	D 2468*	1–8	73°	N/A	Sch 40
ABS plastic pipe Sch 40 and 80	D 1527*	1–12	73°	50–500	Sch 40, 80
ABS sewer pipe and fittings	D 2751	3–12	Up to 180°	N/A	DR 23.5, 35, 42
Solvent cement for ABS plastic pipe and fittings	D 2235	N/A	Matches Pipe/ Fittings or as recommended by cement manufacturer	N/A	N/A
Standard Practice for Safe Handling of Solvent Cements, Primers and Cleaners Used for Joining Thermoplastic Pipe and Fittings	F 402	N/A	N/A	N/A	N/A

*Only available as special order product.

ABS pipe is available in solid core and cellular (foam) core construction in Schedule 40 dimensions. These two forms may be used interchangeably for DWV applications. Cellular core construction involves the simultaneous extrusion of three layers into the pipe wall: a solid outer layer, a foam intermediate layer, and a solid inner layer. The inner surface provides the smooth, continuous surface necessary for satisfactory flow characteristics, and the outer segment develops the beam strength necessary for the product to

FIGURE 2-5 Black ABS pipe requires a special solvent cement. Do not use all-purpose-type cements.

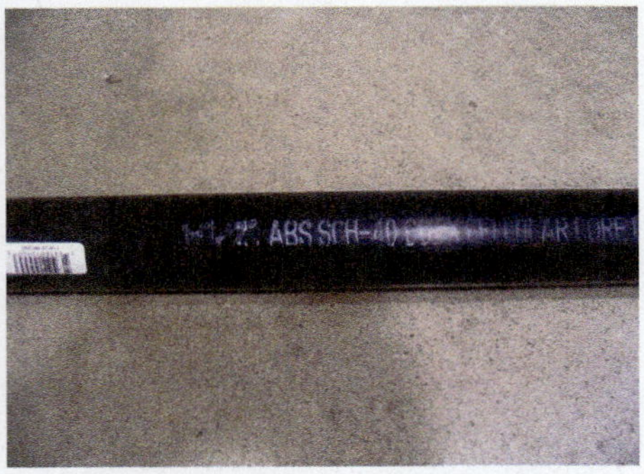

Photo by Ed Moore

behave like a pipe. The closed-cell core holds the outer and inner layers in position with respect to each other but requires less material to do so compared to what is required for a solid wall pipe. The result is a lighter pipe that is satisfactory for DWV applications.

Ultraviolet Radiation

Nearly all plastics can be affected by ultraviolet (UV) radiation. ABS is affected by long-term exposure to UV radiation; carbon black is added to help shield the ABS pipe grades from such radiation. Short-term exposure, such as during construction, is not a problem; however, avoid long-term extended outdoor exposure. Make sure to store materials under cover to minimize long-term exposure to direct sunlight.

Uses

Schedule 40 ABS piping, both solid wall and cellular core, is accepted for DWV systems in all plumbing codes. ABS solid wall pipe is made in Sch 40, Sch 80, and Standard Dimension Ratio (SDR) wall thickness for potable cold water distribution uses, but this application is not in common use today. Table 2-6 provides information about these products, including the ASTM standards that apply.

While ABS materials are resistant to many ordinary chemicals, only the manufacturer of the pipe can recommend suitable uses for special waste systems or industrial process systems. In general, ABS materials are able to contain weak inorganic reagents. Applications that involve petroleum-based products should be avoided. See Chapter 3 for more information on various chemicals and suitable plastic pipe materials.

Compressed Gas Use to Be Avoided

ABS pipe and fitting products are not used in piping systems intended to store and/or convey compressed air or other gases. Furthermore, these piping systems must not be tested with compressed air or other gases unless the procedure being used has been clearly and specifically approved by the manufacturer(s) of the plastic products or system to be tested.

Code Status

ABS pipe and fittings are recognized as acceptable for use in DWV systems in all major model plumbing codes.

Installed Locations

ABS pipe and fittings may be used both above and below ground. Outdoor applications are permitted provided the pipe is further protected from UV light exposure by a water-borne chemically compatible latex paint. Check with the ABS pipe manufacturer or supplier for further information.

Installation Methods

In all cases, careful workmanship and attention to detail are required for a successful long-term installation. Appropriate local code requirements must be followed.

Horizontal Above Ground Support

Support horizontal piping above ground according to Table 2-7 or as required by code using hangers that are wide enough to avoid deforming the pipe at the point of support. Typical clevis hangers ranging from about $\frac{3}{4}$ inches wide in the smallest sizes to more than 1 inch in the largest are usually suitable. If the contents to be carried will be hot (over 120°F), wider saddles would be advisable to reduce the risk of long-term creep. Also consult your local codes.

Frame Construction

ABS pipe may be installed in wall or ceiling framing. The pipe should be protected from nail or screw penetrations by using nail plates wherever the pipe passes through framing members.

Under Slab

Pipe installed under slab is not harmed by direct contact with concrete. When performing under-slab installations, it is important that the pipe be continuously supported along its entire length. Backfill should be free of sharp rocks and other debris that could damage the pipe. Gravel is an acceptable bedding material.

Trenches

Piping installed in trenches in the ground must be placed in continuous bearing and backfilled with sand or granular earth, carefully tamped on each side and to 6 inches

TABLE 2-7 ABS Pipe Support Spacing

Size (in.)	Maximum Spacing (ft.)
$1\frac{1}{4}$–2	3
3–4	4
6–8	5
10–12	6

above the pipe. After the pipe is so bedded, the rest of the trench can be backfilled to the surface.

Vertical Support

Vertical piping should be supported at every floor or every 10 feet, whichever is less. Check for local code requirements that may be more stringent.

Testing

ABS piping installations cannot be tested with compressed gas. Hydrostatic testing is safe and is the recommended method.

Expansion and Contraction

Plastic materials generally have a much larger coefficient of expansion than metals. This fact, however, does not automatically mean that expansion problems are a significant concern for every plastic pipe installation. There are several reasons why the greater coefficient is seldom a problem, including the following:

- The great majority of plastic installations involve relatively short pipe segments where the absolute value of dimension change (with even large temperature change) is not great. In addition, especially for DWV, the complete run of piping is usually not warmed at the same time. That is, a volume of hot water enters the pipe and travels to the discharge end, warming the pipe in a moving wave. There is also usually significant cooling of the water as the wave moves down the pipe.
- Most installations operate in an environment that has very little temperature change—piping in earth or in air-conditioned buildings being two major examples.
- Piping carrying cold water will also see very little temperature change.

For those cases where significant temperature change is a factor, piping dimensional change is accommodated at changes in direction, by piping offsets, by snaking the line (as in a long trench), or by use of expansion joints.

Joining

Socket with Solvent Cement

The most commonly used joining method for ABS pipe and fittings uses solvent cement on the pipe end and on the inside of the fitting socket. The pipe end and socket must be free of dirt, loose particles, and moisture. Place the solvent cement on the outside of the pipe and the inside of the fitting cup immediately before inserting the pipe into the fitting socket.

The inside of the socket is made with a slight taper, with the diameter greater than the pipe OD at the open end of the socket to less than the OD of the pipe at the bottom of the socket. Thus, there is an interference between the outside diameter of the pipe and the inside of the socket approximately midway into the socket. The solvent cement permits the pipe and fitting material to flow sufficiently to allow the pipe to bottom in the socket, and a solid, substantial joint is formed as soon as the cement sets—usually a matter of 1 to 2 minutes depending on conditions.

The solvent cement manufacturer's recommendations should be followed carefully in all details to produce a serviceable joint. For additional information, see ASTM F 402, Standard Practices for Safe Handling of Solvent Cements, Primers, and Cleaners Used for Joining Thermoplastic Pipe and Fittings.

The cement used to join plastic pipe and fittings consists of a solvent appropriate to the plastic being joined, with some of the same plastic type dissolved in the mix. Thus, there can be no universal plastic solvent cement. Each plastic piping material must use the correct solvent cement for that type material.

Although some multipurpose solvent cements are available, most pipe and fitting manufacturers do not recommend their use. Likewise, plumbing codes may not permit such use.

Elastomeric Sleeve Coupling—DWV

While the solvent cement method is the most widely used technique for joining ABS, couplings consisting of an elastomeric sleeve and draw bands also provide a satisfactory joint. These couplings are used to join two plain pipe ends. The coupling most often used is made to conform to CISPI-310-90, and others are available. The user should check local codes for a list of acceptable couplings.

Elastomeric Gasket—Belled Pipe

Some ABS sewer pipe is joined with a hub and spigot assembly that uses an elastomeric gasket to form the seal. The hub, or belled, end of the pipe is formed when the pipe is manufactured, and it contains an elastomeric seal that seats against the pipe wall when the plain pipe end is pushed into the bell. The pipe end must be chamfered and lubricated to facilitate joint assembly.

Socket Fitting Welding

ABS pipe-to-socket joints can be made by fillet welding the top of the socket to the wall of the pipe. The region to be welded is heated to about 500°F to form a puddle and a stick of ABS (about $\frac{1}{8}$ inch square) is worked into the puddle. The puddle is worked all around the joint to form a complete weld. Multiple passes are required. The heat source is a stream of nitrogen or compressed air that is heated in a special electric gun.

IN THE FIELD

Make sure to work in a well-ventilated area to protect your lungs from fumes. Also wear appropriate gloves to protect against burns.

Transition Joints

Suitable techniques and products are available to join ABS to any other piping material. The main methods are the following.

Elastomeric couplings with compression clamps are used to join two plain pipe ends together. These couplings are available for cast iron soil pipe to plastic pipe and for copper tube to plastic pipe. For steel pipe to plastic (both made to Sch 40 dimensions), the standard CISPI coupling described above is satisfactory. Plastic adapter fittings are also made that provide a contour with a raised ridge to provide a better gripping area for the elastomeric coupling.

Threaded adapters are used to join a pipe thread to ABS pipe. The ABS portion may be either spigot (pipe OD size) or socket and the threaded portion may be either male or female NPT threads. An essential requirement for the thread joint is that the thread seal can only be Teflon™ tape or a plastic pipe thread sealant paste specifically tested and approved for ABS by the manufacturer. Any other thread seal material may contain solvents that can attack the plastic and produce failure of the pipe or joint.

Soil pipe hub adapters are used to join ABS DWV pipe to cast iron soil pipe at a hub. One end of the ABS fitting is a socket to join to the ABS piping and the other end is a ferrule that has an enlarged end. The ferrule is placed in the hub, which is then packed with oakum. Lead wool (not molten lead) is then packed in the remaining space to complete the joint.

IN THE FIELD
Always wear protective gloves and a mask when working with lead. Lead can be absorbed through the skin.

Fixture Connections

ABS piping is connected to the drainage side of plumbing fixtures by various special fittings. Water closets are connected to a closet flange. This fitting consists of either a spigot or socket connection to the ABS drainage piping and a flange that sets at the floor level to connect to the water closet by bolts. A hand-formed circle of putty or a wax ring between the flange face and the water closet base provides a durable watertight seal.

Lavatory and sink traps are connected to the ABS piping by trap adapters. These adapters are available in a variety of configurations and sizes. The piping side may be a spigot or socket solvent cement joint connection. The trap connection side usually has some form of compression nut and gland to provide the seal to the trap tubing material.

Marking

The standards for plastic pipe generally require that the product be marked so that it can be readily identified, even if cut in short pieces. Most standards require at least the following items:

- The manufacturer's name or trademark
- The standard to which it conforms
- Pipe size
- Resin type
- DWV, if for drainage

- SDR number or Schedule number
- If the pipe is for potable water, a laboratory seal or mark attesting to suitability for potable water

In addition, a third-party certification mark is often shown. It is not required by the standard, but it may be required by some codes and jurisdictions.

The particular ABS standards differ somewhat on the maximum marking interval distance. Some call for 2 feet, others not more than 1.5 meters (5 feet).

COPPER

Copper is a dense, strong, ductile metal that is resistant to attacks by many chemicals.

A relatively thin-wall copper tube can provide an excellent piping material that is lightweight, strong, and long lasting. The copper most frequently used in plumbing work is referred to as tubing.

Copper tube has both pressure and DWV (drain, waste, and vent) applications. Of all piping material, copper tube has the widest range of plumbing uses. Types K and L are used for water distribution, water services, and many commercial and industrial applications. Type M is used for water distribution and heat-transfer systems such as hot water heating and chilled water cooling. Type DWV is used for drain, waste, and vent. Type ACR is used for refrigerant piping for air conditioning and refrigeration. Most copper tubing is color-coded for quick identification. If color coding is not present, the tube is stamped for identification.

Types of Copper Tube and Use

Type K

Type K is the heaviest wall tubing and is available in hard-drawn temper in diameters from $\frac{1}{4}$ inch to 12 inches and in lengths up to 20 feet. It is also available in soft-temper (annealed) coils up to 2 inches in tube size and up to 100 feet long. The color code for Type K tubing is a green stripe and lettering. Type K copper is typically used for water service and distribution, fire protection systems, solar water heating systems, compressed air, fuel oil, natural and LP gas systems and vacuum systems.

Type L

Type L is the medium thickness wall tubing and is available in the same range of outside diameters and soft and hard configurations as Type K. The color code for Type L tubing is a blue stripe and lettering. Type L copper is typically used for water service and distribution, fire protection systems, solar water heating systems, compressed air, fuel oil, natural and LP gas systems, and vacuum systems.

Type M

Type M is thinner wall tubing available in hard-temper sticks in diameter from $\frac{1}{4}$ inch to 12 inches in 10 feet or 20 feet lengths. Its color code is a red stripe and lettering.

Type M copper is typically used for water service and distribution, fire protection systems, solar water heating systems, compressed air, fuel oil, and vacuum systems.

DWV

DWV is the thinnest wall tubing. It is suitable for drain, waste, and vent only and is available in hard-drawn condition in diameters from $1\frac{1}{4}$ inches to 8 inches in straight lengths of 20 feet. The color code for DWV tubing is a yellow stripe and lettering.

ACR

Approximately the same as Type L (blue lettering) tubing, ACR tubing is shipped from the factory already cleaned and dried and with sealed ends. It is used for refrigeration or medical gases. The tubing is available in 20-foot hard-temper lengths or 50-foot soft-temper coils.

Code Status

Type K, L, and M copper tubing is code approved for water service and hot or cold water distribution. Type K, L, M, and DWV is code approved for building sewer, underground and above ground within the building.

Installation Methods

In all cases, careful workmanship and attention to detail is required for a successful, long-term installation. Appropriate local code requirements must be followed.

Support

According to the 2012 National Standard Plumbing Code, copper tubing must be supported as follows:

- Copper tube ($1\frac{1}{4}$ inches and smaller) every 6 feet horizontally; vertically at each floor or 10 foot intervals
- Copper tube ($1\frac{1}{2}$ inches and larger) every 10 feet horizontally; vertically at each floor or 10 foot intervals

Testing

Testing of the completed piping system can be accomplished by using pressurized air or water. When testing at relatively high pressures air testing is preferred over water testing.

Expansion and Contraction

On excessively long runs or in applications where water temperatures can fluctuate, such as hydronic heating and cooling systems, expansion loops should be installed. Figure 2-6 illustrates a flexible expansion loop.

Joining

Copper tubing can be joined with a variety of methods. The method of choice may depend on a number of factors such as: jobsite conditions, future disassembly requirements, number of joints, and operator skill level.

FIGURE 2-6 The flexible expansion loop will allow for expansion and contraction of the pipe length due to temperature changes of the piping material.

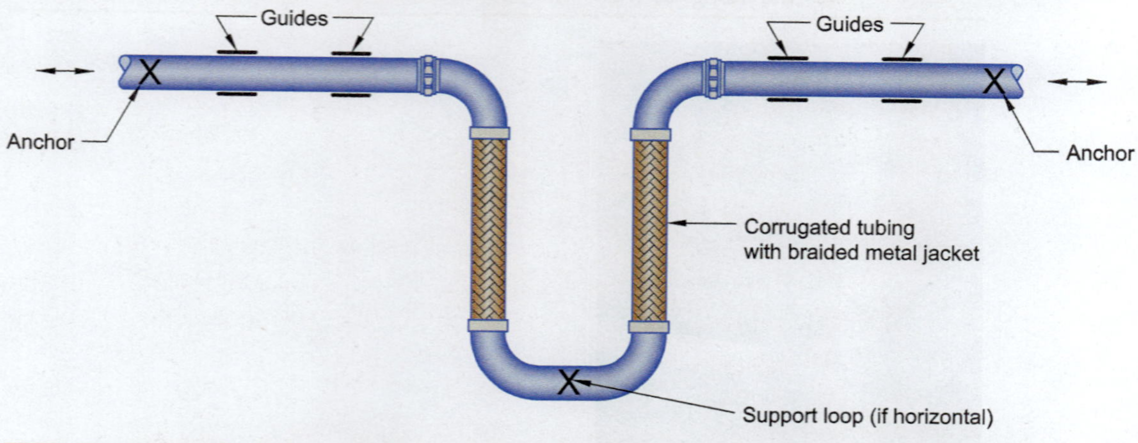

Courtesy of Plumbing–Heating–Cooling–Contractors—National Association

Soldering and Brazing

The soldering and brazing processes involve the process of heating both the pipe and the fitting to a point that a filler material can be melted and drawn into the overlapping area of the joint. The methods used to heat the metals and the filler materials used can vary according to the required strength of the joint and its environment.

Soldering Soldering is typically accomplished with an air-fuel torch that uses one of the following fuel gases: propane, MAPP, acetylene. Most soft solders used for plumbing applications have a melting point below 800°F. Fluxes used for soldering are acid based and help remove any remaining contaminants or oxides leftover from the original cleaning process. In addition to cleaning of the joint, the flux wets the surface to assist the molten solder to flow into the joint. A snug fit is required to provide the capillary action needed to draw the solder in. Figures 2-7 through 2-10 show the steps involved in order to complete the soldering process.

FIGURE 2-7 Using a deburring tool to remove the burr from the inside of the pipe will allow smoother flow.

Photo by Ed Moore

FIGURE 2-8 This open mesh sanding cloth quickly removes any oxidation or contaminants from the outside of the pipe. Be sure to clean the inside of the fittings as well.

Photo by Ed Moore

FIGURE 2-9 Apply a smooth coating of flux to help clean the pipe and help the solder flow more easily into the joint.

Photo by Ed Moore

Brazing Like soldering, brazing involves heating the base metal to a point that it will melt the filler material. In brazing, the filler material is a higher strength material, which normally has a melting point somewhere between 1000°F and 1500°F. Brazing filler material usually either contains phosphorus or silver. Because of the higher melting point, usually an air-acetylene or oxy-acetylene torch is required. Not all filler materials require a flux, but when needed it differs from the fluxes used in soldering in that soldering fluxes are usually water based.

FIGURE 2-10 Solder is fed into the fitting as the torch maintains even heating.

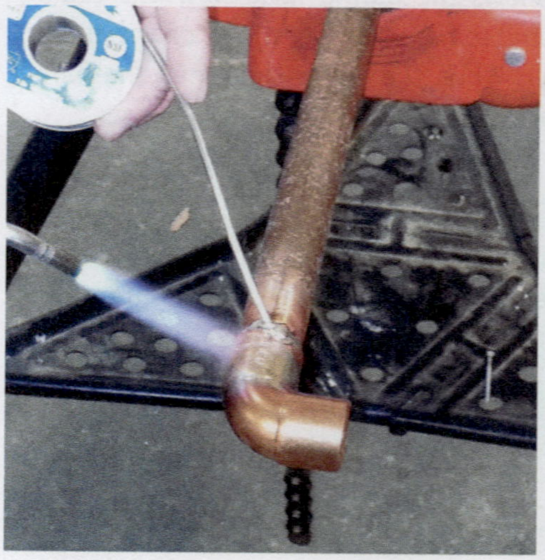

Mechanical Fittings

Whenever there is a possibility that the joint may have to be opened or disassembled, mechanical joints can be used. A mechanical joint is one that does not rely on the bonding of the base metals or require that they be similar. Mechanical joints typically involve two threaded components that will compress or squeeze against the pipe to form a watertight connection. The two most common mechanical joints used in the plumbing industry are the flare fitting and the compression fitting. The flaring process should only be used on annealed or soft tubing.

Flare Fitting

When a flare fitting is used, it is most important to remember to place the nut onto the tube before the tube is flared. Once the flaring nut is installed on the tubing, the end of the tubing should be inserted into the flaring block as shown in Figure 2-11. The end of the tubing should extend approximately $\frac{1}{8}$ inch past the flaring block so that enough material can be flared. If the tubing extends too far, the flare will be too

FIGURE 2-11 The tubing must be firmly secured into the flaring block. The end of the tubing must protrude enough to form a proper flare.

large and the nut will not fit over it. If not enough of the end of the tubing is flared, then the joint will be weak. Figure 2-12 shows the flaring head form the tubing against the flaring block. Figure 2-13 illustrates how the flared end of the tubing will fit against the bevel of the flare fitting. When the nut is tightened the flared end will be forced against this bevel to form a seal.

Press and Push Connectors

Given new regulations that prohibit certain soldering materials as well as those containing lead, new joining methods have been developed that are an alternative

FIGURE 2-12 A proper flare will completely cover the beveled surface of the flaring block.

Photo by Ed Moore

FIGURE 2-13 A proper flare will be the same length as the beveled surface of the fitting. Too much flare will prevent the nut from engaging the fitting.

Photo by Ed Moore

to soldering. Sometimes called "press-to-connect" or "crimp-to-connect" systems, these joining techniques involve proprietary tools that are designed to form fit the fitting to the tubing. As the fitting is crimped onto the tubing, an elastomeric O-ring forms a watertight seal. Figure 2-14 shows a typical handheld crimping tool and Figure 2-15 shows various fittings for use with that tool.

Another type of quick installation fitting that is gaining in popularity is "push-to-connect" fittings. Figure 2-16 shows an example of this type fitting that uses

FIGURE 2-14 Example of a "press-to-connect" tool.

Courtesy of RIDGID®

FIGURE 2-15 Example of "press-to-connect" fittings.

Courtesy of NIBCO®

FIGURE 2-16 Example of "push-to-connect" fittings.

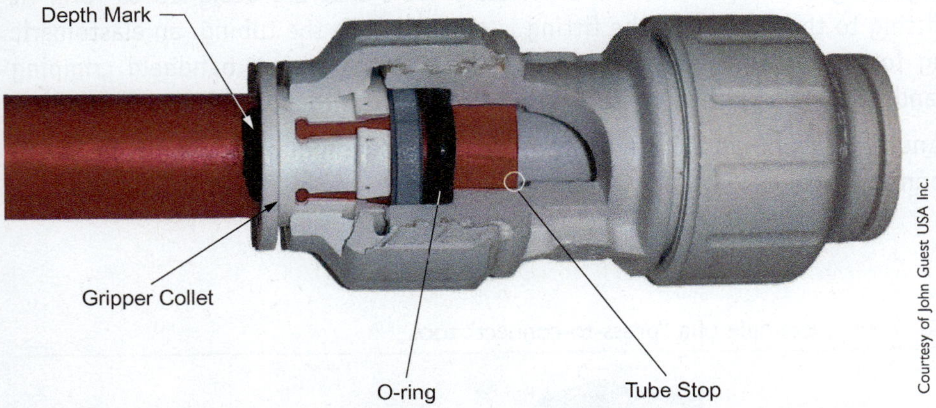

internal barbs that are designed to "bite" into the outside of the pipe as it is pulled out. Like the earlier mentioned fittings, a watertight seal is formed with an elastomeric O-ring and the outside diameter of the pipe. Since this type of fitting is rated for use with copper, PVC, PEX, and PB tubing, it is especially popular for quick repairs on existing systems. While no special tool is required to install this fitting, a special tool may be required for removal.

HUBLESS CAST IRON SOIL PIPE

Hubless or no-hub joints were developed to decrease the assembly time and reduce the dangers associated with the poured lead joints used for cast iron systems. An advantage of this system is that any joint can be readily disassembled for repair, cleaning, or changes. Figure 2-17 shows a typical hubless joint.

Support

According to the 2012 National Standard Plumbing Code, cast iron soil pipe must be supported as follows:

> Cast iron soil pipe—minimum of one hanger per pipe length located within 18 inches of each joint (up to 10 feet maximum pipe length), at changes in direction, and at branch connections. Where pipe is suspended by nonrigid hangers more than 18 inches long, provide lateral support at 40 feet maximum spacing for pipe sizes 5 inches and larger. Lateral support shall consist of either 1) a sway brace or 2) a change in direction or a branch connection that provides the required lateral support.

Testing

Testing of the completed piping system can be accomplished by using pressurized air or water.

FIGURE 2-17 A shielded coupling on hubless cast iron soil pipe.

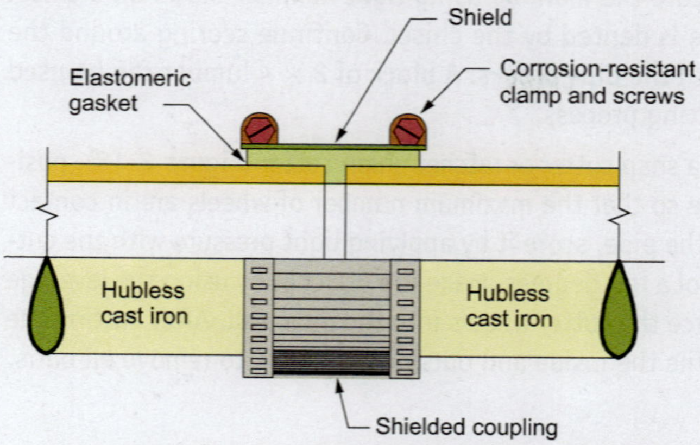

Expansion and Contraction

On excessively long runs or in applications where water temperatures can fluctuate, such as hydronic heating and cooling systems, expansion loops should be installed.

Cutting and Joining

Hubless pipe is cut like hubbed cast iron soil pipe. After the cut ends are filed, the joints are created with a neoprene gasket and stainless steel sleeve and clamps. The clamps are tightened with a torque wrench that will tighten to 60 inch-pounds (some couplings require 80 or 125 inch-pounds). At the appropriate tightness, the wrench clutch will slip, and the technician can be assured that the fitting is tight enough, but not over tightened. The clamp relaxes slightly to a lesser value but will still be water and gastight.

Other variations of the hubless coupling are presently being marketed. Variations such as cast iron supporting sleeves, heavier duty stainless sleeves, interlocking bands and sleeves, and sleeveless couplings are available. Always consult the manufacturer's recommendations for proper installation as some products use special 80 or 125 inch-pound torque wrenches.

HUBBED CAST IRON SOIL PIPE

Cutting

Hubbed cast iron soil pipe is assembled by caulked joints or compression gasket joints. First, the pipe is measured and cut to the required length. The true running length of the pipe is measured from the base of the hub, not the end of the hub. Cast iron soil pipe may be cut with a hammer and cold chisel, snap cutter, ratchet cutter, or power saw.

Saw cuts are made using special abrasive wheel saws designed for cutting cast iron soil pipe. Be sure to follow the manufacturer's instructions when using these powerful saws.

To cut cast iron soil pipe with a hammer and chisel, mark completely around the pipe using a grease pencil. Score the mark by using light hammer blows on a chisel until the entire circumference is dented by the chisel. Continue scoring around the pipe with a stronger force until the pipe breaks. A block of 2 × 4 lumber can be used as a pipe bedding for this cutting process.

To cut cast iron pipe with a snap cutter or ratchet chain cutter (Figure 2-18), position the chain around the pipe so that the maximum number of wheels are in contact with the pipe. Before cutting the pipe, score it by applying light pressure with the cutter and rotating the cutting tool a few degrees. Make the final cut by using the leverage of the cutter mechanism to force the cutter wheels into the pipe wall. After cutting the pipe, peen with a hammer or file the inside and outside diameters to remove all burrs.

Caulked Joints

If a caulked joint is to be made, the spigot (male) end of the pipe is placed in the hub (female) socket, aligned, and two or three rounds of oakum (a rope-like material that swells and seals the joint when wet) are pushed to the bottom of the hub with a yarning iron. The oakum is then tightened in the hub with a packing iron and a hammer. Oakum is again added, pushed in with the yarning iron, and packed with the packing iron until one inch of hub depth remains. The joint is then ready for the addition of molten lead. If a horizontal joint must be poured, a heat-resistant runner is placed adjacent to the hub, with the gate packed for a good pour.

When planning to use a caulked joint, two types of oakum are available. Brown oakum is used for service weight and extra-heavy-duty uses. White oakum is used for water mains and ductile iron piping.

Prior to packing the joint, you should activate the furnace to heat the lead in the pot, as it takes time for the lead to become molten. Lead melts at 622°F and will become goldish-silver in color when it is ready to be poured.

FIGURE 2-18 Example of a cast iron soil pipe cutter.

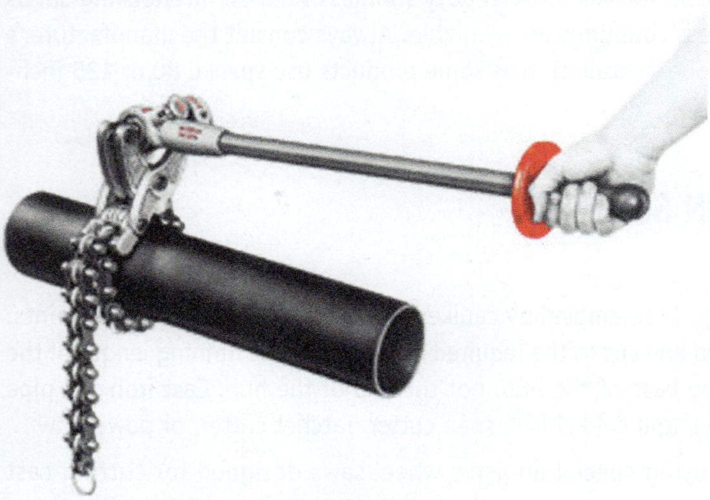

Courtesy of RIDGID®

To check the readiness of the lead for pouring, preheat the ladle by rotating it in the lead. Scoop a small amount of lead into the ladle and swirl it around. If the lead is cold, it will stick to the ladle and must be heated further.

Turn down the heat on the furnace once the lead is hot. This will prevent the lead from burning and caking. Any caked lead should be removed before gathering a ladle of lead. Scoop a full ladle of lead and pour it quickly and smoothly in one continuous stream into the joint until it tops out. Avoid spilling this material, as it will splatter and burn.

After the lead has cooled, set the lead into the hub with inner and outer caulking irons and light hammer blows. Caulking irons are made for the inside diameter and the outside diameter of the lead ring. Caulk the inner ring face first, then the outer ring face. Chisel off any excess lead formations.

This joint holds because the oakum swells once water is applied, causing an extremely tight joint. See Figure 2-19. Plastic pipe can be joined to a cast iron hub using the same method. When plastic is assembled into this type of joint, allow the joint extra time to cool before caulking.

IN THE FIELD

Before using molten lead, be sure that the ladle, oakum, and joint are dry. If water is present, it will explode because the molten lead is so much hotter than the boiling point of water. The force exerted by the steam formed will scatter molten lead in all directions. Keeping your head and face away from the joint as lead is poured reduces the risk of serious injury if the water explodes.

Compression Joints

If a compression (gasket) joint is to be made, a gasket is first placed inside the hub. This gasket must be sized properly for the diameter and weight of pipe being used. Lubricant is applied to the inside of the gasket and on the spigot of the pipe. The pipe end or fitting is pushed into the hub. See Figure 2-20. The pipe or fitting can be driven in with special pulling tools.

FIGURE 2-19 Depiction of a lead caulked joint in cast iron soil pipe.

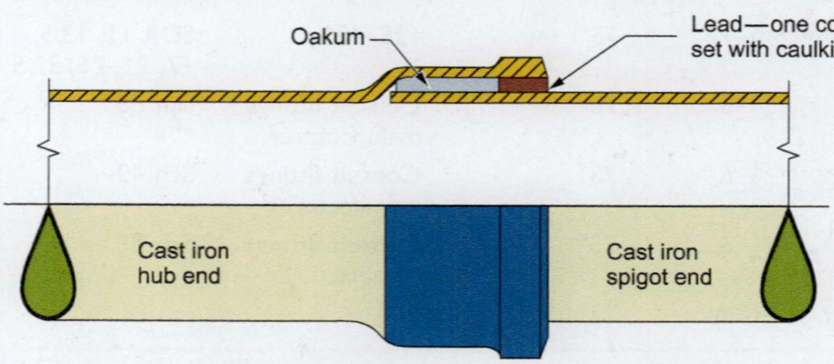

FIGURE 2-20 Depiction of a compression joint in cast iron soil pipe.

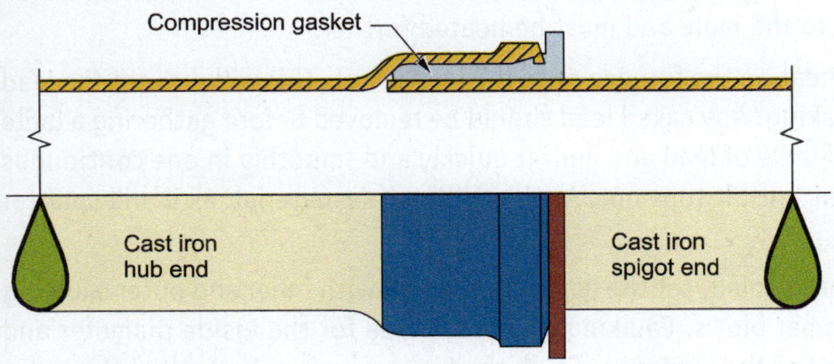

CHLORINATED POLYVINYL CHLORIDE (CPVC) PLASTIC PIPING

Chlorinated polyvinyl chloride (CPVC) is a thermoplastic pipe and fitting material made with CPVC compounds meeting the requirements of ASTM Class 23447 as defined in ASTM Specification D 1784. CPVC applications are for potable water distribution, corrosive fluid handling in industry, and fire suppression systems. It looks similar to PVC except it has a slightly yellowish color. Table 2-8 lists its working characteristics and standards.

Industrial CPVC pipe is manufactured by extrusion in sizes from $\frac{1}{4}$-inch to 12-inch diameter to Sch 40, Sch 80, and SDR (Standard Dimension Ratio) dimensions.

TABLE 2-8 CPVC Pipe Standards

CPVC	ASTM	Size (in.)	Temperature Range (°F)	Rated Pressure (psi)	Dimension Basis
CPVC Hot & Cold Water Distribution System	D2846	$\frac{1}{4}$–2	To 180°	100–400	SDR 11
CPVC Pipe, Sch 40 and 80	F441		73°–180°	130–1,130; 25–280	Sch 40, 80
Standard Practice for Safe Handling of Solvent Cements, Primers, and Cleaners Used for Joining Thermoplastic Pipe and Fittings	F402	N/A	N/A	N/A	N/A
CPVC Pipe SDR-PR	F442	$\frac{1}{4}$–12	73°	125–400	SDR 11, 13.5, 17, 21, 26, 32.5
CPVC Sch 80 Socket Fittings	F439	$\frac{1}{4}$–6	73°	Consult fittings manufacturer	Sch 80
CPVC Sch 40 Socket Fittings	F438	$\frac{1}{4}$–6	73°	Consult fittings manufacturer	Sch 40
CPVC Threaded Fittings for Sch 80	F437	$\frac{1}{4}$–6	73°	Consult fittings Manufacturer	Sch 80
CPVC Solvent Cements	F493	N/A	73°	N/A	

Note: Sch 40 and Sch 80 references apply only to IPS OD piping. SDR references can apply to all of the OD systems, for example, IPS, CTS, AWWA, CI, and Sewer.

CPVC pipe for plumbing systems is manufactured by extrusion in sizes $\frac{1}{4}$-inch through 2-inch copper tube size (CTS) dimensions. The CTS plumbing products are made to copper tube outside diameter dimensions in accordance with ASTM D 2846 specifications and have an SDR 11 wall thickness. The pressure ratings of the CTS SDR 11 systems are 400 psi at 73°F and 100 psi at 180°F. CPVC plumbing pipe is sold in both straight lengths and in small-diameter coils.

Storage and Handling

CPVC pipe and fittings must be stored in such a manner as to prevent physical damage to the materials as well as preventing direct exposure to sunlight. CPVC is affected by long-term exposure to UV radiation. Pigments are added to shield the material from such radiation. Short-term exposure, such as during construction, is not a problem; however, long-term exposure, such as extended outdoor storage, should be avoided. Permanent outdoor installations are permitted provided the pipe is further protected by water-based latex paint.

Pipe lengths and coils should be properly transported to avoid dragging pipe ends and should not be dropped or thrown from trucks or trailers.

Uses

CPVC piping that is suitable for hot and cold water distribution has a 400 psi pressure rating at room temperature and a 100 psi pressure rating at 180°F.

CPVC materials are resistant to many everyday household chemicals, but the manufacturer of the pipe is the only authority to recommend suitable applications. See Chapter 3 for a detailed listing of chemicals and suitable plastic pipe materials and compatible chemicals.

Since CPVC materials do not support combustion, they cannot burn without an external fuel source. This property makes CPVC pipe an attractive alternative to steel and copper pipe for fire sprinkler applications. CPVC fire sprinkler piping systems are approved for light-hazard applications and for use in single- and multifamily dwellings. Installation must be in accordance with the NFPA Section 13, 13D, and 13R.

Compressed Gas Use

CPVC pipe and fitting products cannot be used in piping systems intended to store and/or convey compressed air or other gases. Furthermore, these piping systems cannot be tested with compressed air or other gases unless the procedure being used has been clearly and specifically approved by the manufacturer(s) of the plastic products or system to be tested.

Code Status

Plumbing Applications

CPVC piping for potable hot and cold water distribution systems is recognized in all model plumbing codes.

Plenum Installation

CPVC plumbing pipe is safe for installation in return air plenums; however, the installation must be approved by the local jurisdiction. Even though CPVC is considered a combustible material, it will not burn without a significant external flame source. Once the flame source is removed, CPVC will not sustain combustion. Testing indicates that water-filled CPVC in diameters 3 inches or less will pass the 25/50 flame smoke developed requirements for nonmetallic material in return air plenums.

CPVC fire sprinkler pipe tested and listed in accordance with UL 1887, "Fire Test of Plastic Sprinkler Pipe for Flame and Smoke Characteristics," meets the requirements of NFPA 90A for installation in return air plenums.

Availability

CPVC pipe and fittings are produced by many manufacturers and are available in Sch 40 and Sch 80 dimensions as well as CPVC tubing that is suitable for potable hot and cold water distribution. The tubing is based on copper tube sizes (OD) and IPS pipe (OD) with SDR 11 wall thicknesses.

The ASTM standards and other information for these CPVC piping materials are shown in Table 2-8.

Installed Locations

CPVC water piping may be used in the ground, above ground, in and below concrete slabs, or as water supply and distribution lines. It may be used outdoors if the pipe contains pigments to shield against UV radiation. Painting the assembly with water-based latex paint is one way to develop this protection.

In all cases, good workmanship for the conditions encountered is necessary for a long-term satisfactory installation.

Installation Methods

Horizontal, Above Ground Piping Support

Support horizontal piping above ground according to Table 2-9. Hangers must be wide enough to avoid deforming the pipe at the point of support. Hangers and straps with

TABLE 2-9 Support Distances for CPVC Pipe

Size (CTS) (in.)	73°F	100°F	140°F	180°F
$\frac{1}{2}$	4′	4′	$3\frac{1}{2}$′	3′
$\frac{3}{4}$	5′	$4\frac{1}{2}$′	4′	3′
1	$5\frac{1}{2}$′	5′	$4\frac{1}{2}$′	3′
$1\frac{1}{4}$	6′	$5\frac{1}{2}$′	5′	4′
$1\frac{1}{2}$	$6\frac{1}{2}$	6′	$5\frac{1}{2}$′	4′
2	$7\frac{1}{2}$′	7′	$6\frac{1}{2}$′	4′

sharp or abrasive edges should not be used. Plastic-coated versions of copper hangers are acceptable provided that they allow for movement due to thermal expansion and contraction. However, avoid the use of hangers and supports containing chemicals that are known plasticizers. Typical clevis hangers ranging from about $\frac{3}{4}$ inch wide to more than 1 inch are also suitable.

Frame Construction

When CPVC is installed in walls or ceilings, the pipe should be protected from nail or screw penetrations. Plastic insulators do not need to be used when passing CPVC through wood studs (use oversized holes to accommodate pipe movement). However, when passing through metal studs, some form of protection must be used to protect the pipe from abrasion and to prevent generation of noise. Plastic insulators, rubber grommets, pipe insulation, or similar devices may be used to provide this protection.

CPVC piping systems may penetrate one- and two-hour fire walls, provided that the wall rating is restored. Refer to the installation instructions of the "through penetration firestop system" manufacturer for maintaining the integrity of the wall rating after installation. Only listed firestop systems that are compatible with CPVC may be used.

Slab Construction

In all model plumbing codes, CPVC is approved for under-slab installation. Either straight CPVC pipe can be used with joints or, where joints are not allowed by the plumbing code, continuous coiled CPVC pipe can be used for under-slab installations without joints.

When performing under-slab installations, the CPVC pipe must be evenly supported on a smooth trench or ground. The backfill must be free of sharp rocks or other debris that could damage the pipe. The pipe should be sleeved where it penetrates the slab and at construction joints in the slab.

Manufactured Housing Construction

CPVC is a commonly used plumbing pipe material in manufactured housing. CPVC manufactured in accordance with ASTM D 2846 meets all HUD requirements. The same installation techniques utilized in site-built construction are utilized in this application. See the Manufactured Housing Construction and Safety Standard (MHCSS), Section Number 328.604 for additional details.

Trenches

Piping installed in trenches in the ground must be placed in continuous bearing and backfilled with sand or clean backfill (free of sharp rocks), carefully tamped on each side and to 6 inches above the pipe. After the pipe is properly bedded, the rest of the trench can be backfilled to the surface.

Vertical Support

Vertical piping should be supported at every floor or every 10 feet, whichever is less. Vertical piping must be supported to allow for thermal expansion and contraction. Piping should have a mid-story guide between joists or studs.

Bending

CPVC pipe is available in both straight lengths and coils and can be bent to the manufacturer's specified minimum radii. Consult the pipe manufacturer for bending procedures and recommendations.

Testing

Once an installation is completed and cured per the solvent cement manufacturer's recommendations, the system must be hydrostatically tested. This test should be in accordance with the requirements of the appropriate plumbing code. When pressure testing, the system should be filled with water and all air bled from the highest and farthest points in the installation. If a leak is detected, the joint must be removed and discarded. It is not necessary to fully drain the system.

During subfreezing temperatures, water should be blown out of the lines after testing to prevent damage from freezing.

Thermal Expansion and Contraction

In cases where significant temperature change is a factor, CPVC piping dimensional change is most often accommodated at changes in direction by piping offsets, by snaking the line (as in long trenches), or by the use of expansion joints.

Expansion is mainly a concern on hot water lines. Expansion loops are generally not required on cold water lines. Typically, in single-family construction, the frequency of change in direction eliminates the need for expansion loops in both the hot and cold lines.

Joining Methods

Solvent Cement

The most commonly used joining method for CPVC pipe and fittings uses a solvent cement on the pipe end and in the tapered female socket. Make sure the cement used is specifically rated for CPVC. The pipe end and socket must be free of dirt, loose particles, or moisture.

The CPVC pipe and fittings are assembled after placing solvent cement on the outside of the pipe and the inside of the fitting cup immediately before inserting the pipe into the fitting socket. The inside of the socket is manufactured with a slight taper, with the diameter changing from greater than the pipe OD at the open end of the socket to less than the OD of the pipe at the base of the socket. Thus, there is interference between the outside diameter of the pipe and the inside of the socket approximately midway into the socket. The solvent cement permits the pipe and fitting material to flow sufficiently to allow the pipe to bottom in the socket. A strong joint results as soon as the cement sets (usually 1 to 2 minutes).

The proper procedure for using the solvent cement joining methods is as follows:

1. **Cutting.** The CPVC pipe can be easily cut with a CTS pipe cutter equipped with a cutting wheel designed for plastic pipe. A ratchet cutting tool can also be effective, provided that the cutting blade is kept sharp. Cutting the pipe square

provides optimal bonding area within the joint. If any indication of damage is evident at the pipe end, cut off at least 2 inches beyond the damaged area.

2. **Deburring/Beveling.** Burrs and filings prevent proper contact between the pipe and the fitting during assembly and should be removed from the outside and inside of the pipe. A chamfering tool is recommended for this purpose. Additionally, a slight bevel on the end of the pipe will ease entry of the pipe into the fitting socket, minimizing the chances of pushing the solvent cement to the bottom of the joint.
3. **Fitting preparation.** Wipe any dirt or moisture from the fitting sockets or pipe ends. Check the dry fit of the pipe in the fitting socket. The pipe should make contact with the socket wall one-third to two-thirds of the way into the fitting socket.
4. **Primer application.** If required, a primer should be applied using a dauber approximately half the size of the pipe diameter. Apply primer to the outside of the pipe and the inside of the fitting socket. Do not allow the primer to puddle in the fitting.
5. **Solvent cement application.** Use CPVC cement that conforms to ASTM F 493 specifications. An even coat of cement is applied to the pipe end and within the fitting socket. Do not allow the cement to puddle in the fitting.
6. **Assembly.** Immediately insert the pipe into the fitting socket, rotating $\frac{1}{4}$ to $\frac{1}{2}$ turn while inserting to ensure an even distribution of cement within the joint. Properly align the fitting. Hold the assembly in place for 10 to 20 seconds, allowing the joint to set. An even bead of cement should be evident around the joint. If the bead is not continuous, it may indicate that insufficient cement was applied, in which case remake the joint to avoid potential leaks.

If primer is used, care must be taken to use just enough primer to prepare the contact surfaces. There have been cases where excessive amounts of primer were left on the joint and eventually resulted in joint failure.

Solvent cements are now available that allow one-step joining of CPVC pipe and fittings. This technology eliminates the requirement for the use of primer. Before using one-step cement, check your local code to be sure it is permitted in your area.

Threaded

Schedule 80 CPVC pipe and fittings may be joined by threaded joints. Consult ASTM F 441 and ASTM F 437 for procedures, system constraints, and pressure ratings relative to this joining technique.

PTFE thread tape is compatible with CPVC material and is therefore suitable for sealing threaded connections. If thread-paste sealants are preferred, consult the sealant manufacturer's recommendations for compatibility with CPVC pipe and fittings. Paste-type sealants may contain solvents that damage CPVC pipe and fittings.

Schedule 40 CPVC pipe must not be threaded because too much material is removed. If threaded connections are needed on Schedule 40 CPVC pipe, use threaded adapters that solvent cement to the pipe. Schedule 80 CPVC pipe can be threaded but the pressure rating is reduced by 50 percent. Consult the pipe manufacturer for proper procedures.

Flanges

CPVC can be joined using flanges. While this is most often used for transition joints, flanges can be used throughout a CPVC installation as conditions dictate. Only full-face gaskets should be used, to ensure that the flange faces are pressed together across the full-face area. Washers should be used under the bolt head and under the nut. The flange bolts should be drawn up evenly to avoid overstressing the flange face around the bolt holes and to ensure that the flange faces are drawn together in a parallel fashion.

Socket Fitting—Welding

CPVC pipe-to-socket joints can be reinforced by fillet welding the top of the socket to the outside wall of the pipe after the solvent cement joint has been made. The region to be welded is heated to about 500°F to form a puddle and a stick of CPVC (about $\frac{1}{8}$ inch square) is worked into the puddle. The puddle is worked all around the joint to form a complete weld. Multiple passes are required. The heat source is a stream of nitrogen or compressed air that is heated in a special electric gun.

Transition Joints

Several techniques and products are available to join CPVC to any other piping material. The main methods utilize threaded adapters, compression adapters, and flanges. These methods are described as follows:

- **Threaded adapters.** Threaded adapters may be used to join a pipe thread to CPVC pipe or fittings. The portion that connects to the CPVC may be either socket or spigot (pipe OD size) and the threaded portion may be either male or female NPT threads (male threads are preferred). An essential requirement for the threaded joint is that the thread sealant material conform to the requirements discussed under threaded connections previously in this section.
- **Compression adapters.** Compression adapters are available as couplings (or in other configurations), which seal to the OD of the pipe wall by use of a ring pressing against a gasket sleeve. This arrangement works well to join CPVC to CPVC or other pipe materials.
- **Flanges.** If CPVC flanges are to be bolted to steel flanges, the raised face of the steel flange should be removed so that the full area of the flanges are in contact. Only full-face gaskets should be used to ensure that this requirement is met.

Fixture Connections

Fixture connections are made by various joining and fitting methods. Those most often used are threaded or compression transition fittings to connect to stops or fixture fittings shanks.

Marking

CPVC pipe must be labeled at not more than 1.5-meter (5-foot) intervals as follows:

- The manufacturer's name or trademark
- The standard to which it conforms

- Pipe size
- Resin type or cell class according to ASTM D 1784 (e.g., CPVC 23447)
- Pressure rating
- SDR number or Schedule number
- If the pipe is for potable water, a laboratory seal or mark attesting to suitability for potable water.

When Installing CPVC Piping Systems

Do

- Read the manufacturer's installation instructions.
- Make sure all thread sealants, gasket lubricants, and fire stop materials are compatible with CPVC.
- Keep pipe and fittings in original packaging until needed.
- Use tools specifically designed for use with plastic pipe and fittings. Cut the pipe ends square.
- Deburr and bevel the pipe ends with a chamfering tool.
- Use the proper solvent cement and follow application instructions.
- Rotate the pipe at least $\frac{1}{4}$ turn when bottoming the pipe into the fitting socket.
- Avoid puddling of cement in fittings and pipe.
- Follow the cement manufacturer's recommended cure times prior to pressure testing.
- Allow CPVC tube to have slight movement to permit thermal expansion. Use plastic pipe straps that fully encircle the tube.
- Drill holes $\frac{1}{4}$ inch larger than the outside diameter of the tube when penetrating wood studs.
- Use protective pipe isolators when penetrating steel studs.
- Use metallic clevis or tear-drop hangers when suspending tube from all-thread rod.
- Use compatible sleeving material and tape for under-slab construction.
- Securely tape the top of the sleeve to the pipe for under-slab construction.
- Extend pipe sleeve 12 inches above and below the slab for under-slab construction.
- Backfill and cover underground piping prior to spraying termiticide in preparation for concrete pour.

Do Not

- Do not use petroleum- or solvent-based sealants, lubricants, or fire stop materials.
- Do not use edible oils, such as Crisco, as lubricants.
- Do not use solvent cement that has exceeded its shelf life or has become discolored or thickened.

- Do not pressure test until the recommended joint cure times are met.
- Do not thread, groove, or drill CPVC pipe.
- Do not overtighten or lock down the system.
- Do not install in cold weather without allowing for thermal expansion.
- Do not use tube straps, which tend to restrict expansion and contraction.
- Do not use wood or plastic wedges that strain the tube as it passes through wood studs.
- Do not use pipe isolators as tube passes through wood studs.
- Do not bend CPVC tube around DWV stacks, causing the two materials to bind against each other.
- Do not terminate a run of tube against an immovable object (e.g., a floor joist).
- Do not allow heavy concentrations of termiticides to come into direct and sustained contact with CPVC pipe in under-slab construction.
- Do not inject termiticides into the annular space between the pipe wall and sleeving material.
- Do not spray termiticide, when preparing a slab, without first backfilling over underground piping.
- Do not cut sleeving too short. Sleeving material should extend 12 inches above and below the slab.

POLYBUTYLENE (PB) PLASTIC PIPING

Polybutylene (PB) plastic pipe is no longer available. However, because there was a large amount of it installed, you may encounter it in repair and remodeling work. If at all possible, it should be replaced with current pipe or tubing materials.

PB tubing was popular for water distribution in buildings and gas distribution outside buildings underground. Joints were made with insert fittings or with heat fusion. Figure 2-21 shows PB pipe installed with insert fittings and copper crimp rings.

FIGURE 2-21 Even though PB pipe is no longer used, it can still be found on older jobs.

Photo by Ed Moore

POLYETHYLENE (PE) PLASTIC PIPING

Polyethylene (PE) plastic pipe is manufactured by extrusion in sizes ranging from $\frac{1}{2}$ inch to 63 inches, as shown in Table 2-10. PE is available in rolled coils of various lengths or in straight lengths up to 40 feet. Generally small diameters are coiled and large diameters (OD greater than 6 inches) are in straight lengths. Refer to manufacturer's literature for further information. PE pipe is available in many varieties of wall thicknesses, based on three distinct dimensioning systems:

- Iron Pipe Size Outside Diameter, IPS-OD (SDR)
- Iron Pipe Size Inside Diameter, IPS-ID (SIDR)
- Copper Tube Size Outside Diameter (CTS)

TABLE 2-10 PE Piping Standards

PE	ASTM	Size (in.)	Temperature Range (°F)	SDR (or SIDR) Values	Pressure Rating @ 73°F (psi)
PE Plastic Pipe, Sch. 40	D 2104	$\frac{1}{2}$–6	73°–100°		60–190
PE Plastic Pipe (SIDR-PR) Based on ID	D 2239	$\frac{1}{2}$–6	73°–100°	5.3, 7, 9, 11.5, 15, 19	80–250
PE Standard Spec for Pipe Sch 40 and 80, OD	D 2447	$\frac{1}{2}$–12	73°		50–267
Thermoplastic Gas Pressure Pipe, Tubing and Fittings	D 2513	$\frac{1}{2}$–12	73°–140°	7.3, 9, 11, 13.5, 15.5, 17, 21, 26, 32.5	60–100
Plastic Insert Fittings for PE Plastic Pipe	D 2609	$\frac{1}{2}$–4			
PE Socket Type Fittings for OD Pipe and Tubing	D 2683	$\frac{1}{2}$–4	73°–140°		
PE Plastic Tubing	D 2737	$\frac{1}{2}$–2	73°–100°	7.3, 9, 11	125–200
PE Pipe (SDR-PR) Based on OD	D 3035	$\frac{1}{2}$–24	73°–100°	7, 9, 9.3, 11 13.5, 15.5, 17, 21, 26, 32.5	40–210
PE Butt Heat Fusion Fittings for Pipe and Tubing	D 3261	$\frac{1}{2}$–48, also 90–1, 600 mm	73°–140°	7, 9, 9.5, 11, 11.5, 15.5, 17, 21, 26, 32.5, 41	Not Given
PE Plastic Pipe (SDR-PR) Based on Outside Diameter	F 714	4 and up	73°	7.3, 9, 9.3, 11, 15.5, 17, 21, 26, 32.5, 41	31–254
PE Thermoplastic High Pressure Irrigation Pipeline Systems	F 771	$\frac{1}{2}$–6	73°–100°	SDR 11, 13.5, 17, 21 SIDR 5.3, 7, 9, 11.5, 15, 19	80–200
PE Smooth Wall Pipe for Drainage and Waste Disposal Absorption Fields	F 810	3, 4, 6	73°–100°	N/A	N/A
PE Corrugated Pipe with a Smooth Interior and Fittings	F 892		73°	N/A	N/A
PE Large Diameter Profile Wall Sewer and Drain Pipe	F 894		73°–140°	*	N/A
PE Electrofusion Type Fittings for OD Pipe and Tubing	F 1055		73°–140°	N/A	Not Given
PE Socket Fusion Tools for Socket Fusion Joining	F 1056	$\frac{1}{2}$–4 IPS, $\frac{1}{2}$–13 CTS	N/A	N/A	N/A

*Standard F894 pipes are based on Ring Stiffness Constants (RSC) of 40, 63, 100, and 160. For more details, consult the standard.

PE plastic pipe and fittings are manufactured and installed to conform to various ASTM standards (see Table 2-10). PE piping applications include potable water distribution and service lines, natural gas distribution, various other pressure installations, sewer and drainage piping, and conduit.

PE is a thermoplastic material produced from the polymerization of ethylene and meeting the requirements of Grades PE10, PE24, and PE34, as noted in ASTM D 3350. PE materials can also be described in accordance with the appropriate cell classification as defined in ASTM D 3350, "Standard Specification for PE Plastic Pipe and Fittings Materials."

PE materials used for water service are often designated by a material designation code such as PE 2406 or PE 3408. This material designation code is defined as follows:

1. PE is the abbreviation for polyethylene as defined in accordance with ASTM D 1600, "Standard Terminology Relating to Abbreviations, Acronyms, and Codes for Terms Relating to Plastics."
2. The first digit (2 or 3) represents the density of the PE. The number "2" represents a medium density material in the range of 0.926 grams/cc to 0.940 grams/cc. The number "3" represents a high-density material in the range of 0.941 grams/cc to 0.955 grams/cc.
3. The second digit, "4," represents the slow crack growth requirement of the thermoplastic, which is defined in ASTM D 3350 as either the ESCR (environmental, stress crack) test in ASTM D 1693 or the PENT (polyethylene notched tensile) test in ASTM F 1473. For water service pipe applications, the "4" represents an ESCR, condition C 0.600 hours.
4. The third and fourth digits represent the hydrostatic design stress (HDS) for water at 73°F (23°C) as recommended by the Hydrostatic Stress Board of the Plastics Pipe Institute in PPI TR-4 and divided by one hundred.

PE pipe is available in many forms and colors such as the following:

- Single extrusion colored or black pipe
- Black pipe with co-extruded color striping
- Black or natural pipe with a co-extruded colored layer

Consult the manufacturer's literature and the appropriate ASTM standard for the appropriate form and color for the application.

The American Water Works Association (AWWA) provides the following resources for polyethylene piping:

- AWWA C901—"Standard for Polyethylene (PE) Pressure Pipe and Tubing, $\frac{1}{2}$" through 3" for Water Service"
- AWWA C906—"Standard for Polyethylene (PE) Pressure Pipe and Fittings, 4" through 63" for Water Distribution"

Ultraviolet Radiation

All plastics are affected by UV radiation. Stabilizers are added to polyethylene piping materials to shield them from such radiation. In accordance with ASTM standards, Class C compounds are to contain a minimum of 2 percent carbon black. Properly produced Class C PE pipe may be stored in direct sunlight or installed above ground for extended periods of time. Research studies have demonstrated the long-term durability of PE properly compounded with 2.5 percent carbon black. Results indicate that PE compounds that contain approximately 2.5 percent carbon black will survive at least 30 years of outdoor exposure.

Other colored or natural compounds (previously described in ASTM D 1248 as Class B) do not have a carbon requirement but do require the addition of UV stabilizers to the base resin. Generally speaking, these products can be stored outside and exposed to UV radiation for short periods of time. These products are classified with either a "D" (natural with UV stabilizer) or "E" (colored with UV stabilizer) at the end of the ASTM D 3350 cell classification. The duration of such exposure is typically consistent with that encountered during construction and installation of the piping material. The manufacturer's recommendation should be consulted regarding longer-term exposure and/or long-term storage guidelines.

Uses

PE pipe offers distinct advantages as a piping material. PE's light weight, flexibility, chemical resistance, overall toughness, and longevity make it an ideal piping material for a broad variety of applications such as potable water service or distribution lines, natural gas distribution, lawn sprinklers, sewers, waste disposal, and drainage lines. A complete list of the standards to which these quality piping products are made is presented in Table 2-10.

The selection of materials is critical for water service and distribution piping in locations where there is a likelihood the pipe will be exposed to significant concentrations of pollutants comprised of low-molecular weight petroleum products or organic solvents or their vapors.

PE materials are generally resistant to most ordinary chemicals, but the manufacturer of the pipe is the only authority to recommend suitable uses in the presence of chemically contaminated soils. Consult with the manufacturer regarding the potential for permeation of pipe walls by specific chemical compounds before selecting any materials for use in areas where pollutants may exist.

PE can be used in low temperatures (0°F or colder) without risk of brittle failure. Thus, a major application for certain PE piping formulations is for low-temperature heat-transfer applications such as radiant floor heating, snow melting, ice rinks, geothermal ground source heat pump piping, and compressed air distribution.

Code Status

PE pipe is recognized as acceptable plumbing piping for water services, drainage, and sewer applications in most model plumbing codes. Verify acceptance and installation of PE piping systems with your local code enforcement authority.

Availability

PE pipe and fittings are available from plumbing supply houses and various hardware retailers throughout the United States and Canada. PE pipe is generally less expensive than metallic piping materials.

Installed Locations

PE pipe in pressure applications may be used in ground or above ground. PE pipe installed underground is obviously shielded from the effects of UV light. Above-ground applications are a different matter.

Black PE pipe manufactured in accordance with the requirements of ASTM D 3035, D 2104, D 2447, D 2737, or F 714 contains sufficient carbon black evenly distributed throughout the wall of the pipe to allow its installation in above-ground applications for indefinite periods of time. The engineer typically designs the life of a PE piping system for approximately 50 years. Manufacturers of colored PE pipe should be consulted regarding the suitability of above-ground installation of these products.

No piping should be installed in the ground where it will be exposed to low-molecular weight compounds over a prolonged time period without consideration of the effect that the exposure may have on the piping material. Due to its hydrocarbon structure, PE is susceptible to permeation by some of these low-molecular weight compounds. For this reason, special care should be taken when installing potable water lines through soils, regardless of the type of pipe material under consideration (clay, concrete, PVC, PE, gasketed, etc.). The Plastics Pipe Institute has issued Statement N—Pipe Permeation, which can be studied for further details.

Installation Methods

Storage and Handling

PE pipe, tubing, and fittings should be stored in a way that prevents damage from crushing or piercing, excessive heat, harmful chemicals, or exposure to sunlight for prolonged periods. The manufacturer's recommendations regarding storage should be followed.

PE piping is generally a tough, durable material and is not subject to damage or breakage during normal handling. However, it can be damaged by hard, sharp objects. As with any engineering material, damage can affect the long-term performance of this piping product. Therefore, handling operations, trench installation, and backfill should be performed with reasonable care to prevent excessive scraping, nicks, or gouging of the pipe.

Avoid practices such as dragging coils of pipe or tubing over rough ground and pulling PE pipe through auger or bore holes containing sharp-edged materials, to prevent damage by excessive abrasion and cutting. Uncoiling and other handling should be done without kinking the pipe. If pipe is deeply cut (to a depth greater than 10 percent of its wall thickness) or kinked, the damaged portion should be removed, discarded, and replaced.

In all installations, good workmanship for the conditions encountered is necessary for a long-term satisfactory installation.

Horizontal Above Ground Support

Support horizontal piping above ground according to pipe size, using hangers that are wide enough to avoid a point-loading scenario. For ⅓- to 2-inch PE, maximum support spacing should be 3 feet. For 3- to 6-inch PE, maximum support spacing should be 4 feet.

Typical clevis hangers with a minimum width of one-half the OD of the pipe being installed are generally suitable. Ideal support devices are close-fitting products that are made especially for the pipe, do not crush or abrade the material, and permit movement as conditions may require.

In Trenches

Piping installed in ground trenches must be placed in continuous bearing and backfilled with sand or granular earth, carefully tamped on each side and to 6 inches above the pipe. After the pipe is so bedded, the rest of the trench can be backfilled to the surface.

As a guideline, refer to ASTM D 2774—"Underground Installation of Thermoplastic Pressure Piping." Installation guidelines are also available in the pipe manufacturer's literature.

Large-Diameter Sewers

It is beyond the scope of this text to explore the many considerations that affect the installation of large-diameter (more than 6 inches) piping in the ground. For pipes greater than 6 inches nominal, standard civil engineering methods for the installation of flexible piping products must be employed for the selection of pipe size, wall thickness, bedding and backfill materials, methods, and so on.

Vertical Support

PE vertical piping should be supported at approximately 3-foot centers.

Expansion and Contraction

Plastic materials in general have larger coefficients of expansion than metals. This fact, however, does not automatically mean that expansion problems are a significant concern for every plastic pipe installation.

Since PE piping is used for cold water, buried sewers, or drain fields, there is less chance for significant temperature change to introduce appreciable expansion/contraction problems. For those cases where significant temperature change is a factor, piping dimensional change is accommodated by piping offsets, by snaking the line (as in a long trench), by frequent clamping, or by use of expansion joints.

Joining Methods

PE pipe or tubing can be joined to other PE pipe or fittings or to pipe and appurtenances of other materials using one or more joining systems. The purchaser should verify with the pipe and fittings manufacturer that fittings are capable of restraining PE pipe or tubing from pullout, especially for larger-diameter products with thicker walls. Pressure classes for pipe and fittings should be the same or compatible. Further information and specific procedures may be obtained from the pipe and fittings manufacturer.

Heat Fusion

The most commonly used joining method for PE pipe and fitting is heat fusion. PE pipe and fittings are assembled using butt, saddle, socket, and electrofusion heat fusion methods. These methods involve preparation of surfaces, heating of the surfaces to proper fusion temperatures, and bringing the surfaces together in a prescribed manner to form the fusion bond. ASTM D 2657 is the Standard Practice for Heat Fusion of Polyolefin Pipe and Fittings.

Special tools to provide proper heat and alignment are required for heat fusion connections. These are available from several equipment manufacturers, who can also provide joining procedures. Detailed written procedures and visual aids that can be used to train personnel are available from various pipe and fittings manufacturers. Specific recommendations for time, temperature, and pressure must be obtained from the pipe and fittings suppliers.

The butt fusion technique in its simplest form consists of:

1. Heating the squared ends of two pipes, a pipe and a fitting, or two fittings by holding the squared ends against a heated plate.
2. Removing the plate when the proper melt is obtained.
3. Promptly bringing the ends together.
4. Allowing the joint to cool while maintaining the appropriate applied force.

An alignment jig must be used to obtain and maintain suitable alignment of the ends during the fusion operation.

The saddle fusion technique involves:

1. Melting the concave surface of the base of a saddle fitting while simultaneously melting a matching pattern on the surface of the pipe.
2. Bringing the two melted surfaces together.
3. Allowing the joint to cool while maintaining the appropriate applied force.

The socket fusion technique involves:

1. Simultaneously heating the outside surface of a pipe end and the inside of a fitting socket, which is sized to be smaller than the smallest outside diameter of the pipe.
2. After the proper melt has been generated at each face to be mated, the two components are joined by inserting one component into the other.

The fusion bond is formed at the interface resulting from the interference fit. The melts from the two components flow together and fuse as the joint cools. Optional alignment devices are used to hold the pipe and socket fitting in longitudinal alignment during the joining process, especially with pipe sized 3-inch IPS and larger.

The electrofusion technique involves a resistance wire that is embedded in the fitting. This resistance wire supplies the heat source necessary for fusion.

1. The inside diameter of the fitting allows for pipe insertion.
2. The pipe is clamped in place.

3. Power is supplied to the fittings through a controlled processor.
4. The PE immediately surrounding the wire is melted while the cooler outlying PE serves to contain the melted PE.
5. The pipe, clamped into place, provides the interfacial pressure required for fusion.

Mechanical Fittings

Mechanical fittings provide either a pressure seal alone or a pressure seal and varying degrees of resistance to pullout, including those that hold beyond the tensile yield of the PE pipe. Mechanical fittings may require tightening of a compression nut, tightening of bolts, or merely inserting properly prepared pipe or tubing to the proper stab depth in the fitting. The pipe and fitting manufacturer's recommendations for installation should be followed.

Do not use internal stiffeners that extend beyond the clamp or coupling nut. It is recommended that a solid tubular stiffener be used. The pipe should be cut square using a cutter designed for cutting plastic pipe. It is further recommended that the outside ends of the pipe be chamfered to remove sharp edges that could gouge or cut the gasket when being installed. Chamfering or beveling is a part of the recommended installation procedure for stab fittings.

Mechanical joints using compression fittings are available from $\frac{1}{2}$-inch CTS through 6-inch IPS. Mechanical stab fittings are available from $\frac{1}{2}$-inch CTS through 2-inch IPS. Mechanical transition fittings from plastic to metal pipe are also available in all sizes from $\frac{1}{2}$-inch CTS through 16-inch IPS with some reducing sizes available in smaller diameters (2-inch IPS and less).

Insert Fittings

Insert fittings are available for PE pipe in a variety of styles, including couplings, tees, elbows, and adapters. The pipe end should be prepared for such fittings by cutting the pipe square using a cutter designed for cutting plastic pipe. Two stainless steel clamps are slipped over the end of the pipe. The end of the pipe is forced over the barbs of the fitting until it makes contact with the shoulder of the fitting. The clamps are then tightened to provide a leak-tight connection. Care should be taken to see that the clamp screw positions are offset approximately 180°, as shown in Figure 2-22.

FIGURE 2-22 Tighten band clamps as evenly as possible.

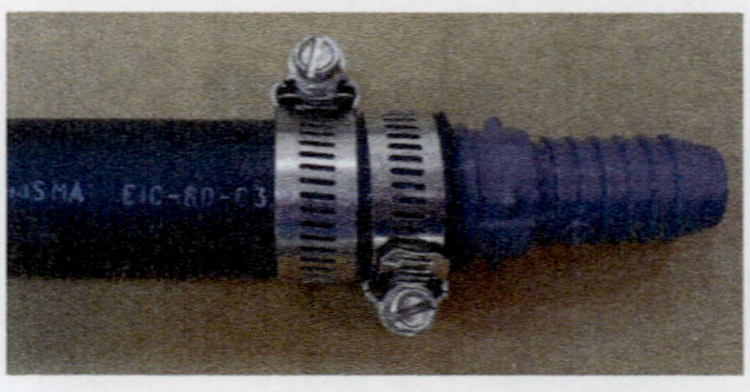

Flaring

Flare joints are generally not recommended with PE tubing. Consult the manufacturer concerning acceptable practice with flare joints.

Threading

Threading cannot be used to join PE pipe, tubing, or fittings for pressure-rated applications.

Solvent Cement

Solvent cement methods cannot be used to join PE pipe, tubing, or fittings.

Marking

In no instance can the marking on pipe be less than that required by the applicable ASTM Standard. The repeat distance for the marking must not be more than 5 inches for all pipe and tubing except the large-diameter drain and sewer pipes.

PE pipe must be labeled as follows:

- The manufacturer's name or trademark
- The standard to which it conforms
- Pipe size
- Material designation code (PE 2406 or PE 3408)
- DWV if for drainage piping
- Pressure rating if applicable
- DR number or Schedule number
- If the pipe is for potable water, a laboratory seal or mark attesting to suitability

PE fittings must be marked as follows:

- The manufacturer's name or trademark
- Material designation code
- Size
- ASTM designation
- Laboratory seal or marking if to be used for potable water

CROSS-LINKED POLYETHYLENE (PEX) PLASTIC PIPING

PEX is a thermoset material made from medium- or high-density cross-linkable polyethylene meeting the requirements of ASTM F 876 and F 877. PEX tubing is used for potable hot and cold water distribution and heat-transfer installations. See Table 2-11.

PEX piping has been used in hot and cold water distribution systems and for hydronic radiant heating in Europe for many years. Introduced into the United States in

TABLE 2-11 PEX Tubing Standards

ASTM Standard	Size (in.)	Temperature Range (°F)	Rated Pressure (psi)
F876	$\frac{1}{4}$–2	73°, 180°, 200°	160, 100, 80
F877	$\frac{1}{2}$–2	73°, 180°	160, 100
F1807	<1	73°, 180°	100 @ 180°F

the 1980s, PEX has replaced PB as the most widely used flexible plumbing piping. PEX tubing is manufactured by extrusion in sizes from $\frac{1}{4}$-inch to 2-inch diameter to copper tube size, OD-controlled (CTS-OD) dimensions. The wall thickness is based on Standard Dimension Ratio 9 (SDR 9) values. It is available in coils and straight lengths.

Uses

PEX tubing can be used in potable water distribution systems, provided it has been tested in accordance with the governing standard, meets the requirements of ANSI/NSF Standard 61, and bears proper certification from a recognized testing agency. A variant of PEX tubing known as PEX-AL-PEX is also widely used for heat-transfer applications. It consists of two layers of PEX separated by a layer of aluminum. The layer of aluminum acts as a barrier to stop the migration of oxygen through the pipe. PEX-AL-PEX is used in both low-temperature applications (radiant floor heating, snow melting, and ice rinks) and distribution piping for temperatures up to 200°F (hot water baseboard, convectors, radiators, etc.).

Code Status

PEX tubing is recognized as acceptable for water distribution piping in most model plumbing codes.

Storage and Handling

PEX tubing can be affected by ultraviolet radiation. Tubing without UV protective packaging must be covered when stored outdoors. Some PEX tubing manufacturers may allow uncovered outdoor storage for a limited time. Check with the manufacturer for exposure limitations.

All plastic piping should be handled to avoid physical abuse. Do not drag, kick, or throw the tubing across any surface. The tubing should not be pulled around sharp corners or any other irregularity. Protect the tubing from sharp edges with sleeving or insulators. As shown in Figure 2-23, nail plates should be used at any place where the tubing could be damaged by screws or nails.

Installation Locations

PEX tubing may be installed below or above ground and in concrete slabs. It may be installed outdoors above ground if shielded completely from sunlight.

FIGURE 2-23 Plastic pipe must be protected from puncture threats.

Photo by Ed Moore

Installation Methods

In all cases, quality workmanship is necessary for a long-term satisfactory installation. Do not use tubing that has been cut, severely scuffed, kinked, or crushed. Cut out and replace damaged sections.

PEX tubing must be placed so that it will not be exposed to sources of high heat. Keep the tubing at least 6 inches horizontally and 12 inches vertically from fluorescent and incandescent light fixtures, heating appliances, electric motors, and furnace or water heater flue vents.

PEX tubing should not be directly connected to water heaters, either electric or gas. Use a minimum 18-inch length of metallic pipe for each connection.

When making transition connections to copper or brass pipe or fittings by soldering or brazing, do the soldering first and allow the joint to cool before making the connection to the tubing. Do not expose PEX tubing to an open flame, as damage can result.

Tubing Support

PEX tubing is usually supported on the face of, or by holes through, joists or studs. Otherwise, support horizontal piping above ground every 32 inches using hangers that are wide enough to avoid deforming the tube at the point of support. Sharp edges must be avoided. Typical clevis hangers or one-hole or two-hole clamps ranging from about $\frac{3}{8}$ inch wide in the smallest sizes to more than 1 inch at the largest are usually suitable. Clamp tubing firmly but not so tightly that it cannot move as it expands and contracts.

PEX piping run vertically should be supported mid-story, at every floor level and again when making significant changes in direction.

Connection to Fixtures

While PEX tubing has some rigidity, it should not be used to support hose bibbs. Attach hose bibbs with a well-anchored drop ear fitting or a length of metallic piping. Fixture stops, control valves, or similar devices may require extra support by means of clamps, threaded adapters, or an enclosure.

Sleeving

PEX tubing must be sleeved when entering or exiting concrete slabs or walls. Sleeving can be foam pipe insulation or a larger size of plastic tubing. If using rigid plastic pipe for sleeving, do not allow the PEX tube to bear against a sharp edge of the sleeve.

Bending

PEX tubing can be formed around corners or obstructions, thereby reducing the need for joints and fittings. The minimum bend radius is eight times the tubing OD for straight lengths or if the bend is in the direction of the coiling bends. The bend radius must be 24 times the tubing OD if the bend has to be opposite the coiling bend (coiled tubing only). Table 2-12 shows the minimum bend radii for PEX tubing. Notice that the bending radius against the coil direction is 3 times that of bending with the coil.

Expansion and Contraction

Plastic materials have a larger coefficient of expansion than metals. PEX piping will expand and contract about 1 inch for every 100 feet of length for each 10°F change in temperature. For tubing sizes 1 inch and smaller, this length change is accommodated by the snaking of tubing around obstacles and by normal slack that is present in the installation of flexible pipe.

For those cases where significant temperature variation is expected, piping dimensional change can be accommodated at changes in direction or with offsets or loops. However, in the normal process of installing flexible PEX tubing, needed slack is generally provided by snaking the piping around obstacles and some sagging

TABLE 2-12 Minimum Bend Radius for PEX Tubing

Tubing Size (in.)	Bend Radius with Coil Direction (in.)	Bend Radius against Coil Direction (in.)
$\frac{3}{8}$	4	12
$\frac{1}{2}$	5	15
$\frac{3}{4}$	7	21
1	9	27
$1\frac{1}{4}$	11	33
$1\frac{1}{2}$	13	39
2	17	51

between supports. Allowing about 7 inches of slack for each 50 feet of run length negates the need for an offset or expansion loop.

Manifold Plumbing Systems

PEX tubing, because of its flexibility, is a good choice for manifold or individual run plumbing distribution systems. In these systems, individual tubing runs proceed from a manifold directly to each fixture in the house. Since PEX tubing is flexible, most if not all directional fittings behind the wall are eliminated.

Individual distribution lines are normally $\frac{3}{8}$ inch or $\frac{1}{2}$ inch. Fixture demand (gpm), length, service pressure, elevation, and required residual pressure determine which size is sufficient.

Groups of lines to a common location are generally bundled for neatness. Hot and cold distribution lines may be bundled together. Color-coded tubing is available to help prevent cross connections. See Chapter 3 for more information on manifold systems.

Joining Methods

In North America, PEX tubing cannot be joined by solvent cement or heat fusion methods. A full line of directional and transition fittings are available. The joint is formed with various mechanical compression methods. The most widely available connection is the insert and crimp ring system produced to the ASTM F 1807 specification. This joining system is supplied by several manufacturers.

In manifold systems, one-piece PEX tubing is installed between the centrally located manifold and the fixture connection. Thus, the only joints are at the ends of the tube. Only a few additional joints are needed to complete the supply line to the header.

PEX tubing must be labeled as follows:

- The manufacturer's name or trademark
- The standard to which it conforms (ASTM F876, F877, or both)
- Tube size and CTS
- Material designation code (PEX0006)
- Pressure/temperature rating(s)
- SDR9
- If the tubing is for potable water, a laboratory seal or mark attesting to suitability for potable water

The marking interval is required to be not more than 5 feet.

POLYVINYL CHLORIDE (PVC) PLASTIC PIPING

PVC is the most widely used plastic piping material. PVC pipe is manufactured by extrusion in sizes and dimension standards shown in Table 2-13 and sold in 10- and 20-foot lengths.

TABLE 2-13 PVC Piping Standard

	Standard				Design Pressure Rating @ 73°F (psi)	
Standard Name	ASTM	Other	Application	Sizes (in.)		Basis for Dimensions
PVC DWV Pipe and Fittings	D 2665	CSA 181.2	DWV	$1\frac{1}{4}$–12	Not rated	IPS Sch 40
PVC Pipe with Foam Core	F 891		DWV	$1\frac{1}{4}$–12, 2–6, 3–18		IPS Sch 40 IPS PS Series
3.25-inch OD PVC DWV Pipe and Fittings	D 2949		DWV			In standard
PVC Pipe Sch 40, 80, 120	D 1785	CSA B137.3	Pressure	$1\frac{1}{8}$–24, $1\frac{1}{8}$–24, $\frac{1}{2}$–12	120–180, 210–1,230, 340–1,010	IPS Sch 40, 80, 120
PVC Pipe Pressure Rated (SDR)	D 2241		Pressure	$\frac{1}{2}$–36	50–315	IPS SDR 13.5, 17, 21, 26, 32.5, 41, 64
AWWA PVC Pressure Pipe, AWWA PVC Water Transmission Pipe		C900, C905	Pressure	4–12, 4–36	Class 200, 150, 100	(2)DR 14, 18, 25
Type PSM PVC Sewer Pipe and Fittings	D 3034		Sewer	4–15		SDR 23.5, 26, 35, 41
Type PS46 PVC Gravity Flow Sewer Pipe and Fittings	F 789		Sewer	4–18		In Standard
PVC Sewer Pipe and Fittings	D 2729		Sewer	2–6		In Standard
Threaded PVC Fittings, Sch 80	D 2464		Pressure	$\frac{1}{8}$–6	50% of pipe	Sch 80
PVC Fittings Sch 40 (Socket and Threaded)	D 2466		Pressure	$\frac{1}{8}$–8		Sch 40
PVC Sch 80 Socket Fittings	D 2467		Pressure	$\frac{1}{8}$–8		Sch 80
PVC Large Diameter Gravity Sewer Pipe and Fittings	F 679		Sewer	18–36		Pipe Stiffness
PVC Large Diameter Ribbed Sewer Pipe and Fitting	F 794		Sewer	4–48		Pipe Stiffness
Making Solvent Cement Joints, PVC	D 2855		All			
Solvent Cement, PVC	D 2564		All			
Joints for IPS PVC with Solvent Cement	D 2672		All			
PVC Primers	F 656		All			
Pressure Joints Using Flexible Elastomeric Seals	D 3139		Supply and distribution lines for water			
Joints for Drain and Sewer Using Elastomeric Seals	D 3212		Drain and Gravity Sewage		25-ft. Head	
Standard Practice for Safe Handling of Solvent Cements, Primers, and Cleaners Used for Joining Thermoplastic Pipe and Fittings		F 402				

Notes: Pressure ratings for Schedule pipe vary by pipe size and by schedule number. Pressure ratings for SDR pipe are constant for all sizes in a single SDR (e.g., SDR 21 pipe is rated 200 psi). Pressure ratings for AWWA pipe include an extra safety factor. Schedule 40 and 80 references apply only to IPS OD pipe. SDR references apply to all OD systems.

PVC plastic pipe is made to conform to various ASTM standards for both pressure and nonpressure applications (also shown in Table 2-13). PVC piping is used for DWV, sewer, water mains, water service lines, irrigation, conduit, and various industrial installations.

PVC is a thermoplastic pipe made with compounds meeting the requirements of ASTM D 1784, Class 12454-B or Class 12454-C for pipe, or those classes plus Class 14333D for fittings. The compound for cellular core material conforms to ASTM D 4396, Class 11432.

PVC DWV pipe is available in both solid wall and cellular core construction; these two forms are used interchangeably for DWV applications. Cellular core construction involves the simultaneous extrusion of three layers of material into the pipe wall: a solid outer layer, a cellular core intermediate layer, and a solid inner layer. The inner layer provides the smooth, continuous channel necessary for satisfactory flow characteristics. The outer layer develops the beam strength necessary for the product to behave like a pipe. The cellular core holds the outer and inner rings in position with each other but requires less material to do so. The result is a lighter, somewhat less costly pipe that is satisfactory for DWV applications. Cellular core pipe is also used for other nonpressure applications.

Storage and Handling

Pipe and fittings should be stored in such a manner as to prevent physical damage to the materials and to keep them clean. Pipe should be properly transported (not dragged) to prevent damage to the pipe ends. Pipe should not be dropped or thrown from trucks.

Temperature Limitations

The maximum use temperature for PVC pressure pipe is 140°F. PVC DWV piping readily withstands the hot and cold water discharges that are normally associated with plumbing fixtures.

As with all thermoplastic piping, pressure ratings are determined and established at 73°F based on testing in accordance with ASTM D 2837 and published in PPI TR-4. Although pressure ratings decrease when the temperature rises, PVC pipe can be used at temperatures greater than 73°F in pressure applications when appropriate temperature de-rating factors are applied. Other factors, such as proper hanger support spacing and the effects of expansion and contraction, need to be considered when working at elevated temperatures. Detailed information about these and other design-related topics are available from pipe manufacturers.

Ultraviolet Radiation

PVC does not readily degrade when exposed to ultraviolet radiation, due to natural UV inhibitors present in the material. Testing and past field experience has shown that when PVC is exposed to sunlight, the pipe will discolor in the area and impact resistance may be lowered slightly, but the majority of PVC's physical properties are not greatly affected. PVC pipe that has been exposed to sunlight for short periods of time, such as during construction, typically will not have problems.

PVC piping may be used in outdoor applications when the piping system is painted with a light-colored water-based acrylic or latex paint that is chemically compatible with PVC. When painted, the effects of UV exposure are significantly reduced. Dark colors can be used, but they will absorb more heat than light colors and will lead to greater expansion and contraction changes of the piping. This can be addressed with proper system design and installation.

Uses

PVC is suitable for DWV, sewers, conveying low-temperature heat-transfer liquids, distribution of potable water, and many agricultural and industrial services. PVC piping is made to IPS sizes with Schedules 40, 80, and 120 wall thicknesses, as well as several Standard Dimension Ratio (SDR) wall thicknesses. Table 2-13 shows the ASTM or other standards, application, available sizes, pressure ratings, and basis for dimensions. While there are many available PVC piping products, the ASTM standard number will be helpful in identifying the pipe you need.

PVC materials are resistant to many ordinary chemicals. In general, PVC is resistant to most acids, bases, salts, oxidants, and weak inorganic reagents. Certain petroleum-based products should be avoided. See Chapter 3 for a detailed listing of chemicals and suitable plastic pipe materials. The product manufacturer should be contacted for suitability.

PVC piping system components are manufactured in a variety of colors, depending on the intended application. A common color scheme is the following:

- White for DWV and some low pressure applications
- White, blue, and dark gray for cold water piping
- Green for sewer service
- Dark gray for industrial pressure applications

This color scheme has an exception: much of the white PVC pipe is dual-rated for DWV and pressure applications.

Compressed Gas Use

PVC pipe and fitting products must not be used in piping systems intended to store and/or convey compressed air or other gases. Furthermore, these piping systems cannot be tested with compressed air or other gases unless the procedure being used has been clearly and specifically approved by the manufacturer(s) of the plastic products or system to be tested.

Code Status

PVC piping is recognized as acceptable material for DWV, sewers, and potable water services and distribution in most model plumbing codes. These codes normally identify acceptable products for specific uses based on the ASTM Standard designation.

Availability

PVC is available from plumbing supply houses, hardware stores, and home improvement centers throughout North America.

Installed Locations

PVC pipe may be used in the ground or above ground. It may be used outdoors if the pipe contains stabilizers to shield against UV radiation or if it is painted with a water-based latex paint.

Installation Methods

In all cases, knowledge of products being used and good workmanship practices for handling, joining, and installation are required to achieve a satisfactory long-term installation. Appropriate local code requirements and the manufacturer's installation instructions must be followed.

Horizontal Above Ground Support

Proper hanger and support recommendations may be different for pressure systems versus nonpressure systems. Local code requirements must be followed. Table 2-14 shows recommendations for pipe support spacing on horizontal PVC DWV piping above ground. Support recommendations for PVC pressure systems are shown in Table 2-15.

IN THE FIELD

The 2012 National Standard Plumbing Code, International Plumbing Code, and Uniform Plumbing Code each limit the maximum horizontal pipe support spacing to 4 feet on all PVC pipe.

Proper support selection and placement are critical to prevent excessive weight loading or bending stress, to minimize pipe sag, and to address the effects of thermal expansion and contraction. Proper support spacing is dependent on pipe size, the location of concentrated loads, and the operating temperatures of the system. Higher temperatures require additional supports. If the contents to be carried will occasionally be hot (over 120°F), wider

TABLE 2-14 PVC DWV Pipe Support Spacing

Size (in.)	Maximum Spacing (ft.)
$1\frac{1}{4}$–2	3
3–4	4
6–8	5
10–12	6

TABLE 2-15 PVC Pressure Pipe Support Spacing (ft.)

Pipe Size (in.)	Schedule 40 Temperature (°F)					Schedule 80 Temperature (°F)				
	60	80	100	120	140	60	80	100	120	140
$\frac{1}{4}$	4	$3\frac{1}{2}$	$3\frac{1}{2}$	2	2	4	4	$3\frac{1}{2}$	$2\frac{1}{2}$	2
$\frac{3}{8}$	4	4	$3\frac{1}{2}$	$2\frac{1}{2}$	$2\frac{1}{2}$	$4\frac{1}{2}$	$4\frac{1}{2}$	4	$2\frac{1}{2}$	$2\frac{1}{2}$
$\frac{1}{2}$	$4\frac{1}{2}$	$4\frac{1}{2}$	4	$2\frac{1}{2}$	$2\frac{1}{2}$	5	$4\frac{1}{2}$	$4\frac{1}{2}$	3	$2\frac{1}{2}$
$\frac{3}{4}$	5	$4\frac{1}{2}$	4	$2\frac{1}{2}$	$2\frac{1}{2}$	$5\frac{1}{2}$	5	$4\frac{1}{2}$	3	$2\frac{1}{2}$
1	$5\frac{1}{2}$	5	$4\frac{1}{2}$	3	$2\frac{1}{2}$	6	$5\frac{1}{2}$	5	$3\frac{1}{2}$	3
$1\frac{1}{4}$	$5\frac{1}{2}$	$5\frac{1}{2}$	5	3	3	6	6	$5\frac{1}{2}$	$3\frac{1}{2}$	3
$1\frac{1}{2}$	6	$5\frac{1}{2}$	5	$3\frac{1}{2}$	3	$6\frac{1}{2}$	6	$5\frac{1}{2}$	$3\frac{1}{2}$	$3\frac{1}{2}$
2	6	$5\frac{1}{2}$	5	$3\frac{1}{2}$	3	7	$6\frac{1}{2}$	6	4	$3\frac{1}{2}$
$2\frac{1}{2}$	7	$6\frac{1}{2}$	6	4	$3\frac{1}{2}$	$7\frac{1}{2}$	$7\frac{1}{2}$	$6\frac{1}{2}$	$4\frac{1}{2}$	4
3	7	7	6	4	$3\frac{1}{2}$	8	$7\frac{1}{2}$	7	$4\frac{1}{2}$	4
$3\frac{1}{2}$	$7\frac{1}{2}$	7	$6\frac{1}{2}$	4	4	$8\frac{1}{2}$	8	$7\frac{1}{2}$	5	$4\frac{1}{2}$
4	$7\frac{1}{2}$	7	$6\frac{1}{2}$	$4\frac{1}{2}$	4	9	$8\frac{1}{2}$	$7\frac{1}{2}$	5	$4\frac{1}{2}$
6	$8\frac{1}{2}$	8	$7\frac{1}{2}$	5	$4\frac{1}{2}$	10	$9\frac{1}{2}$	9	6	5
8	9	$8\frac{1}{2}$	8	5	$4\frac{1}{2}$	11	$10\frac{1}{2}$	$9\frac{1}{2}$	$6\frac{1}{2}$	$5\frac{1}{2}$
10	10	9	$8\frac{1}{2}$	$5\frac{1}{2}$	5	12	11	10	7	6
12	$11\frac{1}{2}$	$10\frac{1}{2}$	$9\frac{1}{2}$	$6\frac{1}{2}$	$5\frac{1}{2}$	13	12	$10\frac{1}{2}$	$7\frac{1}{2}$	$6\frac{1}{2}$
14	12	11	10	7	6	$13\frac{1}{2}$	13	11	8	7
16	$12\frac{1}{2}$	$11\frac{1}{2}$	$10\frac{1}{2}$	$7\frac{1}{2}$	$6\frac{1}{2}$	14	$13\frac{1}{2}$	$11\frac{1}{2}$	$8\frac{1}{2}$	$7\frac{1}{2}$
18	$13\frac{1}{2}$	12	11	8	7	$14\frac{1}{2}$	14	12	11	9
20	14	$12\frac{1}{2}$	$11\frac{1}{2}$	10	$8\frac{1}{2}$	$15\frac{1}{2}$	$14\frac{1}{2}$	$12\frac{1}{2}$	$11\frac{1}{2}$	$9\frac{1}{2}$
24	15	13	$12\frac{1}{2}$	11	$9\frac{1}{2}$	17	15	14	$12\frac{1}{2}$	$10\frac{1}{2}$

saddles are recommended to reduce the effects of long-term creep. When operating at or near the maximum temperature limits, it may be more economical to provide continuous support for the piping.

Many hangers designed for metal pipe are suitable for use with PVC. However, the hangers and supports used must provide an adequate load-bearing surface that is smooth and free of rough or sharp edges that could damage the pipe. Typical clevis hangers ranging in width from about $\frac{3}{4}$ inch to more than 1 inch are usually suitable. Movement caused by the effects of expansion and contraction as well as movements from pressure fluctuations must be considered to ensure proper hanger selection and placement. Hangers and supports used must not compress the pipe, which could prevent movement and cause damage. Also, avoid the use of hangers and supports containing chemicals that are known to degrade plastics.

Frame Construction

PVC pipe may be installed in wall or ceiling framing. The pipe should be protected from nail or screw penetrations by using nail plates wherever the pipe passes through framing members.

Trenches

Piping installed in ground trenches must be placed in continuous bearing and backfilled with sand or granular earth, carefully tamped on each side and to 6 inches above the pipe. After the pipe is so bedded, the rest of the trench can be backfilled to the surface. Small-diameter (up to 2-inch) pipe should be snaked from side to side of the trench to minimize expansion or contraction problems, especially if the installation is made in hot weather.

Large-Diameter Sewers

For pipes 10 inches and larger, standard civil engineering methods must be used for the selection of pipe size, wall thickness, bedding and backfill materials, methods, and so on.

Vertical Support

Vertical piping should be supported at every floor or every 10 feet, whichever is less, unless the design engineer has made other provisions to allow for expansion and contraction. The 2012 International Plumbing Code adds the further restriction that pipe sizes 2 inches and smaller have mid-story guides. Always confirm support requirements using your local code.

Bending

Some small-diameter pipe is available in coils, and such pipe can be bent. Most PVC pipe, however, should not be bent. Fittings must be used for changes of direction.

Testing Piping Installations

Once an installation is completed and the cement has had time to cure per the solvent cement manufacturer's recommendations, the system must be hydrostatically tested. This test should be in accordance with the requirements of the appropriate plumbing code. When pressure testing, the system should be filled with water and all air bled from the highest and farthest points in the installation. If a leak is detected, the joint must be removed and discarded. It is not necessary to drain the system fully if the affected fittings can be isolated for the required work. During subfreezing temperatures, water should be drained out of the lines after testing to prevent damage from freezing.

Expansion and Contraction

Plastic materials in general have larger coefficients of expansion than metals. This fact, however, does not automatically mean that expansion problems are a significant concern for every plastic pipe installation, for the following reasons:

1. The great majority of PVC pipe installations involve relatively short pipe segments, where the absolute value of dimension change (with even large temperature change) is not great. In addition, especially for DWV, conditions are usually such that the complete run of piping is not warmed at the same time. That is, a volume of hot water enters the pipe and travels to the discharge

end, warming the pipe in a moving wave, with the pipe cooling again as the wave passes.

2. Most installations operate in an environment that has minimal temperature change—piping in earth or in air-conditioned buildings being two major examples.
3. Piping carrying cold water will also see very little temperature change.

For those cases where significant temperature change is a factor, piping dimensional change is accommodated at changes in direction, by piping offsets, by snaking the line (as in a long trench), or by use of expansion joints.

Joining Methods

Solvent Cement

The most commonly used joining method for PVC pipe and fittings uses solvent cement on the external pipe end and in the internal socket. Use only solvent cement manufactured specifically for PVC and the size of piping being joined. Generally, large-diameter pipe and fittings require heavier-bodied cement with greater gap-filling properties, compared to small-diameter piping. In addition, the appropriate-size applicator must be used to ensure sufficient cement is applied in a timely manner. By far, the majority of field-related failures are the result of improper solvent cementing procedures. It is critical that the cement manufacturer's instructions be followed and applied to obtain leak-free joints.

The pipe end and socket must be free of dirt, loose particles, and moisture. A primer is brushed on the pipe end and the inside face of the socket. Excess application of primer should be avoided, as excessive amounts that are not removed can result in eventual joint failure.

The PVC pipe and fitting are then assembled after placing solvent cement on the outside of the pipe and the inside of the fitting socket immediately before inserting the pipe into the fitting socket.

The inside of the socket is manufactured with a slight taper, with the diameter greater than the pipe OD at the open end of the socket and less than the OD of the pipe at the bottom of the socket. Thus, there is an interference between the outside diameter of the pipe and the inside of the socket approximately midway into the socket.

The solvent cement permits the pipe and fitting material to flow sufficiently to allow the pipe to bottom in the socket, and a solid, substantial joint is formed as soon as the cement sets. For sizes 4 inches and smaller, the joint will have "handling strength" in 1 or 2 minutes under normal working conditions. For larger sizes and extreme weather conditions, longer set times are required.

The solvent cement manufacturer's instructions should be followed carefully in all details to produce a serviceable joint. For additional information, see ASTM F 402, Standard Practice for Safe Handling of Solvent Cements, Primers, and Cleaners Used for Joining Thermoplastic Pipe and Fittings.

Flanging

Flanges are used where the piping system may be subject to periodic disassembly and are commonly used as a transition fitting to connect PVC to other piping materials or large valve bodies. PVC flanges consist of a bolting ring and center hub that is solvent cemented on to the PVC pipe. This provides a surface for connection to another flange of the same size and bolt pattern. The two flanges are sealed by placing a gasket between the flanges (bolting rings) and tightening the bolts in a manner to ensure equal force all around the flanges.

Class 150 PVC flanges have a maximum pressure capability of 150 psi at 73°F. Thermoplastic flanges have bolt hole patterns that are compatible with those found in ASME B16.5, Flanges and Flanged Fittings. Flanges are available in two basic styles: Van Stone (two-piece) and companion (one-piece).

Van Stone flanges allow alignment of the bolt holes after assembly because the bolt ring is separate from the hub piece. Solid one-piece companion flanges are not capable of this flexibility and therefore require thoughtful planning prior to assembly.

Full-face gaskets of $\frac{1}{8}$-inch thickness are used to seal the flanges. Gasket hardness is recommended to be between 50 and 80 durometers. Care must be taken when joining flanges together, whether bolting plastic to plastic or metal to plastic. Any undue stress from misalignment, unequal tightening, or over-torquing can cause flange failure.

Some metal and plastic butterfly valves and some flanges have raised areas near the inside sealing surface. This creates a ring gasket effect that is not usually recommended. To accommodate these types of valves and flanges, it is normally recommended that the lowest recommended assembly torque be applied. Use extreme caution when connecting to a raised face flange because the unsupported area produces a bending stress on the flange ring.

When bolting flanges together, use proper-sized flat washers under the bolt heads and nuts. Failure to do so can cause premature flange failure from excessive local compressive stress.

Elastomeric Gasket—Hub and Spigot

Some PVC sewer pipe and some pressure water pipe are joined with a hub and spigot assembly that uses an elastomeric gasket to form the seal.

Thermal Welding

Thermal welding is a method of joining PVC piping components used in industrial applications, for shop fabrications and repairing a leaking joint. The pipe and fitting are joined by hot gas welding at the fillet formed by the junction of the fitting socket entrance and the pipe. Hot gas welding (which is similar to gas welding with metals except that hot gas is used for melting instead of a direct flame) consists of simultaneously melting the surface of a plastic filler rod and the surfaces of the base material in the fillet area while forcing the softened rod into the softened fillet.

Welding with thermoplastics involves only surface melting because plastics, unlike metals, must never be puddled. Therefore, the resulting weld is not as strong as the parent pipe and fitting material. This being the case, fillet welding is recommended as a repair technique for minor leaks only.

Welding temperature for PVC thermoplastic materials is 500°F to 550°F. All welding should be conducted by personnel adequately trained in the art of hot air welding of thermoplastics.

Transition Joints

Suitable techniques and products are available to join PVC to any other piping material. The main methods are the following:

- Elastomeric couplings with compression clamps are used to join two pipe ends together. These couplings are available for cast iron soil pipe to plastic and for copper tube to plastic. For steel pipe to plastic (both made to Sch 40 dimensions), the standard CISPI hubless coupling is satisfactory. Plastic adapter fittings are also made that provide a contour with a raised ridge to provide a better gripping area for the elastomeric coupling.
- Threaded adapters are used to join a component with standard tapered pipe thread to PVC pipe. The PVC portion may be either spigot (pipe OD size) or socket, and the threaded portion may be either external or internal NPT threads. PTFE tape is compatible with all plastic piping materials. If paste thread sealants are preferred, consult the sealant manufacturer's instructions for compatibility with the plastic being used. Some paste-type sealants may contain solvents that can damage certain plastic materials.

Caution must be used during assembly to prevent over-torquing of the PVC components. One or two turns beyond finger tight is all that is required to obtain a leak-free seal once the thread sealant has been applied (usually two to three wraps of PTFE tape in the direction of the threads).

It should be noted that threaded PVC joints do not carry the same pressure rating as solvent cemented joints and their use at temperatures above 110°F could result in leakage. Threaded PVC pipe and threaded PVC fittings should not be used in high-pressure systems where a leak could endanger personnel.

Thermoplastic external threads are susceptible to breaking from bending stress. This may be caused by two things:

1. The stress concentrations produced by the sharp notch of the thread form
2. The reduction in strength due to the removal of material when making the thread

Extra care must be taken to prevent side loads or bending stress when using thermoplastic external threads.

Thermoplastic fittings with internal taper pipe threads are susceptible to breaking from excessive hoop stress caused by wedge effect of the tapered external thread. This is especially true when a metal external taper pipe thread is tightened into a thermoplastic internal taper pipe thread. Generally, stress is shared equally between the two components when joining taper pipe threads. However, since metal has much greater mechanical strength properties compared to plastic, it does not compress when tightened. This shifts all the stress to the PVC material component. For this reason, some plumbing codes prohibit or limit the use of internal threaded plastic pipe fittings.

Soil pipe hub adapters are used to join PVC pipe to cast iron soil pipe at a hub. One end of the PVC fitting consists of a socket to join to the PVC piping and the other end is a ferrule that has an enlarged end. The ferrule is placed in the hub, which is then packed with oakum. Lead wool (not molten lead) is then packed in to complete the joint.

Fixture Connections

PVC piping is connected to the waste side of plumbing fixtures by various special fittings. Water closets are connected to a closet flange. This fitting consists of either a spigot or socket connection to the PVC waste piping and a flange that sets at the floor level to connect to the water closet by bolts. The connection is prevented from leaking by use of putty or a wax ring between the flange face and the water closet base.

Lavatory and sink traps are connected to the PVC piping by trap adapters. These adapters are available in a variety of configurations and sizes. The piping may be external or internal. The trap connection side usually has some form of compression nut and gland to provide the seal to the trap tubing material.

Marking

The outside of PVC pipe must be labeled as follows:

- The manufacturer's name or trademark
- The standard to which it conforms
- Pipe size
- Material designation code (e.g., PVC 12454, or PVC 1120, or PVC Type 1, Grade 1)
- DWV, if for drainage
- Pressure rating, if for pressure
- SDR number or Schedule number
- If the pipe is for potable water, a laboratory seal or mark attesting to suitability for potable water

The required marking repeat interval on pipe ranges from 2 feet to 5 feet.

When Installing PVC Piping Systems

Do

- Read the manufacturer's installation instructions.
- Follow recommended safe-work practices.
- Follow local code requirements.
- Keep pipe and fittings in original packaging until needed.
- Cover pipe and fittings with an opaque tarp if stored outdoors.

- Follow proper handling procedures.
- Inspect pipe for damage prior to use.
- Use tools specifically designed for use with plastic pipe.
- Use a drop cloth to protect finishes in the work area.
- Make certain that lubricants, fire stop materials, and sealants are chemically compatible with PVC.
- Cut the pipe ends square.
- Bevel and deburr the pipe ends with a chamfering tool.
- Use the proper primer and solvent cement and follow the manufacturer's application instructions.
- Use the proper size applicator for the pipe being joined.
- Rotate the pipe $\frac{1}{4}$ turn when bottoming pipe in fitting socket.
- Avoid puddling of cement in fittings and pipe.
- Use PTFE tape or approved paste thread sealant on all threaded connections.
- Assemble threaded joints carefully to avoid cross threading.
- Allow for movement due to expansion and contraction.
- Use hangers designed for use with plastic that will not damage the piping.
- Follow proper hanger support spacing requirements.
- Protect from nails, screws, and abrasive surfaces.
- Fill lines slowly and bleed the air from the system prior to pressure testing.
- Follow the manufacturer's recommended cure times prior to pressure testing.
- Test in accordance with local codes.
- Use only glycerin and water solutions for freeze protection when applicable.

Do Not

- Do not drop pipe or allow objects to be dropped onto plastic pipe.
- Do not use solvent cement near sources of heat, open flame, or when smoking.
- Do not use solvent cement that has exceeded its shelf life or has become discolored or jelled.
- Do not allow solvent cement to puddle in the pipe or fittings.
- Do not use products containing fractures, splits, or gouges.
- Do not use ratchet cutters below 50°F.
- Do not overtighten threaded connections.
- Do not install in cold weather without allowing for expansion.
- Do not pressure test until recommended cure times are met.
- Do not pressure test with compressed air or any other gas.
- Do not use in compressed gas or air piping systems.
- Do not use glycol-based solutions as antifreeze.
- Do not install next to heat-producing sources.
- Do not use PVC pipe or fittings to provide structural support.

STEEL PIPE

Steel pipe is the heaviest and strongest piping material available. Its uses include DWV, water service and distribution, heating and cooling, snow melting and snow making, radiant panel heating, fire protection, and gas distribution.

Steel pipe is either galvanized or black. Black pipe has a light coat of black paint applied to protect it in storage. Steel pipe is available in lengths up to 21 feet (longer lengths are available by special order) and in a wide variety of wall thicknesses. Galvanized pipe is steel pipe that has a protective metal coating applied so that it is less prone to corrosion. Galvanized pipe is not allowed to be used in fuel gas systems.

Steel pipe is made to dimensional standards called Schedules. In the size ranges normally encountered ($\frac{1}{8}$ inch to 4 inches), standard weight pipe is called Schedule 40 and extra heavy pipe is Schedule 80. For ease of combining materials, many plastic materials are made to Schedule 40 or Schedule 80 dimensions.

Code Status

Galvanized pipe is suitable for cold and hot water distribution but not for water service. Galvanized steel pipe is allowed for use as DWV as long as it is above ground and within the building. Ductile iron gravity sewer pipe is allowed for sewer pipes outside the building.

Support

According to the 2012 National Standard Plumbing Code, steel pipe must be supported as follows:

> Steel threaded pipe: $\frac{3}{4}$ inch size and smaller,10 feet intervals; one inch size and larger, 12 feet intervals

Testing

Testing of the completed piping system can be accomplished by using pressurized air or water. When testing at relatively high pressures air over water testing is preferred.

Expansion and Contraction

On excessively long runs or in applications where water temperatures can fluctuate, such as hydronic heating and cooling systems, expansion loops should be installed.

Joining

Steel pipe and its various fittings are the model for the plumbing industry. Nearly all the other materials used in plumbing work are substitutes for steel pipe and are evaluated based on how they compare with steel pipe.

Steel pipe is used on larger jobs for water and waste systems, fire protection sprinkler systems, heating and cooling systems, and snowmaking and snow-melting systems. Steel pipe also has many applications for process piping installations.

Steel pipe is the strongest and toughest piping material made. It also has the lowest coefficient of expansion with temperature change, is better able to withstand cyclic stress change, and is most resistant to mechanical or thermal shock. It can be machined and welded. In short, mechanically it is a superior product. However, substitute materials are usually satisfactory because steel characteristics far exceed the physical requirements for typical piping applications.

Malleable Iron Fittings

Malleable iron fittings are used for pressure applications and are available in standard and extra-strong weights, galvanized or painted black. The fittings are available in a wide variety of ells, 45s, tees, and reducers. Malleable iron fittings are low cost and very long lasting.

Welding Fittings

Welding fittings are used with black pipe to form very strong systems. Welding is a method of assembly in which two adjoining pieces of metallic materials are fused together by heating the material to the melting point and adding a filler material to complete the joint. Steel pipe and fittings can be welded together through the electric arc or gas welding process.

Grooved Fittings

Grooved fittings have grooves placed near the fitting end to accept a clamping device to hold the fitting to the pipe. Similarly, the pipe will have a groove formed near the end of the pipe. The clamping device will span the gap between the end of the pipe and the fitting to secure the two together. A gasket will form a liquid tight seal. Figure 2-24 shows a cross section of this type of joint.

FIGURE 2-24 Notice that each pipe has a groove formed near the end for joining purposes.

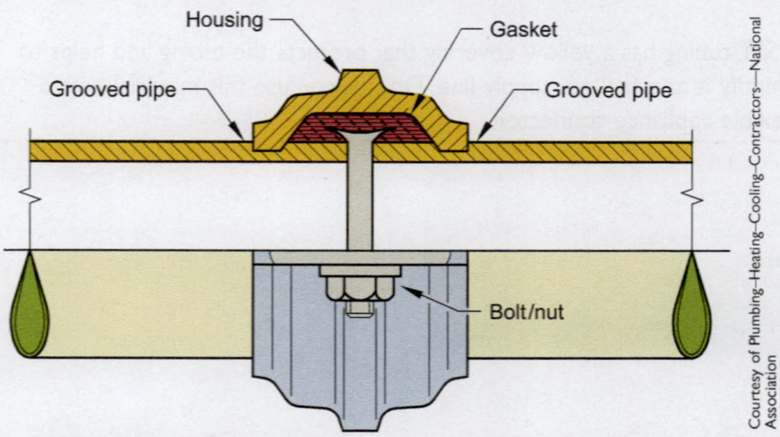

CORRUGATED STAINLESS STEEL TUBING

Gas Piping

Corrugated stainless steel tubing (CSST) is relatively semiflexible stainless steel tubing that is primarily used for fuel gas piping. The tubing is a thin-walled 300 series stainless steel that attains its flexibility by having corrugations along the wall. While this tubing is considered to be semiflexible, do not confuse it with a flexible appliance connector. CSST is usually made of a heavier gauge (thicker) material and cannot withstand multiple flexing cycles. Most manufacturers of CSST do not approve it being directly connected to movable appliances. Figure 2-25 shows a section of CSST tubing with its approved connector. Unlike a flexible appliance connector, the stainless steel tubing used for piping the system is covered with a bright yellow polyethylene jacket that contains the following information.

- The manufacturer's name
- The maximum working pressure
- The pipe size
- The testing standards and approval agencies

It is very important to notice that all CSST manufacturers gain flexibility by varying the specific dimensions of their tubing; one manufacturer may have deeper corrugations while another has wider ones. For this reason, it is crucial to never use fittings from one manufacturer with another one's tubing. All major manufacturers have fittings that have a standard national tapered pipe thread on one end so it is possible to use any manufacturer's product with another type of piping material; but the connection must be with an approved adapter.

When sizing CSST, it is important to use the EHD sizing and not the nominal pipe size. EHD is the "Equivalent Hydraulic Diameter," which is a standardized flow comparison. As stated earlier, each manufacturer uses different spacing and depths of their corrugations so naturally the internal diameter and the flow resistance for each pipe is different. The EHD of the tubing allows sizing charts to give a more accurate indication and comparison of the tubing performance. Always use the manufacturer's sizing charts to accurately size each piping system.

FIGURE 2-25 CSST tubing has a yellow covering that protects the tubing and helps to identify it as a fuel gas supply line. Do not confuse this material with a flexible appliance connector.

Photo by Ed Moore

The major manufacturer websites for CSST products offer installation manuals and other important product information. Visit them at:

Wardflex—http://www.wardflex.com
Gastite—http://www.gastite.com
Tracpipe—http://www.tracpipe.com

REVIEW QUESTIONS

True or False

1. __________ Plastic materials will not conduct electricity.
2. __________ As pipe size increases, the pressure rating increases.
3. __________ Plastic pipe sizes are either Outside Diameter or Inside Diameter controlled.
4. __________ Universal plastic pipe cements are preferred by the manufacturers of plastic pipe.
5. __________ Most plastic piping must be protected from prolonged exposure to sunlight.
6. __________ All CSST tubing is manufactured to the same dimensions.
7. __________ Cast iron pipe comes standard in 20 feet lengths.
8. __________ Type L tubing has thicker walls than Type M.
9. __________ In general, as the tubing diameter increases, the maximum piping support distance can increase.
10. __________ Plastic pipe expands and contracts more than steel.

Chapter 3

LEARNING OBJECTIVES

The student will:

- Identify special applications for thermoplastic piping materials in fire-resistive construction systems.
- Describe the properties of general chemical resistance.

Special Applications of Plastic Pipe and Fittings

PLASTIC PIPING IN FIRE-RESISTIVE CONSTRUCTION

Plastics and Combustion

Plastics are organic materials that, like many other building materials, are combustible and can be made to burn in a fire situation. Included in this group are shingles, carpeting, wood wall studs, paper-faced insulation batts, wire coatings, furniture, wall coverings, and so on. A source of heat and oxygen are essential to the combustion of plastic pipe materials, which in themselves have self-ignition temperatures generally exceeding 650°F. This temperature is well above that of ordinary combustible materials such as wood, which ignites between 260°F and 500°F. See Table 3-1.

It has been estimated that plastic piping systems represent less than 1 percent of the total combustible products in most building structures and much of that piping is located behind fire-resistant walls. Studies show that a typical two-story house contains about 150 pounds of plastic piping material, compared to 40,000 to 50,000 pounds of other combustible materials such as structural lumber and furniture.

In a fire situation temperatures can exceed 1,000°F. At those temperatures, combustibles will burn. Nearly all burning substances can emit toxic gases when burned under uncontrolled conditions. Investigations of documented fire fatalities show that carbon monoxide gas is commonly formed by burning and is the primary toxic combustion product. Therefore, while plastic pipes are combustible and can be consumed in a fire, because of their low fire mass compared to all the other combustibles in a building structure, they supply a very small portion of the toxic gases released in an out-of-control building fire situation.

IN THE FIELD

In most states, every home with fuel burning appliances must be equipped with a carbon monoxide detector. Warning: Not all smoke detectors will detect carbon monoxide.

TABLE 3-1 Ignition Temperature of Plastic Pipe Materials and Other Common Building Materials

Material	Ignition Temperature (°F)
PTFE	986°
PB	914°
CPVC	900°
PVDF	790°
ABS	780°
PVC	735°
PE	660°
PEX	660°
Douglas Fir	500°
Cotton	490°
Paper	445°
Wood	401°

Note: PVC, CPVC, and PVDF do not support combustion. They require an external flame source to support combustion at temperatures below 50°F. Ignition temperature is determined by testing specimens of the material. These are self-ignition values based on ASTM D 1929 test method.

When installed properly, plastic piping is safe and meets or exceeds applicable building code requirements. It is important for the plumber to understand that not all plastics are the same and that each material family and product has certain installation requirements that may be slightly different.

In addition to being installed mostly behind gypsum wallboard, which protects it from any fire in the occupied areas, plastic piping is used for pressure, gravity flow, and drain, waste, and vent (DWV) systems. Plastic hot/cold water distribution and fire sprinkler systems are not likely to contribute fuel to a fire or to cause it to spread. As a matter of fact, there have been some incidents in which heat from a fire caused a plastic water pipe to rupture and release a spray of water that extinguished the fire.

Building codes seek to strike a balance between utility, cost, and safety of a building. They specify, among a long list of items, sections that deal with fire and preventing the spread of fire and/or smoke for specified periods of time. Thus, various construction assemblies for walls, floors, and ceilings are described as 1-hour, 2-hour, or 4-hour rated. These references mean that for the stated time, a fire on one side of the barrier (floor, wall, ceiling) will not spread to the other side.

One-hour rated walls, ceiling, and floors are required by building codes in more circumstances than 2-hour or 4-hour ratings. The concept of the 1-hour requirement is that it gives building occupants time to evacuate a building in the event of a fire. That being the case, the escape route(s) from a building (usually hallways and corridors) must have 1-hour rated walls, floors, and ceilings.

Two-hour and 4-hour construction designs are most often used to separate portions of the structure that have different owners, or that could have different owners, such as a building wall built on the property line.

Penetration of Rated Assemblies

The actual fire resistance capability of wall, ceiling, and floor sections is established by subjecting large sections (10′ × 10′ walls or 12′ × 15′ floors) to ASTM E 119 tests. In these tests, a finished assembly section containing the pipe system is positioned in an atmospheric pressure furnace as a wall or floor and tested under controlled time-temperature conditions for the rating time period of 1, 2, 3, or 4 hours. The assembly is observed during the test to ensure that the temperature rise on the unexposed side does not exceed limits and that flame does not pass through the assembly during the test. Typically, fire has been observed to spread in buildings from one area to another through the collapse of walls or by openings through which flame and hot gases can pass.

Because utilities are essential for human occupancy of a building, it is necessary to put openings through the fire-rated walls of the structure to run utilities such as water; drain, waste, and vent lines; HVAC; and electrical. However, to maintain the fire resistance of a building, these penetrations must be done in such a way that the fire rating of the assembly is not destroyed or reduced. This is accomplished by the use of fire-stop systems and devices that are available commercially and include such things as assemblies, sealant, foams, putties, or wraps around the pipe or opening, some of which may expand under a fire condition so as to close the opening entirely. Figure 3-1 shows an example of fire-rated expandable foam that can be used for pipe penetrations.

FIGURE 3-1 Expandable foam can be used to seal pipe penetrations. This brand is listed for fire-rated walls.

Photo by Ed Moore

The ASTM E 814 test method for Fire Tests of Through Penetration Fire Stops is used to evaluate the performance of such systems and devices. This is accomplished by using the same time/temperature furnace conditions of E 119 but with a smaller test assembly holding the pipe penetration sealing system or device in place.

In addition, the furnace is under a positive pressure and the fire-stop system is sprayed with a fire hose at an appropriate time in the test to see how the assembly resists the force of the water stream from the fire hose. More detailed information on testing procedures and fire-stops is given later in this chapter.

In summary, plastics are combustible materials that can be consumed in an out-of-control fire. Because of their location, they are generally involved in the later stages of a building fire. Due to the relatively low amount of fire mass, they are responsible for a very small portion of the total emissions given off during a fire.

Building codes specify the fire ratings required in the building structure. All penetrations must be fire-stopped using approved methods.

Plastic pipe products meet all applicable building code requirements and fire ratings when installed properly. It is recommended that piping installers become familiar with fire-stop techniques, systems, and materials, and how to properly select and install them on the various pipe materials so that the job meets the requirements of the building codes.

MASTER COMPARISON OF THERMOPLASTIC PIPING MATERIALS

Chapter 2 focused on the capabilities of the most common plastic and metallic materials used for piping. In this chapter, Table 3-2 lays out the pertinent considerations involved in comparing plastic piping materials and products for a given application. The table shows the application first, the plastic material, the ASTM or other standards for piping products that apply, joining methods, dimensional standards, and sizes available for the application.

Plastic Pipe Selection

For a given application, the best piping choice for the building owner is the one with the lowest life-cycle cost for the installation. While life-cycle cost is an easy concept,

TABLE 3-2 Plastic Piping Material

Plumbing and Related Uses	ABS	CPVC	PB*	PE	PEX	PP	PVC
Drain, waste, and vent	X						X
Hot and cold water		X	X		X		
Fire sprinkler piping							
Outside sewers and drains	X						X
Septic fields—subsoil				X			X
Tubular waste	X					X	X
Water piping	X*	X	X	X	X		X

*PB is no longer allowed to be used, but it can still be found in older construction.

the individual elements are not always easy to measure because they must include all operation and maintenance costs over the life of the system.

Since most owners do not have enough experience to evaluate the long-term part of this concept, they may ask for help from their designer or contractor to arrive at a decision; or, in many cases, they simply decide on the basis of first cost. While the initial cost of a plastic piping installation will normally compare favorably with other materials, initial cost should be reviewed with the long-range consequences in mind.

For permanent piping installations, the most important requirement for the piping is that it shall be of adequate flow capacity for all likely demands and be capable of withstanding the material to be carried—chemically, thermally, and mechanically. Plastic piping has excellent all-around ability to withstand all of these types of stress for a wide range of applications in modern buildings.

The charts in this chapter can be used to assist the designer and contractor in picking a suitable plastic material for the application as influenced by local availability of products, skill and experience of available work force, requirements for specialized tooling (if any), preferred joint type, pressures to be contained, and temperatures to be encountered.

After the preceding factors are evaluated, the piping material can be selected on the basis of cost, provided that there are multiple choices.

Although most of the products are approved by most plumbing codes, the designer, the contractor, and the installer should know which plastic piping products their local codes allow.

Typical Applications

Table 3-2 shows that ABS and PVC are suited for drain, waste, and vent piping applications. Note that there are two material choices and each material is available in the form of IPS-OD pipe with both solid wall and cellular core wall. There is also a standard for a special size pipe that is 3.25-inch OD so that it will fit within a 2″ × 4″ wall.

In a similar fashion, the three hot and cold water pipe materials are available in CTS tube sizes and they are supplemented by some IPS pipe sizes.

The table also shows possible material choices for sewer, drain field, fire sprinkler, septic field, tubular waste, and water piping.

Industrial Special Applications

For these job types, the possible selections from the tables in this chapter are reviewed against chemical exposure data. It is important that recommendations be obtained from the plastic pipe manufacturer when selecting piping materials to transport any chemical that the processes in a building might require.

Heating and Cooling Applications

These projects frequently require a range of pipe sizes. In such cases, the tables should be scanned for materials that are available in the required sizes and temperature rating that are sufficient for the system operating pressures. This first screening should generate a fairly short list, which can then be reviewed for availability and cost.

In the past, many cooling systems operated with chilled water at temperatures of 40°F and above. However, recent developments have led to ice storage systems that have fluid temperatures below 32°F and therefore the fluid's freezing temperature must be lowered.

Sizing Plastic Piping for Plumbing Systems

Plumbing systems deliver potable water from a source to the point of use. They must also contain and remove water-borne waste from the point of use and deliver it to a point of disposal. The first and foremost requirement for a plumbing system is that the quality of the potable water must not be compromised by any foreign material, including anything that can be extracted from the piping materials, the valves, or any of the appurtenances. In order to have this assurance, all these materials and even the water contact materials used in water treatment plants and water tanks need to qualify for PW (potable water) contact based on the tests and requirement of ANSI/NSF Std 61. All plastic PW piping products carry the PW mark and meet the Std 61 requirements.

The fixtures that have water supply and waste pipe connections must be designed and installed to prevent any possibility of cross connections and contamination of the potable water supply system. The DWV system must carry away from the plumbing fixtures not only the water-borne waste that (usually) originates from the potable water discharged at plumbing fixtures, but any other material or process used or encountered on the premises.

Additional requirements that follow closely are that the system has sufficient water supply capacity and sufficient waste-removal capacity to serve the maximum probable needs of the occupants of the building.

Water Piping

All model plumbing codes require that both the hot and cold water distribution systems within buildings use only plastic piping that is suitable for hot water applications. The reason for this restriction is to ensure that cold water–only material cannot be installed by mistake in the hot water distribution piping.

Water service piping bringing water to the building from wells or mains is permitted if they are suitable for cold water only.

Waste Piping

All model codes require adequate DWV systems for all buildings wherein water distribution systems and fixtures are installed. Most plumbing codes recognize plastic pipe for all types of waste piping, drains, vents, and rainwater piping.

DETERMINATION OF LOADS IN PLUMBING SYSTEMS

As described in Chapter 2, plumbing fixtures are characterized by intermittent use. A water closet or urinal is flushed, a certain requirement for water flow is imposed on the system for a short period of time, and a certain waste load is imposed on the drain system for a somewhat different time period. When a shower is on, the water load and waste load are practically identical in flow rate and timing, but filling (or draining) a bathtub, kitchen sink, or lavatory result in different supply and drainage flow rates and very different times of occurrence.

With any plumbing fixture, when a use cycle is completed, there is a time lapse before the fixture can be used again. This time between uses is the dominant factor in determining the average flow to or from a plumbing fixture.

Therefore, we are interested in two kinds of flow rates to or from a fixture:

- The instantaneous flow for the fixture. These flows must be accommodated in the piping to and from the fixture.
- The average flow over time of successive uses of the fixture.

These flows—instantaneous and average—are used to determine the impact of more than one fixture on the system. How these various fixtures and the flows that are developed combine to produce a total system demand is the problem that must be solved to design a plumbing system.

Water Supply Fixture Units (WSFU)

To make possible the summation of the impact on the system of several plumbing fixtures—whether the same or different types—Professor Roy B. Hunter of the University of Iowa assigned a series of index numbers for different fixtures that can be added together and compared to a curve he developed to determine the maximum probable demand that the group of fixtures can impose on the water supply system. Pipe sizes are then selected from tables that show flow capacity of piping as a function of pipe size and allowable pressure drop. Professor Hunter's work was done in the 1920s and 1930s.

These index numbers are called water supply fixture units (WSFU). Further research and field testing in the 1990s by Professor Tom Konen of Stevens Institute has refined these numbers to appropriate values for plumbing fixtures as manufactured now and in the recent past. These values are presented in Table 7-7 of this text.

Drainage Fixture Units (DFU)

Dr. Hunter's work also included the development of index numbers for drainage calculations. These numbers are called drainage fixture units (DFU). Drainage fixture unit values for typical plumbing fixtures are shown in Chapter 11. For a group of plumbing fixtures, the DFUs are combined and pipe sizes are selected from tables that show capacity (in DFUs) as a function of pipe size and slope (for gravity drain lines).

Hunter's Curve

Chapter 7 presents Hunter's Curve in tabular form. The correlation between WSFU and flow rate is expressed in gallons per minute. Notice that there are two columns—the higher one is for systems supplying principally flushometer water closets and the lower curve is for systems supplying all other fixture types. At 1,000 WSFU, these two branches merge.

Maximum Probable Demand

The early concept of fixture unit values made two distinctions to recognize that people impose loads on a plumbing system, not plumbing fixtures. Those distinctions were to put installations into two categories: public and private. Public values were higher than the private values because it was recognized that generally more people had access to plumbing fixtures in public settings.

The results of the research (both field surveys and computer models) of Dr. Konen indicated the need for more divisions of the types of people loading. Four categories of use were selected, as follows:

- Private residential
- Multifamily residential
- Nonresidential
- Heavy use

The goal of this study was to enable the determination of a close estimate of maximum probable demand—the water flow or drainage load that would not be exceeded more than about 1 percent of the time. In almost all cases the calculated values are never exceeded, and most of the time the system load is much less than the estimated maximum.

Water Pipe Sizing

There are two basic design concepts for water piping. The common method used for many years was to provide large mains supplying smaller branches. The mains reduce in size along the way as the loads are taken off on the way to the ends of the system. The WSFU load served by each segment of the piping system is calculated, and the design flow is selected. The pipe sizes are then selected from a table similar to Table 3-3.

Another method that is gaining in popularity uses individual runs to each fixture (described in Chapter 7). These pipe runs originate in a header (or manifold) located at some convenient place. The header may contain valves for each run so that each fixture can be isolated from the system for service as needed.

Drain Pipe Sizing

Drain pipes are sized by DFU load, whether they are vertical or horizontal. If the pipe is horizontal, the slope of the line is factored into the required pipe size. Table 3-4 shows the relationships.

Vent Pipe Sizing

Vent pipes are sized by DFU loading on the drain being vented, the stack diameter, and the length of the vent pipe. Refer to Chapter 17 for more detail.

TABLE 3-3 Water Flow Pressure Drop in psi/100 ft.

Schedule 40 Plastic Pipe (PE, PVC, & ABS) (Smooth)										
		4′/sec				8′/sec				
Normal Size	Inside Diameter	Flow gpm	WSFU (A)	WSFU (B)	PD	Flow gpm	WSFU (A)	WSFU (B)	PD	Flow at 10 Psi per 1000
½	0.622	3.8			7	7.6	10		24	4.8
¾	0.824	6.6	8		5.1	13.3	18		17.5	9.5
1	1.049	10.8	15		3.7	21.5	34		13	19
1¼	1.38	18.6	27		2.7	37.3	75	21	9.5	
1½	1.61	25.4	41		2.3	50.7	125	51	8	
2	2.067	41.8	92	32	1.7	83.6	290	165	6	
2½	2.469	59.6	172	72	14	119.3	490	390	4.8	
3	3.068	92.1	330	210	1.1	184.3	850	810	3.7	
4	4.026	158.7	680	620	0.8	317.5	1,900	1,900	2.7	
CPVC Tubing SDR 11 (ASTM D 2846)										
½	0.485	2.3			16.6	5.5			65	
¾	713	4.9			9.9	9.8			40	
1	0.921	8.3	11		4.6	16.6	25		18	
1¼	1.125	12.4			3.5	24.8	40		14	
1½	1.329	17.3	26		2.8	34.6	69	19	11	
2	1.739	29.6	51	13	2.1	59.2	171	72	8.4	
Schedule 80 Plastic Pipe (PVC, CPVC) (Smooth)										
½	0.546	2.9			6.9	5.8			27.6	
¾	0.742	5.4			4.8	10.8	15		19	
1	0.957	9.4	12	3.4	18.8	23			13.6	
1¼	1.278	16	24		2.4	32	58	17	9.6	
1½	1.5	22	34		2	44.1	100	36	8	
2	1.939	36.8	73	23	1.5	73.6	250	120	6	

Note: Column A is for systems without flushometer valves. Column B is for systems with flushometer valves.

TABLE 3-4 Building Drains and Sewers[1]

Maximum Number of Drainage Fixture Units (DFU) That May Be Connected to Any Portion of the Building Drain or the Building Sewer.

Pipe Size-Inches	Slope Per Foot $\frac{1}{16}$-Inch	Slope Per Foot $\frac{1}{8}$-Inch	Slope Per Foot $\frac{1}{4}$-Inch	Slope Per Foot $\frac{1}{2}$-Inch
2			21	26
3		36[2]	42[2]	50[2]
4		180	216	250
5		390	480	575
6		700	840	1,000
8	1,400	1,600	1,920	2,300
10	2,500	2,900	3,500	4,200
12	3,900	4,600	5,600	6,700
15	7,000	8,300	10,000	12,000

Notes:

1. On-site sewers that serve more than one building may be sized according to the current standards and specifications of the Authority Having Jurisdiction for public sewers.
2. See Section 11.5.6.c & d for the restrictions on the number of water closets installed on 3″ building drains and sewers.

Source: 2012 National Standard Plumbing Code.

Storm Drainage

Table 3-5 shows the capacity of vertical drain pipes expressed in terms of roof area served for several rainfall rates. Table 3-6 shows the capacity of horizontal storm drains at various pipe slopes and rainfall rates.

TABLE 3-5 Vertical Drainpipe Capacity

Diameter of Conductor or Leader (in.)	Design Flow in Conductor (gpm)	Allowable Projected Roof Area at Various Rates of Rainfall				
		1″	2″	4″	5″	6″
2	23	2,176	1,088	544	435	363
$2\frac{1}{2}$	41	3,948	1,974	987	790	658
3	67	6,440	3,220	1,610	1,288	1,073
4	144	13,840	6,920	3,460	2,768	2,307
5	261	25,120	12,560	6,280	5,024	4,187
6	424	40,800	20,400	10,200	8,160	6,800
8	913	88,000	44,000	22,000	17,600	14,667

Note: Consult with local code officials for the value to use for a particular location.

TABLE 3-6 Horizontal Storm Drain Capacity

Diameter of Conductor or Leader (in.)	Design Flow in Conductor (gpm)	Allowable Projected Roof Area at Various Rates of Rainfall per Hour (sq. ft.)					
		1″	2″	3″	4″	5″	6″
Slope $\frac{1}{16}$ in./ft.							
5	100	9,600	4,800	3,200	2,400	1,920	1,600
6	160	15,440	7,720	5,147	3,860	3,088	2,575
8	340	32,720	16,360	10,907	8,180	6,544	5,450
10	620	59,680	29,840	19,893	14,920	11,936	9,950
12	1,000	96,000	48,000	32,000	24,000	19,200	16,000
Slope $\frac{1}{8}$ in./ft.							
3	34	3,290	1,645	1,097	822	658	550
4	78	7,520	3,760	2,507	1,880	1,504	1,250
5	139	13,360	6,680	4,453	3,340	2,672	2,230
6	222	21,400	10,700	7,133	5,350	4,280	3,570
8	478	46,000	23,000	15,333	11,500	9,200	7,670
10	860	82,800	41,400	27,600	20,700	16,560	13,800
12	1,384	133,200	66,600	44,400	33,300	26,640	22,200
15	2,473	238,000	119,000	79,333	59,500	47,600	39,670
Slope $\frac{1}{4}$ in./ft.							
2	17	1,632	816	544	408	326	272
3	48	4,640	2,320	1,547	1,160	928	775
4	110	10,600	5,300	3,533	2,650	2,120	1,770
5	196	18,880	9,440	6,293	4,720	3,776	3,150
6	314	30,200	15,100	10,067	7,550	6,040	5,035
8	677	65,200	32,600	21,733	16,300	13,040	10,870
10	1,214	116,800	58,400	38,933	29,200	23,360	19,470
12	1,953	188,000	94,000	62,667	47,000	37,600	31,335
15	3,491	336,000	168,000	112,000	84,000	67,200	56,000

Typical design rainfall rates in the United States vary from 2 inches to 4 inches per hour in most regions, with the Gulf Coast, Florida, and the southern Atlantic coast having areas as high as 9 inches/hour. Refer to *Plumbing 301* for more information concerning sizing of storm drains.

VENT TERMINALS

Vents in sanitary waste (DWV) systems serve two purposes:

- To allow sewer gases to travel up the stack and vent to the atmosphere above the roof.
- To allow air to enter the system so that pressure variations caused by drainage flows through the system do not draw the water seals out of the traps in the system. The circulation of air has the added benefit of starting the aerobic decomposition of waste before it reaches the waste treatment plant.

The drainage system must have at least one main vent stack that will terminate above the roof line. Extensive systems may have additional roof vent penetrations plus interconnecting vent piping to serve all the fixture drains.

Air Admittance Valves

Depending on the size of the system and the number of fixtures, the vent piping can be simplified and reduced by the use of air admittance valves (AAVs). An AAV is a one-way valve that allows air to enter the plumbing drainage system when negative pressure develops in the system but closes by gravity and seals the vent terminals under other conditions so that sewer gases do not escape into the building environment. AAVs, sometimes called "Studor Vents," are available in two types: those rated for use on individual, branch, and circuit vents, and those rated for stacks. It is important to match the correct one for the application. Systems designed with AAVs should have at least one open pipe vent terminating to the outdoor atmosphere. AAVs are designed and tested to meet performance standards that were developed to ensure acceptable performance of AAVs over the lifetime of a piping system. Figure 3-2 shows an AAV that can serve up to a 2-inch vent.

Uses

AAVs are installed to serve as vent terminals for individual, branch, or stack vents. They are most commonly used for solving issues that surround the venting of kitchen island sinks.

Code Status

While the 2012 International Plumbing Codes allow the usage of AAVs, the 2012 National Standard Plumbing Code does not address them. Mostly this is due to the debate of using "mechanical devices" in the DWV system. Some local plumbing codes have been changed to permit their usage. Check your local plumbing

FIGURE 3-2 Air admittance valves, like this one, have greatly simplified rough-in of island kitchen sinks.

Photo by Ed Moore

code to determine whether AAVs are approved. If the local plumbing code requires a listing or testing by a recognized agency, the agency's mark must be visible on the product.

Standards

There are two standards that cover AAVs:

- ASSE Standard 1050: Performance Requirements for Stack Type Air Admittance Valves for Plumbing DWV Systems
- ASSE/ANSI Standard 1051: Performance Requirements for Individual and Branch Type Air Admittance Valves for Plumbing Drainage Systems

Products Available

AAVs are available in sizes that fit vents from $1\frac{1}{4}$ inch to 4 inches. They can be used with any kind of piping material. In many cases, AAVs can be purchased with a fitting that allows them to be attached with either a mechanical threaded joint or solvent cement joint.

Installation

Most manufacturers have very detailed instructions on how to install AAVs. Contractors should refer to these instructions and to any provisions in the local plumbing codes to ensure that the installation meets the necessary requirements. The following are key guidelines for successful installation of AAVs:

- When installed on single fixtures or branches, AAVs should be located a minimum of 4 inches above the weir of the fixture trap. Attic installations must be 6 inches above any ceiling insulation.
- AAVs must be located to allow adequate air to enter the valve. When located in a wall space or attic space, ventilation openings must be provided, opening into the space. Location of the AAV in a vanity cabinet or sink cabinet is acceptable.
- AAVs must be installed in the vertical upright position. The maximum offset from the vertical position cannot exceed 15 degrees.
- AAVs must be connected to drains with a vertical connection to maintain an unblocked opening in the piping to the AAV.
- Stack type AAVs are acceptable only in engineered drainage systems. The AAV must be at least 6 inches above the highest flood level rim of the fixtures being vented in stack applications.
- When a horizontal branch connects to a stack more than four branch intervals from the top of the stack, a relief vent must be provided.
- A minimum of one vent must extend outdoors to the open air for every building plumbing drainage system. For drainage systems connected to private sewage disposal systems, this vent should be located as close as possible to the connection between the building drain and the building sewer.
- AAVs should be installed after the drainage system rough-in test.
- AAVs should not be used as a vent terminal for any sump vent.
- AAVs should not be used to vent a special waste or chemical waste system.
- AAVs should not be located in a supply or return plenum.
- The maximum height of drainage stack being vented by a stack type AAV should be six branch intervals.
- When exposure to extreme temperatures is possible, protection of the AAV with insulation material is recommended. Care should be taken not to block the flow of air into the AAV.
- Refer to the manufacturer's recommendations for proper sizing of AAVs.

PLASTIC PIPING IN NONPOTABLE WATER SYSTEMS

Plastic piping may be used in systems other than the water service, water distribution, and DWV applications that have been the focus of this chapter. In many of these applications, plastic piping is the preferred piping because of its unique combination of strength, flexibility, and chemical resistance.

The following descriptions are brief overviews of the various applications and do not contain all of the information that might be required for design of a proper system. Before installing systems in these applications, consult with the piping manufacturer or equipment manufacturer for specifics related to these special applications.

Radiant Heat

Flexible plastic tubing such as PEX-AL-PEX is a natural choice for radiant heating systems. One type of radiant system, in-floor, consists of tubing loops cast directly into concrete, lightweight gypsum concrete, or attached to the underside of subflooring. Since the loops are generally 200 feet to 500 feet in length, and the tubing is spaced from 6 inches to 18 inches apart, flexible plastic tubing is easily installed into these systems. Loops can be installed with no joints located in or under the floor. It is customary to have a loop for each section of the building—a bathroom, for example. In turn, the flow of heated water will be controlled by a thermostat-driven control valve. Other types of radiant systems utilizing baseboard radiators can also be supplied by plastic tubing. Figure 3-3 shows typical construction for radiant tubing in a concrete slab.

When the tubing is encased within the concrete floor, such as in Figure 3-3, the tubing can be attached directly to reinforcement mesh or may be stapled or otherwise attached to the top of the subfloor or underfloor insulation (when used) prior to placement of the concrete.

For systems without concrete, the tubing is attached directly to the underside of the subfloor with clamps or staples. Generally, a radiant barrier and insulation are installed below the tubing to reflect the radiant heat back into the heated space. Some systems utilize a thin aluminum plate with a preformed groove for the tubing. Figures 3-4 and 3-5 show different examples of radiant floors.

FIGURE 3-3 Tube spacing is based on the amount of heat loss for that location. Spacing is usually closer near outside walls.

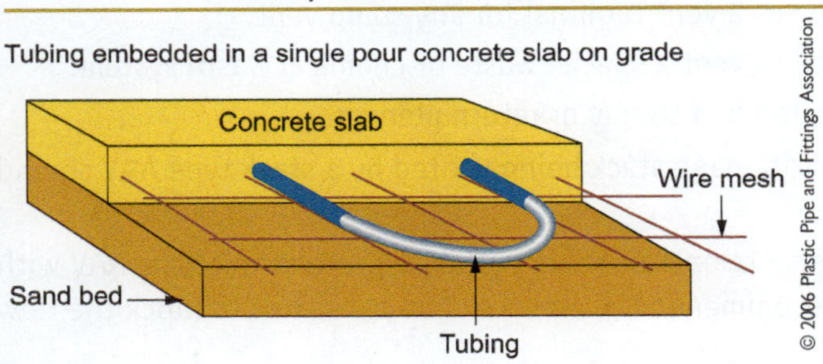

FIGURE 3-4 Furring strips are used as spacers and to secure the finished floor. Pay particular attention to nail locations so as not to puncture the tubing.

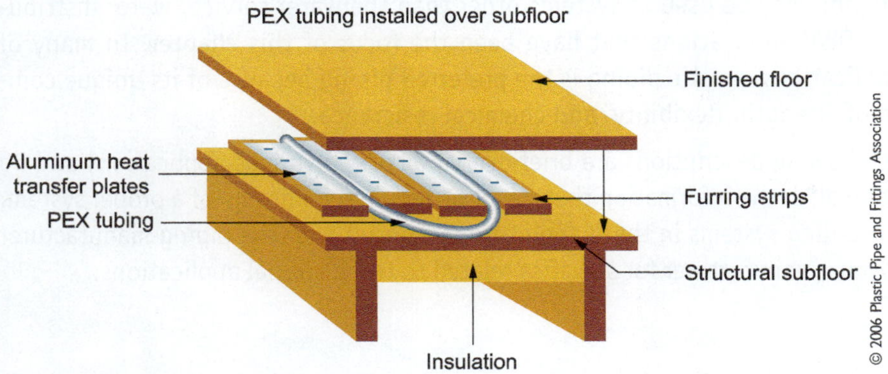

FIGURE 3-5 This installation method is especially well suited for remodel jobs where the finished floor is already in place.

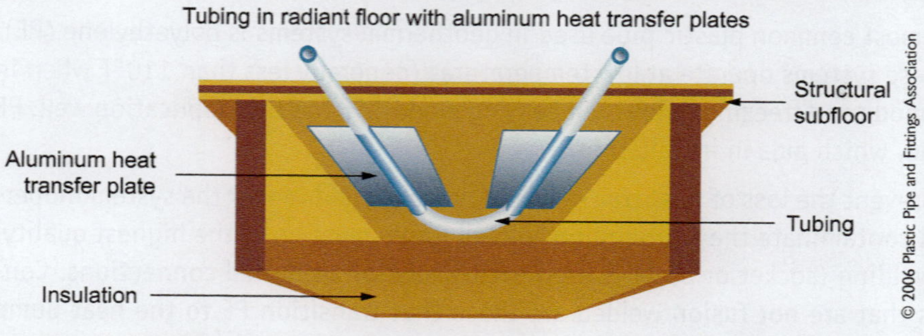

Note: The insulation should be in contact with the piping and aluminum plate.

The concept of in-floor systems is simple. Warm to hot water (as required by the system and heating load) is circulated in the tubing. The heat from the circulating water is transferred either directly to standard or lightweight concrete or to the space under the subfloor. The entire floor now becomes a heating mass that warms people and objects in the room primarily by radiant transfer and, to a much lesser degree, by conduction.

Without going into a detailed discussion about the dynamics behind radiant heat transfer principles, a simple description is that warm objects (above absolute zero) give off radiant infrared heat waves from all sides. A warmer object will transfer energy to a colder object. Radiant energy waves are the way that the sun's energy is transferred across millions of miles of empty space. Since our bodies are great receptors of radiant energy (we are easily warmed by the sun), heating by radiation is comfortable and efficient. For more information concerning transfer of heat, refer to *Plumbing 301*.

Most in-floor systems operate at water temperatures of 85°F to 140°F. Radiator-type systems can operate at up to 200°F. The maximum rating of the tubing and the maximum system temperature must be checked before installing plastic tubing into radiator systems. To allow for some temperature overrun of the heating system during especially cold times, the system should be designed to operate at 20°F less than the maximum temperature rating of the tubing. These systems must be left dry or protected with antifreeze if installation is to take place when freezing temperatures may occur.

Snow Melting Systems

Snow melting systems work in the same manner as radiant heating systems. However, since systems must be capable of heating the concrete without the benefit of an insulated structure, the heat output (BTUs/hour) is generally 3 to 6 times higher than for indoor systems. Tubing spacing is usually 12 inches or less to avoid melting in strips. These systems must be protected from freezing by the addition of antifreeze.

Geothermal Ground Source Systems

Geothermal systems use the near-constant temperature of the earth (below about 5 feet deep) as a heat sink/source for cooling and heating systems. These systems generally use long lengths of plastic pipe that are laid horizontally or in vertical bore holes through which an antifreeze solution is circulated. Depending on whether the system is in the cooling or heating mode, heat will either be transferred to the earth or the solution will be heated by the earth. This solution is then circulated back to a heat pump unit where the energy transfer is completed.

The most common plastic pipe used in geothermal systems is polyethylene (PE). Since these systems operate at low temperatures (generally less than 110°F when in cooling mode and freezing or lower in heating mode), PE fits this application well. PE is flexible, which aids in installation.

To prevent the loss of heat transfer fluid, which would render the system inoperable and contaminate the surrounding soil, all joints must be of the highest quality. Fusion welding (socket or butt) is used exclusively for all buried connections. Connections that are not fusion welded are those that transition PE to the heat pump units above ground inside the building.

All piping currently in use is IPS-OD HDPE (high-density polyethylene). When repairing or adding on to an existing system, it must be determined what material has been installed.

Ice Rinks

Ice rink systems are installed in similar fashion to hydronic heating systems, but they are used to freeze the ice. A cooled antifreeze solution is circulated through loops on close centers. The typical material used is PE. PE has excellent low-temperature properties that make it ideal for these types of systems.

Several connection methods are used depending on the type of supply header system employed. Fusion, mechanical and barb, and hose clamp fittings are common. Since there is currently no standard for these systems, IPS-OD, IPS-ID, and CTS-OD piping are all being used. When repairing one of these systems, the size basis must be determined.

Freezer Warehouses

These systems are installed to keep the ground underneath a freezer warehouse from freezing. Piping is laid in the earth under a concrete slab, and brine solution is circulated through the pipe to keep the ground under the floor at a constant temperature. The most common material used for these systems is PE, but many systems were installed with polybutylene. Since there are no standards in place for these systems, different size bases are being utilized as well as several connection methods.

Water Source Heat Pumps

Plastic piping is ideal for these systems since they operate at water temperatures of 60°F to 110°F. CPVC, PE, and PVC piping are all suitable for this application. These systems often include a heater and a cooling tower to keep the water within the operating range, but the major system advantage is that they can absorb heat from the sunny side of a building and transfer it to the cool side as demanded by individual occupants.

Gas Piping

Although ASTM D 2513—Thermoplastic Gas Pressure Pipe, Tubing and Fittings—has provisions for the use of PE and CPVC materials, PE piping is the primary product for this application. The design stress for plastic pipe used in distribution of natural gas and petroleum fuels is regulated by the U.S. Department of Transportation as published in Part 192, Title 49 of the Code of Federal Regulations. The American Gas Association Plastic Materials Committee, the Fuel Gas Division of PPI, and members of ASTM Committee F-17 are cooperating with the ASME Gas Piping Technology Committee to provide assistance in selecting safe design stress levels for various kinds of plastic pipe. The National Fuel Gas Code limits the use of plastic gas pipe to below-ground installations outside the building. Plastic gas pipe may go above ground in a metal-encased meter riser.

Under Slab Warm Air Duct

Where forced warm air furnaces are used with concrete slab floor homes, the heating ducts are often embedded in gravel below the slab. PVC pipe has been used for years as the duct to carry the warm air from counter-flow furnaces to the registers around the perimeter of the building.

Chilled Water Cooling Systems

For years, PVC piping has been used for chilled water cooling systems that operated at water temperatures of 40°F to 45°F. New systems have been developed that operate at temperatures below 32°F, and therefore chemicals must be added to lower the freezing temperature of the liquid used. These are sometimes referred to as ice storage systems because the design provides for chiller operation at off-peak hours when power rates are lower. Then the units actually make some ice that will provide the cooling capacity needed during the day.

CORROSIVE FUME HANDLING SYSTEMS

PVC and CPVC extruded piping is used for the construction of corrosive-resistant fume-exhaust systems. Systems of this type are typically found in the metals finishing industry, electronics industry, anodizing industry, and laboratory work.

Many industrial operations (metal finishing, plating, etc.) produce acidic fumes, VOCs, heavy metals, and other corrosive or hazardous materials during the manufacturing process. The corrosive fumes are removed by an air handling system, which involves pulling the fumes off the processing tanks via exhaust hoods that act as collection points to the system.

The majority of duct systems handle fumes from multiple tanks, using smaller-diameter duct piping, which in turn is connected to a large-diameter main trunk line. The fumes are pulled through the system by corrosive-resistant fans and blowers and, depending on the application, are then forced through in-line fume scrubbers, mist eliminators, air stripping columns, or other equipment designed to clean the air and collect the contaminants present in the fumes for recycling, additional treatment, or proper disposal. Since many of the fumes and other by-products generated by these processes are too aggressive for metallic air handling systems, the use of thermoplastic duct provides a viable alternative for many applications.

INDUSTRIAL APPLICATIONS

Plastic piping systems are used extensively in many diverse industrial applications due to the inherent light weight, corrosion resistance, long service life, and lower overall installed costs of these materials when compared to metallic systems. The exceptional chemical resistance of these products makes them ideal for use in many aggressive applications where collection, transfer, and mixing of chemicals are required.

Most plastic piping products are produced to Schedule 80 dimensions for these types of applications due to the pressure requirements and added safety measures needed. These products can typically be found in the following industrial applications: chemical processing industries, electroplating, pulp and paper manufacturing, high purity applications, ultra pure water, computer equipment processing, aeration systems, atomic energy, hot and cold corrosive fluid transfer, newspaper industry, industrial waste treatment brines, plant water distribution, chemical laboratories, color industries, petrochemicals, photographic industry, military applications, precious metals recovery, battery manufacturing, fertilizer manufacturing, aircraft industries, textile industries, mining, and other industrial applications involving corrosive fluid transfer.

Other common applications for plastic piping include: food processing, aquatic farming (lobster and fish hatcheries, etc.), pharmaceutical, filtering systems, chilled water meat packing, hot and cold potable water systems, well piping, irrigation, agricultural, groundwater remediation projects, monitoring wells, desalinization projects, commercial drainage applications, roof drains, containment ponds, water and wastewater treatment, solid waste treatment (landfills), and sewage treatment.

CHEMICAL RESISTANCE

Plumbing Piping Systems

There are several piping materials used in plumbing systems; the systems themselves operate under different conditions. For instance, gravity flow systems are subject to both internal and external conditions and a wide variety of solutions or mixtures in addition to the normal sanitary wastes. These can include—in small quantities—drain cleaners, bleach solutions, paint thinner, paint stripper, spot remover, biodetergent ammonia, nail polish remover, and detergents.

While all plumbing codes prohibit the discharge of most of these reagents into the sanitary waste system, small or diluted quantities can be expected to reach the DWV and the sewer piping system. Based on more than 30 years of experience, the plastic sanitary waste and plastic sewer piping materials covered by the standards listed in Chapter 2 have the chemical resistance characteristics needed for these systems. In general, experience has shown that all of these plastic materials are equal to or better than metals in terms of chemical resistance for gravity flow plumbing piping.

For plumbing pressure piping, water distribution is the primary system. The plastic materials used and listed for water distribution are not affected by corrosive or aggressive water. These materials are also not affected by the external environments found in most buildings.

For underground installation of plastic piping, all the materials used for both pressure and nonpressure piping are resistant to all kinds of aggressive or hot soils and cinders. There is one limitation: plastic potable water piping should not be installed in soils contaminated with organic hydrocarbons such as gasoline leakage from underground tanks.

Other Piping Systems

Even in environments outside of traditional plumbing, plastic piping systems have shown great resistance to many of the chemicals encountered in both pressure and gravity flow piping systems. Chemical resistance tables are available that indicate the resistance of each material to a variety of chemical solutions that may be encountered in a wide variety of industrial and commercial applications. These tables often indicate that resistance can vary by solution concentration and by temperature. In addition, stress and pressure can also be factors.

Because there are so many chemicals and mixtures of chemicals used in industrial piping applications, piping system designers and engineers must use care and judgment in selecting the best plastic piping material for each application. In addition to consulting the pipe manufacturer, designers can also rely on their own experience and the experience of others. Because many industrial piping systems are operated under well-defined and controlled conditions, and because they are also likely to have shorter service life requirements, material selection for these systems differs from those of plumbing systems that are usually expected to have a service life of 20 to 50 years.

REVIEW QUESTIONS

True or False

1. __________ Because of the low fire mass in a typical building, thermoplastics are small contributors to a structural fire.
2. __________ Since plastic pipe is fire resistant, there is no need to caulk penetrations through a fire-rated wall.
3. __________ The steps used in sizing plumbing systems for plastic pipe are different than sizing for metal pipe.
4. __________ The lower the ignition temperature of a product, the safer it is.
5. __________ Air admittance valves should be located a minimum of 4 inches above the weir of the fixture trap.

Chapter 4

LEARNING OBJECTIVES

The student will:

- Describe the general concepts of welding.
- Differentiate between the types of welding gases and safety issues related to handling welding cylinders.
- Describe the types of gas welding equipment and explain safety issues for gas welding equipment.
- Discuss the proper use of fillers and fluxes for soldering, brazing, and welding.
- Explain safety procedures and safety issues related to welding.

Introduction to Welding

Welding is a skilled technique for permanently joining metals and sometimes thermoplastics. Most often, welding is performed by melting the edges of the metal pieces to be joined and adding a molten filler material that cools to form a strong permanent joint. This chapter will introduce the basics of gas welding, the equipment, and safety precautions as they pertain to the field of plumbing. Additional material concerning shielded-metal arc welding and weld testing can be found in Appendix A.

HISTORY OF WELDING

Welding as we think of it today was developed in the late 1800s and used a special technique known as forge welding. Forge welding has been used for centuries by blacksmiths to join metals through heating and hammering metals into shape.

During the late 1800s and early 1900s, welding technology advanced rapidly as arc welding processes and oxy-acetylene processes were invented and developed. The two world wars in the first half of the twentieth century fueled the need for armament production that was reliable and inexpensive. Many companies were created to manufacture welding machines and supplies to meet the demand for the technology.

In the second half of the twentieth century, more welding techniques were developed, including shielded-metal arc welding and gas-metal arc welding.

Welding applications today are everywhere you go. When you drive over a bridge, the metal components in the bridge structure are welded. When you fly on an airplane, much of the sheet metal on the aircraft has been joined by welding.

Welding applications can be found in the manufacturing, fabrication, repair services, maintenance, transportation, oil, and mining industries. Many artists and sculptors use welding in their work.

PRACTICE OF WELDING

Welding is a specialized trade requiring specific training and certification. Because it involves working with high temperatures, electricity, and gases, special care must be taken when working in this field. Welding is a highly scientific field and requires extensive knowledge of metal composition and characteristics as well as electrical theory. Appendix A of this text addresses the basic knowledge required of a welder.

IN THE FIELD
Always know what you are welding on. Workers are killed or injured every year because they were welding on vessels containing chemicals or gases, or that were under pressure.

WELDING STANDARDS

Much pipe welding must conform with one of two widely used sets of standards. One set is the ASME (American Society of Mechanical Engineers) power boiler code, pressure vessel code, and welding standards. The other set is the ANSI B31 standards for power piping. Pipe welding that conforms to either of these standards is referred to as code welding or code work.

In addition, the American Welding Society (AWS) has its own set of standards for the industry. These standards are comprehensive and can be found on the AWS website at http://www.aws.org.

GAS WELDING EQUIPMENT AND SAFETY

Welding Gases

Many welding fuel gases are available in pressurized cylinders. These gases include oxygen, acetylene, propane, methane (natural gas), and MAPP™ gas. Hydrogen is also available in cylinders, but its welding application is limited because it is extremely hazardous and expensive. Other gases are used in welding processes to control the atmosphere above the weld. These gases are argon, helium, nitrogen, and carbon dioxide. Caution: while these last few gases are not flammable, they can cause suffocation if they displace all of the air in the work environment.

Acetylene

Acetylene is the most widely used fuel gas for welding and cutting steel. When combined with 100 percent oxygen, the oxy-acetylene flame temperature reaches approximately 6,300°F.

Acetylene burned in air without any mixing has a lazy, luminous flame that gives off a great deal of carbon as soot. When properly mixed with oxygen or air, it has an intense blue flame. Acetylene with air is used in the Prestolite™ torch or the Turbo Torch™, both of which produce enough heat for soldering or brazing.

One of the early problems with acetylene was that it is very unstable when pressurized over 15 psi. At pressures above 15 psi, acetylene can explode even without an ignition source. This 15 psi limit greatly reduced the amount of acetylene that could be stored in a tank. To overcome this problem, acetylene is stored in cylinders that are filled with a porous material and acetone. The acetylene dissolves in the acetone, which helps stabilize the gas for safe storage. A full cylinder of acetylene has a pressure of about 225 psi. Cylinders are available in capacities from 8 lbs. to 232 lbs. The larger cylinders may be painted black, orange, or red; smaller cylinders are often gray with red labels. Due to this variation, workers should always verify the type of gas by the printed label, not the color of the cylinder.

IN THE FIELD

When working with acetylene cylinders, the top valve should be opened only $\frac{1}{4}$ to $\frac{1}{2}$ turn so that it can be turned off quickly in case of emergency. Acetylene tanks should always be in an upright position. Acetylene is an extremely combustible gas and leaks could cause a fire or an explosion.

Oxygen

Oxygen is an odorless, tasteless, colorless gas. In itself, it is not combustible. However, it is the element that combines with other elements in all combustion processes. Many substances (such as oil) that are safe in air can be very flammable in the presence of pure oxygen.

Commercial oxygen is obtained from the fractional distillation of liquid air and is sold in a variety of convenient quantities. Green is the standard color for oxygen cylinders. Sizes best suited for portable jobsite use are the 70 $ft.^3$, 118 $ft.^3$, and 230 $ft.^3$ cylinders, where the pressure of a full oxygen container is about 2,200 psi at 70°F. Because of this high pressure, the valve on an oxygen cylinder is a high-pressure valve and should be fully opened and back-seated to minimize stem packing leakage.

Cylinders should always be transported capped to avoid breaking off a valve and creating a hazardous situation. Always follow proper safety procedures when dealing with gas cylinders.

IN THE FIELD

If a pressurized cylinder were to fall over and break the top valve off, the cylinder would take off like a rocket. In some cases, cylinders have been known to shear through a concrete wall. Always have cylinders capped when not in use.

Propane

Propane gas is a hydrocarbon usually present in oil or gas wells. It is separated from the main petroleum product and supplied to distributors. When subjected to higher than atmospheric pressure and room temperatures, propane is a liquid. This allows large quantities of propane to be stored in cylinders.

The propane-oxygen flame burns at approximately 5,190°F, which is cooler than oxy-acetylene, so it is not suitable for welding. Propane does develop enough heat to be desirable for soldering and brazing applications.

MAPP™

MAPP™ gas is a special man-made fuel gas that is more stable than acetylene. When mixed with oxygen, it burns at approximately 5,301°F, cooler than acetylene, but hotter than propane. It can be stored at higher pressures and is safer than acetylene or propane, so it is well suited to certain applications.

Safe Handling of Welding Cylinders

When dealing with pressurized fuel cylinders, safety must be the first consideration. Review Table 4-1 for cylinder handling rules and methods. Safety is a legal requirement, with OSHA specifying rules for proper handling of these cylinders. Treat these vessels with care and respect.

GAS WELDING EQUIPMENT

The equipment required for gas welding and cutting operations is the simplest and least expensive of all welding apparatus. The items needed include the following:

Pressure Regulator

Pressure regulators receive gas from the supply cylinders and deliver it to the hoses at a closely controlled pressure.

Regulators are precision devices that take a wide-ranging inlet pressure and deliver a nearly constant, adjustable outlet pressure. An adjusting screw is used to set the desired output. Best practice requires that the adjusting screw be backed out completely when the regulator is not in service. The area where the adjusting screw

TABLE 4-1 Safe Handling of Cylinders

	Transport	Storage and Use	Cautions
All cylinders	Use transport caps at all times in transit. Use a hand truck to transport.	Chain horizontally to work table, truck bed, or wall. Keep away from electrical current. Close off valve when not in use. Keep full and empty containers separated.	Do not use as a roller. Do not strike or drop. Do not lift by the transport cap. Do not force the control valve. Do not tamper with fuse plug.
Oxygen cylinders		Store away from oil, grease, and similar combustibles, as well as areas where oil may drip on cylinders. Clean oil or grease near the assembly immediately.	Do not grease or oil any connection. Do not test connections with an oil-based soap solution. Do not use oxygen for testing gas lines, running air tools, or feeding internal combustion engines.
Fuel cylinders		Store away from welding and cutting areas. Always store cylinders in an upright position. Open valve only $\frac{1}{4}$ to $\frac{1}{2}$ turn. Use only for intended purpose.	

FIGURE 4-1 This cutaway shows the working components of a gas welding regulator.

Photo by Ed Moore

threads into the regulator is referred to as the bonnet. Typically, the bonnet area is the weakest point on the regulator, so never stand in front of the regulator when opening it. If a regulator fails to operate properly, NEVER attempt to repair it! Send it to a qualified repair shop or have it replaced. Figure 4-1 shows a cutaway of a pressure regulator.

The tank connection of an oxygen regulator is female, right-hand thread. The hose-connection thread on an oxygen regulator is male, with a right-hand thread, while the tank connection of an acetylene regulator is female, left-hand thread. The hose connection on an acetylene regulator is male, with a left-hand thread.

Gauges

Gauges make it easy to monitor the gas remaining in the supply cylinders and to monitor the pressure delivered for the work. For this reason, oxy-fuel assemblies have two gauges: one to indicate the pressure in the tank and one to indicate the operating pressure in the hoses and torch head. Figure 4-2 shows both.

Gauges, like regulators, are precision instruments, so care must be taken when handling them. The backs of the gauges are designed to blow out if the pressure tube inside the case ruptures.

Hoses

Hoses connect the output of the regulators to the torch handle, which controls the flow of gases to the tip. The connecting hoses, which are usually sold in linked sets, are green for oxygen (with right-hand connectors) and red (or orange) for acetylene (with left-hand connectors).

FIGURE 4-2 Pressure gauges showing both the tank pressure and the torch working pressure.

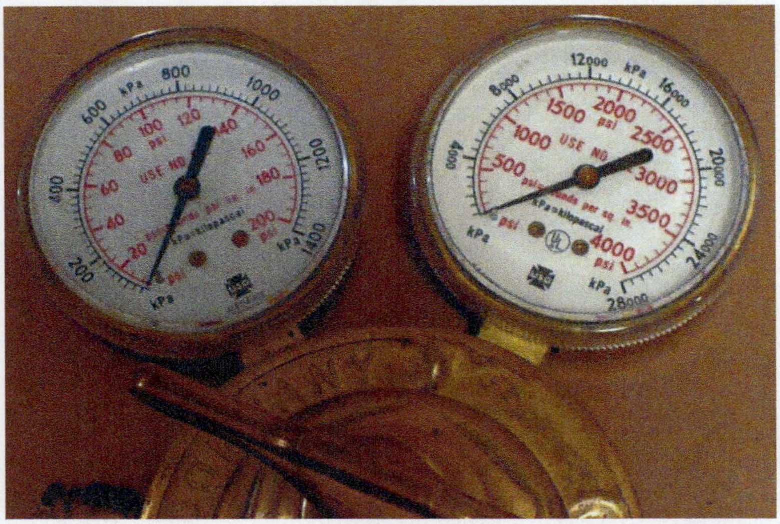

Photo by Ed Moore

Hoses are available in many different diameters and length and should be selected based on the volume of gas flow and length required. It is best to use a small, light section for the last few feet to the torch to facilitate handling and reduce operator fatigue.

IN THE FIELD

Hoses are made of a rubber compound and can easily be melted. Make certain to route hoses away from hot materials, sharp objects, or the spray of molten metal from the cutting process.

Torches

The torch consists of three major components: the handle (or mixing chamber), the tip, and the regulating valve. On one end of the handle the hose connects to a control valve that allows the user to fine-tune the gas flow and thereby control the size of the flame. If two compressed gases are used, such as an oxygen/acetylene torch, then two valves will be present and the handle will act like a mixing chamber. The mixing chamber allows the gases to combine before they proceed onto the tip. If only a single gas is being used, ambient air will be combined with the fuel gas at the other end before it enters the tip.

Check valves should be installed between the torch handle and the regulator to prevent a flashback. A flashback is when the flame burns at a faster rate than the gas is being supplied. When this occurs, the flame will travel backward toward the hoses, regulator, and the tank. This can cause the hoses to melt, a fire, or an explosion.

Torch Tips

A range of tip sizes is available, and the appropriate tip size is selected for the welding task to be done. When the task is cutting, a cutting torch assembly is required.

This assembly includes a tip with several small openings for the gas to exit. A ring of small oxy-acetylene flames and a center hole for pure oxygen delivery, a lever-operated valve to control the center oxygen, and separate valves to control the delivery of the oxygen that burns with the acetylene. When using the cutting attachment, the oxygen valve on the torch handle is wide open and all oxygen control is accomplished by the valve closest to the tip. Figure 4-3 shows an oxy-acetylene cutting head.

The torch tips are sized by a numbering system that varies from manufacturer to manufacturer. Generally, the larger the number, the greater the gas flow and flame size. Check your particular brand for the recommended tip size for the welding to be accomplished. Safety procedures for handling this equipment are given in Table 4-2.

FIGURE 4-3 Typical oxy-acetylene cutting torch. The oxygen valve closest to the hoses should be opened all the way when cutting.

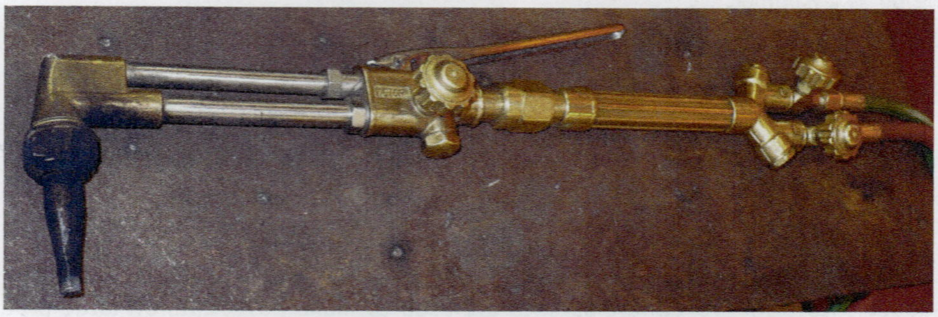

Photo by Ed Moore

TABLE 4-2 Safety and Handling of Equipment

Equipment	Handling Suggestions	Safety Suggestions
Regulator	Keep regulator off when not in use. Be sure to turn adjustment screw completely off prior to start-up. Failure to do so may burst the regulator and cause injury to the operator. Inspect all nuts and connections prior to using the regulators. Stay within recommended pressures. Have only qualified organizations repair faulty regulators.	Handle properly. Do not force connections of hoses and regulators. Use the proper regulator on the proper cylinder. Do not lubricate with oil or oil-based solutions.
Gauges	Handle carefully. Replace any broken gauge glasses.	Keep glass protected from breakage. Do not attempt to repair a defective gauge.
Hoses	Test hoses for leaks by immersing in water at normal working pressure. Leaky hoses should be repaired or replaced. Keep hose connections tight. Do not repair a hose with tape. Keep the hose properly rolled and stored when not in use. Spring-type hose reels are also available.	Use only one type of gas in the hose. A mixture of different gases in the hose could result in an explosion. Avoid dragging the hose on a greasy floor. Avoid circumstances where a hose can be rolled upon, kinked, burned, or subjected to falling objects or hot metals.

(*continued*)

TABLE 4-2 Safety and Handling of Equipment (Continued)

Equipment	Handling Suggestions	Safety Suggestions
Torch handles	Handle properly.	Keep all connections to the torch tight. Test them accordingly. Use flashback check valves. Any torch suspected of being faulty should be repaired by the torch manufacturer or its authorized repair facility. Avoid any contact with greasy or oily substances.
Torch tips	Do not remove a tip from the torch when it is hot. Allow it to cool first. Use the manufacturer's wrench, not pliers, to remove the torch tip. Keep the tip hole clean at all times using a tip cleaner. A wood block may be used to clean the deposits from outside of the tip. Never use the tip as a hammer.	Restrictions in the tip may cause back pressure and leakage. Handle tips carefully.

Auxiliary devices for gas welding equipment include tip cleaners (used to remove slag and spatter from the welding or cutting tips) and strikers for safely lighting the torch. See Figure 4-4.

Figure 4-5 shows a pipe clamp that is used to maintain alignment until welds can be made. Cutoff and beveling machines are used to make bevel cuts on large pipe with the proper angle on the face to prepare the pipe for welding.

FIGURE 4-4 The friction lighter (shown in the top of the figure) is the safest method to light a fuel gas torch. The tip cleaner (shown in bottom of figure) is used to remove slag and other debris that can clog the gas openings in a tip.

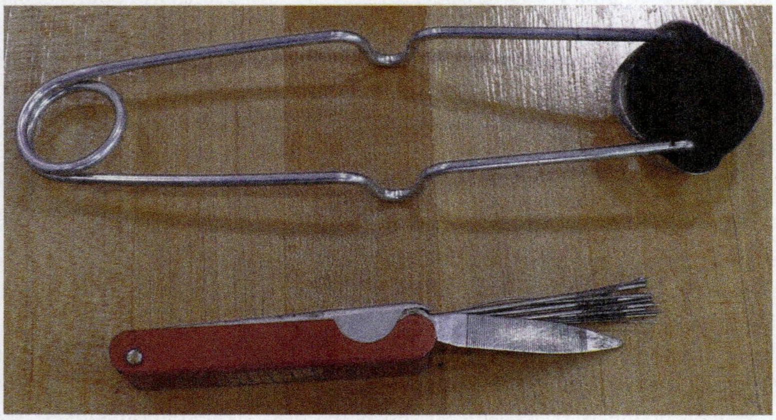

Photo by Ed Moore

FIGURE 4-5 Special clamps are available for positioning the pipe to be welded.

FILLERS AND FLUXES FOR SOLDERING AND BRAZING

Brazing and soldering fluxes are available in liquid, paste, powder, or dry crystal form. They may be self-cleaning or may require additional joint cleaning. Most are available in either oil- or water-based form. To minimize continued flux corrosion after joints are made, non-self-cleaning, water-based fluxes are recommended for soldering. The system should also be flushed thoroughly and the tube exterior wiped clean with a wet cloth after soldering to minimize continued corrosion.

Welding fluxes are intended to clean surfaces, dissolve (or remove) oxides, and release trapped gases and slag particles. Each of these operations is different for different materials, so there is no universal flux material. Always verify the application of a flux before use.

In many soldering and brazing operations, a flux is used to aid in making the joint. The flux is used to keep the joining surfaces clean after the surfaces have been cleaned manually and to facilitate the solder or brazing material's wetting and adhering to the joint surfaces. The flux also prevents oxidation while the joint is being heated and keeps the oxides away from the filler material. As the filler material melts, it easily displaces the flux.

Fluxes are available that are considered to be highly corrosive, medium corrosive, and noncorrosive. Specific properties are these:

Highly Corrosive

- Highly corrosive (acid or salty)
- Used on aluminum or steel

Medium Corrosive

- Fluxing ability short-lived at soldering temperature

- Easily removed after soldering
- Best for copper and copper alloys

Noncorrosive

- Used for silver, nickel, copper, brass, and bronze alloys

WELDING RODS

The filler materials available for these various operations have a wide variety of physical characteristics. Many rods are coated with either a protective coating or a flux. For example, steel rods are copper-coated to keep them from rusting, while some types of aluminum rods are coated with a flux.

Welding filler rod is available in diameters from $\frac{1}{16}$ inch to $\frac{3}{8}$ inch usually 36 inches long. The American Welding Society (AWS) classifies rods for various services and strengths. For example, an RG-80 rod would be interpreted as follows:

- R is for weld rod.
- G is for gas.
- 80 is for 80,000 psi tensile strength.

See Table 4-3 for a broad range of filler materials, pertinent properties, and recommended applications.

SAFETY AND SAFETY EQUIPMENT

Fire and Burn Hazards

Many situations can be hazardous when using open flames in soldering, brazing, welding, or cutting procedures. The most obvious danger is fire and all its variations. Burns, explosions, asphyxiation, and eyesight damage are also possible. To minimize these risks, observe the following minimum precautions:

- Keep the torch away from flammable materials or vapors.
- Keep work in hazardous areas to a minimum—prefabricate subassemblies in the shop whenever possible.
- Secure the work on fireproof tables when prefabricating.
- Cover and protect flammable or finished surfaces.
- Have adequate and correct fire extinguishers close by.
- Do not use flames or sparks around containers of flammable materials or containers that formerly contained such materials.
- Do not use higher oxygen cutting pressure than necessary to accomplish the job. Higher pressure will waste oxygen and produce a much wider scatter of cutting sparks. Be sure the cutting spark stream will not strike other persons, flammable materials, or the welding equipment.

TABLE 4-3 Filler Materials Used for Joining Metals[1]

Operation	Composition of Filler Material	Completely Solid	Completely Liquid	Usage and Notes
Soldering	Tin-Lead[2] $\frac{50}{50}$	362°F (183°C)	418°F (214°C)	Only on nonpotable water applications
	Tin-Antimony-Lead[2] 20/1/79	363°F (184°C)	418°F (214°C)	Higher pressure systems
	Tin-Antimony-Lead[2] 40/2/58	365°F (185°C)	418°F (214°C)	Not intended for zinc-based metal alloys
	Tin-Antimony $\frac{95}{5}$	452°F (233°C)	418°F (214°C)	Food handling systems
	Tin-Zinc $\frac{80}{20}$	390°F (199°C)	418°F (214°C)	Aluminum
	Tin-Zinc $\frac{70}{30}$	390°F (199°C)	418°F (214°C)	Aluminum
	Tin-Zinc $\frac{30}{70}$	390°F (199°C)	418°F (214°C)	Aluminum
	Zinc-Aluminum $\frac{95}{5}$	720°F (382°C)	418°F (214°C)	High temperature and strength are traits of this corrosion-resisting filler. Note that this material has no pasty range condition—it is either a solid or a liquid, never a semi-liquid.
	Cadmium-Zinc $\frac{82.5}{17.5}$	509°F (265°C)	418°F (214°C)	Used to join wide-clearance joints
	Cadmium-Zinc $\frac{10}{90}$	509°F (265°C)	418°F (214°C)	Used to join wide-clearance joints
	Cadmium-Silver $\frac{95}{5}$	640°F (338°C)	418°F (214°C)	High-strength butt joints
Brazing	Silver alloy (13 different alloys)	1,116°–1,434°F (602°–779°C)	1,444°–1,636°F (618°–891°C)	Joining both ferrous and nonferrous materials; free-flowing; flux required; better results with close-tolerance fit. See manufacturer for specific traits.
	Aluminum-Silicon alloys (4 alloys)	970°–1,070°F (521°–577°C)	1,080°–1,135°F (582°–613°C)	Aluminum brazing
	Magnesium alloys (4 alloys)	770°–830°F (410°–443°C)	1,050°–1,110°F (565°–599°C)	Magnesium brazing
	Gold alloys (4 alloys)	1,635°–1,815°F (891°–991°C)	1,635°–1,885°F (891°–1,029°C)	Electronic and missile components
	Copper and Copper-Zinc alloys (3 and 2 alloys)	1,980°F (1,082°C)	1,980°F (1,082°C)	Ferrous metals and nickel; not for copper or stainless steel alloys
	Copper-Phosphorus alloys (5 alloys)	1,190°–1,310°F (643°–710°C)	1,135°–1,650°F (724°–899°C)	Close tolerances required; primarily for copper alloys
	Nickel alloys (7 alloys)	1,610°–1,975°F (877°–1,079°C)	1,610°–2,075°F (877°–1,135°C)	Food equipment and other applications where high resistance to corrosion is needed
Welding	RG-45 Low Carbon Steel	2,300°–2,700°F (1,260°–1,482°C)	2,300°–2,700°F (1,260°–1,482°C)	Holding strength 45,000–50,000 psi range for general purpose use
	RG-60 Low Alloy Steel	2,400°–2,700°F (1,316°–1,482°C)	2,400°–2,700°F (1,316°–1,482°C)	Piping for process plants, power plants, and severe condition systems; holding strength in the 50,000 to 60,000 psi range; excellent for carbon steel, wrought iron, and low alloy steels
	RG-65 Low Alloy Steel	2,400°–2,700°F (1,316°–1,482°C)	2,400°–2,700°F (1,316°–1,482°C)	Carbon and low alloy pipe, plate, and sheet

[1]Additional information is available from the manufacturers of these products.

[2]Lead-based solders and fluxes (greater than 0.2% lead) are prohibited for use on potable water supply systems according to federal law.

- Do not leave a lighted torch unattended.
- Have a safety spotter (or fire watch) anytime you are welding or cutting in a pipe chase or other area where sparks can fall out of sight. Some method of communication between the spotter and the welder is necessary.
- Plan to minimize fire hazards.
- Check hoses and valves for signs of leakage and replace any defective items. Inspect building walls to be sure that the flame has not ignited any surface. Keep a fire extinguisher handy. Thoroughly extinguish all smoldering surfaces. Never leave the jobsite until you are absolutely certain that no fire hazard exists.
- Watch for drafts that may affect flame direction.
- Avoid working in confined spaces. Have a coworker spot for you. Have plenty of ventilation and an escape route available at all times.
- Test the air for combustibles in confined spaces prior to entry.
- Always position yourself so that molten or hot metal will not fall on you.

Constantly guard against burns and burn potential. Burns can be caused by the actual torch flame as well as by sparks, discarded hot metal, or hot equipment. To protect yourself, wear proper protective coverings including gloves, apron, leggings, jacket, cap, and goggles. To protect others, properly shield below the work area to keep hot material from falling on other workers or on hazardous materials. Shield the work area from onlookers.

Guard against explosions by keeping cylinders out of confined areas and do not leave a pressurized torch in a confined area. A minor leak could produce explosive gas concentrations. Demand the highest level of equipment maintenance. Repair or replace any defective or marginal equipment.

When lighting, always open and light the acetylene first, then open the oxygen. When shutting off, close the oxygen first, then the acetylene. This sequence is recommended by the American Welding Society because it reduces the risk of the higher-pressure oxygen backflowing into the lower-pressure acetylene hose.

Fire Watch

Whenever the work being performed has the potential to produce an ignition source or excess heat that is capable of causing a fire, a fire watch must be provided. A fire watch is a properly trained individual who will assess the jobsite conditions before any work begins, monitor the work site during the hot work operation and make sure the work site is safe after the work has concluded. It is important that person's only responsibility during the hot work is to monitor the site; they should not be doing other tasks. The fire watch individual should also have a way to constantly communicate with the individuals doing the work so that they can be warned of any dangerous situations that might occur. The fire watch individual should also have the power to stop the work if they believe any danger is present.

Many companies have what is known as a "hot work permit," which is a permit from a supervisor to do the work and usually contains a checklist of the following information or action items:

- Date of job.
- Location and type of work being performed.
- Area within 35 feet of work has to be cleaned of all flammable or explosive material.
- List of all PPE needed to perform the work.
- All conveyors and ducts that might carry a spark are shut down.
- Proper ventilation is available.
- All opening in floors and walls are sealed to prevent sparks from existing.
- Proper fire extinguishers are present and means of sounding an alarm are possible.

Asphyxiation

Asphyxiation hazards must always be considered. Normal combustion produces carbon dioxide and water vapor. An improper flame produces carbon monoxide as well. Materials often found in fluxes can be problems in confined spaces: fluorides, beryllium, mercury, lead, and similar substances are sometimes encountered. Continuous exposure to these materials can lead to chemical pneumonia and cause severe illness. Permanent injury to your lungs can also result.

IN THE FIELD

Cutting or welding galvanized or coated material will produce toxic fumes from the zinc and other metals. These fumes can produce nausea and headaches after brief exposure. Work in a well-ventilated area.

Open workplaces are preferable, but if it is necessary to work in a confined space, provide positive ventilation and frequent rest periods for the welder.

Eye and Face Protection

Eye protection is essential for brazing, welding, cutting, and soldering. Goggles should be used that fit over the eyes and keep particles from entering from the side. Sunglasses are not a satisfactory substitute for proper welding goggles.

If welding or cutting overhead, use a clear face shield over the proper goggles to provide satisfactory protection. Also make sure to wear a fire-resistant cap to protect your head from sparks and slag.

The following goggle shade numbers are recommended as a minimum. The higher the number, the darker the shade.

- Soldering: #2
- Brazing: #3 or #4

- Cutting: #3 to #5
- Welding: #4 to #8

Operations involving electric-arc processes require greater eye protection. See Chapter 5 for recommended types and sizes.

Safety Procedures

Proper use and storage of equipment is essential for safety:

- Open and close the cylinder valve once or twice quickly prior to installing the regulator connector to clear any loose dirt.
- Never force a regulator onto a cylinder valve. The regulator should screw on several turns by hand before needing a wrench for final tightening.
- Transport cylinders with protective caps in place.
- After use, close cylinder valves and exhaust pressure in acetylene and oxygen regulators and hoses.
- If the flow of either gas is irregular, stop the welding/cutting process until the regulators can be readjusted or replaced.
- For new hoses, blow the talc dust from the inside before connecting for use.
- Segregate or mark hot surfaces until they have time to cool.
- Use proper scaffolding or aerial lifts when working above the floor.
- Use a life belt and lifeline on a technician welding in a confined space.
- First aid and fire extinguishing equipment should be available on all jobsites and should be well marked for immediate access.
- Emergency numbers should be readily available.
- All welding equipment should be stored in a secure tool box any time that the equipment is not in use.
- When storing tanks on the jobsite, be sure acetylene and oxygen tanks are stored at least 20 feet apart. This is OSHA regulated.
- And last, but very important: Do not tamper with pressurized gases at any time and under any circumstances. The misuse of pressurized gases is incredibly dangerous. These are flammable materials.

If you have an interest in learning more about welding, ask your instructor, employer, or local career and technical school where you can find training in your area. Again, welding deals with extremely high temperatures as well as with gases and electricity, and specialized training is a must if you wish to perform welding work as a plumber.

REVIEW QUESTIONS

Fill in the Blank and Short Answer

1. Acetylene is dissolved in __________ to stabilize the gas for safe storage.
2. What brought about the first major advancements in welding?
3. Name two industries that rely on welding.
4. The HIGHER/LOWER the shade number, the darker the lens. (circle one)
5. When using an oxy-acetylene cutting head on a torch, the oxygen valve closest to the hoses should be opened how far?

Chapter 5

LEARNING OBJECTIVE

The student will:

- Describe the basic techniques of soldering, brazing, and flame-cutting.

Soldering, Brazing, Cutting, and Gas Welding

DETERMINING THE APPROPRIATE JOINING METHOD

In general, joints between steel and copper can be made with heat and filler material by soldering and brazing as well as by welding. Many other metals can be joined by these processes. The following are some considerations that dictate the joining method:

- Pipe material
- Severity and length of service
- Fluid inside the pipe
- Operating temperatures and pressures
- Expansion, contraction, and vibration in the system

For less-demanding service with copper material, soldering is satisfactory. If the joint is to withstand high stress at high temperature, copper or steel systems with brazed or welded joints are indicated. The most severe service requires steel pipe with welded joints.

However, if copper pipe is to be joined to steel, brazing is the best joining method to use. The strength of the joint is affected by the pipe material and filler material.

Soldering Defined

Soldering is defined in the *American Welding Society (AWS) Handbook* as a joining method in which coalescence is produced by heating, generally to a temperature below 1,000°F (538°C), and using a nonferrous filler metal (solder) that has a melting point below that of its base metal. The filler metal fills the joint cavity by capillary action.

Brazing Defined

Brazing is defined as a group of welding processes in which coalescence is produced by heating to a suitable temperature above 1,000°F and by using a nonferrous filler metal with a melting point below that of the base metal.

SOLDERING

As described in *Plumbing 101*, the soldering method of making a joint involves first heating the tube, then the socket fitting, to a temperature high enough to melt solder (but not the pipe or fitting) and have the molten solder fill the annular space between the pipe and the fitting. If the tube and fittings have been properly cleaned and fluxed, this technique will provide excellent joints for low-temperature, low-pressure applications (250°F, 100 psi).

Other considerations that affect the efficiency of making such joints are these:

1. Proper size torch tip.
2. Use of an oversized tip has a tendency to overheat the materials and cause excess surface oxidation. An undersized tip tends to produce a cold joint in which the solder does not wet the tube-fitting surfaces, resulting in a much smaller solder shear area available to develop the joint strength. Multiple orifice tips are available for large-diameter tube soldering.
3. Proper fuel pressure. Too large a flame (excess regulator pressure) can cause overheating of the materials and may blow gases into the joint. Too small a flame (not enough regulator pressure) will give results similar to improper tip size and slow production rate. The heat should be applied briefly to the tubing, then to the fitting. This process produces the minimum gap between tube and fitting and thus improves joint strength.

Approved Solders and Fluxes

Federal law and local codes prohibit the use of solders and fluxes that contain more than 0.2 percent lead on potable water lines.

BRAZING

Brazing is similar to soldering but requires more demanding techniques to produce satisfactory joints. Brazing requires filler materials that melt at temperatures above 1,000°F. The most common filler materials are brass and a family of fillers that are based on alloys of silver.

The preparations for brazing are similar to those for soldering: square cut tubing, reamed tube end, well-cleaned surfaces, close-fitting tube and female sockets, use of appropriate flux, and a flame that will develop the required temperature.

Brazing fluxes go through several visual changes that indicate the temperature of the work:

212°F	Water boils out
600°F–750°F	Flux bubbles vigorously
800°F–850°F	Flux melts
1,100°F	Flux is quiet and clear (be ready to add filler)
1,600°F	Flux breaks down

Between the clear, quiet phase, and complete breakdown, filler material must be added to the joint. At this point, the filler material should flow smoothly to fill the area between pipe and fitting. Adding too much more heat will cause the joint to overheat.

Brazing Technique

Use the flame to heat the joint so that the joint melts the filler. If you melt the filler with the flame, the molten filler will drop onto the joint and may or may not heat it enough to wet and fill the joint.

Since the joint materials have to be above 1,000°F, the torch must be applied all around the joint to assure that the filler wets and adheres to the entire joint area.

A brief summary of the brazing technique follows:

1. Properly clean the tube and fitting.
2. Apply the proper flux. When adding flux to large-diameter fittings, flux both inside and outside the filler cup. The flux on the outside of the cup will aid you in knowing when to add filler.
3. Support the pipe and fittings properly before brazing.
4. Do not overheat the joint. Overheating will cause oxidation and inhibit producing a proper joint.
5. Do not underheat the joint. Underheating prevents melting and wetting of the filler. (The filler metal beads on the pipe if underheated.)
6. Braze quickly. Once the proper temperature is reached, the filler metal should be applied immediately for the highest quality joint.
7. Flush the inside of the pipe after brazing.

Note: for refrigeration and medical gas piping, it is essential that the inside of the piping does not oxidize during the brazing operation. To prevent oxidation from occurring, an inert gas is passed through the pipe during the brazing operation. The inert gas commonly used is nitrogen. This gas must be purged later to prevent line contamination. Medical gas certification such as that offered through the National Inspection, Testing and Certification Corporation (NITC) or other organizations as dictated in the National Fire Prevention Association (NFPA) Standard 99 is required to work on medical gas systems.

WELDING

Unlike soldering or brazing, welding requires that the base materials themselves melt to join. A filler material is sometimes added to maintain a certain thickness to the weld. Many materials can, at least theoretically, be welded by the torch process. Especially when applied to steel, welded joints are strong, long lasting, and develop the most streamline flow paths. The technician needs special training to produce welded joints efficiently, but the effort is worth it. With additional coursework and training, welding certifications are available.

FLAME CUTTING

Steel and iron can be cut using an oxy-acetylene torch. A special torch tip, which has a small circle of six oxy-acetylene flames, is used in the cutting operation. Inside the circle is an opening that conveys pure oxygen to the work when the operator activates a valve connected to the torch. Figure 5-1 shows a close-up of an oxy-acetylene cutting tip.

After the iron or steel is brought to the melting point, the oxygen valve in the torch tip is opened and the molten material is blown away. In the hands of a practiced worker, the torch can be moved at a continuous rate to cut very efficiently and cleanly.

FIGURE 5-1 Notice that the oxygen hole for blowing away the molten metal is larger than the other holes on an oxy-acetylene cutting torch tip.

Cutting Procedure

Preparation and Safety Precautions

Preparation and safety precautions for cutting include the following:

- Secure the work so that it can't fall before or after the cutting is accomplished.
- Keep flammable materials away from the work and out of the probable spray path of the cutting sparks.
- Advise other workers of your intentions. Rope off the work area, especially when working overhead.
- Mark the cutting line on the work, using soapstone or some other light-colored marker.
- Set up the cylinders, regulators, hoses, and torch as described in Chapter 4.
- Select the proper torch tip and gas pressures for the thickness and type of the metal based on the manufacturer's recommendations.
- Use proper goggles, protective clothing, and coverings.

Cutting Process

The cutting process itself is executed as follows:

1. Light the acetylene with a striker.
2. Slowly add oxygen to clean up the flame. Too rapid a change will extinguish the acetylene flame.
3. Increase the oxygen until the blue-white outer cone disappears. You now have the hottest flame for the most efficient preheating for cutting.
4. Apply the flame to the edge of the metal. Hold the tip so that the smaller blue cone is approximately $\frac{1}{8}$ inch to $\frac{1}{4}$ inch above the metal. If possible, rest your arms

FIGURE 5-2 When the torch is held properly, the operator can produce a smooth cut without becoming overly tired.

Photo by Ed Moore

on a fixed object; this will eliminate unnecessary movement and produce a smoother cut. See Figure 5-2.

5. When the work is brought to melting temperature, direct pure oxygen onto the work, by depressing the oxygen lever on the torch handle, and the molten iron or steel will burn away. **Caution:** when starting a hole in the middle of the metal, some of the molten metal can be blown back at you.

Kerf

The void produced in the work by this cutting process is called a *kerf*. The best kerf is one in which the cut is accurate and square to the top of the metal.

The width of the kerf is affected by the size of the cutting tip, oxygen pressure, and the speed of motion of the torch. Figure 5-3 shows a picture of poor kerf made in a piece of plate steel.

FIGURE 5-3 The empty void left after the cutting process is called the kerf. In this case, the kerf is poor because it is not accurate and square. Remember to factor in the kerf when cutting; otherwise, the part being cut may be too short.

Photo by Ed Moore

Drag

Drag is the term used to describe the lines formed on the work face by the variations in cutting gas and manual manipulation. With low oxygen pressure or too fast a cut, these lines become pronounced.

When cutting along a straight path, hold the torch tip $\frac{1}{8}$ inch to $\frac{1}{4}$ inch above the work and direct the flame slightly toward the direction of travel so that the metal is preheated by the flame. If the torch tip is held too close, the top edge of the kerf will be rounded over instead of a sharp corner. If a perpendicular cut edge is desired, hold the flame perpendicular to the plane of the work at right angles to the line of the cut. If a bevel angle is desired, the flame must be held at the proper angle to achieve the required bevel angle. Figure 5-4 shows the drag lines produced by a skilled welder.

The best cuts are achieved by observing the following:

- Use the proper tip. Undersizing produces rough cuts; oversizing wastes gas and overheats the material.
- Use recommended gas pressures for satisfactory cuts without waste.
- Control torch with smooth, regular motions.
- Keep torch tips clean.
- Keep equipment clean and dirt-free.
- Do not use regulators that deliver gas erratically. Erratic delivery may signal an empty tank.

PLASMA ARC CUTTING

Plasma arc cutting uses an electric arc and compressed air to cut metals. This method can be used to cut metals that cannot be cut by oxy-acetylene processes, such as stainless steel. One outstanding characteristic of the plasma method is that thin materials can be cut at such a high rate that the edges are cool enough to touch when the cutting is completed. In addition, this type of cutting increases productivity because of its faster speed. Figure 5-5 shows a plasma cutting torch.

OXY-ACETYLENE FLAT WELDING

The first requirement for welding steel is the ability to create an area of molten material, called the *puddle*, where two pieces of metal can be joined together. Steels have a significant plastic range where the material is molten but viscous. At the

FIGURE 5-4 When done properly, the cut should be smooth without any big gouges or defects.

Photo by Ed Moore

FIGURE 5-5 The plasma cutter can cut any material that conducts electricity very quickly.

Photo by Ed Moore

high-temperature end of the plastic range, molten steel is as free-running as water and welding is not possible. It is important to be able to tell the difference between hot metal and molten metal when welding.

The oxy-acetylene process develops the puddle by using a neutral flame. As the melting temperature is approached, the flame must be moved in a circular pattern so that the melt area does not get too hot. Thus, the temperature (which controls the viscosity of the melt) is controlled by the initial flame size and the speed and size of the circular sweep of the flame.

See Appendix A for specific instructions and exercises for learning to form a puddle and adding filler for welds.

PERSONAL PROTECTION

Proper protective devices are necessary in any welding activity. Goggles, face shield and helmet, and gloves must be used. Also consider using leggings, foot coverings, and aprons if the exposure warrants it. Additional dangers other than the burn itself can include falling off a ladder or platform and receiving additional injury.

Never work directly below a welding task where molten metal can fall on you. A molten bead can burn through your clothing and skin.

Always wear a welding shield to weld. The flashing of the arc can cause serious burns to your eyes and skin in a very short period of time.

REVIEW QUESTIONS

True or False

1. __________ Solder joints are suitable for copper tubing applications for fluids up to 250°F and 100 psi.
2. __________ A properly made brazed joint is stronger than a soldered joint.
3. __________ Soldering is preferred when joints are buried.
4. __________ An inert gas must be passed through the tubing systems when brazing to prevent oxidation of the inside surfaces of the material.
5. __________ You should light a torch with a striker, never a cigarette lighter.

Chapter 6

LEARNING OBJECTIVES

The student will:

- Differentiate among the types of materials used in water supply piping and contrast the advantages and disadvantages of each type.
- Describe the characteristics, methods of treatment, advantages, and disadvantages of the three types of water sources.
- Identify and describe the principal building water supply systems, including water system service mains and building distribution piping.
- Describe methods of developing or maintaining system pressure in building piping.

Piping Materials, Sources, and Distribution for Potable Water

STANDARDS

As the technology of mankind has advanced, the materials used to transport water have become more sophisticated. From hollowed logs in prehistoric times to the range of materials available today, we can see the progress of resourceful people adapting and using materials known at the time the installations were made.

In order to identify the best material for a given task, a system of standards to define water piping products and installation methods has been developed. These standards are developed by committees of experienced individuals representing all sections of the industry and the public. In this way, a maximum of experience is applied in a minimum time to write a document, available to everyone, which describes in detail a material to be used for a particular water piping application.

This process means that most of the materials in use today are manufactured to conform to standards that are appropriate for that particular material. The only exceptions are new products or new applications of established products. Such exceptions are usually tried out on a job-by-job basis.

The principal organizations involved in developing piping standards are:

ASSE	American Society of Sanitary Engineering
ASTM	American Society for Testing and Materials
ASME	American Society of Mechanical Engineers
AWWA	American Water Works Association
IAPMO	International Association of Plumbing and Mechanical Officials
NSF	National Sanitation Foundation
PDI	Plumbing and Drainage Institute

For standards that have widespread application or use, the sponsoring organization may go to the American National Standards Institute (ANSI) for that organization's endorsement. Being classified as an American National Standard means that the standard has followed proper procedures for fair and equitable development.

For many years, the federal government also produced standards written by various agencies. Of particular interest were the Commercial Standards (CS) of the Department of Commerce and the Federal Specifications (FS) of the General Services Administration. In recent years, however, these standards have not stayed current with modern materials and methods. To stay current, federal authorities now reference consensus voluntary standards published by the organizations shown above. American National Standards are usually the most acceptable to authorities.

When a manufacturer sees a new application or room for improvement to an existing product, they will typically try to design a product to fill the requirement of the customer. By working with contractors on the development, valid data can be obtained under actual job conditions. If this early experience is promising, standards are developed and extensive sales efforts are made to get acceptance for the product throughout the industry. In some cases, products that have been used in other countries have been adopted based on their performance overseas. This variation is not absolutely foolproof, but it comes close. Several years of satisfactory experience with a product or method in Europe is a good indication that the product or method will probably be successful in the United States. For a more complete list of standards that apply to plumbing materials, refer to Chapter 3 of the 2012 National Standard Plumbing Code.

INSTALLATION CONSIDERATIONS

The first installation consideration is an economic one: the installed cost of the material. To choose among materials that have similar uses, the designer will usually select the one with the least-installed cost. This judgment is not easy to make because of the many factors that change from job to job. Since labor costs are a large portion of the installation cost, factors such as the time and ease of installation are very important. In all cases, we are making our selection based on our best judgment.

IN THE FIELD

Several manufacturers have developed new products that use mechanical devices to crimp special copper fittings onto copper pipe (see Figure 6-1). Each fitting is equipped with an O-ring that makes a pressure-tight seal once the crimping process is complete (see Figure 6-2). Although the crimping device and fittings are more costly, considerable amounts of time can be saved during installation.

The second task is to determine the importance of material longevity for the job at hand and select a material that best fits the requirement. In cases where a failure of the system would result in any loss in productivity, such as a factory-stopping production or a street being closed while concrete or pavement are removed for a repair, an increase in product cost could be justified

FIGURE 6-1 A compression tool uses electrical powered jaws to crimp the fittings onto the copper pipe, saving time on installation versus other methods.

Photo by Ed Moore

FIGURE 6-2 Press fittings have an internal O-ring that creates a seal.

Photo by Ed Moore

by a promise of longer system life. In many cases, the validity of a choice can only be determined by time. It is possible that many poor choices have been made in the past, so never simply buy into a "that's the way we did it in the past" mentality unless you know all of the job conditions and all of the available options. Also remember that plumbing and mechanical codes change based on many factors, so always check for code compliance before making any decision.

Ultimately, any material succeeds or fails in the marketplace. Economical materials continue to be sold; uneconomical materials disappear.

IN THE FIELD

Remember, water service is the pipe from the water main or other source of potable water supply to the water distribution system of the building served. Water distribution piping is the piping within the building or on the premises that conveys water from the water service pipe to the point of use.

MATERIAL CHARACTERISTICS

A review of water piping materials in current use will illustrate the concepts discussed above.

Asbestos-Cement Pipe

Asbestos-cement pipe is no longer listed as an approved water service material in the 2012 National Standard Plumbing Code. Since this was an extremely popular choice for such a long time, it is very possible to find it in existing systems. When properly installed and flushed, asbestos-cement pipe does not easily corrode or introduce contaminating materials into the water supply. It has very long life, adequate burst strength, and resistance to crushing loads. It has low impact strength, but this is not significant for buried pipes. The joints are easily assembled and leak-tight. However, there was always a danger that hazardous airborne asbestos particles may be released during the fabrication and installation.

IN THE FIELD

Only work with asbestos-cement pipe when you have been properly trained in the handling of asbestos material. It should only be cut by breaking it with a hammer or cutting it with a wheel-type cutter, and only after it has been wetted with potable water. This helps reduce the amount of airborne fibers. Once removed, the fragments should be properly sealed and labeled as dangerous materials.

Plastic Pipe

When properly installed and flushed, plastic pipe materials do not corrode or introduce materials into the water supply. Plastic pipe has long service life. It also has adequate burst strength and resistance to crushing. Plastic pipe offers low water absorption, a variety of joints, and chemical resistances for various applications. Several types of plastic pipe are available, as shown by the following examples:

Polyethylene (PE)

Polyethylene pipe (PE) is a flexible plastic pipe, which is available in long rolls. Since it is not suitable for hot-water use, it is typically used for water service and lawn sprinkler systems. Normally it is black, but it is available in yellow for use in the gas industry. In plumbing applications, it is typically joined with hose barbs and band clamps, as shown in Figure 6-3. The gas industry relies on a heat fusion joint.

FIGURE 6-3 Mechanical joints for PE pipe use a hose barb and two stainless steel band clamps.

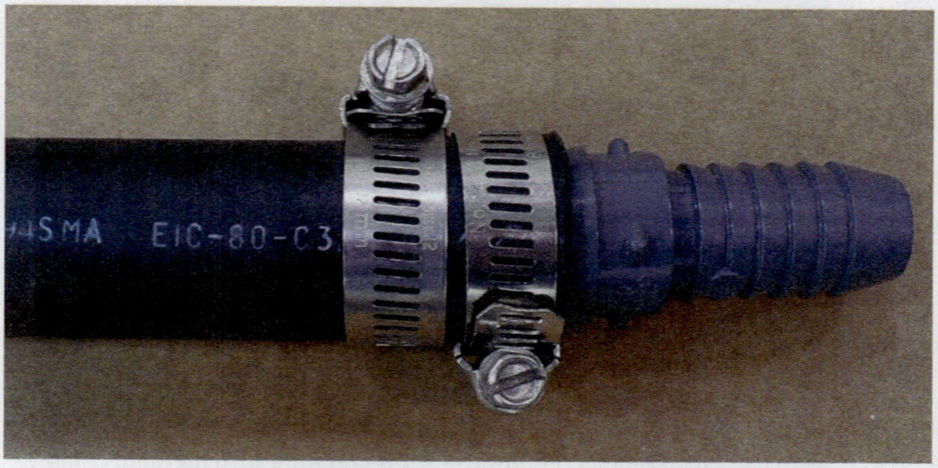

Chlorinated Polyvinyl Chloride (CPVC)

Chlorinated polyvinyl chloride (CPVC) is a yellowish plastic pipe that is available in long lengths, typically 20 feet. It is commonly used in hot and cold water distribution piping, but is also allowed for water service pipe. This material is assembled with solvent-cemented joints, using glue that is specific to CPVC pipe. Do not use PVC or all-purpose glues (see Figure 6-4). Schedule 80 pipe may be threaded.

FIGURE 6-4 CPVC pipe is joined with cement joints using solvent cement specifically designed for CPVC.

Polyvinyl Chloride (PVC)

Polyvinyl chloride (PVC) is a white plastic pipe that is available in long lengths, typically 20 feet. It is no longer allowed to be used for cold water distribution but is still commonly found installed as water service pipe. Solvent-cemented joints are used for assembly, although threaded joints are also used with Schedule 80 material.

Polybutylene (PB)

Note: this product is no longer available.

While polybutylene (PB) is not available at this time, there is a large amount of it installed that you may encounter in repair and remodeling work. If possible, it should be replaced with current pipe or tubing materials.

PB is a bluish-gray pipe that was popular for water services and hot and cold distribution piping. Insert fittings, mechanical fittings, heat fused joints, or stab type fittings were used in assembly. Figure 6-5 shows PB pipe with metallic compression rings.

Cross-Linked Polyethylene (PEX)

This special form of flexible polyethylene material has been treated at the molecular level so that the polymers are cross-linked for added strength, flexibility, and tolerance for elevated temperatures. It is available in either 20-foot straight lengths or longer rolls. It is also available in white, red, or blue. Brass mechanical fittings are used to join the pipe; these fittings are typically held in place either by copper or stainless steel bands or by the pipe's elastic properties. Figure 6-6 shows both metallic band clamps; Figure 6-7 shows a brand that uses a plastic sleeve to hold the fitting in place.

Polyethylene-Aluminum-Polyethylene (PEX-AL-PEX)

Polyethylene-aluminum-polyethylene (PEX-AL-PEX) is a laminated piping composed of two layers of cross-linked polyethylene (PEX) surrounding a thin tubing of

FIGURE 6-5 Polybutylene (PB) is a bluish-gray pipe that is no longer used due to numerous failures. It should be replaced when found.

Photo by Ed Moore

FIGURE 6-6 PEX pipe joints may be held in place with either copper or stainless steel bands. Copper bands are typically less expensive, but require a crimping tool for each pipe size. The crimping tool for stainless steel bands is suitable for pipe sizes from $\frac{3}{8}$ inch to 1 inch.

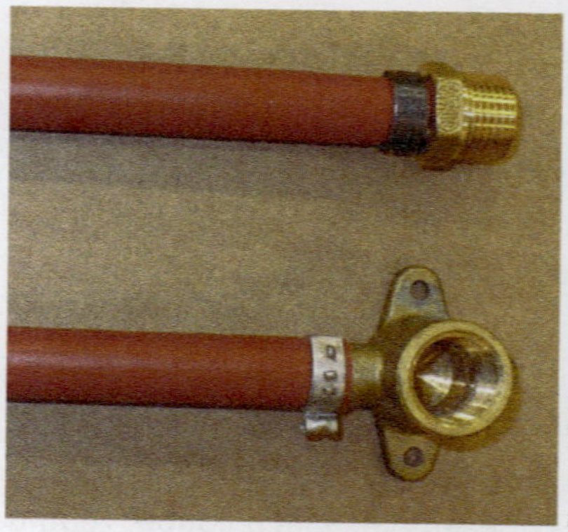
Photo by Ed Moore

FIGURE 6-7 This brand of PEX takes advantage of the pipe's elastic memory and a cold expansion process for installing fittings. Notice the additional PEX band on each fitting end.

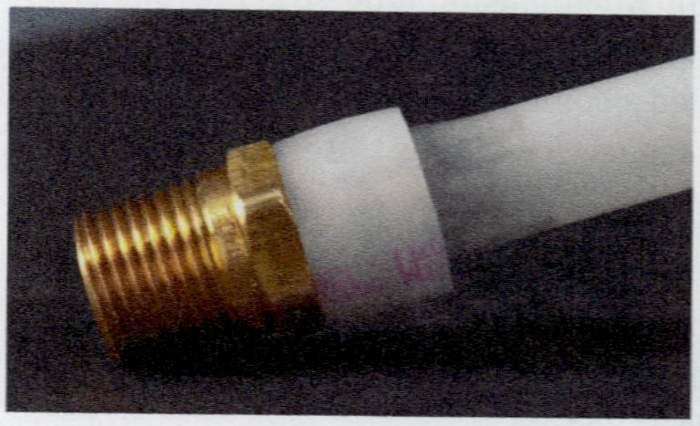

aluminum. The aluminum adds stiffness to this semirigid tubing, allowing it to retain its shape when bent with special tools. It comes in long rolls and is used extensively in the radiant floor heating industry because of its long life and tolerance to exposure to concrete. Like PEX, connections with PEX-AL-PEX are made with brass mechanical fittings.

Glass Pipe

Relatively fragile and costly, glass pipe does not corrode or introduce contaminating material into the water supply. It is assembled using mechanical joints with rubber washers or O-rings. Glass is used for distilled water or other special systems. Glass pipe is not permitted under some codes.

Metallic Pipe

Metal pipes are generally stronger and heavier than the materials listed above. They may be applied to a greater range of temperatures and pressures than the nonmetallic types but are typically more difficult to work with. Specific characteristics and applications include the following:

Brass

Brass pipe can be used for water services and distribution piping. It is corrosion-resistant and uses threaded or soldered joints. Brass pipe is now seldom used because of its high cost. It is more commonly found in compressed air systems.

Copper

Copper tubing used for water distribution is available in both soft and hard form and is categorized as Type M, L, or K. All three types have the same outside diameter, which is always $\frac{1}{8}$ inch larger than the stated size. For example, $\frac{1}{2}$ inch copper tubing has an actual outside diameter of $\frac{5}{8}$ inch. Type K has the thickest walls and the smallest inside diameter, while Type M has the thinnest inside walls and largest inside diameter. Soft copper tubing is available in 20-foot to 100-foot coils and is commonly used to run natural gas to appliances and refrigeration lines. Soft tubing can be bent to save fittings, which means that it works very well in situations where the tubing needs to make several turns. It not only saves time but will also cut down on your chances of developing a leak.

Hard copper tubing, on the other hand, is most commonly available in 10-foot and 20-foot lengths, and is very convenient when making long straight runs or for stub-outs, such as a tub spout. It also holds up much better than soft tubing in high-traffic areas. The copper industry has adopted a color code for the labeling of each of the different types of copper: Type K uses green lettering, Type L uses blue, and Type M uses red.

While both hard and soft copper can be brazed or soldered, only soft copper should be flared for joining pipe. Compression fittings are available for both hard and soft copper. Figure 6-8 shows a relatively new fitting that uses an O-ring and barbs that dig into the copper pipe when installed.

FIGURE 6-8 This fitting has internal barbs that dig into the pipe when installed.

Photo by Ed Moore

Cast Iron

Ductile cast iron pipe is used for water mains. It is rugged and noncorroding. Cast iron water mains are also popular but are somewhat less rugged than ductile iron.

Steel (Galvanized)

Galvanized steel was the most widely used material until the middle of the twentieth century, and is still popular in industrial applications. It is very rugged and the strongest material (mechanically) but is somewhat more subject to corrosion than other metals. It is also more difficult to work with than many new materials. Consequently, steel is used less frequently in today's market.

Lead

Lead piping for water services was the material of choice early in the twentieth century, but it is no longer approved by the major codes. Lead continued to be used for water services until the 1960s and 1970s. Joints in lead pipe are made with ellipsoids of solder (by a process called wiping) and are difficult and time-consuming to make.

Due to environmental and health considerations, lead is no longer used in new installations. When encountered, it is usually replaced with less toxic, easier-to-use materials.

Summary

If other materials are in use in your area, or if you encounter unusual materials in older buildings, ask your instructor or employer about these materials.

Special installation precautions may be required in certain nonmetallic systems. For example, the electrical system in a building cannot be grounded to the plumbing system when the water service is nonmetallic. Alternate grounding procedures will be necessary. Also, metallic tracer wire may be required by local code when installing plastic water service tubing and pipe.

In summary, water piping materials are selected from materials that have been manufactured to industry standards, using the material that is best suited and has the least installed cost for the job. The least-installed-cost judgment is made by evaluating the exact application, cost of material, ease (or difficulty) of working with the material, skill and experience of the crew available to do the work, durability of the material, and expected life required for the completed installation. When replacing metallic water pipe with nonmetallic pipe, make sure electrical systems are grounded.

SOURCES AND TREATMENT OF WATER

Water can be obtained from various sources, depending upon the amount needed. Larger cities, which require large volumes of water, may use lakes, rivers, or reservoirs as the point of supply. Smaller cities can utilize fields of wells. An individual or private system must be developed for any building located where it cannot be served by a larger system.

The first requirement of a community or public system is that it must provide sufficient water at adequate pressure to meet the needs of the buildings or communities that it serves. Minimum pressures are set forth in local codes and are based on the loads demanded within the building. Both the National Standard Plumbing Code and International Plumbing Code limit the maximum pressure to 80 psi. If the reservoir or other source is located above the system, gravity may provide enough pressure for the system, but in most cases, pumps are required to maintain flow and pressure.

Large systems have the following advantages:

1. Adequate quantity and pressures are usually assured.
2. Water treatment and water quality are required to be monitored regularly.
3. Piping systems are more likely to be monitored and maintained regularly.

Disadvantages of a large system include the following:

1. A cross-connection can affect more people.
2. There is a tendency to waste water if there is an impression of an abundant supply.

Individual or Private Systems

Private systems can use any of the following water sources with varying levels of success:

Ponds and Streams

Ponds and streams are the least desirable source of water because of the need for considerable treatment and continuous monitoring.

Drilled Well

A drilled well is obtained by turning a specialized drill bit into the ground, generally 100 feet or more. It is the most desirable source of water, because it is usually sunk into deep strata, where a minimum of treatment is required. The well casing, usually 4 inches to 10 inches in diameter, is sealed at the top to minimize chances of surface contamination reaching the deep water level.

Driven Well

A driven well is obtained by literally pounding the well pipe into the ground. Therefore, it is not suitable for rocky strata. It is usually shallower than a drilled well, generally less than 100 feet. The well diameter is usually $1\frac{1}{4}$ inches to 2 inches. It is a satisfactory source if the water at the shallower level is pure.

Dug Well

Historically, this was the typical well of the pioneers. It requires the least equipment, since it is usually dug by hand. It consists of a large-diameter, shallow hole, which may be lined with brick or large pipe. It is easily contaminated by objectionable materials on the surface of the ground.

Bored Well

To make a bored well, a large hole is bored by an auger and then cased. It is sealed on top, reducing the chances of surface contamination reaching the water source. The well diameter is considerably larger, from 3 feet to 4 feet, and the well is usually less than 100 feet deep.

Public/Private Systems

Private systems make it possible to utilize a site that is not served by a large or public system. However, water quality in a private system is uncertain, and the amount of water can be severely limited. Installation and maintenance costs are high, and treatment and monitoring costs can be high.

Because of the limitations of the private system, a public water supply system and a private system should never be connected together. Since the water quality in a private system is less reliable than in the typical large system, the large system would be at risk of widespread contamination if the private system supply became fouled. Nearly all plumbing codes prohibit such interconnections.

WATER TREATMENT

Natural water always contains impurities. Raindrops form on dust particles, and falling rain becomes rich in dissolved gases. After percolating into the various geological levels in the earth, small amounts of the minerals in the earth are dissolved by the water, which carries the materials in solution. Shallow water can also have bacteria present. Thus, the water available to us has many added ingredients, some relatively harmless and some that have to be removed or neutralized.

Potable water is produced by treatment at or near the water source to control conditions that are dangerous to us. Conditions that must be controlled include the following:

Suspended Solids

Suspended solids can cause the water to have a certain color and *turbidity*, or cloudiness. These solids are eliminated by holding the water in settling tanks until the solids fall to the bottom. Sometimes chemicals, such as aluminum sulfate, are added to cause the smaller solids to mass together and settle faster; this process is known as coagulation or flocculation.

pH Value

In 1909, the concept of pH was introduced by S. P. L. Sorensen, a Danish chemist. The term *pH* is believed to come from the French term *potentiel hydrogène*, meaning the potential of hydrogen. The number is an index from 0 to 14, relating to the acidity (corrosiveness) or alkalinity (scaling potential) of the water. A reading of exactly 7 is neutral, neither acidic nor alkaline. A pH below 7 is acidic, and the lower the number, the more acidic the water. A pH above 7 is alkaline, and the larger the number, the more alkaline the water. Alkalis are also called bases. The value for potable water should be in the range of 7 to 8 to assure noncorrosive and nonscaling water. Water that is corrosive can cause pipes and other plumbing fixtures to corrode or otherwise fail prematurely.

It is important to remember that pH is an intensity measure, not a quantity measure, just like a thermometer can tell how hot a room is but cannot tell how much heat must be removed to cool the room. It is also important to know that the pH scale is based on an exponential function. In simple terms, a pH reading of 4 is ten times as acidic as a reading of 5 and one hundred times as acidic as a 6.

Carbon Dioxide (CO_2) Content

Carbon dioxide is absorbed by rainwater as it falls from the sky. This carbon dioxide is dissolved into the water, making it acidic. The CO_2 can be reduced by deaeration or heating.

Dissolved Air Content

The oxygen in air is dissolved into water, making it corrosive and therefore harmful to the piping system. Oxygen is reduced either by heating the water or through a process known as vacuum deaeration.

Hardness

As water is absorbed into the ground and works its way into the water table, it picks up minerals from the soil. These additional minerals make the water hard and abrasive, which can lead to premature failure of the piping components. Hard water is treated by softening, usually near the point of use rather than at the source.

Bacterial Content and Effects of Water Temperature

Chlorination is the most effective control for waterborne bacteria. Elevated water temperature increases the degree of the problem of chemical attack from acids or bases and of carbonate ("lime") scaling. Water at a temperature above 180°F is considered to be sterilizing for most bacteria, so water that hot is beneficial for that problem. Additional information concerning water treatment can be found in *Plumbing 301*.

IN THE FIELD

Hot water used for potable water should be limited to 120°F to limit the possibility of scalding.

Flushing and Sterilization

New or significantly modified piping should be flushed thoroughly to remove dirt, pipe chips or shavings, and other foreign particles. After thorough flushing, the piping should be sterilized.

Sterilization can be accomplished by following either of these methods:

- Fill the system with a solution of 50 parts per million (ppm) of chlorine for 24 hours.
- Fill the system with a solution of 200 ppm of chlorine for 3 hours.

After the elapsed time, it is essential to flush the system and fill it with potable water. A water sample should then be taken and tested. If necessary, the sterilization must be repeated until bacteria cannot be detected. Tables 6-1 and 6-2 contain a brief outline of water conditions, detrimental effects, and corrective methods. Additional information can be found at the Water Quality Association website (http://www.wqa.org).

TABLE 6-1 Detrimental Water Conditions

	Corrosion	Erosion	Scaling	Clouding	Taste Problems	Precipitation Forming
Acids	X		X	X	X	
Solids in Suspension		X		X		X
Salt	X			X	X	
CO_2	X	X				
Oxygen	X	X				
Oil-Grease				X	X	X
Sewage (Organic Matter)	X			X	X	X
Calcium Compounds			X			X
Magnesium Compounds			X			X
Sodium Compounds					X	
Ferrous Compounds			X	X	X	X

TABLE 6-2 Treatments for Detrimental Water Conditions

	Settle, Precipitate, Filter	Soften Chemical	Neutralize	Heat or Deaerate	Deionize	Reverse Osmosis	Aerate	Distill	Chlorinate
Acids			X		X				
Solids in Suspension	X								
Salt					X	X		X	
CO_2				X				X	
Oxygen				X				X	
Oil-Grease	X								
Sewage (Organic Matter)	X					X	X	X	X
Calcium Compounds	X	X			X	X		X	
Magnesium Compounds	X	X			X	X		X	
Sodium Compounds					X	X		X	
Ferrous Compounds	X	X			X	X		X	

PUBLIC WATER DISTRIBUTION SYSTEMS

Public or municipal water distribution systems serve the majority of the population in this country in the twenty-first century. The components of a public water system are discussed below.

Distribution

After the public water purveyor has developed a water supply, treated the water to potable standards, and provided the necessary pumps for distribution, the water must be delivered to the users through a community piping system.

Typical public water systems consist of large-diameter distribution lines and smaller trunk lines connected to local mains. The local mains supply individual building service pipes.

A classification of the parts of such large systems is given below.

Large Distribution Lines

These lines are large-diameter pipes that carry water from the treatment plant to the next major part of the distribution system, trunk lines.

Trunk Lines

Trunk lines carry the water from the distribution lines to smaller local mains.

Local Mains

Local mains are the intermediate supply pipes between the trunk lines and the individual building water service pipes.

Building Water Service Pipes

The building water service pipe connects the water main (or other source of potable water supply and/or a private well) to the water piping system of the building served.

Distribution Layout

The main distribution scheme of a public system may be laid out in many different ways. The most common arrangements are defined with respect to the terms above.

Loop

In the loop system, large distribution lines are installed in loops that cover part or all of the city. Local mains are connected to the loops.

Grid System

In a grid system, the large distribution lines and trunk lines are installed in a checkerboard or grid layout with local mains connected to the grid.

Tree System

In a tree system, large distribution lines and trunk lines extend outward from the source, decreasing in size as the distance from the source increases. Local mains connect to the branches of the tree.

Loop and grid systems provide greater reliability and capacity as parallel flow paths keep pressure drops small. In the event of a system problem, the problem area can be isolated and most of the system will continue to function. System extensions can also best be made to the grid or loop system because of the low-pressure-drop characteristic.

Components and Operation of a Public Water System

Water pressure and line sizes may vary throughout a system depending on the components and functions needed from the water system. Table 6-3 indicates water pressures and line sizes that are considered adequate for the uses shown without oversizing the system.

Water Main Installations

Outside water distribution systems are usually installed by the utility company, but plumbing contractors frequently install extensions to serve industrial buildings or subdivisions in a city. There are basic considerations for installing underground water mains (main distribution lines, trunk lines, water mains, or building water service pipes). These concepts include:

- Follow all rules of safe trenching and shoring to protect workers and the public.
- Provide solid, continuous bedding for the tubing or pipe.
- Place the pipe at depths below the frost line.

TABLE 6-3 Typical Components and Operation

Category	Normal Values
Residential Areas	40–60 psi
Other Building Types	60–80 psi
Water Velocity in Mains	10 fps
Minimum Pipe Sizes	
Water Service	$\frac{3}{4}$ inch
Fire Line	5 inches to 6 inches to hydrant
Minimum Flow Capacity	3 inches to building systems
Fire Hydrant	600 gpm
Line Velocity	2 fps

Note: "psi" is pounds per square inch; "fps" is feet per second; "gpm" is gallons per minute.

- Protect the pipe from corrosive soils with protective coatings or use pipe material impervious to the corrosive condition. More information is available in *Plumbing 301*.
- Test the line carefully. Remember that some joints can be blown apart and a large-diameter line could stand up if not partially backfilled before testing.
- Use proper backfill material (no rocks, frozen earth, or destructive debris) over the pipe.
- Follow all codes regarding placement of water service piping and spacing from sewer piping. See Figure 6-9.
- Use thrust blocks to support pipe and fittings at changes of direction.

Pressurization Methods

It is necessary for the entire water distribution system to be pressurized to keep water flowing at all points of need. Pressure is maintained by one or more of the following methods:

Gravity Feed

Gravity feed may be used to maintain pressure if the water source is higher than the public system. For every vertical foot of water column, there will be approximately 0.433 psi of pressure generated.

Elevated Tanks Versus Pressurized Reservoir

Elevated tanks, open to the atmosphere, are filled by the operation of system pumps. The arrangement is then the same as gravity feed. Various control arrangements are used to control the pumping to minimize pressure variation.

Closed reservoirs are pressurized by system pumps, which develop the required pressure. The pressurized reservoir then provides a large volume of water to supply the system needs. These tanks do not have to be built on long legs to maintain system pressure, but they must be built to withstand the required system pressure at the

FIGURE 6-9 A minimum separation distance of 12 inches must be maintained at all times for underground pipes. If pipes must cross, it is preferred that the water service pipe be on top.

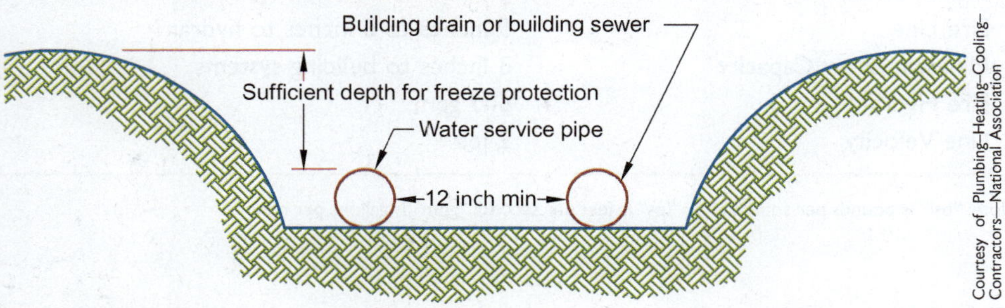

site. Thus, it is an economic decision whether the low-pressure elevated tank costs less than the lower high-pressure tank.

Multiple Pumps

By careful selection of pumps, it is possible to maintain system pressure by turning pumps on and off as system flow (and system pressure) changes. The use of variable speed drives has greatly increased the efficiency of these types of systems by allowing the pumps and motors to be run at lower speeds during periods of low demand. This reduces the stress and energy surges of starting and stopping large-horsepower motors.

System Water Pressure Range

The system water pressure range should be between 30 and 75 psi static (not flowing). Higher pressures lead to rapid deterioration of faucets and ballcocks as well as excessive water waste in unrestricted outlets. Pressures higher than 75 psi can also produce excessive flow velocities, in turn causing noise or even pipe failure.

Pressures lower than 30 psi will usually produce inadequate flow and result in unsatisfactory performance of many plumbing devices. Buildings over three stories will not receive adequate flow with building supply pressures less than 30 psi.

Most codes require pressure-reducing valves to reduce service pressures to fixtures. The output of such valves is normally set at about 50 psi, but most devices can be adjusted to a range of pressures. See Figure 6-10. Refer to local codes for engineered systems that may require greater psi pressure.

FIGURE 6-10 Typical water pressure reducing valve.

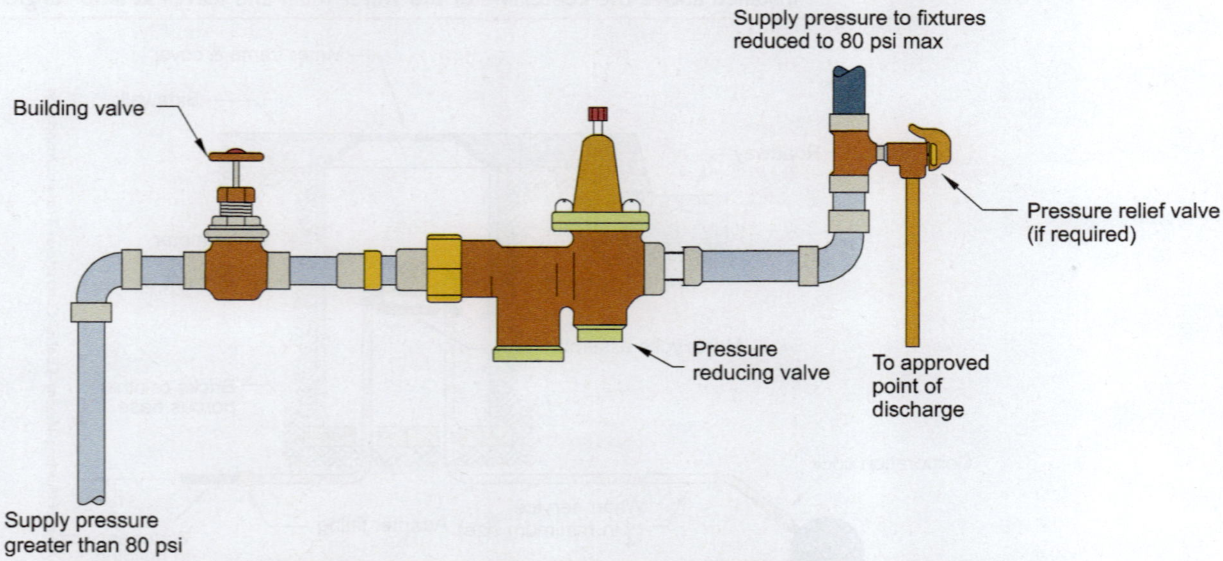

Courtesy of Plumbing–Heating–Cooling–Contractors—National Association

Building Services

A connection from a water main to a building or house service is made by a corporation cock, available from $\frac{3}{4}$ inch to 2 inches. These valves are connected to the main by a tapping machine, which can drill, tap, and install the cock while the main is under pressure. If the needed service calls for a relatively large-diameter pipe to be run from a relatively small-sized main, such as a 2-inch service from a 5-inch main, a manifold installation is used. As the name suggests, several smaller cocks are tapped into the main and tied into the common building service. This technique protects the strength of the main while providing the flow required for the building.

As discussed in Chapter 9 of *Plumbing 101*, the area of a circle is calculated by squaring the radius or diameter. Therefore, it would take four 2-inch diameter pipes to equal the same area as one 4-inch diameter pipe.

Larger taps are made by bolting a valved saddle tee on the main, drilling the main through the valve, withdrawing the drill, and closing the valve. This installation is done while the main is pressurized in service. See Figure 6-11 for a typical public water service.

Corporation Cock

The corporation cock is the initial valve tapped into the water main. The water service (minimum $\frac{3}{4}$ inch) connects to the corporation cock and then extends to the curb box, which is located just after the property line.

Curb Cock and Curb Cock Box

Once the water service line has entered the property, there needs to be a method to shut off the water for repairs or installation of the remainder of the system; that is

FIGURE 6-11 Typical water service connection. Notice how the corporation cock is installed above the centerline of the water main and leaves at a 45° angle.

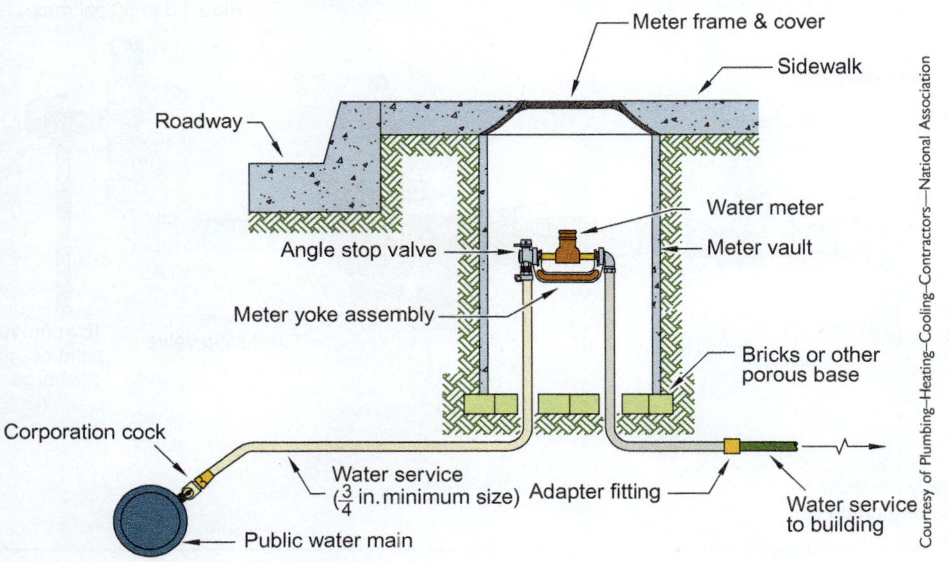

the purpose of the curb cock. Since the water service line is buried, a box or combination of a box and pipe must extend from the water service line to the finish grade so that the valve is readily accessible.

The curb cock box has a foot, base, adjustment sleeve, and cover plate. The foot and base interlock around the curb cock valve. The adjustment sleeve allows for setting the cover plate to grade level. The cover plate keeps dirt and debris from entering the box. A long rod or curb key is used to open or close the valve.

Meter Yoke, Meter, and Backflow Preventer

Most water services are metered to determine usage for billing purposes. The meter yoke is used to hold the meter in place. The meter can be below ground in a pit or in the building.

If the meter is below ground, it must be easily accessible for replacement or reading. Valves are usually placed on both sides of the meter in the building to shut off water flow if meter replacement is necessary. Meters are often provided with a remote-reading register outside the building to allow the meter to be read more efficiently. Some newer meters store usage data until read by a handheld computer. Other devices can be connected to telephone lines for truly remote readings and are useful in areas where security or convenience is of concern. Figure 6-12 shows a typical residential water meter equipped with remote meter reading ability.

FIGURE 6-12 Typical water meter with remote meter reading capabilities.

Photo by Ed Moore

As a means to protect the community water supply, many water purveyors are requiring some means of backflow prevention. A residential dual check is one of the more commonly used valves to prevent the water from leaving the building and returning to the community water supply. Depending on the type of hazards that may exist in the building or the type of backflow, different backflow devices may be required. Section 10.5.3 of the 2012 National Standard Plumbing Code lists suitable backflow protection based on the mechanism of backflow and the hazard it presents.

Building Shutoff Valve

As the name implies, the building shutoff valve is responsible for shutting off the water that feeds into the building. This valve is placed immediately inside the building, or in a curb box just outside the building for quick shutoff. Tag this valve for identification if its purpose is not obvious. It should be readily accessible at all times.

Ball valves, which shut off with a 90° rotation of the handle, have become the preferred valve for this application. A stop and waste valve with a bleed feature is commonly recommended to drain the water from the vertical line if alterations are necessary. It is important not to install a valve of this sort in a pit because it opens the downstream piping to the environment when in the closed position. If the valve was placed in a pit and the area became flooded, contaminants could enter the piping system. Most codes require a full-port service valve.

Pressure-Reducing Valves

Whenever the water supply pressure is higher than 80 psi, a pressure-reducing valve is required to be installed in the water service line before entering the building. There are two main types of pressure-reducing valves, direct acting and pilot operated. Pilot operated valves are for large water mains and are typically made of ductile iron. They operate through a pilot tube that connects to the downstream side of the valve. The pilot tube sends downstream water pressure to an externally mounted diaphragm valve. The diaphragm valve will drive the main valve open or close depending on whether the downstream pressure is above or below the valve set point.

On smaller applications, a direct acting valve similar to the one shown in Figure 6-13 is used. Direct acting valves get their name because the water in the main pipe is in direct contact with the diaphragm that senses the pressure. In this arrangement, the incoming water is directed into an inner chamber, which is opened and closed by a globe style seat. When the downstream pressure reaches a pressure higher than the valve set point, the water pressure will override the spring pressure and cause the valve to close. At this point, the valve will remain closed as long as the downstream pressure is higher than the set point. Once the pressure-reducing valve

FIGURE 6-13 Typical direct acting pressure-reducing valve. Note that these valves have an arrow indicating direction of flow and a metal nameplate that lists its operating range.

Photo by Ed Moore

is closed, water will be trapped within the building. As the water is heated, it will expand and increase system pressure even higher. This is the reason why most municipalities require an expansion tank to be installed whenever a pressure-reducing valve is required. The expansion tank is usually located near the water heater, since it is usually the cause of the expansion.

Pressure-Reducing Valve Sizing

A common mistake is to automatically size the pressure-reducing valve based on the pipe size when it should be sized based on the design pressure and flow. An oversized pressure-reducing valve will have to position the valve disk extremely close to the valve seat to handle low flow conditions. Since the water is now moving through a small area, it will travel across the valve seat very quickly and can erode material away. The cutting away of the valve seat will have the appearance of a small groove as though it was cut with a wire, hence the term *wire drawing*. This condition will cause the valve to fail prematurely and be noisy in operation. You should consult the valve manufacturer's literature to properly size the valve.

When there is a large fluctuation in the expected water flow, such as in an apartment complex or an industrial setting, it may be necessary to install two valves in parallel. Instead of one valve that is sized for the maximum flow of the building, the two smaller valves can be sized to a portion of the flow. These smaller valves will be wide open during peak flow times but can throttle down to low flow conditions better than one larger valve. For more even flow, keep the length of piping as close to the same as possible.

Since all pressure-reducing valves have moving parts and small passages, they will eventually need service or replacement. For this reason, it is advisable to install a union in series with the valve so the system can easily be disassembled for repairs.

Strainers

Whenever the piping system contains control valves, moveable parts or is suspected of having small particles such as dirt, silt, or other suspended solids, strainers should be installed. Most valve and pump manufacturers install some form of strainer on the inlet of their products. This protects the inner working parts and extends life. Strainers are typically sized based on the diameter pipe they are being installed on and can be configured as either wye-patterned, bucket type or inline. Once the strainer body has been sized, the proper strainer mesh must be selected. The strainer mesh is usually a screen made of either stainless steel or brass wire. Wire mesh strainers are typically numbered to indicate the numbers of wires in a given distance. This means that a number 20 mesh strainer would have larger openings than a number 40. This is opposite from the sizing of filters, which are sized based on the smallest particle they can effectively stop. In a filtering system, a 5 micron filter would stop particles larger than 5 microns while a 20 micron filter will only stop those 20 microns and larger. So why not pick the smallest strainer or filter possible? As filters and strainers become more restrictive, there will be a larger pressure drop across them. It should also be pointed out that as filters and strainers become dirty, they will filter even finer and develop even more pressure drop. This means that less pressure will be available downstream. Figure 6-14 shows a typical wye-pattern strainer.

FIGURE 6-14 This in-line strainer has a removable stainless steel wire mesh strainer. Note the direction of flow.

Photo by Ed Moore

Hose Bibb

Hose bibbs are known by many names: sill cock, pave, and spigot to name a few. As the name hose bibb implies, the primary purpose of the fitting is to allow the connection of a hose to the potable water system. Since the other end of the hose could be exposed to a number of hazards, it is crucial that some form of backflow protection be incorporated into the fitting. Most new hose bibbs have built-in vacuum breakers and mounting flanges to help protect the potable water and to secure them to the building. Figure 6-15 shows a hose bibb with a PEX connector and a built-in vacuum breaker. The only hose-end valve that does not require backflow protection is the boiler drain. Boiler drains are typically located near the bottom of water heaters, boilers, and other pressurized storage tanks. Boiler drains are primarily used to service the unit or to periodically flush it out.

As of January 4, 2014, the Safe Water Drinking Act bans the use of any valves or fittings with more than a weighted average of 0.25 percent lead. This is a dramatic decrease from the previous 8 percent standard.

PRIVATE WATER DISTRIBUTION SYSTEMS

Private water distribution systems provide water from wells and springs by pumping. Pressures are usually stabilized by a pressurized tank. The pump is activated when pressure drops below a set level and continues to run until peak pressure in the tank is restored.

A private water distribution system is a scale model of a public system. The water line from the pump is similar to the water service line and feeds the fixtures accordingly. There is no meter, meter yoke, curb cock box, or curb key valve installed on a private single-user system. However, a pressure gauge, a relief valve, a control switch, and related appurtenances are necessary components for such systems.

A valve should be installed in the pipe to the building delivery piping so that the condition and operation of the well, the pump, and the tank can be observed. This

FIGURE 6-15 This hose bibb has a built-in vacuum breaker and is designed for use with PEX piping.

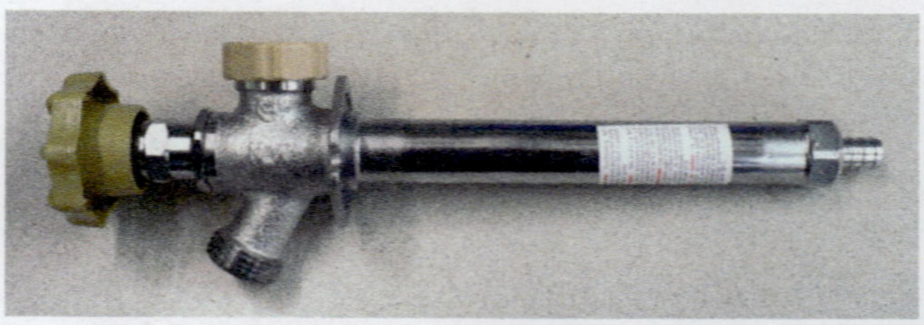

Photo by Ed Moore

will also serve as a way to isolate the water service pipe from the rest of the house in the event that the pump needs to be replaced.

IN THE FIELD

A specially made brass manifold (see Figure 6-16) is available to mount the controls necessary to control the well pump and monitor the system. This can be a great time saver.

As in a public system, the valves used for water services, at water meters, or at private pressure tanks should be the full-port type.

NONPOTABLE WATER DISTRIBUTION SYSTEMS

Some jurisdictions permit the installation of nonpotable water distribution systems using gray water to supply water to urinals or water closets for water conservation purposes. These piping systems are sized similar to potable systems and are supplied by water from a tank-and-pump arrangement.

Such piping systems must be continuously marked along their lengths as nonpotable to avoid the possibility of a cross-connection between a nonpotable and potable supply system.

SUMMARY

Installation of water mains is the first step in providing a serviceable public water distribution system. It is important that the guidelines mentioned above as well as all local codes be followed to install a safe, long lasting water distribution system.

Public water distribution systems are pressurized by gravity, elevated tanks, or pressurized tanks.

The desired range of water pressure in the building service is 30–75 psi, although some private systems operate in the 20–40 psi range.

FIGURE 6-16 Specially made manifolds can be a great time saver by eliminating individual joints.

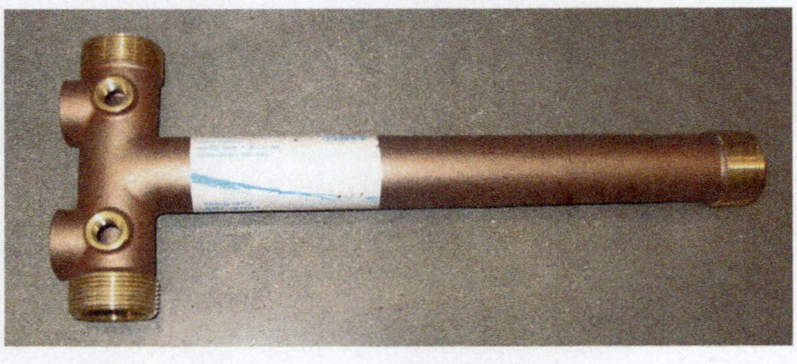

Photo by Ed Moore

REVIEW QUESTIONS

True or False

1. __________ PE pipe is used on either hot or cold piping.
2. __________ Water service is the section of piping that runs from the water meter to the water distribution section.
3. __________ A low pH in water means the substance is very acidic.
4. __________ A residential potable water system should not exceed 80 psi pressure.
5. __________ One treatment for calcium compounds is filtering.
6. __________ Pressure-reducing valves close if the upstream pressure becomes too high.
7. __________ A strainer should be installed upstream of any sensitive valves.
8. __________ A corporation cock is normally installed just inside the building.
9. __________ A drilled well is preferred over a dug well.
10. __________ *Turbidity* is a term that refers to cloudy or colored water.

Chapter 7

LEARNING OBJECTIVES

The student will:

- Summarize basic pressurization methods for large systems and for building systems.
- Discuss issues related to water supply systems in buildings, including concepts of water flow rate, water supply fixture units (WSFU), and pipe sizing.
- Discuss the differences between main-and-branch piping systems and individual run piping systems.

Water Pipe Sizing: Main-and-Branch and Individual Run Systems

THE GOAL OF BUILDING PIPING DESIGN

The intent of any water piping design for a building is to select appropriate pipe sizes to ensure that there will be satisfactory delivery of water at safe, useful temperatures to the plumbing fixtures in the building. A significant portion of this chapter is designed to provide the information and methods to develop such piping designs.

WATER PRESSURE

Before we discuss how water flow rates and pipe sizing are determined, we need to consider the concept of water pressure. Pressure is developed when force is exerted over an area. In the most typical case, we are looking only at the force acting on an area that is 1′ × 1′, which is commonly referred to as a unit area. This will give a unit of measure of pounds per square inch, which is abbreviated as psi.

If water is poured into a cube that is 1-foot high, 1-foot wide, and 1-foot long, it weighs 62.4 pounds. All of this weight exerts a force on the bottom of the container. Because the area of the bottom is 12″ × 12″, it has an area of 144 square inches. The pressure at the bottom of the container can then be calculated as:

$$\text{Pressure} = \frac{62.4 \text{ pounds}}{144 \text{ square inches}} = 0.433 \text{ pounds per square inch (psi)}$$

Because this pressure is based on a unit area, the pressure at the bottom of any container will be 0.433 psi as long as there is one vertical foot of water depth. Therefore, for every one-foot increase in elevation, the water pressure will increase 0.433 psi. This concept can be used to determine the pressure at any depth, regardless of the shape of the container. See Figure 7-1.

Simply multiply the vertical distance measured in feet from the water surface down to the point of measurement by 0.433. For example, if the water is 10 feet deep, the pressure at the bottom will be 4.33 psi (10 × 0.433).

FIGURE 7-1 When the depth of water is the same, the pressure at the bottom of each container is the same regardless of shape.

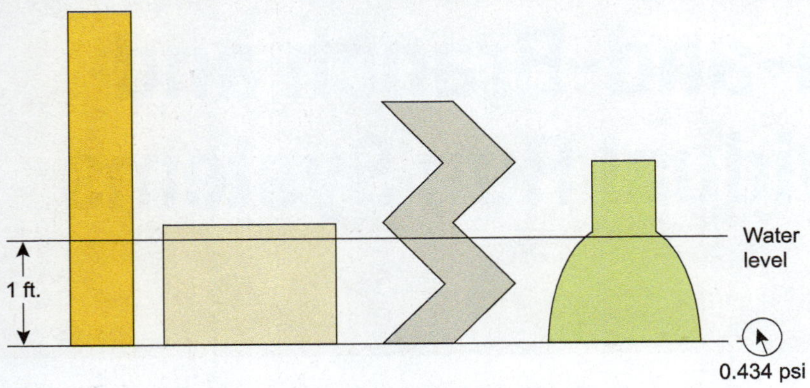

EXAMPLE 1

A $1\frac{1}{2}$ inch pipe travels vertically from the ground floor of a building to the ninth floor. Assuming 10 feet per floor, what would be the pressure at the bottom of the pipe, when filled with water?

$$\text{Pressure (ground floor)} = \frac{0.433 \text{ psi}}{1 \text{ ft. wc}}\left(\frac{9 \text{ floors}}{1}\right)\left(\frac{10 \text{ ft.}}{\text{floor}}\right)$$

$$\text{Pressure (ground floor)} = 0.433 \times 90 = 38.97 \text{ psi}$$

Notice that the diameter of the pipe had nothing to do with the static pressure. It will only affect pressure when there is water flowing.

When dealing with pumps and other hydronic systems, the terms *head pressure* or *head* are frequently used. These terms are usually expressed in feet of water column (ft. wc, as seen in the equation above) and are related to pressure in psi by the formula:

$$\text{Pressure (psi)} = 0.433 \times \text{Height of the water column (ft.)}$$

Until now, we have been talking about how a vertical column of water can generate pressure. The converse is true as well. A certain amount of pressure is needed to support a column of water. Changing the above formula around, it is possible to determine how tall a column of water will be supported by a known pressure.

EXAMPLE 2

A plumbing technician wants to run water to the top of a structure. The water pressure at the ground floor to the structure is 80 psi. What is the maximum height the water will rise in the structure?

$$\text{Max height of water} = 80 \text{ psi}\left(\frac{1 \text{ ft. wc}}{0.433 \text{ psi}}\right)$$

$$\text{Max height of water} = 184.76 \text{ ft.}$$

Note: At 184.76 ft., there will be no pressure left to do work. The water would be stationary in the pipe. Increasing the diameter of the pipe would not change this fact.

Maintaining Water Pressure in the Community

As discussed in Chapter 6, public water distribution systems maintain nearly constant water pressures in the mains by using large elevated tanks or pressurized tanks placed at necessary points in the system. An alternative to elevated tanks is to use pumps to adjust delivery flow and pressure as the system load varies. The elevated tank method is more common, as it requires less complex pump and control equipment. Elevated tanks also provide an available supply of water in case of emergency and provide a flywheel effect for the system, stabilizing pressure variations. Figure 7-2 shows a newly constructed elevated water tank to help deliver water to a new subdivision.

The following sections describe further variations of these systems.

Forced-Feed System

Forced-feed or direct-feed systems consist of a series of pumping stations located throughout the grid or loop distribution system. The load pattern, based upon present and predicted future needs, establishes the design criteria and the location of the pumping stations.

The primary supply pumping station is located at the water treatment plant. Auxiliary pumping stations are designed to satisfy system needs. Displacement pumps are usually used in auxiliary stations because they build up pressure as long as they operate. Note that these auxiliary pumps do not add to available water volume, since they do not draw water from wells or reservoirs—they only boost the pressure of the water that is already in the system.

FIGURE 7-2 This newly constructed water tower helps supply water to new housing development.

Photo by Ed Moore

One disadvantage of the forced-feed system is that the water source capacity must always exceed system demand. If the water supply is inadequate, even briefly, pressure drops may occur, which can cause vacuum conditions in some mains and increase the possibility of backflow from customers' premises into the water mains.

Another disadvantage may be erratic response to variations in demand for water. The uncertainty of these variations means that personnel and/or expensive, sophisticated equipment must monitor and control the pumps' output.

Gravity-Feed System

In a gravity-feed system, the water is placed at a higher elevation than the final point of distribution to provide the pressure necessary for delivery to point of use. A typical utility system would include more than one storage tank to best serve the loads on the system.

Reservoirs, lakes, or rivers are sources of water. Covered tanks are the preferred devices to store water after it has been brought up to potable standards at the treatment plant. Supply and demand are easily monitored by taking water level readings in the tank.

When pumps are required to raise the water to the elevation needed to maintain the gravity head, centrifugal pumps are usually selected because they can move larger volumes of water at low pressures, compared to other types of pumps.

Considerations of Flow in Public Water Distribution Systems

Public, municipal, or utility water piping systems are usually designed by civil engineers. The criteria for these large systems are different from those for a distribution system in a building. The principal concern is for fire emergencies, with a somewhat lesser concern for adequacy of the layout to be able to supply the next wave of subdivision developments. The fire issue is resolved by recording the flowing pressure in the street mains when a very large flow is being taken from one or more hydrants. This test is of interest to insurance companies that are being asked to insure a proposed commercial or industrial facility. A typical test might be to flow 1,000 or more gallons per minute and to have the system pressure in the street mains stay above 40 psi. In many areas, fire hydrants are color coded according to the pressures they can exert.

If acceptable values are not observed, the management of the water utility could elect to increase the capacity of the system so as not to lose the proposed real estate improvement—which, in most cases, would mean jobs for local citizens and tax income to the community.

Considerations of Flow in Building Systems

The vast majority of plumbers work primarily with water flow within a building system. With some appropriate background information, they should be able to address most problems that occur in these complicated systems.

Pressure Fluctuation

Pressure in the building supply system may change because of problems external to the building. For example, street pressure variations can be caused by unusually heavy demands elsewhere, such as firefighting or a break in the water main. Variations can also be caused by loss of power or pump failure at the pumping station. Lack of rainfall or snow can also reduce the available pressure if the water level in the source reservoir falls significantly below normal levels.

Building pressure variations also may be caused by internal problems, such as undersized supply designs, excessive pipe friction losses, or fitting and meter friction losses. Building flow variations may be caused by unusual or emergency demands or changes in occupancy or use. Problems can be short- or long-term, with easy or difficult solutions.

Pressure, Velocity, Friction, and Flow

Proper design of the plumbing system is necessary for adequate service. Proper sizing requires understanding the concepts of pressure, velocity, friction, and flow. While undersized piping can cause problems such as low operating pressure, noisy pipes, and excessive friction wear of pipe inside diameter; oversized pipes create higher installation costs, waste material resources, and increase hot water consumption.

Pressure

The pressure of the water in the piping system causes the water to move through any opening in the system (faucet, valve, or leak). Whatever the supply pressure is, it will be used up by the pressure drop caused by pipe friction (due to flow), by head loss due to elevation in the building, and by pressure drop at the water outlet.

Velocity

Velocity is the speed of the water as it passes through the pipe while traveling downstream. Velocity is determined by dividing the quantity of water by the cross-sectional area of the pipe. As the cross-sectional area increases, the velocity of the water decreases for a given pressure.

Velocity is typically measured in feet per second (fps). Velocities of water within pipes should never exceed 10 fps to guard against noise and rapid pipe deterioration. Some pipe and tubing manufacturers recommend velocities in the 4–6 fps range. Be sure to check these maximum values prior to sizing a system.

Friction

Friction is the resistance to flow of materials through the system. Friction can be affected by the piping components and how they are arranged. The more friction in a system, the harder it will be to move material through it. The amount of friction can vary based on the different piping materials and different pipe sizes used. Pipe with a rough inside wall or a small diameter will have more friction compared to larger-diameter, smooth-wall pipe. If water travels too fast through the system, the amount of friction will increase as well. Copper and plastic have less resistance than steel pipe because the inside walls of copper tubing and plastic pipe are smoother than the inside wall of galvanized steel pipe. Internal changes in diameter, such as a burr from cutting or the use of internal fittings, will also cause an increase in friction losses.

Friction losses must be considered when designing a piping system. Losses due to friction reduce the pressure available at the fixture fitting. Tables of friction loss related to pipe sizes and flow rates are included later in this chapter.

Flow

Flow is the amount of water that passes a particular point in a certain amount of time. If you know how much water is coming out of a pipe and the amount of time it took, then you can calculate the flow rate. You may notice that, while Table 7-1 lists some of the more common units of flow, each unit is simply a volume divided by time.

EXAMPLE 3

A 5-gallon container is placed under an open faucet. It takes 2 minutes to fill the container. The flow rate at the faucet is calculated as follows:

$$\text{Flow rate} = \frac{\text{Volume}}{\text{time}}$$

$$\text{Flow rate} = \frac{5 \text{ gallons}}{2 \text{ minutes}}$$

$$\text{Flow rate} = 2.5 \text{ gallons per minute (gpm)}$$

If the answer was needed in cubic feet per minute, use the conversion fact that 1 cubic foot equals 7.48 gallons.

$$\text{Flow rate (cfm)} = \left(\frac{2.5 \text{ gallons}}{\text{min}}\right)\left(\frac{1 \text{ cubic ft.}}{7.48 \text{ gallons}}\right) = \begin{array}{l}0.334 \text{ cubic feet} \\ \text{per minute (cfm)}\end{array}$$

Or the answer could be found in cubic inches per minute, based on the fact that there are 231 cubic inches to a gallon.

$$\text{Flow rate (cim)} = \left(\frac{2.5 \text{ gallons}}{\text{min}}\right)\left(\frac{231 \text{ cubic inches}}{1 \text{ gallon}}\right) = \begin{array}{l}577.5 \text{ cubic inches} \\ \text{per minute (cim)}\end{array}$$

Because the rate at which the water travels in the pipe is so important to the life of the pipe, the pressure losses to the system, and the noise made by the system, it is necessary to know the velocity at which the water travels in the pipe. Example 4 shows how the velocity of the fluid in a pipe can be determined, based on its flow rate and cross-sectional area.

TABLE 7-1 Common Units of Measure

English Units	Metric Units
Gallons per minute (gpm)	Liters per second (lps)
Gallons per hour (gph)	Cubic meters per hour (cmh)
Cubic feet per minute (cfm)	Cubic meters per minute (cmm)
Cubic feet per second (cfs)	Cubic meters per second (cms)

EXAMPLE 4

If the 0.334 cfm of water in Example 3 were traveling through a $\frac{3}{4}$ inch pipe to the faucet, what would be the velocity of the water in the pipe?

To be able to calculate feet per second, we must first calculate the cross-sectional area of the pipe. Then we can move on to calculate the velocity in feet per second.

Remember that the cross-sectional area of a tube is calculated using the formula

$$\text{Area of a circle} = \pi \times r^2 \text{ or } \pi\left(\frac{d}{2}\right)^2$$

Since we know that the answer will contain the unit "feet," it is easier to change inches into feet now so all of the units of measure will be the same. Therefore, the 0.75 inch diameter can be divided by 12 inches per foot, which equals 0.0625 foot. Now the area can be found by dividing the diameter by 2 to get the radius and applying the formula for area.

$\text{Area of a circle} = \pi \times r^2$
$\text{Area of a circle} = 3.14 \times (0.0313)^2$
$\text{Area of a circle} = 3.14 \times 0.000979 \text{ ft.}^2 = 0.0031 \text{ ft.}^2$

Now use the formula that shows the relation of velocity, flow rate, and area:

$$\text{Velocity (fpm)} = \frac{\text{Flow rate (cfm)}}{\text{cross-sectional area (ft.}^2\text{)}}$$

$$\text{Velocity (fpm)} = \frac{0.334 \text{ ft.}^3}{1 \text{ min}} \times \frac{1}{0.0031 \text{ ft.}^2}$$

$\text{Velocity (fpm)} = 107.47 \text{ fpm}$ [in order to change minutes to seconds, divide by 60]

$$\text{Velocity (fps)} = \frac{107.47 \text{ ft.}}{1 \text{ minute}} \times \left(\frac{1 \text{ minute}}{60 \text{ seconds}}\right) = 1.80 \text{ fps}$$

Since velocities of water in pipes should be kept below 10 fps, this should pose no problems.

DISTRIBUTION PIPING SIZED BY AVAILABLE PRESSURE

Once an adequate building supply system is assured, the building distribution piping can be sized. The design flow for each section of the piping layout can be determined. It is important to remember that by the time the water reaches the highest, most remote fixture, it must still have at least the minimum pressure that the fixture requires for the fixture to function properly. If the distribution piping is too small, then an excessive amount of pressure will be lost due to flow restrictions. If the piping is too large, there will be an undue expense on the job.

The first step is to determine the amount of pressure that can be lost as the water travels through the longest circuit. To determine this value, add the minimum pressure required at the highest fixture ($P_{\text{fix-min}}$), the pressure loss due to elevation

(P_{elev}) at the highest fixture, and the pressure loss through the water meter (P_{meter}). Subtract this amount from the available pressure at the water main (P_{wm}).

$$\text{Pressure available for flow loss } (P_{pd}) = (P_{wm}) - (P_{fix-min}) - (P_{elev}) - (P_{meter})$$

Divide the resulting pressure available for flow loss by the equivalent length of the longest circuit, which is the distance from the water main (or other source) to the most distant fixture. Most designers estimate the equivalent length as the actual length of piping multiplied by 1.5 to account for fittings in the line.

$$\text{Pressure drop available (per foot)} = \left(\frac{P_{pd}}{\text{Equivalent length of pipe}}\right)$$

This quotient is the available pressure to move water through the piping in psi/ft. By multiplying the pressure drop available (per foot) by 100, the column pressure drop (psi/100 ft.) can readily be used to size pipe.

Use a pipe size/water flow table (see Table 7-9 to select the pipe size for each section of the circuit being studied). Note that the pipe size must be selected so that the allowable pressure drop is not exceeded and the water velocity limit is not exceeded—whichever condition calls for the bigger pipe size is the condition that must be honored.

The shorter or less elevated branches in the building can then be sized. Most designers use the same limiting considerations for these less-demanding circuits as for the longest (or highest) circuit. This is a conservative practice, but it usually does not lead to significant oversizing.

Hot and cold water piping sections are sized by the same method.

EXAMPLE 5

A piping system is connected to the supply main with 100 feet of Type L copper tube. A $\frac{3}{4}$ inch meter is installed, and the highest fixture, which requires 12 psi minimum pressure, is 30 feet above the main. Select a pipe size that will allow 10 gpm to flow without exceeding either the 8 fps velocity limit or the pressure drop per foot limit. Table 7-2 can be used to calculate the pressures associated.

Equivalent length of longest branch = 100 × 1.5 = 150 ft. (1.5 to allow for fittings)

$$\text{Pressure drop per foot} = \left(\frac{25\text{ psi}}{150\text{ ft.}}\right) = 0.17\text{ psi/ft.}$$

$$\text{Pressure drop per } 100' = 0.17 \times 100 = 17\text{ psi}/100'$$

TABLE 7-2 Pressure Available After All Pressure Losses

Initial pressure	+65 psi
Minimum pressure required at highest fixture	−12 psi
Loss because of elevation (30 ft. × 0.434 psi/ft.)	−13 psi
Loss through meter (30 gpm, $\frac{3}{4}$ in.)	−15 psi
Balance available for pressure drop due to flow	25 psi

Table 7-9 shows that a $\frac{3}{4}$ inch line (type L smooth wall) would be able to handle 12.1 gpm without exceeding 8 fps. It also does not exceed the 17 psi/100 ft. limitation. Any smaller pipe would have to have a greater velocity to achieve the same flow rate, and there would be a higher pressure drop in the process. Any larger pipe would allow the same flow rate but at a lower pressure drop and at a lower velocity.

WATER SUPPLY FIXTURE UNITS

The least-certain element in sizing a piping system is the determination of maximum water flow in all parts of the system. We know the maximum supply flow to a water closet, bathtub faucet, and a lavatory faucet, but what is the probability that all will be in use at the same time? Or, on a larger scale, what is the probability that more than one bathroom will be used at a time? More than ten bathrooms? More than ten apartments?

These are important questions. If we overestimate the values, we oversize the system, at a cost penalty to the owner. If we underestimate the values, the system will not deliver enough water to all points of need, which can be an inconvenience or even represent an economic loss to the owner if they rely on water flow for production. For a number of years, it has generally been accepted as an industry practice to identify maximum flows for certain fixtures. Table 7-3 shows the maximum flow rates that are allowed by law for some of the more common fixtures.

EXAMPLE 6

Figure 7-3 shows an isometric view of a simple residential plumbing plan. Use Table 7-3 to determine the total gpm for the structure. Table 7-4 shows the tabulated values.

TABLE 7-3 Maximum Demand at Individual Water Outlets

Type of Outlet	Maximum Demand (gpm)
Metering Lavatory Faucet	0.25 gal/cycle
Self-Closing Lavatory Faucet	0.5
Drinking Fountain Jet	0.75
Ordinary Lavatory Faucet	2.5
Shower Head, $\frac{1}{2}$ inch	2.5
Laundry Faucet, $\frac{1}{2}$ inch	2.5
Ballcock in Water Closet Flush Tank	3.0
Dishwashing Machine (domestic)	4.0
Laundry Machine (8 or 16 lbs.)	4.0
Sink Faucet, $\frac{3}{8}$ inch or $\frac{1}{2}$ inch	4.5
Bath Faucet, $\frac{1}{2}$ inch	5.0
Hose Bibb or Sillcock ($\frac{1}{2}$ in.)	5.0
Sink Faucet, $\frac{3}{4}$ inch	6.0
$\frac{1}{2}$ inch Flush Valve (15 psi flow pressure)	15.0
1 inch Flush Valve (15 psi flow pressure)	27.0
1 inch Flush Valve (25 psi flow pressure)	35.0

FIGURE 7-3 Piping layout of fixtures within a residential building. Piping that supplies more than one fixture is lettered (A–N).

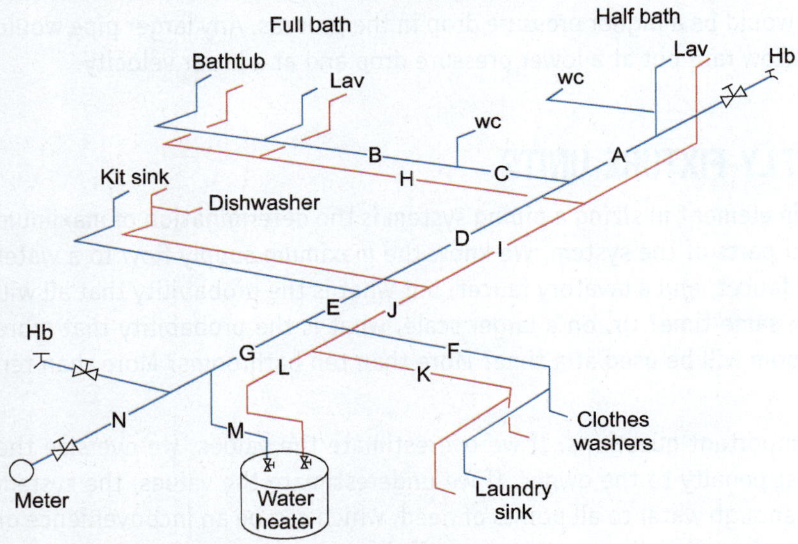

TABLE 7-4 Water Demands Based on Maximum Flow Rates Allowed

Quantity	Fixtures (Residential)	Demand (gpm)	Total (gpm)
2	Lavatory faucets	2.0 × 2	4.0
1	$\frac{3}{8}$ inch sink faucet	4.5	4.5
1	Bathtub faucet	5.0	5.0
1	Laundry faucet	5.0	5.0
2	Ballcocks in flush tanks	3.0 × 2	6.0
1	Dishwasher	4.0	4.0
1	Clothes washer	4.0	4.0
2	Hose bibs	5.0 × 2	10.0
		Total	42.5

Since this summation would be based on all fixtures running at the same time, and this is very unlikely, the piping would be oversized for the vast majority of the time. The method that has been developed to solve this problem uses the concept of water supply fixture units (WSFU). All fixtures are assigned WSFU values, which are index numbers meant to weigh the values of the fixtures based on how often and for how long they are used.

The water supply fixture unit is based on the following traits:

- Actual flow required
- Duration of use of the fixture
- Time between uses

To design a system, the WSFU for all individual fixtures are totaled using the values from Table 7-5, which is taken from Table B.5.2 of the National Standard Plumbing

Code 2012. The table lists common plumbing fixtures and WSFU values for many fixtures in different applications. Note that these tables change frequently as codes are updated. Always check the most current version of your local code when using WSFU tables to size piping.

The information given in Table 7-5 was developed by research done by Stevens Institute, originally published in 1993 and refined in 1996. The research identified four general categories of fixture applications with significantly different factors for diversity. These differences are recognized by assigning different WSFU values to fixtures depending on where they are installed. The categories are the following:

- Single-dwelling units
- Multiple-dwelling units
- Other than dwelling units
- Heavy-use assembly (encompassing most commercial and industrial uses)

TABLE 7-5 Water Supply Fixture Units (WSFU) and Minimum Fixture Branch Pipe Sizes

	Minimum Branch Pipe Size	Individual Dwelling Units	Serving 3 or More Dwelling Units	Other than Dwelling Units	Heavy-Use Assembly
Bathroom Groups Having 1.6 gpf Water Closets Other than the Flushometer Valve Type					
Half-Bath or Powder Room		3.5	2.5		
1 Bathroom Group		5.0	3.5		
1-$\frac{1}{2}$ Bathroom Groups		6.0	4.0		
2 Bathroom Groups		7.0	4.5		
2-$\frac{1}{2}$ Bathroom Groups		8.0	5.0		
3 Bathroom Groups		9.0	5.5		
Each Additional Half-Bath		0.5	0.5		
Each Additional Bathroom Group		1.0	1.0		
Bathroom Groups Having 3.5 gpf (or higher) Gravity Tank Water Closets					
Half-Bath or Powder Room		4.0	3.0		
1 Bathroom Group		6.0	5.0		
1-$\frac{1}{2}$ Bathroom Groups		8.0	5.5		
2 Bathroom Groups		10.0	6.0		
2-$\frac{1}{2}$ Bathroom Groups		11.0	6.5		
3 Bathroom Groups		12.0	7.0		
Each Additional Half-Bath		0.5	0.5		
Each Additional Bathroom Group		1.0	1.0		
Other Groups of Fixtures					
Bathroom Group with 1.6 GPF Flushometer Valve		6.0	4.0		
Bathroom Group with 3.5 GPF (or higher) Flushometer Valve		8.0	6.0		
Kitchen Group with Sink and Dishwasher		2.0	1.5		
Laundry Group with Sink and Clothes Washer		5.0	3.0		

(*continued*)

TABLE 7-5 Water Supply Fixture Units (WSFU) and Minimum Fixture Branch Pipe Sizes (Continued)

	Minimum Branch Pipe Size	Individual Dwelling Units	Serving 3 or More Dwelling Units	Other than Dwelling Units	Heavy-Use Assembly
Individual Fixtures					
Bar Sink	3/8″	1.0	0.5		
Bathtub or Combination Bath/Shower	1/2″	4.0	3.5		
Bidet	1/2″	1.0	0.5		
Clothes Washer, Domestic	1/2″	4.0	2.5	4.0	
Dishwasher, Domestic	1/2″	1.5	1.0	1.5	
Drinking Fountain or Water Cooler	3/8″			0.5	0.75
Hose Bibb (first)	1/2″	2.5	2.5	2.5	
Hose Bibb (each additional)	1/2″	1.0	1.0	1.0	
Kitchen Sink, Domestic	1/2″	1.5	1.0	1.5	
Laundry Sink	1/2″	2.0	1.0	2.0	
Lavatory	3/8″	1.0	0.5	1.0	1.0
Service Sink or Mop Basin	1/2″			3.0	
Shower	1/2″	2.0	2.0	2.0	
Shower, continuous use	1/2″			5.0	
Urinal, 1.0 GPF	3/4″			4.0	5.0
Urinal, greater than 1.0 GPF	3/4″			5.0	6.0
Water Closet, 1.6 GPF Gravity Tank	1/2″	2.5	2.5	2.5	4.0
Water Closet, 1.6 GPF Flushometer Tank	1/2″	2.5	2.5	2.5	3.5
Water Closet, 1.6 GPF Flushometer Valve	1″	5.0	5.0	5.0	8.0
Water Closet, 3.5 GPF Gravity Tank	1/2″	3.0	3.0	5.5	7.0
Water Closet, 3.5 GPF Flushometer Valve	1″	7.0	7.0	8.0	10.0
Whirlpool Bath or Combination Bath/Shower	1/2″	4.0	4.0		

Notes:

1. A Bathroom Group, for the purposes of this Table, consists of not more than one water closet, up to two lavatories, and either one bathtub, one bath/ shower combination, or one shower stall. Other fixtures within the bathing facility shall be counted separately to determine the total water supply fixture unit load.
2. A Half-Bath or Powder Room, for the purposes of this Table, consists of one water closet and one lavatory.
3. For unlisted fixtures, refer to a listed fixture having a similar flow and frequency of use.
4. The listed fixture unit values for Bathroom Groups and Individual Fixtures represent their load on the cold water service. The separate cold water and hot water fixture unit values for fixtures having both cold and hot water connections shall each be taken as 3/4 of the listed total value for the individual fixture.
5. When WSFU values are added to determine the demand on the water distribution system or portions thereof, round the sum to the nearest whole number before referring to Table 10.14.2B for the corresponding gallons per minute (gpm) flow. WSFU values of 0.5 or more should be rounded up to the next higher whole number (9.5 = 10 WSFU). Values of 0.4 or less should be rounded down to the next lower whole number (9.4 = 9 WSFU).
6. The listed minimum supply branch pipe sizes for individual fixtures are the nominal (I.D.) pipe size in inches.
7. "Other Than Dwelling Units" applies to business, commercial, industrial, and assembly occupancies other than those defined under "Heavy-Use Assembly." Included are the public and common areas in hotels, motels, and multi-dwelling buildings.
8. "Heavy-Use Assembly" applies to toilet facilities in occupancies that place heavy, but intermittent, time-based demands on the water supply system, such as schools, auditoriums, stadiums, race courses, transportation terminals, theaters, and similar occupancies where queuing is likely to occur during periods of peak use.
9. For fixtures or supply connections likely to impose continuous flow demands, determine their required flow in gallons per minute (gpm) and add it separately to the demand (in gpm) for the distribution system or portion thereof.

Source: Table 10.4.2A 2012 National Standard Plumbing Code.

In an individual dwelling, a lavatory would have a WSFU total rating of 1.0, while in a structure for three or more dwelling units, its total rating would be 0.5. In a setting such as a public school, the same lavatory would be rated at 1.0. In each case, the actual flow to the fixture remains the same, but the duration of use and the time between uses could vary. It should also be pointed out that when a fixture has both a cold water and a hot water supply, the individual supply (cold or hot) will be 75 percent of the total for the fixture. Also remember that the pipe sizes listed are the minimum allowable; larger sizes may be necessary if the supply pressure at the fixture is low or the length to the fixture is excessively long.

TABLE 7-6 Converting Demand in WSFU to GPM[1]

WSFU	GPM Flush Tanks[2]	GPM Flush Valves[3]	WSFU	GPM Flush Tanks[2]	GPM Flush Valves[3]
3	3		120	49	74
4	4		140	53	78
5	4.5	22	160	57	83
6	5	23	180	61	87
7	6	24	200	65	91
8	7	25	225	70	95
9	7.5	26	250	75	100
10	8	27	300	85	110
11	8.5	28	400	105	125
12	9	29	500	125	140
13	10	29.5	750	170	175
14	10.5	30	1000	210	210
15	11	31	1250	240	240
16	12	32	1500	270	270
17	12.5	33	1750	300	300
18	13	33.5	2000	325	325
19	13.5	34	2500	380	380
20	14	35	3000	435	435
25	17	38	4000	525	525
30	20	41	5000	600	600
40	25	47	6000	650	650
50	29	51	7000	700	700
60	33	55	8000	730	730
80	39	62	9000	760	760
100	44	68	10,000	790	790

Notes:

1. This table converts water supply demands in water supply fixture units (WSFU) to required water flow in gallons per minute (GPM) for the purpose of pipe sizing.
2. This column applies to portions of piping systems where the water closets are the flush tank type (gravity or pressure) or there are no water closets, and to hot water piping.
3. This column applies to portions of piping systems where the water closets are the flush valve type.

Source: Table 10.14.2B 2012 National Standard Plumbing Code.

While it seems possible that there are many other installation categories where the WSFU values would be different, the research done at Stevens showed that these categories captured the range of usage of typical plumbing fixtures and that combining the WSFU values gave a good correlation of short-term water consumption for a large variety of plumbing projects.

After the WSFU for each fixture is listed, the necessary pipe size must be determined. In most cases, it is handy to measure or calculate the flow rate for each section of the piping system in gallons per minute (gpm). Table 7-6 converts the water demand in WSFU to gpm, depending on whether the system contains flush valves or flush tanks.

EXAMPLE 7

How many gallons per minute would a domestic clothes washer in an individual dwelling be rated for?

According to Table 7-3, the clothes washer, domestic, would have a total of 4.0 WSFU. When the 4.0 WSFU is applied to Table 7-6, it is found to equate to 4 gpm.

The piping system shown in Figure 7-3 can be sized based on the water supply fixture units, which would be more representative of the normal conditions. Figure 7-4 shows Figure 7-3 redrawn to include WSFU totals on the various pipe sections. The pipe size to each individual fixture will be based on the minimum allowable pipe listed in Table 7-5; larger sizes can be run, based on pressure or economical reasons. Pipes that supply more than one fixture are based on Table 7-8. Table 7-7 shows a tabulation of pipe sizes and demands.

FIGURE 7-4 Water supply fixture units have been added to the layout from Figure 7-4. Gpm values are used to size lettered pipes. All pipe sizes are listed in Table 7-7.

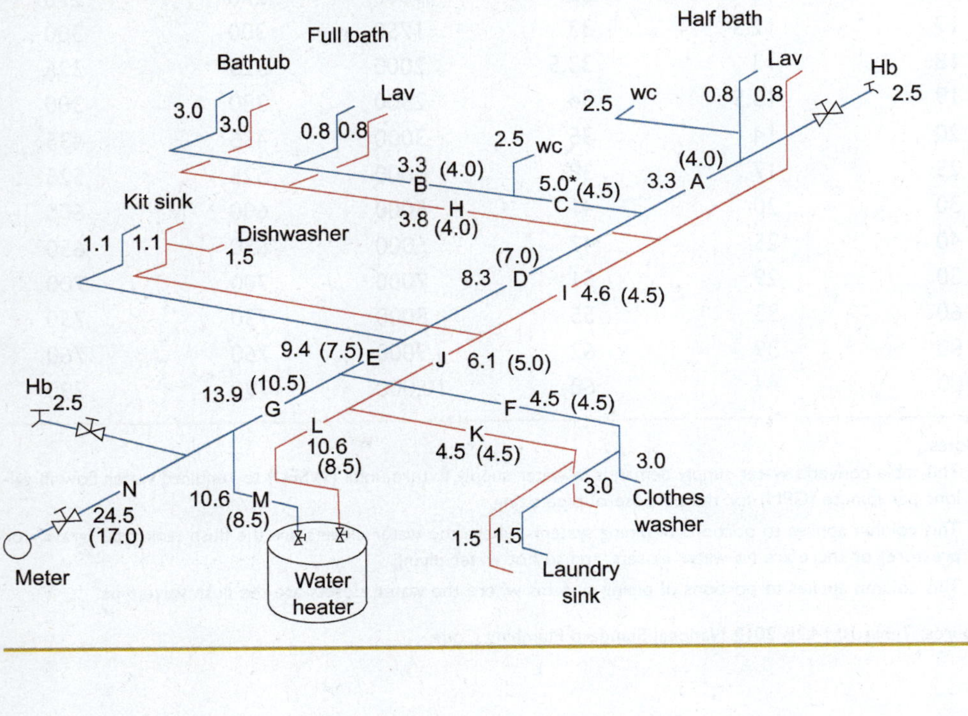

TABLE 7-7 Pipe Sizing with WSFU and gpm

Fixture or Pipe	Hot			Cold		
	WSFU	gpm	Pipe Dia (in.)	WSFU	gpm	Pipe Dia (in.)
Half Bath, Lavatory	0.8		$\frac{3}{8}$	0.8		$\frac{3}{8}$
Half Bath, Water Closet				2.5		$\frac{1}{2}$
Hose Bibb				2.5		$\frac{1}{2}$
Pipe A				3.3	4.0	$\frac{1}{2}$
Full Bath, Shower/Tub	3.0		$\frac{1}{2}$	3.0		$\frac{1}{2}$
Full Bath, Lavatory	0.8		$\frac{3}{8}$	0.8		$\frac{3}{8}$
Pipe B				3.3	4.0	$\frac{1}{2}$
Full Bath, Water Closet				2.5		$\frac{1}{2}$
Pipe C				5.0*	4.5	$\frac{1}{2}$
Pipe D				8.3	7.0	$\frac{3}{4}$
Kitchen Sink	1.1		$\frac{1}{2}$	1.1		$\frac{1}{2}$
Pipe E				9.4	7.5	$\frac{3}{4}$
Washing Machine	3.0		$\frac{1}{2}$	3.0		$\frac{1}{2}$
Laundry Sink	1.5		$\frac{1}{2}$	1.5		$\frac{1}{2}$
Pipe F				4.5	4.5	$\frac{1}{2}$
Pipe G				13.9	10.5	$\frac{3}{4}$
Pipe H (full bath group)	3.8	4.0	$\frac{1}{2}$			
Pipe I	4.6	4.5	$\frac{1}{2}$			
Dishwasher	1.5		$\frac{1}{2}$			
Pipe J	6.1	5.0	$\frac{1}{2}$			
Pipe K	4.5	4.5	$\frac{1}{2}$			
Pipe L	10.6	8.5	$\frac{3}{4}$			
Pipe M	10.6	8.5	$\frac{3}{4}$			
Pipe N				24.5	17.0	1

*The actual value of 6.3 WSFU can be reduced, when considered as a group.

Note that the flows of the hose bibbs (sometimes referred to as lawn faucets) are not included. This is a common practice, especially in communities that prohibit the use of lawn sprinklers during peak times. In cases where the hose bibbs are considered (steady flows), the first hose bibb in a section is considered to be an additional 2.5 WSFU. Each additional hose bibb would add 1 WSFU to the piping section. In other words, we would normally not increase pipe size in a dwelling unit to allow for hose bibb flow. In industrial or commercial buildings, steady flows due to processes such as equipment cooling must be recognized and included in the design. Maximum probable peak flow in any section of a piping system is difficult to determine, especially for very small systems like the example given in Figure 7-3.

Table 7-7 also shows the minimum size pipe that should be used to connect the various fixtures. These minimum-sized pipes should be brought to within 30 inches of the fixture connected.

TABLE 7-8 Type L Copper Tube Water Flow Velocity

	4 fps				5 fps				8 fps				
Tube Size (in.)	WSFU (tanks)	WSFU (valves)	Flow (gpm)	Pd psi/100 ft.	WSFU (tanks)	WSFU (valves)	Flow (gpm)	Pd psi/100 ft.	WSFU (tanks)	WSFU (valves)	Flow (gpm)	Pd psi/100 ft.	Tube Size (in.)
$\frac{1}{8}$			1.8	7.8	2		2.1	11.7	4		3.6	28.0	$\frac{1}{8}$
$\frac{1}{2}$	3		2.9	5.9	4		3.6	8.9	7		5.8	21.3	$\frac{1}{2}$
$\frac{3}{4}$	7		6.0	3.9	9		7.3	5.8	16		12.1	13.9	$\frac{3}{4}$
1	14		10.3	2.8	18		12.9	4.3	31		20.6	10.2	1
$1\frac{1}{4}$	23		15.7	2.2	29		19.5	3.3	56	15	31.3	8.0	$1\frac{1}{4}$
$1\frac{1}{2}$	34	5	22.2	1.8	47	11	27.7	2.7	101	36	44.4	6.5	$1\frac{1}{2}$
2	79	26	38.6	1.3	117	43	48.2	2.0	261	136	77.2	4.7	2
$2\frac{1}{2}$	173	73	59.5	1.0	247	120	74.4	1.5	470	360	119.0	3.7	$2\frac{1}{2}$
3	300	170	84.9	0.8	406	281	106.2	1.3	749	713	169.9	3.0	3
4	635	567	149.3	0.6	854	833	186.7	0.9	1739	1739	298.7	2.2	4
5	1189	1189	232.7	0.5	1674	1674	290.9	0.7	3338	3338	465.5	1.7	5
6	2087	2087	334.6	0.4	2847	2847	418.2	0.6	6382	6382	669.1	1.4	6

Sizing Based on Velocity Limitations and Pressure Drop

A common method of pipe sizing is to focus on the velocity of water that flows through the pipe. It is considered good engineering practice to limit the velocity of water in a system to less than 8 feet per second. This eliminates noise and excessive wear on the materials. If the system contains quick opening valves, the velocity is typically limited to less than 4 fps, to avoid water hammer or other excessive pressure surges. Table 7-9, sizing tables based on velocity limitation, lists the flow capacity, WSFU rating, and pressure drop (in psi/100 ft.) at two velocities (4 fps and 8 fps) for several pipe materials.

TABLE 7-9 Sizing Tables Based on Velocity Limitation

		Velocity = 4 feet per second				Velocity = 8 feet per second			
Nominal Size (in.)	Actual ID (in.)	Flow (gpm)	Load (WSFU)* Column A	Load (WSFU)** Column B	Friction (psi/100)*	Flow (gpm)	Load (WSFU)* Column A	Load (WSFU)** Column B	Friction (psi/100)*
Copper and Brass Pipe, Standard Pipe Size (Smooth)									
1/2	0.625	3.8			6.8	7.7	10		24.0
3/4	0.822	6.6	8		5.0	13.2	18		18.0
1	1.062	11.0	15		3.7	22.1	34		13.3
1 1/4	1.368	18.3	27		2.7	36.6	70	21	9.7
1 1/2	1.600	25.1	40	10	2.3	50.1	125	51	8.2
2	2.062	41.6	92	31	1.7	83.2	290	15	6.1
2 1/2	2.500	61.2	180	80	1.3	122.4	490	390	3.9
3	3.062	91.8	330	210	1.1	183.6	850	810	3.9
4	4.000	156.7	680	620	0.8	313.4	1900	1900	2.8
Copper Water Tube, Type M (Smooth)									
1/8	0.450	2.0			9.6	3.9			34.0
1/2	0.569	3.2			6.4	6.3	8		26.0
3/4	0.811	6.4	8		5.0	12.9	18		18.0
1	1.055	10.9	15		3.6	21.8	34		13.0
1 1/4	1.291	16.3	24		2.9	32.6	60	15	10.0
1 1/2	1.527	22.8	35		2.4	45.6	110	38	8.5
For 2 Inches and Larger, Use Values for Type L Copper									
Copper Water Tube, Type L (Smooth)									
3/8	0.430	1.8			11.0	3.6			39.0
1/2	0.545	2.9			8.1	5.8	7		29.0
3/4	0.785	6.0	7		5.3	12.1	17		19.0
1	1.025	10.3	14		4.0	20.6	30		14.0
1 1/4	1.265	15.7	23		3.0	31.3	55	15	11.0
1 1/2	1.505	22.2	35		2.5	44.4	100	36	8.7
2	1.985	38.6	80	26	1.8	77.2	270	140	6.2
2 1/2	2.465	59.5	170	70	1.4	110.0	470	360	5.0
3	2.945	84.9	300	170	1.1	169.9	750	730	4.0
4	3.905	149.3	625	575	0.0	298.7	1750	1750	2.8

(continued)

TABLE 7-9 Sizing Tables Based on Velocity Limitation (Continued)

		Velocity = 4 feet per second				Velocity = 8 feet per second			
Nominal Size (in.)	Actual ID (in.)	Flow (gpm)	Load (WSFU)* Column A	Load (WSFU)** Column B	Friction (psi/100)*	Flow (gpm)	Load (WSFU)* Column A	Load (WSFU)** Column B	Friction (psi/100)*
Copper Water Tube, Type K (Smooth)									
$\frac{3}{8}$	0.402	1.6			11.0	3.2			36.0
$\frac{1}{2}$	0.527	2.7			8.2	5.4	6		30.0
$\frac{3}{4}$	0.745	5.4	6		5.6	10.9	15		20.0
1	0.995	9.7	13		4.1	19.4	29		14.0
$1\frac{1}{4}$	1.245	15.2	24		3.1	30.4	55	15	12.0
$1\frac{1}{2}$	1.481	21.5	33		2.6	43.0	97	35	9.0
For 2 Inches and Larger, Use Values for Type L Copper									
Galvanized Iron and Steel Pipe, Standard Pipe Size (Fairly Rough)									
$\frac{1}{2}$	0.622	3.8			8.0	7.6	10		31.0
$\frac{3}{4}$	0.824	6.6	8		6.0	13.3	18		22.0
1	1.049	10.8	15		4.5	21.5	34		17.0
$1\frac{1}{4}$	1.380	18.6	27		3.4	37.3	75	21	13.0
$1\frac{1}{2}$	1.610	25.4	41		2.8	50.7	125	51	11.0
2	2.067	41.8	92	32	2.2	83.6	290	165	8.1
$2\frac{1}{2}$	2.469	59.6	172	72	1.8	119.3	490	390	6.8
3	3.068	92.1	330	210	1.4	184.3	850	810	5.4
4	4.026	158.7	680	620	1.1	317.5	1900	1900	4.0
Schedule 40 Plastic Pipe (PE, PVC, & ABS) (Smooth)									
$\frac{1}{2}$	0.622	3.8			7.0	7.6	10		24.0
$\frac{3}{4}$	0.824	6.6	8		5.1	13.3	18		17.5
1	1.049	10.8	15		3.7	21.5	34		13.0
$1\frac{1}{4}$	1.380	18.6	27		2.7	37.2	75	21	9.5
$1\frac{1}{2}$	1.610	25.4	41		2.3	50.7	125	51	8.0
2	2.067	41.8	92	32	1.7	83.6	290	165	6.0
$2\frac{1}{2}$	2.469	59.6	172	72	1.4	119.3	490	390	4.8
3	3.068	92.1	330	210	1.1	184.3	850	810	3.7
4	4.026	158.7	680	620	0.8	317.5	1900	1900	2.7
Schedule 80 Plastic Pipe (PVC, CPVC) (Smooth)									
$\frac{1}{2}$	0.546	2.9			6.9	5.8	7		22.7
$\frac{3}{4}$	0.742	5.4			4.8	10.8	15		16.0
1	0.957	9.4	12		3.4	17.9	23		11.2
$1\frac{1}{4}$	1.278	16.0	24		2.4	32.0	58	17	11.2
$1\frac{1}{2}$	1.500	22.0	34		2.0	44.1	100	36	6.5
2	1.939	36.87	730	230	1.5	73.65	2400	1200	5.0

*Column A applies to piping that does not supply flushometer valves. Use this table for tank-type water closets.

**Column B applies to piping that supplies flushometer valves.

TABLE 7-10 Allowance in Equivalent Length of Pipe for Friction Loss in Valves and Threaded Fittings

	Equivalent Feet of Pipe for Various Sizes							
Fitting or Valve	$\frac{1}{2}$ inch	$\frac{3}{4}$ inch	1 inch	$1\frac{1}{4}$ inches	$1\frac{1}{4}$ inches	2 inches	$2\frac{1}{2}$ inches	3 inches
45°	1.2	1.5	1.8	2.4	3.0	4.0	5.0	6.0
90°	2.0	2.5	3.0	4.0	5.0	7.0	8.0	10.0
Tee, Run	0.6	0.0	0.9	1.2	1.5	2.0	2.5	3.0
Tee, Branch	3.0	4.0	5.0	6.0	7.0	10.0	12.0	15.0
Gate Valve	0.4	0.5	0.6	0.8	1.0	1.3	1.6	2.0
Balancing Valve	0.8	1.1	1.5	1.9	2.2	3.0	3.7	4.5
Plug-Type Cock	0.8	1.1	1.5	1.9	2.2	3.0	3.7	4.5
Check Valve—Swing	5.6	8.4	11.2	14.0	16.8	22.4	28.0	33.6
Globe Valve	15.0	20.0	25.0	35.0	45.0	55.0	65.0	80.0
Angle Valve	8.0	12.0	15.0	18.0	22.0	28.0	34.0	40.0

TABLE 7-11 Allowance in Equivalent Length of Tube for Friction Loss in Valves and Threaded Fittings

Fitting or Valve	$\frac{1}{2}$ inch	$\frac{3}{4}$ inch	1 inch	$1\frac{1}{4}$ inches	$1\frac{1}{2}$ inches	2 inches	$2\frac{1}{2}$ inches	3 inches
45° Elbow—Wrought	0.5	0.5	1.0	1.0	2.0	2.0	3.0	4.0
90° Elbow—Wrought	0.5	1.0	1.0	2.0	2.0	2.0	2.0	3.0
Tee, Run—Wrought	0.5	0.5	0.5	0.5	1.0	1.0	2.0	
Tee, Branch—Wrought	1.0	2.0	3.0	4.0	5.0	7.0	9.0	
45° Elbow—Cast	0.5	1.0	2.0	2.0	3.0	5.0	8.0	11.0
90° Elbow—Cast	1.0	2.0	4.0	5.0	8.0	11.0	14.0	18.0
Tee, Run—Cast	0.5	0.5	0.5	1.0	1.0	2.0	2.0	2.0
Tee, Branch—Cast	2.0	3.0	5.0	7.0	9.0	12.0	16.0	20.0
Compression Stop	13.0	21.0	30.0					
Globe Valve				53.0	66.0	90.0		
Gate Valve			1.0	1.0	2.0	2.0	2.0	2.0

If a building design or available pressure seems marginal for satisfactory performance, the system can be checked by using Tables 7-10 and 7-11.

EXAMPLE 8

If 650 feet of 2-inch galvanized pipe carries water at a velocity of 8 fps, what is the outlet pressure if the inlet is 60 psi and the pipe is horizontal?

Table 7-9 shows 8.1 psi drop per 100 feet of 2-inch galvanized pipe at a flow of 8 fps.

$$\text{Outlet pressure} = 60 \text{ psi} - \frac{8.1 \text{ psi}}{100 \text{ ft.}}(650 \text{ ft.})$$

$$\text{Outlet pressure} = 60 \text{ psi} - 52.6 \text{ psi}$$

$$\text{Outlet pressure} = 7.4 \text{ psi}$$

If the velocity is 4 fps, the outlet pressure is as follows:

$$\text{Outlet pressure} = 60 \text{ psi} - \frac{2.2 \text{ psi}}{100 \text{ ft.}}(650 \text{ ft.})$$

$$\text{Outlet pressure} = 60 \text{ psi} - 14.3 \text{ psi}$$

EXAMPLE 8
continued

Outlet pressure = 45.7 psi

Note that Table 7-9 indicates a flow rate of 83.6 gpm when the water velocity is 8 fps. The flow rate drops to 41.8 gpm when the water velocity is 4 fps. Pressure drop is reduced greatly as flow (or velocity) is reduced.

If the piping changes elevation in the building, there is also a pressure change due to the elevation change, as described at the beginning of this chapter. The pressure change is subtracted if the point of use is above the supply and added if the point of use is below the supply.

Fitting and Valve Allowance

In the previous example, the 650 feet of pipe was assumed to be straight. Obviously, there would have to be couplings to join the pipe, but it turns out that a coupling offers very little resistance to flow. Unfortunately, in the real world, few piping systems are completely straight. Piping systems contain valves, tees, and elbows, each of which offers varying degrees of resistance to flow. If the piping contains a significant number of fittings, the length of the pipe must be increased by the equivalent length of the fittings. Tables 7-10 and 7-11 list these values.

EXAMPLE 9

What is the pressure drop of a horizontal piping system that includes the following items if the velocity is 4 fps?

In order to find the pressure drop through the system, the equivalent length of all pipe, fittings, and valves must be determined. Use Table 7-12 to determine a total equivalent length, and then calculate the associated pressure drop.

At 4 fps, $\frac{3}{4}$ inch galvanized pipe has a friction value of 6.0 psi per 100 ft.

$$\text{Pressure drop} = \left(\frac{6\text{ psi}}{100\text{ ft.}}\right) \times (70.7\text{ ft.})$$

Pressure drop = 4.24 psi

EXAMPLE 10

What is the pressure drop in the previous example if the discharge end of the pipe is 20 feet higher than the inlet?

Total pressure drop = Pressure drop due to flow + pressure drop due to elevation

$$\text{Pressure drop} = 4.24 + \left(\frac{0.433\text{ psi}}{1\text{ ft. of elevation}}\right) \times (20\text{ ft.})$$

Pressure drop = 4.24 + 8.66

Pressure drop = 12.9 psi total drop

INDIVIDUAL RUN SYSTEMS

Up until now, all of the piping examples have been drawn in what is commonly known as the main-and-branch layout. This design uses a large supply pipe to route water from the incoming supply to a cluster of fixtures, at which point it branches off with smaller pipes. These smaller pipes can divide again until they reach the individual fixture that they

TABLE 7-12 Equivalent Length

Quantity	Pipe or Fitting	Unit Equivalent Length	Equivalent Length
50 ft.	$\frac{3}{4}$ inch galvanized pipe		50 ft.
4	45° ell	1.5	6
2	90° ell	2.5	5
1	Tee	0.8	0.8
1	Gate valve	0.5	0.5
1	Swing check	8.4	8.4
Total Equivalent Length (ft.) =			70.7 ft.

serve. Another possible design layout is known as the individual run system. An individual run system, also known as a manifold system, uses a dedicated pipe or tube to supply water to each faucet or fixture fitting from a header that is located conveniently in the building. Thus, there are no tees in this system—each tube from the header serves only one faucet. Figure 7-5 shows an isometric sketch of this type system.

As indicated by Figure 7-6, there is a valve for each takeoff from the header. Some local codes permit this valve to serve as the shutoff for the branch, so no stop is required at the location of the fixture. Many consumers like having the shutoff valves in a single location because they can easily shut off a branch should the need arise. However, without a shutoff at the fixture, a person may be required to travel some distance before the water can be turned off. Imagine an upstairs toilet overflowing when the only water shutoff valve is downstairs in a utility room. These valved headers can easily be labeled to facilitate identification. Figure 7-6 shows a typical valved manifold or header assembly.

Even though the most popular choice of piping material is plastic tubing taken from large coils, materials suitable for other potable water supply can be used. The long coils make it possible not to have fittings in the tube from the header to the

FIGURE 7-5 This isometric sketch of a manifold system shows how each fixture is on a dedicated run.

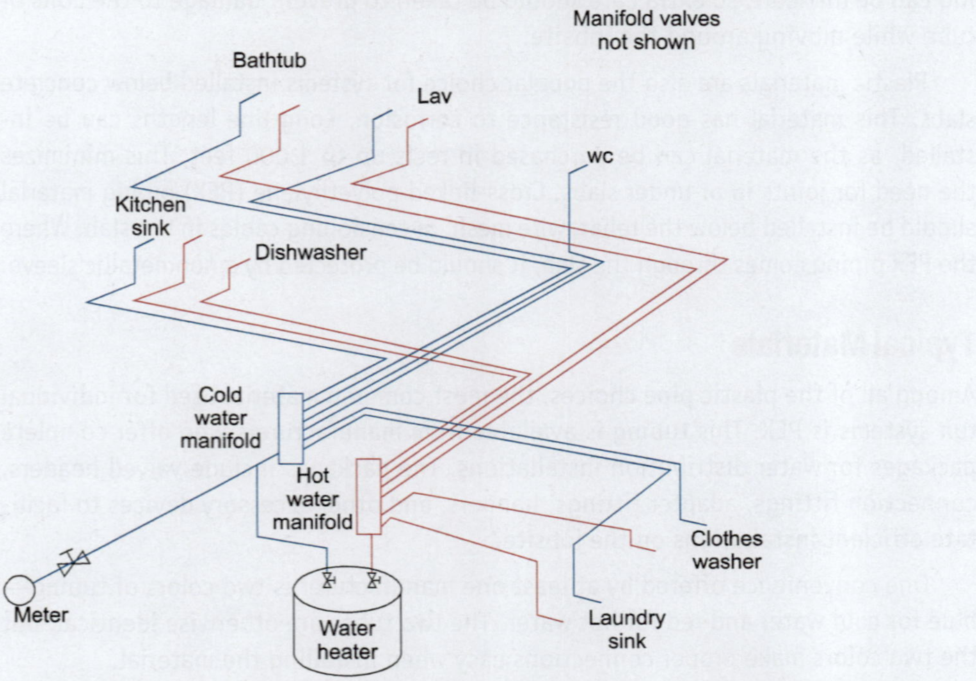

FIGURE 7-6 In this individual run system, each branch run (takeoff) has a shut-off valve.

Photo by Ed Moore

faucet, so the possibility of future leaks is minimized. Handling large coils of tubing can be difficult, so extra care should be taken to prevent damage to the coils of tube while moving around the jobsite.

Plastic materials are also the popular choice for systems installed below concrete slabs. This material has good resistance to corrosion. Long line lengths can be installed, as the material can be purchased in reels up to 1,000 feet. This minimizes the need for joints in or under slabs. Cross-linked polyethylene (PEX) piping material should be installed below the rebar, wire mesh, or tensioning cables in the slab. Where the PEX piping comes through the slab, it should be protected by a nonmetallic sleeve.

Typical Materials

Among all of the plastic pipe choices, the most common material used for individual run systems is PEX. This tubing is available from manufacturers who offer complete packages for water distribution installations. The packages include valved headers, connection fittings, adapter fittings, hangers, and other accessory devices to facilitate efficient installations on the jobsite.

One convenience offered by at least one manufacturer is two colors of tubing—blue for cold water and red for hot water. The two tubes are otherwise identical, but the two colors make proper connections easy when installing the material.

FIGURE 7-7 This brass manifold can serve as a remote manifold. It allows three $\frac{1}{2}$-inch pipes to be supplied by one $\frac{3}{4}$-inch pipe.

Photo by Ed Moore

Connections are made with fittings inserted into the tubing. Some systems expand the tubing to accommodate the fitting; after a brief time the tubing shrinks back to grip the tube. Other systems utilize a crimp ring around the tube end to compress the tubing and grip the fitting. Additional design options utilize compression-type fittings, typically brass, which have an internal barb and a compression ring, which is tightened with a compression nut to grip the tubing. Many types of valves and fittings are available for this material to make a complete installation (see Figure 7-7).

Sizing

Individual run or parallel run systems are usually sized similar to the conventional main-and-branch design described earlier in this chapter. Some codes, such as the 2012 International Plumbing Code, allow individual distribution lines with a developed length of 60 feet or less to be reduced by one nominal size. This reduction is allowed only if the available pressure is 35 psi or higher. On runs longer than 60 feet, the installer should be aware of how an increase in flow rate (gpm) will increase the pressure drop over the length of the pipe. Check your local code for requirements specific to your area. Table 7-13 shows pressure drops for PEX tubing based on flow rates (gpm) and tubing diameter. Normally, the water heater is fed by a $\frac{3}{4}$ inch line from the cold side of the header and a $\frac{3}{4}$ inch line back from the heater to the hot water side of the main system header.

In the event that the available water pressure at the building is only slightly more than the minimum required, these tubing sizes can be increased one size to help ensure satisfactory service.

TABLE 7-13 Pressure Drop in PEX Tubing

	$\frac{3}{8}$ inch		$\frac{1}{2}$ inch		$\frac{3}{4}$ inch		1 inch	
Nominal Size (ID) gpm	Psi drop per 100 ft.	Velocity	Psi drop per 100 ft.	Velocity	Psi drop per 100 ft.	Velocity	Psi drop per 100 ft.	Velocity
1	7.0	3.33	1.6	1.81	0.3	0.96	0.1	0.55
2	25.4	6.67	5.8	3.62	1.1	1.81	0.3	1.10
3	53.9	10.00	12.2	5.43	2.3	2.72	0.7	1.65
4	91.8	13.34	20.8	7.24	3.9	3.63	1.1	2.19
5			31.4	9.05	5.9	4.54	1.7	2.74

Source: Data from Plastic Pipe and Fittings Association, 2006.

Most manufacturers of individual run systems have detailed sizing and installation guides. Consult your local representative or supplier to find out what is available. With some time taken to understand the system applications, a successful installation can be ensured. The Plastic Pipe Institute has a Design Guide available at http://www.plasticpipe.org, which has tables that can be used to determine the flow rate and pressure drops associated with each design.

Operating Advantages

The individual run system greatly reduces the likelihood of surprise temperature changes in one circuit (e.g., the shower) when water is drawn from another faucet or when a toilet is flushed. This is true because the pressure drop up to the header is usually small and the flows beyond the header are independent of each other.

Since the line going to an individual faucet or shower is usually smaller (in diameter) than the trunk line supplying the entire bathroom group, less water is wasted while waiting for hot water to arrive. The time that is required to have hot water reach an outlet is reduced because only the small tubing has to be filled with hot water, rather than a larger-sized main and then the final branch. This is the case in conventional main-and-branch systems. However, it is important to recognize that getting hot water to your lavatory faucet does nothing to make hot water available at the adjacent bathtub. Thus, the advantage of quicker delivery to one faucet is offset by the need to wait at any other faucets in the same room.

When using plastic tubing, there is less heat loss or gain in the system. The material has a better insulating value than metal piping. This allows the water to maintain its temperature longer between draws.

A very important consideration is the reduced chance of water hammer. This is true for two reasons: There is only a small amount of water flowing in the tube, and the nonmetallic pipe helps to absorb the shock wave since the piping material yields very slightly to the pressure input from the effect of a quick-closing valve.

Another advantage of plastic material is its inherent corrosion resistance. While no material is impervious to all attacks, plastics offer perhaps the most defenses. When installed in the slab or underground, this resistance helps to provide a long life of leak-free operation.

System Disadvantages

The most obvious disadvantage to the individual run systems is the increased amount of pipe needed to complete an installation. On a typical ranch-style house, an average of 1.8 times more pipe is needed than that of a typical main-and-branch design. The installer should compare this increase in cost from the additional piping material to the decrease in cost of purchasing and installing fittings. On smaller homes, or homes with closely grouped plumbing fixtures, this cost would be less significant. Another disadvantage to the individual run systems is the increased pressure drop on longer runs. In situations where only the fixture farthest from the water source was run, the individual run system had a larger pressure drop than would be found in a similar-sized house with a main-and-branch design. However, when additional fixtures were run at the same time, the individual run system was able to supply more fixtures with stable pressure.

Cost Considerations

Contractors who regularly use this system believe that the installed cost of the water distribution system of their typical job is lower than the cost of an equivalent system using the main-and-branch method described previously. Significantly shorter installation time and the reduced chance of leaks help to increase productivity, providing these reported cost savings.

Vendors and manufacturers' representatives at trade shows are good sources for information about savings. Consult these helpful resources when you are able. They can provide a wealth of information for you.

Other Uses

Frequently, the manifold type of design is used in hydronic heating systems. The biggest difference in these systems is that each run leaving the supply manifold will come back to a return manifold, thereby creating a loop. An automatically controlled valve is added to each loop so that the temperature of the loop can be controlled by increasing or reducing the flow of water in that loop. Figure 7-8 shows a conceptual drawing of this type of system.

For radiant heat panels, the most popular material choice is PEX or polyethylene-aluminum-polyethylene (PEX-AL-PEX). Its corrosion resistance makes it a natural selection for embedding in floors, walls, or ceilings.

One of the problems faced with hydronic heating systems is eliminating the amount of oxygen that is diffused through the pipe. When oxygen is allowed to enter a hot water system, it increases the amount of corrosion that occurs to the iron materials in that system. To help with this problem, manufacturers have developed PEX-AL-PEX, which is tubing that has a layer of aluminum sandwiched between two layers of PEX material. This extra layer of aluminum acts as an oxygen barrier.

These systems are not only being used for comfort heating but have also been applied to other outdoor problems. The difficulty of snow and ice removal on

FIGURE 7-8 Hydronic heating systems use a controlled flow of hot water to heat an area. Each loop has a separate control valve that is controlled by a thermostat.

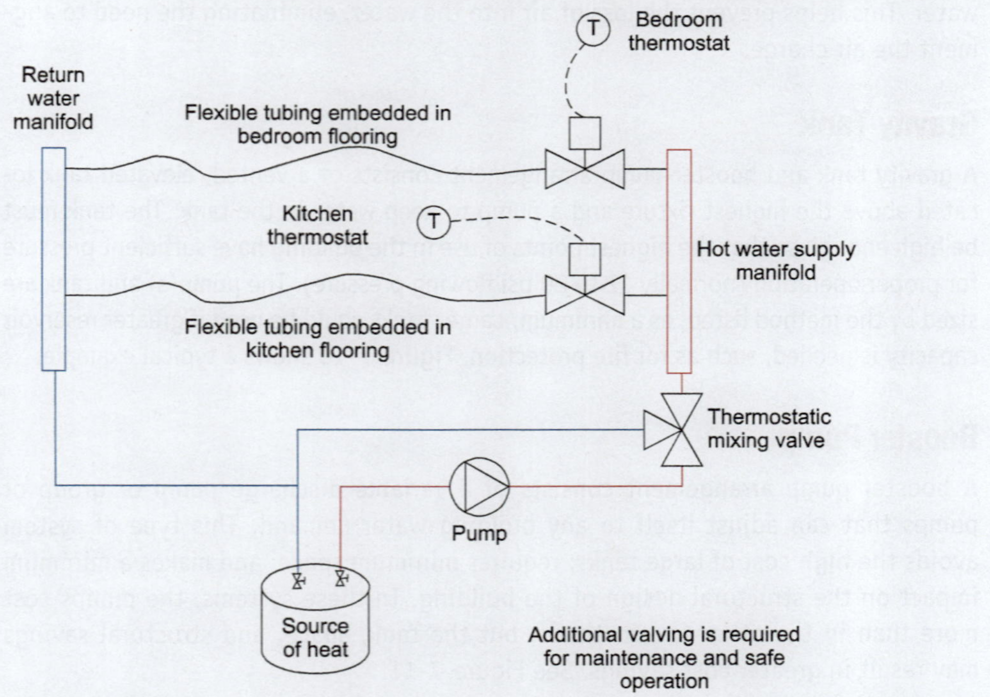

walkways and steps can be minimized by embedding tubing in the pavement. This can also help minimize the number of ice-related accidents in public places. These loop design systems have also been installed under turf to prolong the growing season by keeping the roots warm. This can be a tremendous advantage to greenhouses and golf courses, allowing them to increase their productive times.

PRESSURE BOOSTING SYSTEMS

In tall buildings or in areas with public systems that have low pressure, the supply pressure to the building may be unable to deliver sufficient water to furnish all needs. Three methods have been devised to assure adequate water supply in such cases:

- Hydropneumatic tanks
- Elevated gravity tanks
- Booster pumps to develop the necessary water pressure

Table 7-14 lists advantages and disadvantages for each system type.

Hydropneumatic Tank

A hydropneumatic tank and pump arrangement consists of a closed tank and one or more pumps to pressurize the tank. The tank contains water and compressed air. The air inside the tank compresses as the pumps fill the tank. If there is a water demand during the off cycle of the pumps, the compressed air will push the water out of the tank and into the system. The tank is sized to deliver water to the building for 5 to 10 minutes without requiring pump operation. Once started, the pumps will operate for a similar period of time before turning off. The rule of thumb is to select a pump rated in gallons per minute (gpm) equal to the peak demand and a tank size in gallons equal to ten times the gpm rating. If needed, the air cushion in the tank can be augmented by a compressed air source. Figure 7-9 shows a typical system.

A newer variation of this system uses a bladder membrane to separate the air and water. This helps prevent the loss of air into the water, eliminating the need to augment the air charge.

Gravity Tank

A gravity tank and booster pump arrangement consists of a vented, elevated tank located above the highest fixture and a pump to keep water in the tank. The tank must be high enough so that the highest points of use in the building have sufficient pressure for proper operation (normally 12 to 20 psi flowing pressure). The pump(s) and tank are sized by the method listed, as a minimum. Larger tanks could be used if greater reservoir capacity is needed, such as for fire protection. Figure 7-10 shows a typical example.

Booster Pump

A booster pump arrangement consists of a variable discharge pump or group of pumps that can adjust itself to any building water demand. This type of system avoids the high cost of large tanks, requires minimum space, and makes a minimum impact on the structural design of the building. In these systems, the pumps cost more than in the other two methods, but the tank, space, and structural savings may result in greater cost savings. See Figure 7-11.

TABLE 7-14 Characteristics of Water Pressurizing Methods

System	Advantages	Disadvantages
Hydropneumatic	Tank may be equipped with a bladder membrane (to storage air pressure). Tank does not have to be elevated. Tank can be buried underground to save space. Tank may be located anywhere near the building. Pump operation is required only at periods of high demand.	Tanks without air bladders are more subject to corrosion. The tank is more costly than the gravity system. Pumps of a higher head are required compared to gravity systems. More sophisticated controls and more expensive equipment as opposed to gravity system. System pressure variation of 20 psi may occur. Valuable floor space may be required by this system.
Gravity	Very reliable, as a reserve capacity is available even if power failure occurs. The system is simple. Least cost to operate. Minimum pressure variations in the system. Less energy is required as pumps only operate to replenish the water level in the tank. Fewer pressure regulators are required on the system. Additional water is available for fire protection.	Expense is required for support at the top of the building, leak protection, insulation, and enclosure. Tank rupture could lead to expensive damage. Relief pans and drains are recommended. Since tank is open to atmosphere, there is potential for contamination.
Booster System	Could be less expensive initially. Less floor space is required, with the majority of floor space and added structural burden placed on basement or lower floors.	More sophisticated controls and more expensive equipment as opposed to a gravity system. Oversizing the pumps may result in higher costs both initially and during operation. At least one pump of the multistage system must be continuously operating. No emergency reserve is available in times of failure of the pump or the incoming supply. The system is naturally noisy as the pump operates regularly. Should a breakdown occur in the controls of the pump system, a factory representative is usually needed to repair the defect.

FIGURE 7-9 Hydropneumatic systems use compressed air to maintain pressures. Newer model tanks have a bladder to reduce loss of air.

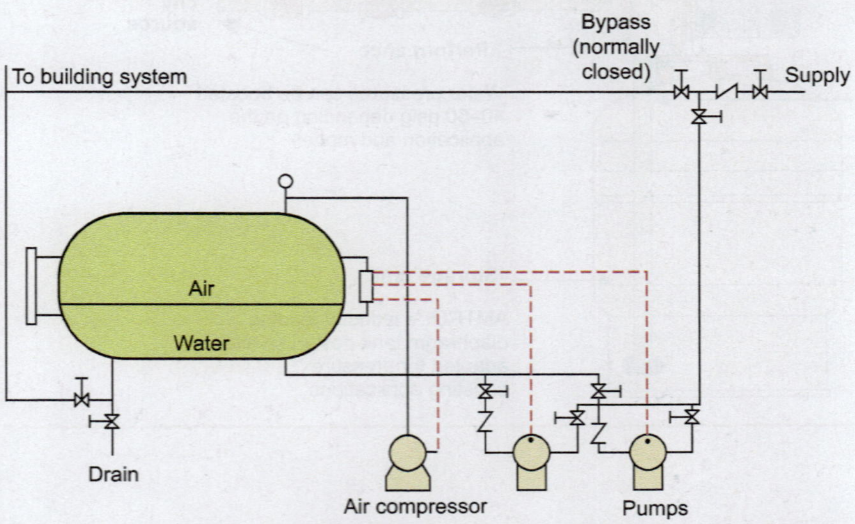

FIGURE 7-10 Gravity tanks provide pressure due to being elevated above the point of use.

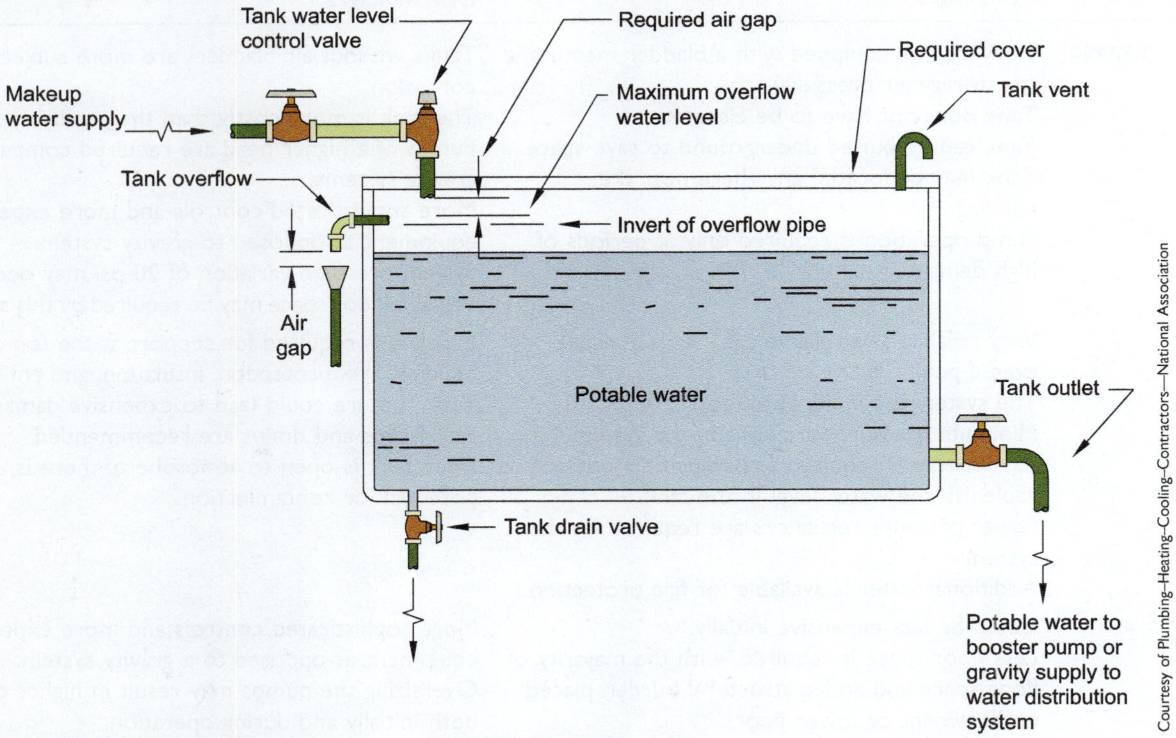

FIGURE 7-11 Pressure booster systems use a water pump to increase the pressure to the downstream system. The pump does not add water but simply increases the available pressure.

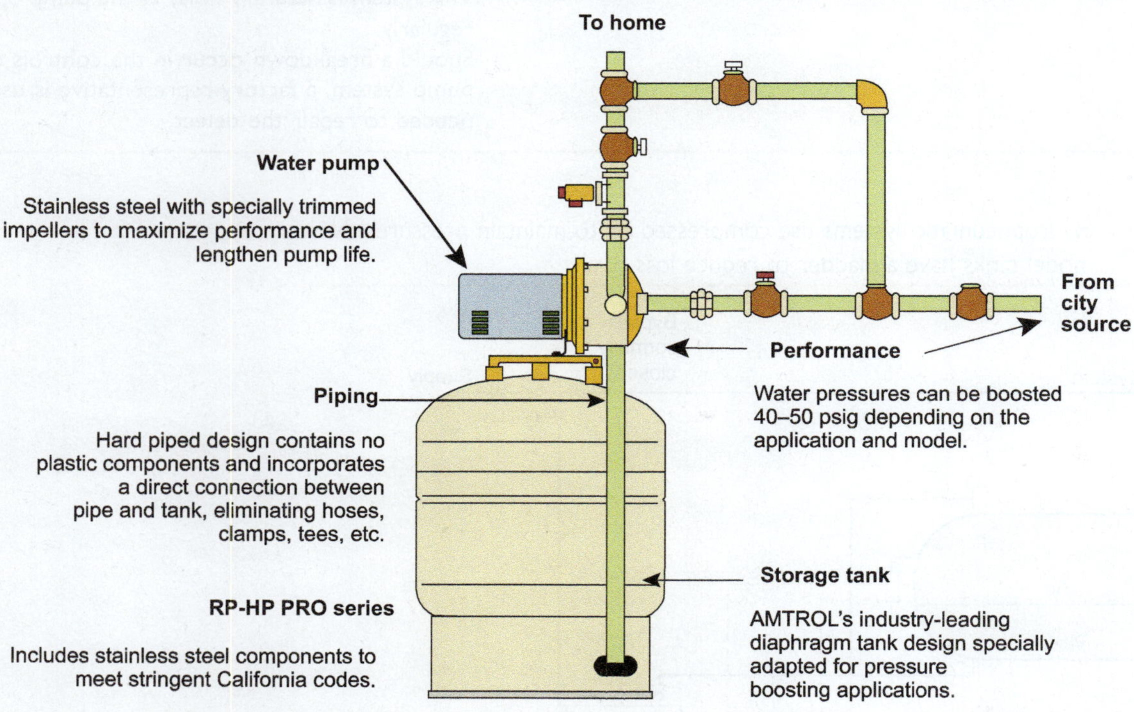

A surge tank may be located upstream of the booster pumps if the availability of the water supply is intermittent. This tank prevents damage to the pumps, as a supply of water is always present.

REVIEW QUESTIONS

True or False

1. _________ Pressure at the base of a column of water is the same as it is at the top, as long as the water is not moving.
2. _________ Elevated water tanks often are the best way to maintain pressure in a system.
3. _________ Variations in water pressure in a building may be caused by heavy demand on the public mains and/or by undersized components in the building system.
4. _________ The piping manufacturer's recommended maximum flow velocity through the pipe should be known and honored.
5. _________ In a water system, it is necessary to estimate the maximum probable flow at all points in the system.
6. _________ The water supply fixture unit concept is used to estimate maximum probable demand in piping systems.
7. _________ The header (or manifold) is usually equipped with individual valves for each tube.
8. _________ The individual run system uses a tube that extends from the header to a single faucet or fixture fitting.
9. _________ Temperature changes due to usage at other fixtures are reduced by using a manifold and individual runs.
10. _________ PEX-AL-PEX is PEX piping that is only used with aluminum fittings.
11. _________ All individual run systems for domestic water supply will contain a main supply header and a return header, which forms a loop.

Short Answer

12. Name three methods of pressurizing the water supply in a building.
13. Using Table 7-9, determine the following values:

    _________ Friction loss for 1 inch Type M copper, 4 fps velocity

    _________ Friction loss for 2 inches galvanized, 8 fps velocity

    _________ Gpm for 1 inch Type L copper, 14 psi/100 ft. pressure loss

    _________ Fps velocity for 1½" Type K copper, 2.6 psi/100 ft. pressure loss

14. Using Table 7-10, what are the equivalent feet of pipe for the branch of a 1-inch tee?
15. Using Table 7-5, what is the WSFU rating for a domestic kitchen sink in an individual dwelling unit?

Chapter 8

LEARNING OBJECTIVES

The student will:

- Define hot water.
- Explain the impact of codes on the requirements for hot water.
- Describe the key components of a hydronic heating system.
- Describe the construction and components of storage-type, direct-fired water heaters.
- Identify the safety precautions and steps to replace water heaters.
- Identify common causes of complaints concerning water heaters.
- Describe the characteristics of tankless water heaters, indirect water heaters, and circulating systems.
- Describe the factors necessary to size water heaters for residential settings.

Water Heaters and Hot Water Distribution

DEFINITIONS

Basic Principle No. 3 of the National Standard Plumbing Code states that "hot water shall be supplied to all plumbing fixtures which normally need or require hot water for their proper use and function."

A properly designed hot water supply system should be designed to conform to local code requirements, meet the needs of the customer, and provide a reasonably long service life. The system should also be designed to minimize initial installation costs but still be accessible and maintainable. While some codes are specific as to what constitutes hot water, others are vague. Hot water is defined in the International Plumbing Code as water at a temperature greater than 110°F (49°C). The 2012 National Standard Plumbing Code simply defines hot water as "potable water that is heated to a required temperature for its intended use."

In buildings other than dwelling units, tempered water only is permitted to be delivered to fixtures that normally would be supplied with hot and cold water. Tempered water is defined by the 2012 National Standard Plumbing Code as "a mixture of hot and cold water to reach a desired temperature for its intended use."

Unless otherwise specified by local code, health department, or other authorities having jurisdiction, water temperatures should be in the general ranges specified in Table 8-1.

WATER HEATER TYPES

Water is heated by one of two methods: direct or indirect. The direct method uses a primary fuel to heat the water in its container. The indirect method uses steam, hot water, or some other heated fluid to heat the water in a heat exchanger coil. Primary fuels include fuel gases, fuel oils, and electricity.

TABLE 8-1 Hot Water Temperature Ranges

Building Type	Use	Temperature Range (°F)	Temperature Range (°C)
Dwelling Units	Showers and bathtubs	85–110	29–43
	Other normal use	110–120	43–49
	Maximum temperature	120	49
Restaurants	General purpose	140	60
	Sanitizing	180	83
Hospitals	Laundries	165	74
	Photo processing	68	20
	Sterilizing	180–190	82–88
	Bathing	85–110	29–43

Note: The above temperatures are simple guidelines. Temperature ranges are very controversial, so any decisions concerning their selection should be verified with the health department or other authorities having jurisdiction.

Several industry standards apply to water heaters. In most jurisdictions in the United States and Canada, units that exceed 100 gallons, 200 MBH (1 MBH = 1,000 BTU per hour) input, or 200°F operating temperature must be constructed according to the ASME Boiler and Pressure Vessel Code. Other heater standards promulgated by the American Gas Association (AGA) or Underwriters Laboratories (UL) describe the design and performance limits of smaller water heaters.

Direct-Fired

In a direct-fired water heater, the water is heated by a flame or electric heating element. These water heaters can be tank-type for storage or minimum-volume–type heaters that can be arranged to heat the contents of an adjacent storage tank. Minimum-volume direct-fired water heaters can also be used as instantaneous heaters—units that turn on and heat the water to operating temperature only when there is flow through the heater.

Tank-type water heaters (also called storage heaters) provide a quantity of available hot water for a wide range of flow rates in the system. See Figure 8-1. For a given system-required capacity, the tank-type water heater usually will be less costly to install. However, if the storage heater is insufficient to meet demand and gets cool, it may require considerable time to recover the temperature.

Tankless water heaters heat the water when flow occurs (see Figure 8-2). When such water heaters supply a constant load or an on-off load that does not vary much when flowing, they can deliver a satisfactory temperature to the load. When the load is made up of many fixtures with random usage, the temperature-delivery pattern of this type of water heater will be variable, possibly beyond acceptable limits. Tankless water heaters have minimum standby losses and require minimum space for installation, but they usually require higher rates of energy input and are more susceptible to scaling problems if the water contains scaling chemicals.

FIGURE 8-1 Typical gas-fired tank-type water heaters store heated water until needed by the user.

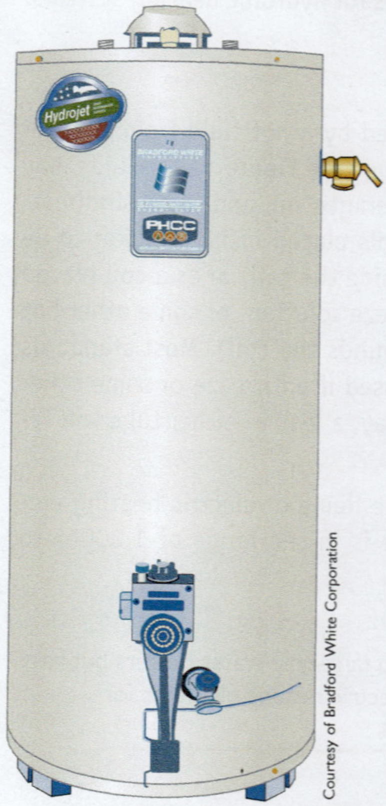

Courtesy of Bradford White Corporation

FIGURE 8-2 Tankless water heaters are much smaller and do not have the standby losses common to tank-type water heaters.

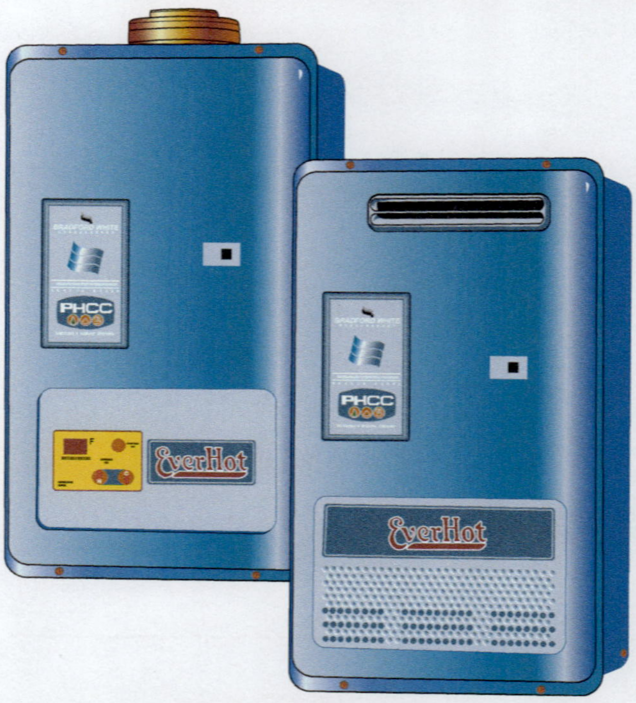

Courtesy of Bradford White Corporation

Because tankless water heaters have such high heating capacities, they are starting to be used for space heating applications. Some manufacturers have built-in circulation pumps that work well with storage tanks for hydronic heating systems.

Indirect-Fired

In an indirect-fired water heater, the water is heated by a hot fluid from another source, such as a boiler or solar water heating system (see Figure 8-3). Fluids may be hot water, steam, hot antifreeze solutions, refrigerants, oil, and so on. Indirect-fired water heaters can take the form of heating coils placed inside a boiler to be heated by boiler water (the water to be heated is inside the coil) or as a coil placed within a storage tank with steam, hot water, antifreeze solution, or some other hot liquid within the coil (the water to be heated surrounds the coil). Most standards now require that a double-wall heat exchanger be used if antifreeze or some other nonpotable water is used to transfer heat. In that way, a leak in either tube will result in a noticeable leak, not a contamination.

Because the temperature difference between the flame or electric heating element and the heated water in a direct-fired heater is in the range of 1,000°F to

FIGURE 8-3 Indirect-fired water heaters look similar to tank-type water heaters but have an internal heat exchanger. Some have electric heating elements for backup heat.

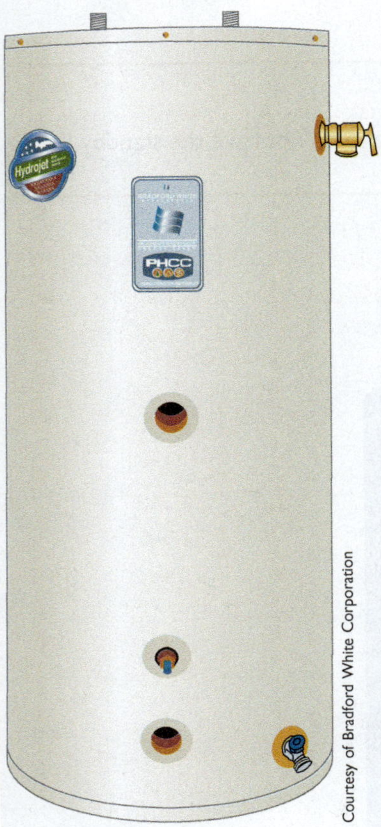

Courtesy of Bradford White Corporation

2,500°F, direct-fired units are physically smaller than indirect-fired types of the same rating, where the temperature difference may be as low as 50°F or 100°F. With this lower temperature difference between the heating fluid and the fluid to be heated, much more surface area must be available to heat a given volume of water.

Scaling

A problem that occurs in many areas is fouling of water heaters with scale buildup. This is a rare phenomenon in chemistry—very few chemicals are less soluble in hot water than in cold water. Unfortunately, one family of these chemicals is the carbonates found in well water in many areas of the country. These chemicals precipitate out of solution when the water is heated and form a scale on the surface of the water heater. The scale insulates the contained water from the heating source in either the direct or indirect type of unit. Because of the high temperatures in the direct-fired type, this insulating scale can lead to metal fatigue, local thermal overstress, or burned-out tanks or coil tubes. The indirect types can scale to the point that flow is stopped. Any scale reduces the ability of any type of heater to produce hot water. One noticeable symptom of scale buildup is a rumbling noise created when the water in contact with the heating surface turns to steam and breaks through the scale buildup. Many newer tankless water heaters monitor the water output temperature and the set-point temperature to determine if scaling could be present. If so, an error code will be generated to warn the customer of an impending danger.

Circulation of Hot Water

Hot water is conveyed to the fixtures in a building in a piping system similar to the cold water distribution system. Good design practice dictates that whenever the distance to the farthest fixture exceeds 100 feet, a hot water circulation system or some other means should be provided to maintain heated water temperature within 25 feet of any heated water outlet. The circulation system maintains hot water in the mains at all occupied times. In this way, hot water arrives at any faucet with little or no delay after the faucet is opened and water waste is minimized since there is very little waiting time for hot water to come to the fixture.

Water Heater Controls and Markings

Water heaters must be equipped with an operating thermostat, a relief valve or valves to limit temperature-and-pressure values to safe levels (see Figure 8-4), and an energy cutoff device (which requires manual reset) to interrupt all energy to the heater if the water temperature exceeds the factory-set value. If the unit burns a fuel, then a flame-failure safety system is also required.

MIXED WATER TEMPERATURE CONTROL

The 2012 National Standard Plumbing Code, 2012 International Plumbing Code, and 2012 Uniform Plumbing Code require that showers and bath/shower combinations

FIGURE 8-4 This standard temperature and pressure (T&P) control valve shows signs of corrosion and scale build up. Since this can cause a malfunction, this valve must be checked annually for proper function.

have individual balanced pressure, thermostatic, or combination automatic compensating valves that limit the maximum temperature of water to less than 120°F. These valves will reduce the possibility of anyone being scalded when there is a reduction of pressure and flow in the cold water line. In older valves, a reduction of cold water flow would cause an increase in the amount of hot water flowing from the faucet. Figure 8-5 shows a newer shower valve that has internal compensation as well as a rotation limiter to reduce the amount of hot water that can be delivered.

In cases where the hot water is also used to heat the building, the temperature of the water can go as high as 140°F. This water must be tempered before it can be used for domestic hot water purposes. Figure 8-6 shows a mixing valve used to temper the hot water output of a solar water heating system before it is sent into the house.

FIGURE 8-5 Newer shower and tub faucets come with rotation stops that limit the maximum amount of hot water that can be delivered.

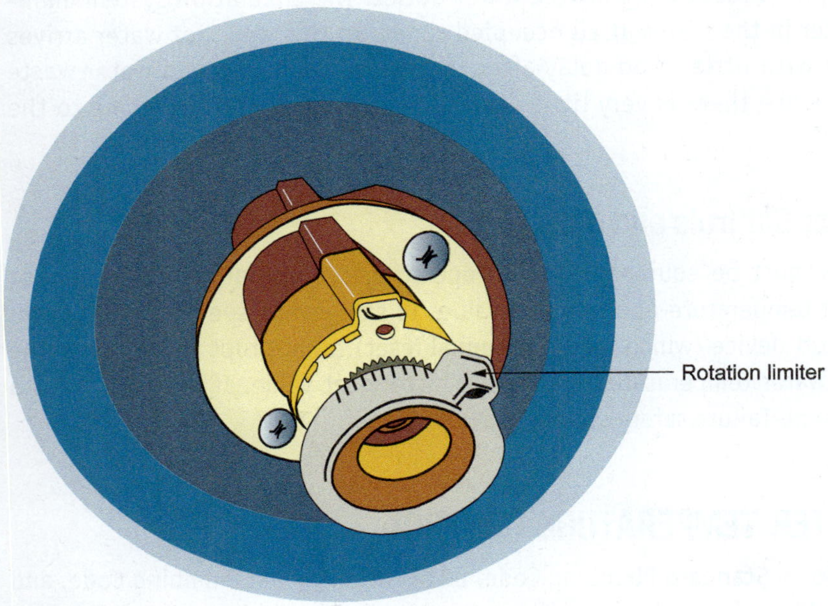

2. Place thread sealant on the male thread that enters the tank tapping.
3. Shut off gas and water to the tank. Be certain to instruct the occupants of the building not to operate any faucets during this brief task.
4. Disconnect gas piping from supply pipe, tubing to pilot and main burner, and thermocouple.
5. Air-lock the tank by making certain all hot water faucets are closed, opening the drain valve, and waiting until the water stops flowing (only a small amount will drain out). Caution: if you lose vacuum, you may cause water damage to property and/or injury to yourself and others around you.
6. Remove the original gas control and quickly insert the new gas control. Air will be drawn in the control opening and only a small amount of water will flow from the drain.
7. Tighten the new control into proper position and reconnect the gas lines and thermocouple.
8. Check for gas leaks with an approved leak indicator solution.
9. Open water valve(s).
10. Relight the water heater. Verify proper operation of pilot-out safety system and temperature control.

On larger water heaters, you will find that the safety and operating functions will be done by a group of devices rather than a single unit.

Separate controls would be selected in the design of a particular water heater for any of several reasons:

- Single control of sufficient capacity might not be available.
- Higher operating temperature required than provided for in single units. Different flame safety system required on the larger water heater.
- Operating temperature sensing point physically some distance away from safety limit sensing point.
- Use of commercial- or industrial-type (rather than residential-type) devices.

For whatever reasons, such units are in use. It is usually easier to service these control systems because each component has only one function to perform and replacement components are usually available. Some water heaters are equipped with a spark-ignited pilot that lights at the beginning of each heating cycle. When the water temperature reaches the thermostat set-point, the main burner and pilot are shut off. In this way, fuel is saved by not having the pilot burning at all times.

The immersion bulb of most water heater thermostats is coated with a thin, smooth plastic film to minimize the chance of scale accumulation on the bulb. Such scale buildup would result in overheating the water because the sensing bulb would be insulated from the actual water temperature.

Occasionally, pilot flame interruptions occur because of a problem with the thermocouple or its installation. Thermocouples can deteriorate; as they do, they produce less of an electrical signal to hold the gas safety system latched. Other common problems that can occur are as follows: the cold end connection to the gas control can become loose or corroded, which reduces the amount of voltage applied to the control valve; the hot end of the thermocouple can shift out of the pilot flame; or the pilot flame might be too small to produce enough heat.

If the thermocouple has to be replaced, obtain the proper replacement, remove the old thermocouple, and install the new one. Be sure that at least the top $\frac{1}{2}$ inch extends through the pilot flame and that the flame surrounds the thermocouple tip. Test the unit for performance, making sure that the main burner shuts off if the pilot light goes out. Refer to Table 8-2 for additional troubleshooting aids.

TABLE 8-2 Troubleshooting—Gas Water Heaters

Complaint	Possible Cause	Service Tip
No Hot Water	Pilot extinguished	Relight heater. Check and adjust pilot if necessary. Check thermocouple output and dropout reading. Clear pilot burner.
	Gas supply off	Restore gas supply and relight all appliances.
	Gas plug cock closed	Open cock fully and relight pilot.
	Control knob not in "on" position	Turn knob to "on" position and relight.
	ECO has cut off pilot	Allow tank to cool. Relight heater. Dial down setting 10°F. Thermostat may be defective.
Water Not Hot Enough	Control dial set too low	Dial control to higher setting; avoid scalding temperatures.
	Control defective	Replace control if it is out of calibration. Test by turning to higher setting. Avoid scalding temperatures.
	Temperature differential too high	Replace the control if the differential exceeds 25°F.
Insufficient Hot Water	Control dial set too low	Dial control to higher setting; avoid scalding temperatures.
	Control defective	Replace control if it is out of calibration.
	Gas plug cock not fully open	Turn plug cock to open position.
	Knob on dial not in "on" position	Turn control knob to "on" position.
	Excessive demand	Check for change in hot water use. Review methods of diversifying demands to accommodate heater operation. Size heater properly to accommodate demand required.
	Heater fouled with scale	Remove scale or replace heater.
	Temperature differential too high	Replace the control if the differential exceeds 25°F.
	Heater appears to have a slower recovery	Check gas input. If needed, adjust gas pressure or replace main burner orifice. Heater may have been installed in warmer months when water supply may have been warmer.
	Flue baffle not in flue	Install flue baffle.
	Dip tube missing or in "hot" side	Install dip tube in cold side of heater.
Water Too Hot/Heater Too Noisy	Heater stratification (stacking)—this occurs when water is drawn off slowly from the heater and the cold water entering the base of the tank keeps the gas control on; the water at the top of the tank becomes extremely hot	Check for leaky faucets.
	Sizzle: normal condensation	Normal condition when heating a cold tank.
	Rumble: sediment build-up	Sediment may be drained or may have to be removed using deliming solution. The sediment collects at the base of the heater, and as the flame heats the tank base, the water trapped below the sediment becomes superheated and flashes to steam, especially when water is drawn off.

(continued)

TABLE 8-2 Troubleshooting—Gas Water Heaters (Continued)

Complaint	Possible Cause	Service Tip
	Ticking: expansion or contraction of flue, tank, or piping	Normal condition.
	Combustion sounds: burner using too much primary air	Install and regulate air shutter. If occurring in wintertime, dial down control based on occupants' demands. If gas heater is so equipped, adjust air shutter to blue condition but not noisy or lifting off burner ports. Install a water mixing valve if necessary.
	Combustion sounds: wrong burner orifices	Check to assure that the orifices match the gas firing rate.
Water Leaks	Joint leakage	Repair faulty installation.
	Condensation	Insulate pipes if necessary.
	Relief valve discharge	Relief valve may be wrong rating. Install water hammer arrestors so that relief valve does not open due to pressure surges in building system. Continuous leakage should be checked for signs of valve failure. Check for the installation of a dual check valve or similar backflow device in the piping to the heater; if it exists, install a pressure relieving device or an expansion tank in line to the heater. Do not allow the relief valve to cycle regularly to solve this expansion problem.
	Drain cock leaking	If valve will not close, turn valve wide open to allow for sediment discharge, which may be preventing proper closure. If valve will still not close, replace it.
	Condensation may cause occasional dripping noises on burner	Normal condition when heating a cold tank.
Gas Odors	Gas valve or piping connections	Check for leaks.
	Vent blockage	Check for backdraft, blockages, excessive sooting, or improperly installed vent piping.
	Overfiring	Adjust input.
	Draft diverter not installed	Install draft diverter.

GAS-FIRED WATER HEATERS MANUFACTURED AFTER JANUARY 1, 2004

Accidental flammable vapor ignition incidents occur every year. Some of these situations involve persons cleaning greasy machine parts or attempting to clean or remove flooring with gasoline. Another common situation involves people spilling gasoline in a basement or garage. Sometimes gasoline has been illegally stored in glass containers that were accidentally broken. These events have had horrible results when the gasoline vapors were ignited by water heater pilots or main burners. People in the vicinity of these explosions were seriously burned. Some were even killed. Of course, large property damage almost always resulted as well.

For many years, the defense against this hazard was the requirement in most plumbing codes that water heaters had to be installed 18 inches above the floor, in basements, or garages. Several water heater manufacturers attacked the problem and developed solutions that were passive, meaning that the installer does not have to contribute to the solution. New regulations have been adopted: now, the burner chamber is inaccessible to the user (access covers are held in place with many screws, etc.) and the combustion air reaches the burner chamber through a

FIGURE 8-7 The flame arrestor screen located on the bottom of a gas-fired water heater prevents a flame from exiting the combustion chamber.

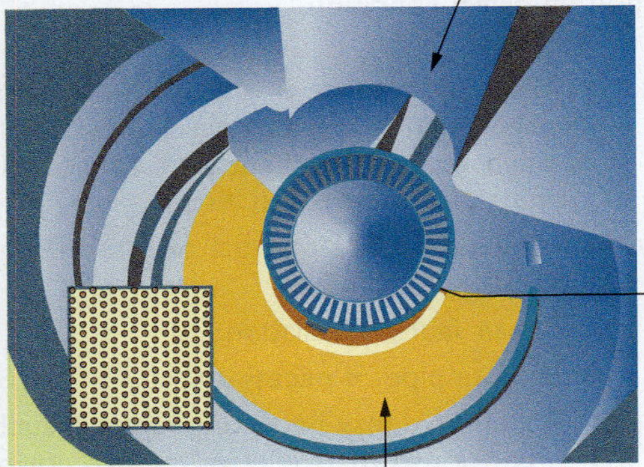

grating that allows a combustible atmosphere to reach the fire but quenches any flame that attempts to pass back beyond it. See Figure 8-7.

Check your local code for specific installation requirements. Your local code may still require the 18-inch height for installation. It often takes some time for codes to recognize product developments that make those requirements obsolete.

All of the control functions described above are incorporated into these new water heaters. The principal difference is that it is not possible to light the pilot flame manually. Either the pilot is lit automatically on each cycle, or it can be started with a pushbutton that energizes a piezo-electric igniter like those often found on outdoor grills.

ELECTRIC ELEMENTS

An electric water heater utilizes resistance heating elements to warm the water. These elements can be directly submerged in the heated water or, for low-wattage units, wrapped around the tank wall to heat the water by conduction through the tank. Direct immersion water heater elements are more efficient and available in higher wattages than the wraparound version, but the tank tapping needed to pass the immersion heater through the tank wall and into the water is a potential leak point.

The most common electric water heater has two elements: one near the top of the water heater, the other near the bottom. Each element has a thermostat mounted at the same level as the element. The controls are wired so that if the top of the tank is cooler than the upper thermostat, the upper element is activated and the lower element is switched off. When the upper thermostat is satisfied, the upper element is off and the lower element is permitted to operate subject to the control of the lower thermostat.

The result is that a water heater completely full of cold water heats only the upper portion of the tank so that hot water is available quickly. Once this upper section is heated, the lower element turns on to heat the remaining water in the tank. Note that only one water heater element can be on at any time. Most electric utilities require this mode of operation to reduce a huge demand of electricity all at once. Most heating elements will be damaged if they are energized in air. Therefore, be sure not to turn on the power to an electric water heater unless you are certain that it is full of water.

Heating Element Failures

Heating element failures are generally determined to be one of the following problems:

- The element is electrically open. This means that the heating element (wire) has broken, so no current can pass and heating cannot be accomplished.
- A sign of this type of failure is that the correct amount of voltage will be applied to the element, but there will be no current flow.
- Water enters the heating element, providing a circuit path to ground. Depending upon the exact circumstances, this situation can produce continued heating even when the thermostat signals to turn off the heat. This usually results in tripping the ECO or causes fuses or circuit breakers to open. It can also energize the jacket of the appliance, so you feel a shock when you touch the jacket. This is why equipment grounding conductors are so important.
- The heating element touches the sheath, thus grounding the electric supply. This condition will also cause circuit breakers or fuses to open.

Electric Heater Thermostats

Thermostats are usually arranged to sense the water temperature through the tank wall. The upper thermostat is single pole, double throw, and the lower thermostat is single pole, single throw.

The normal residential adjustment is to set both the upper and lower thermostat at about 120°F. In this way, the water heater will be able to furnish hot water to the building under most operating conditions. These recommended settings minimize the chance of dangerous scalding injuries.

The energy cutoff device for an electric water heater is a manual-reset thermal element that is factory-set to trip at about 180°F. All energy to the heater is shut off if this device trips. Figure 8-8 shows a top thermostat and ECO unit typical of a residential water heater.

IN THE FIELD

Lowering the thermostat setting to below 120°F can allow Legionella bacteria to grow. Legionella bacteria can cause Legionnaires' disease, a harmful form of pneumonia.

FIGURE 8-8 The top thermostat and ECO for an electric water heater are combined into one unit.

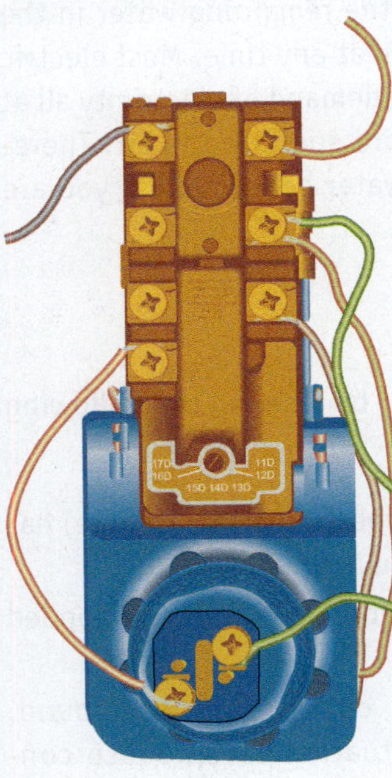

Relief Valves

A combination temperature-and-pressure relief valve or separate temperature-and-pressure relief valves must be installed on every water heater—except small instantaneous hot water dispensers—as the ultimate safety protective devices. The temperature sensor on the relief valve must be located in the top six inches of the heater. This device should be periodically inspected to determine that it is free-working. The handle can be opened manually; the valve should re-close upon release. The relief valve discharge pipe should discharge through an air gap. This discharge pipe should be the full size of the relief valve drain outlet. The piping should not be trapped or valved, and the outlet end should not be threaded. The outlet should be piped to an approved point of discharge, where escaping steam or hot water will not create a safety hazard. Customers should be instructed to call their plumber if they notice flow from the relief valve. The relief valve piping should be rated for use in hot water discharges or should comply with the ASME A112.4.1 standard.

OIL BURNERS

Oil-fired water heaters are generally similar to gas water heaters except that more volume in the burner chamber is needed to burn fuel oil properly, compared to gas. Most oil burners combine a motor-driven blower and oil pump to provide combustion air and oil under pressure for satisfactory air/oil mixing. This mixing is essential to efficient, clean combustion.

Oil burners utilize an electric spark to initiate the fire, so no standby pilot flame is required. It should be noted that this electric spark is generated with a very high voltage. This voltage will be higher than most voltmeters can safely measure, so do not attempt to test it. All burners incorporate some type of flame-failure sensing system so that operation is stopped if flame is not present when it is called for. These flame-failure controls usually require manual resetting if they are activated. Because atmospheric air conditions can affect the firing rate of most oil burners, special equipment is needed to adjust the combustion air.

Oil-fired water heaters also incorporate thermostats, relief valves, and energy cutoff devices similar to those found on other water heaters.

NAMEPLATE INFORMATION

No matter what energy source the water heater uses, check to be sure that the heater has an ASME, AGA, or UL certification on the label to be assured that the appliance is constructed according to industry standards. The information on the label will also allow you to determine that the water heater has the proper storage capacity, heating rate, and working pressure rating. Also noted on the label will be the manufacturer's trademark, model, and serial number.

INSTANTANEOUS WATER HEATERS

Instantaneous water heaters, also called tankless water heaters, provide hot water to the system fixtures without using a storage tank. A variety of products are available, including direct-fired models that heat the water directly from their own energy input and indirect-fired models that use the heat energy from another source to heat the water. Most manufacturers make models that are compatible with electrical, fuel gas, oil, and solar energy.

Direct-Fired

Direct-fired instantaneous units use either a fuel-fired or electric element to heat the water. These units operate whenever there is water flow and shut off when the water flow stops. Figure 8-9 shows a small direct-fired electric water heater that is located in a bathroom vanity. This allows for small amounts of water to be delivered almost instantly.

IN THE FIELD

By installing a point-of-use instantaneous water heater on long runs, the amount of water wasted while waiting for it to warm up can be greatly reduced.

Direct-fired tankless water heaters are suited for many hot water demand applications. Sizes are available in a wide range of flow rates. Larger commercial units are well suited to swimming pool, industrial, or dishwasher loads. New residential models are available with modulating firing rates so they adapt well to various load conditions. Small direct-fired instantaneous water heaters are found in hot water beverage dispensers. Figure 8-10 shows a wall-mounted direct-fired water heater that supplies hot water for the entire house.

FIGURE 8-9 This point-of-use tankless water heater (electric) serves a lavatory located a long distance away from the main water heater.

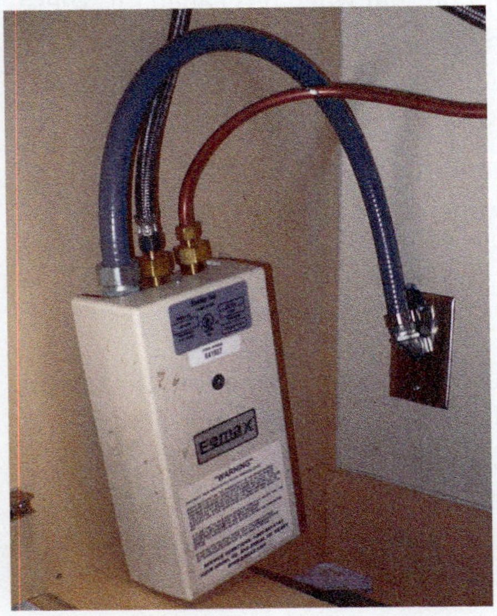

Photo by Ed Moore

FIGURE 8-10 This gas-fired tankless water heater will modulate its firing rate based on the amount of hot water required.

Photo by Ed Moore

Indirect-Fired

Indirect-fired units use heat from another source to heat the water. Steam or hot water is passed through a coil or containment tank that is, in turn, directly connected to the potable water source. If the steam or hot water is not potable, a double-wall heat exchanger is used to prevent contamination in the event of a leak. In some cases, the coil may contain potable water and be submerged in the building heating boiler. In this way, the water heater requires no floor space. Because the boiler capacity is large compared with the hot water needs, ample hot water is usually available.

Instantaneous indirect-fired water heaters save on space and investment in a storage tank, but they represent added operating cost if the heating boiler has to be fired to serve only this relatively small load. They also require added investment and maintenance in automatic mixing valves if the range of hot water flows is fairly wide.

Figure 8-11 shows a coil and tank piped to a hot water boiler to heat domestic hot water. Boiler water circulates through the tank and around the coil to heat the domestic hot water. Note that the boiler water path is from the top of the boiler, through the heat exchanger, and into the return at the bottom of the boiler. With this arrangement, boiler water will circulate by gravity (no pump is needed) when domestic water takes heat from this water through the wall of the heating coil. The return connection is made directly to the boiler, not the return header. In this way, natural convection produces the boiler water flow through the heat exchanger.

Because of high energy costs, indirect-fired water heaters are not often selected for general year-round use in residential dwelling applications. However, in businesses such as hospitals that require reliable hot water or steam year-round, they are very desirable.

Controls and Accessories

Whenever an indirect-fired water heater is installed, additional controls and accessories are required to help maintain a safe water temperature and pressure. In many cases, the unit that acts as a heat source is capable of supplying water at

FIGURE 8-11 This indirect-fired water heater relies on natural convection for circulation.

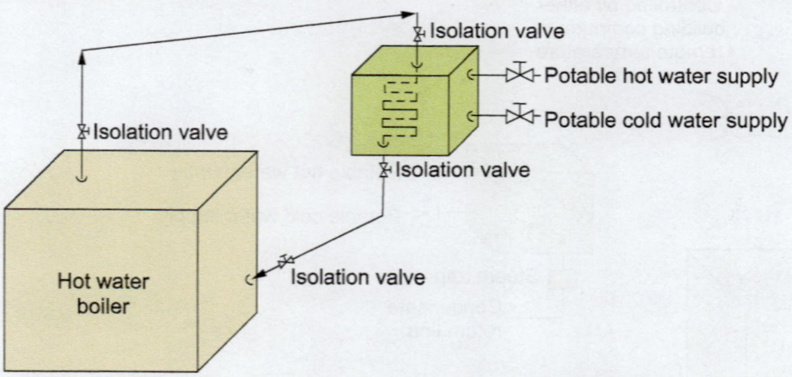

temperature levels that are unsafe for potable systems. Additional valving is also needed for preventive maintenance. Examples of this include:

- Gate or ball valves and tees in the boiler piping so that the exterior of the coil can be flushed or treated with scale-removing chemicals.
- Similar valves and tees for the freshwater side of the heat exchanger.
- A flow control valve installed in the space heating supply line so that overheating the building will not occur when the boiler is kept hot to provide domestic hot water.
- A relief valve installed to maintain safe pressure and temperature levels.

In some cases, the submerged coil is located inside the boiler. This eliminates the need for additional floor space but must have been considered when the boiler was purchased. As before, cold water enters through the bottom of the coil and hot water exits at the top. A flow control valve, located at the discharge of the coil, is included in the flow path to regulate the flow of water so that enough transition time is available to assure full temperature rise.

These systems perform satisfactorily in all seasons because the boiler is arranged to maintain a sufficiently high water temperature in warm weather to heat the domestic hot water. However, these units are not satisfactory if the domestic water has a scaling characteristic.

Figure 8-12 shows a further modification, in which the hot water or steam is taken from a header and directed to a heat exchanger. An additional control valve will be needed to maintain the temperature of the potable water. The control valve can be controlled either through a remote bulb thermostat mounted on the tank or through a building automation computer.

When certain firing rates or gallon capacities are exceeded, the water heater is considered a boiler. This may require additional controls and/or inspections.

FIGURE 8-12 Some indirect-fired water heaters use steam control valves that monitor the hot water temperature through either an aquastat or a building automation system.

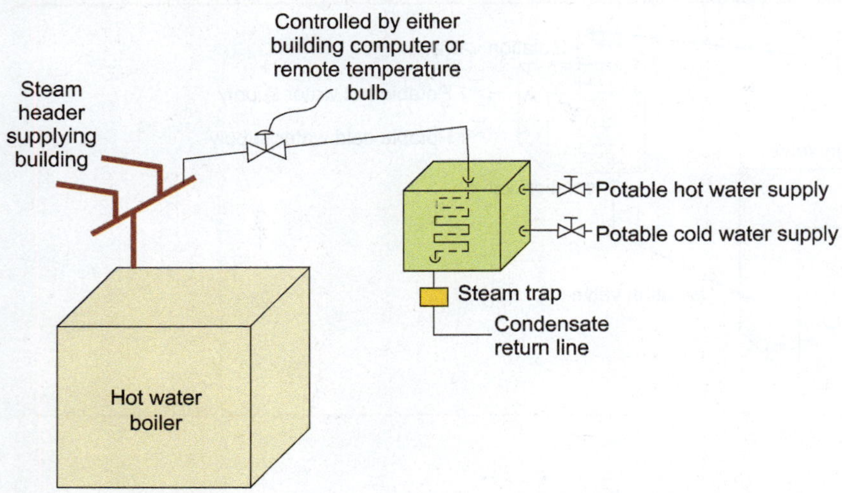

It should be noted that in a complete boiler installation, the above drawings are not complete in all details.

RESIDENTIAL SIZING

There are several different methods used to properly size water heaters. Many water heater manufacturers, such as Bradford White (http://www.bradfordwhite.com/rightspec.asp) and Rheem (http://www.rheem.com/products/water_heating/tankless/ez_spec_sizing/), have software or sizing guides that aid in the selection process. The information needed will depend on whether the water heater has a storage tank or not. A term commonly used in the sizing of water heaters is *first hour rating*. The first hour rating is the amount of hot water, in gallons, the heater can supply per hour when the tank is initially full of hot water. The first hour rating will be listed on the Energy Guide label as "Capacity (first hour rating)." It is affected by the tank capacity, type of fuel used, and the size of the burner or element. The Federal Trade Commission requires an Energy Guide label on all new conventional storage water heaters but not on heat pump water heaters. The following is an example of how both can be sized for a residential building.

Storage Tank Type

A very simple, popular sizing process uses a chart that bases hot water consumption on the number of bedrooms and the number of baths in a house. The theory is that the more bedrooms a house has, the more people it can contain—thereby more demand. Also the more bathrooms, the greater the peak demand will be for a short period of time. Table 8-3 shows an example of how these tables are set up; remember it is always best to use the table or method described by the specific water heater manufacturer.

TABLE 8-3 Water Heater Sizing for Example 1

Number of Baths	1 to 1.5			2 to 2.5				3 to 3.5			
Number of Bedrooms	1	2	3	2	3	4	5	3	4	5	6
GAS											
Storage, gal	20	30	30	30	40	40	50	40	50	50	50
1000 Btu/hr input	27	36	36	36	36	38	47	38	38	47	50
1-hr draw, gal	43	60	60	60	70	72	90	50	66	66	80
Recovery, gph	23	30	30	30	30	32	40	32	32	40	42
ELECTRIC											
Storage, gal	20	30	40	40	50	50	66	50	66	66	80
kW input	2.5	3.5	4.5	4.5	5.5	5.5	5.5	5.5	5.5	5.5	5.5
1-hr draw, gal	30	44	58	58	72	72	88	72	88	88	102
Recovery, gph	10	14	18	18	22	22	22	22	22	22	22
OIL											
Storage, gal	30	30	30	30	30	30	30	30	30	30	30
1000 Btu/hr input	70	70	70	70	70	70	70	70	70	70	70
1-hr draw, gal	89	89	89	89	89	89	89	89	89	89	89
Recovery, gph	59	59	59	59	59	59	59	59	59	59	59

EXAMPLE 1

What size gas water heater would be required for a three-bedroom house with two full bathrooms?

Using Table 8-3, a 40-gallon water heater with a 36,000 DTU/hr input and a first hour rate of 70 gph would be needed.

While the chart is simple to use, it assumes that standard-size fixtures and faucets are used. If larger whirlpool bathtubs or higher flow–rate shower heads are used, the first hour rate and/or tank size may need to be increased.

Tankless Type

Since there is no initial stored hot water to draw from, tankless or demand-type water heaters need to be sized based on the flow rate expected and the water temperature increase needed. In order to accomplish this, it is necessary first to determine which hot water demands will occur at the same time during the peak use period. This demand will give the flow rate, or gallons per minute, that the water heater will need to be rated for. For example, if it were expected that one shower (2.0 gpm) and one lavatory (2.0 gpm) would be used at the same time, then the water heater would need to be rated at a 4.0 gpm flow rate. Note: these flow rates are 75 percent of the values taken from Table B.3 (Maximum Demand at Individual Water Outlets) of the 2012 National Standard Plumbing Code. In areas where lower flow rate fixtures are required, the flow rates would be less.

Next, the amount of heat input needed to handle the demand must be calculated by taking this flow rate and the expected temperature rise needed and using the formulas in Example 2. The expected temperature rise is the temperature at which the hot water needs to be supplied, minus the incoming water temperature. Incoming water temperature will change depending on the geographical location. In northern areas, it may be as low as 38°F, but in southern areas it may be as high as 70°F.

EXAMPLE 2

What would be the Btu/hr input for an 80 percent efficient tankless water heater that is supplied with incoming 60°F water, if the desired outlet temperature is 120°F at 4.0 gpm or 240 gph?

For gas:

$$\text{Btu/hr input} = \frac{\text{flow rate} \times \text{temperature rise} \times 8.33\ \text{lbs/gallon}}{\text{thermal efficiency}}$$

$$\text{Btu/hr input} = \frac{240\ \text{gph} \times (120 - 60) \times 8.33}{0.80} = 149{,}940\ \text{Btu/hr}$$

For a 98 percent electric water heater:

$$\text{kW input} = \frac{\text{flow rate} \times \text{temperature rise} \times 8.33\ \text{lbs/gallon}}{\text{thermal efficiency} \times 3{,}412\ \text{Btu/kW}}$$

$$\text{Btu/hr input} = \frac{240\ \text{gph} \times (120 - 60) \times 8.33}{0.98 \times 3412} = 35.8\ \text{kW}$$

Increasing Capacity

When the water heater does not keep up with the building's instantaneous demand for hot water, one or more of the following conditions may be the cause:

- Increase in the amount or duration of building demand
- Reduction in capacity of the water heater due to reduction of tubing diameter from fouling, scaling, or corrosion
- Simultaneous occurrence of formerly separately timed loads

In many cases, there are several possible solutions. One solution is to place a smaller storage-type or demand-type water heater at one of the farthest fixtures. Sometimes a considerable amount of water and energy is being lost due to long runs of hot water that cool off during nonuse periods. These long runs have to be replenished with hot water before the fixture can serve its purpose. Figure 8-13 shows how a small point-of-use water heater can be installed to handle a lavatory. Here the smaller point-of-use water heater will only run until the hot water from the larger heater has replaced the cold water in the supply pipe. The placement of the second water heater may depend on lengths of runs, space available, and other economic factors.

Circulating Systems

Circulating systems are installed in hot water systems to maintain hot water in the mains so that no faucet is more than a short distance from a hot main. In this way, hot water is available quickly at any faucet and water waste is held to a minimum. A hot water return pipe is run parallel to the hot water supply main. Near the end of the main (or several branches) the return line connects to the main. This allows a return path back to the water heater to allow circulation. This circulation can be gravity or pumped depending on the system size and design. Figure 8-14 shows a circulation loop installed to serve a remote fixture.

FIGURE 8-13 By placing a small water heater close to a remote fixture, hot water will instantly be present at the outlet.

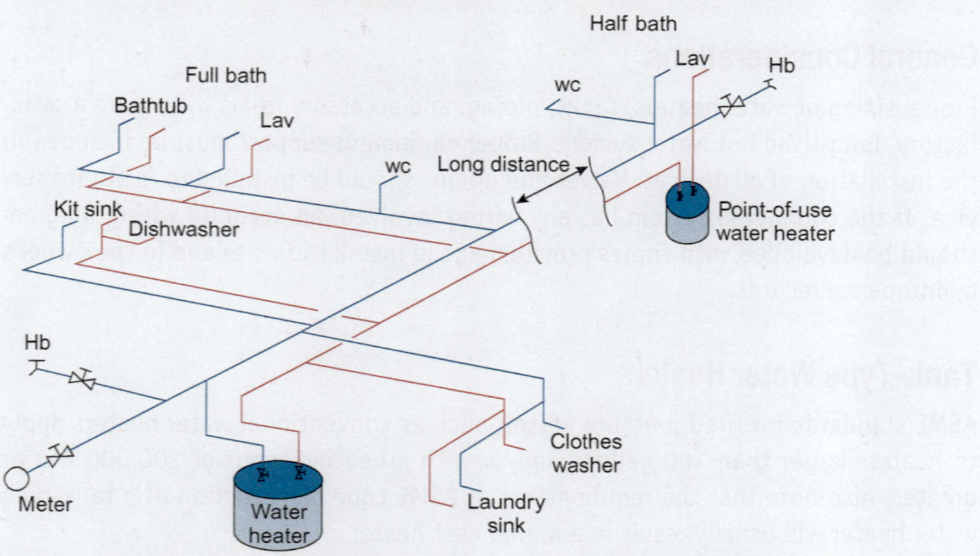

FIGURE 8-14 Circulation loops must be well insulated and thermostatically or time controlled.

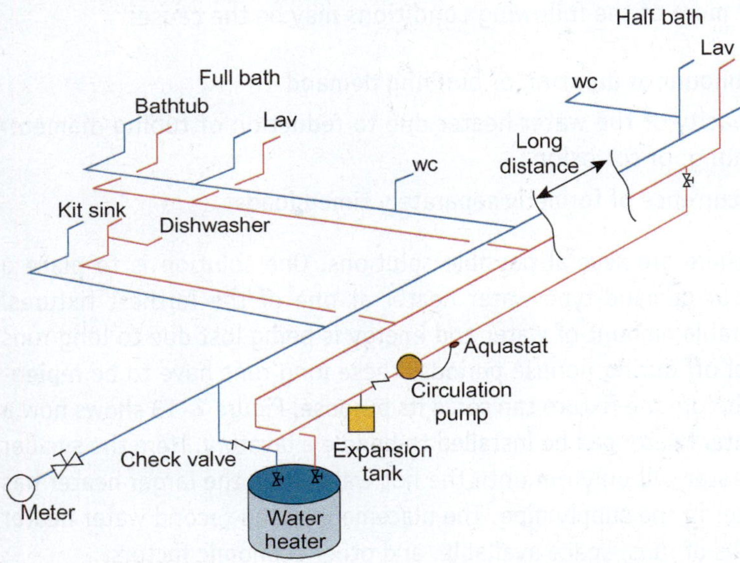

To minimize energy loss from the hot mains, most systems are equipped with timers or aquastats that measure water line temperature and operate the circulating pump only during occupied times. Additionally, the piping is insulated. Remember that an expansion tank is required for thermal expansion in many cases.

IN THE FIELD
Section 10.15 of the 2012 National Standard Plumbing Code requires that, where the developed length from the source of the heated water to the farthest fixture exceeds 100 feet, some means of maintaining the water temperature to within 25 feet of the fixture is required. This can be in the form of a circulation loop, electric heat tracing, or use of a point-of-use water heater.

General Considerations

Proper sizing of water heaters, tanks, piping, and accessory items will assure a satisfactory, long-lived hot water system. Proper equipment support must be included in the installation of all devices. Valves and unions should be installed to facilitate service. If the completed system has any wiring involved, an accurate wiring diagram should be developed with copies provided at the installation site and in the owner's maintenance records.

Tank-Type Water Heaters

ASME standards for fired pressure vessels such as conventional water heaters apply to heaters larger than 100 gallons and/or with a heating input of 200,000 Btu or greater. Also note that the requirements of ASME code construction of a tank-type water heater will usually result in a higher cost heater.

Storage tank or tank-type water heaters should be equipped with sufficient valves for maintenance and control and with inspection openings so that the inside of the tank can be examined. Smaller tanks use $1\frac{1}{2}$-inch or 2-inch pipe outlets with nipples and caps as the preferred closure.

Larger tanks use gasketed hand-holes and covers for these inspection openings. Very large tanks will have openings large enough for a person to pass through so that tanks can be inspected and/or repaired internally.

The completed installation of water heater, piping, and storage tank should be insulated to minimize heat loss.

When large water heating units are replaced, be careful to follow guidelines for the proper removal of asbestos. Contract with certified asbestos removal contractors for abatement and removal of this material.

It should be noted that the figures shown are not complete in all details for a proper boiler installation. The figures show the plumbing connections.

HYDRONIC HEATING SYSTEMS

Overview

There are many different methods of heating a building; some involve heating the air, some heat water, some create steam, and others rely on the solar radiation to heat the building components. This section focuses on the use of hot water to heat the building. Hydronic heat means that water is used to convey the heat. The most common system layout involves a heating source that will heat the water, and a piping system that will carry the water from the heating source to the areas of the building that needs heat. To make hydronic systems more safe and energy efficient, pumps, controls, and specialized heating terminals are used. Since hydronic heat uses relatively small diameter pipes, compared to the large duct systems used in forced air, it has the distinct advantage of not consuming large amounts of building space. It is also very quiet and can be easily adjusted to supply different amounts of heat to different parts of the building. The drawback to most hydronic systems is that the building still requires a system to circulate fresh air and to condition the air. Conditioning the air means that the humidity must be controlled, the air must be cleaned (filtered) and cooled to provide cooling for the building. Newer energy-efficient houses are so well constructed that they must also have fresh air provided for the occupants. Figure 8-15 shows a typical hydronic system that provides domestic hot water and space heating. It is important to point out that per code, 120°F water is the maximum temperature for domestic hot water, yet some codes allow hot water for heating purposes to reach 140°F. In this situation, precautions must be taken to protect the building occupants.

According to the Uniform Mechanical Code, the maximum pressure design of a piping system for hydronic heating is 160 psi, and the highest working temperature is 250°F. Higher temperature-and-pressure systems are designed for special heating applications that usually do not involve comfort heating. Most comfort heating systems are low-pressure systems operating at less than 15 psi and below 180°F.

FIGURE 8-15 Hydronic heating system diagram showing three zones and domestic hot water (DHW).

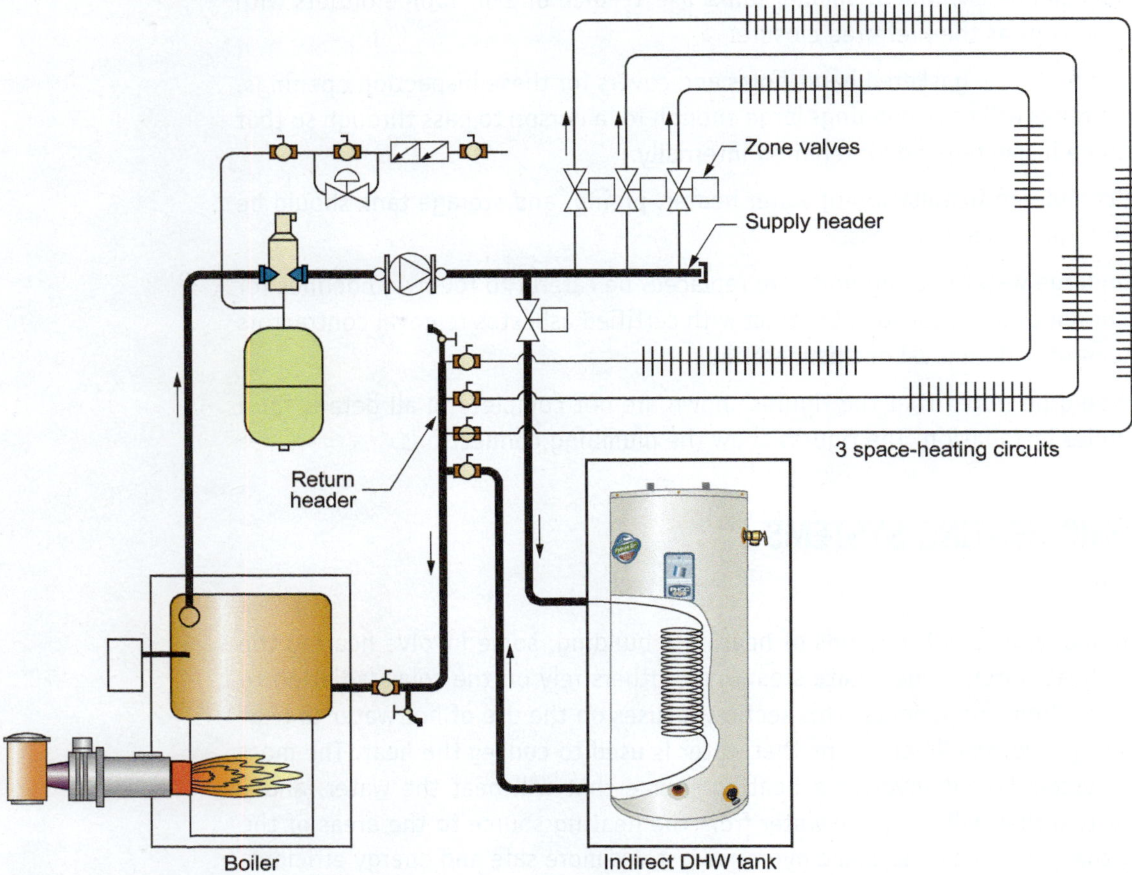

Modes of Heat Transfer

Buildings gain or lose heat by three primary modes of heat transfer: radiation, conduction, and convection. In the summer months, heat is trying to enter the building and during the winter months the heat is trying to escape the building. A properly designed heating system is sized so that it can efficiently provide heat to all parts of the building during those winter months. Because the heating load will vary throughout the day, many hydronic systems are designed so that they gradually make adjustments throughout the day so that the building maintains a constant temperature. These three modes of heat transfer also enable the heating system to replenish the heat being lost to the cooler outdoors. The following is a brief explanation of how the different modes of heat transfer work throughout the building and the heating system.

Radiation

Most people know that the earth is warmed by the solar radiation from the sun. The energy from the sun travels millions of miles through space and only when it is absorbed by an object is the heat revealed. Radiation does not require a medium to transfer heat; it can move through a vacuum. What most people don't realize is that any object can radiate heat or the lack of heat. If you place your hands near (but don't touch) a cup of hot drink, you can feel the heat radiating off the object. If you do the same with a cup of ice, you feel cold because your body is radiating heat from your body into the cooler cup of ice. On a sunny day, the sun will shine through a window and heat up any object that it shines on. Hydronic systems use similar modes of heat transfer, such as warming up the entire floor. Radiant floors have warm water circulating beneath them and as the floors heat up they radiate heat into the room.

Conduction

Conduction is the primary mode of heat transfer in and out of a building. Conduction is heat energy being transferred through the molecules of a material. This heat can also go through a series of objects that are in contact with each other. In the example of the radiant floor, the heat from the water is conducted into the pipe and then into the concrete floor. The concrete then conducts the heat into the finished floor. There are two consideration factors that determine how well heat will conduct through a series of object, such as a floor or wall—how well the individual materials conduct heat and how well they are in contact with each other. Some materials like copper are good conductors of heat while other materials are not. Polystyrene is an example of a material that does not conduct heat very well. As a general rule, metals conduct heat much faster than plastics.

Convection

Convection is a method of heat transfer that involves heat moving through fluid mediums. An example of convection would be how air warmed by a radiator will rise and circulate within a room. As warm air rises, cooler air will move in to take its place and once warmed, it too will rise.

System Sizing

Before a system can be sized, important data must be collected. This data includes information about the building and how it is to be used. Where the building is located will greatly affect the heating requirements of the building. A building in New York will have different heating requirements than if the exact same building were placed in Miami, Florida. The square footage of the building and the type of construction will also greatly affect the heat loss of the building. Walls, floors, and ceiling made with metal studs and building components would have a far greater temperature loss than ones built with wood frame construction. The type and amount of insulation will also greatly affect the thermal efficiency of the building. Another consideration that affects the heat load of the building is how the building will be used. A building with a large amount of computers or lights will not need as much heating as a building that is used for storage. The system designer typically will select a design water temperature and then size each of the terminal units to provide the necessary heat for each section of the building. Manufacturers of terminal units will rate the heating output of their unit according to the incoming water temperature and a specific flow rate. Today most contractors use software that takes the information listed above and calculates the heat loss for each area of the building. Figure 8-16 shows a screenshot of typical load calculation software.

FIGURE 8-16 Computerized design software can quickly produce information needed for component sizing.

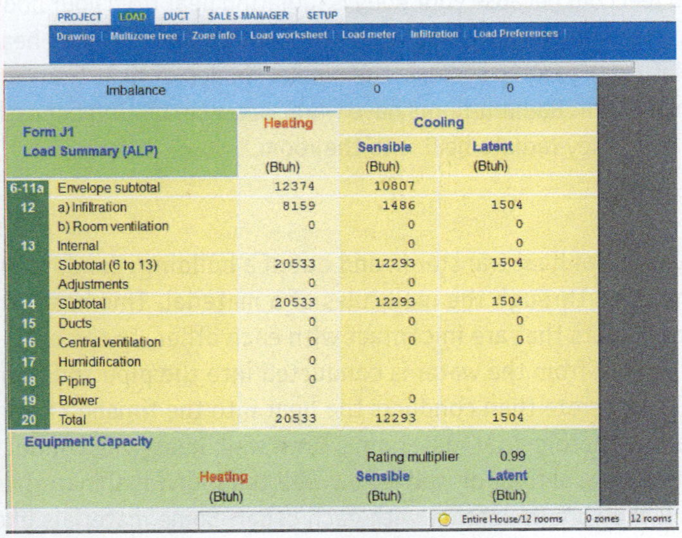

Source: Wrightsoft Corporation.

Heating Terminals

Once the heat loss for the building has been determined, the system designer would determine which type of heating terminal unit to use. The heating terminal units can be radiant floor heating, baseboard heating units, free standing fan coil units, duct mounted coils, or any combination of the above.

Baseboard Heating Units and Free Standing Fan Coils

Baseboard heating units and free standing fan coils are good in applications where more heating is required in one area of the building than another. The designer will select the appropriate number of units or the size of the units depending on how much heat loss that part of the building is expected to have. Manufacturers of terminal units will rate the heating output of their unit according to the incoming water temperature and a specific flow rate. The individual units can be spaced out according to how the building is being used and where there are the largest heat losses. These units are also more applicable in renovations of existing buildings, where it would be difficult to modify the flooring system. Both the baseboard and fan coil units are typically controlled by a thermostat that is mounted in the space. The thermostat can either control the fan or a control valve that controls the flow of water through the unit. See Figure 8-17.

Radiant Floor Heating

When a building is being constructed or if there is access beneath the existing floor, radiant floor heating systems work well. A radiant floor heating system has the advantage of placing the heat the closest to the occupants and can evenly distribute the heat through the building. Like the baseboard heating units and the fan coils, it

FIGURE 8-17 Thermostats can be used to control the circulating fan, the control valve, or both.

Photo by Ed Moore

also can be set up for zone control. Figure 8-18 shows an example of how a hydronic floor system can be laid out with individual zones that are each controlled by a thermostat. In this situation, the building occupants can adjust each area to fit their needs. Figure 8-19 shows a cutaway of a radiant floor heating system.

FIGURE 8-18 When the heating demands of the system vary, individual zones can be established to provide a higher level of comfort and efficiency.

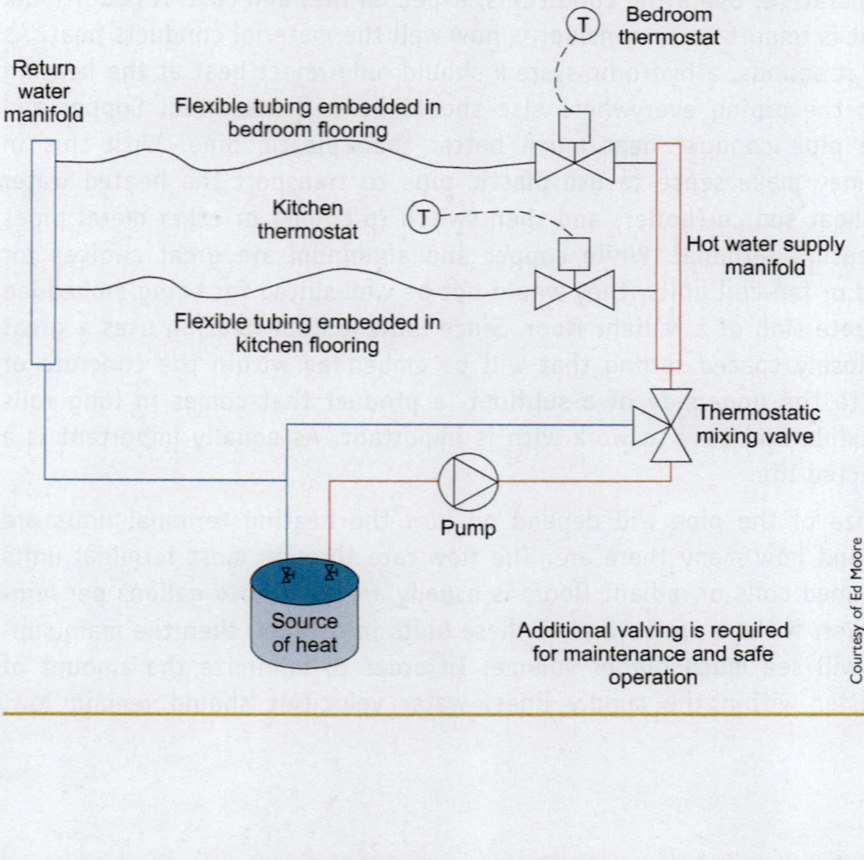

Courtesy of Ed Moore

FIGURE 8-19 The tubing of the hydronic heating system must be in contact with the flooring to assure good heat transfer. The insulation should be on the side opposite the living space.

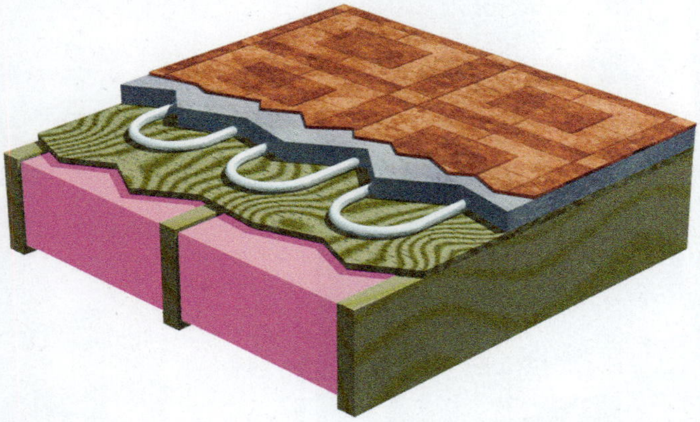

Components

Hydronic systems are comprised of many different components that all serve a distinct function. The following are some of the most basic components that are found in hydronic systems. Their size or level of control may vary depending on the size of the system, but their purpose is primarily the same.

Piping

There are a great number of piping materials available to plumb a hydronic system. Like any other system, some important factors to consider are pressure, temperature, operating conditions, expected life, and cost. An additional factor that is important to consider is how well the material conducts heat. As simple as it sounds, a hydronic system should only reject heat at the heating terminals; the piping everywhere else should be well insulated. Copper and aluminum pipe conduct heat much better than plastic pipe. With this in mind, it may make sense to use plastic pipe to transport the heated water from the heat source (boiler) and then switch to copper or other metal pipes at the heating terminal. While copper and aluminum are great choices for baseboard or fan coil units, they would not be well suited for being embedded in a concrete slab of a radiant floor. Since radiant floor heating uses a great deal of closely spaced tubing that will be embedded within the concrete or attached to the underside of a subfloor, a product that comes in long rolls and is flexible and easy to work with is important. As equally important is a long expected life.

The size of the pipe will depend on how the heating terminal units are arranged and how many there are. The flow rate through most terminal units such as finned coils or radiant floors is usually around 1 to 4 gallons per minute; however, if there are several of these units in parallel, then the main supply lines will see much higher volume. In order to minimize the amount of heat transfer within the supply lines, water velocities should remain low, approximately 4 to 5 feet per second. This will allow smooth laminar flow and less turbulence. The more turbulence, the more heat transfers through the wall of the pipe. Keeping to conservative flow rates will also reduce the amount of pressure drop through the pipes, which means there will be less pressure for the pump to overcome.

Thermal Expansion

As the water in the system heats up it will expand; and since water is incompressible there must be provisions made to allow for this, otherwise dangerously high pressures could exist. This expansion is usually provided for with an expansion tank. The actual size of the tank will depend on the amount of water in the system, the expected temperature fluctuations, and the system operating pressure. Manufacturers of expansion tanks provide charts to help determine the proper tank or number of tanks needed.

As mentioned earlier, the system operating pressure is usually low, around 12 to 15 pounds per square inch. The system pressure is directly related to the difference in heights of the piping within the system. As explained in *Plumbing 101*, as water is raised to an elevated height, pressure is developed at the base of the system. For every foot in elevation of the piping system, there will be approximately 0.433 psi generated. The pressure can be found through the following formula. The addition of 5 psi is a buffer to make sure that there will always be a positive pressure on the system.

$$P_e = \left(H\,ft. \times .433\frac{psi}{ft.}\right) + 5\ psi$$

$P_e =$ Pre-charge pressure in the expansion tank

$H =$ The highest point in the piping system, above the opening for the expansion tank

Another important consideration for the expansion tank is its location within the system. The expansion tank should be located near the suction side of the circulating pump. When the circulation pump starts, it will draw water from the low-pressure (suction) side and move it to the high-pressure (discharge) side. As the water moves from the suction line, a pressure drop would normally exist. Since the expansion tank has a positive pre-charge pressure, it will temporarily feed some of the water stored in it into the suction line. This will help maintain a positive pressure. If a negative pressure was allowed to exist, any air trapped in the system would form bubbles. As the bubbles move through the pump, they would go from the negative pressure side to positive pressure side and rapidly collapse. If they were in contact with the side of the pump, the surface tension between the pump surface and the air bubble would tear pieces of material away. This rapid collapse would cause pitting on the inside surface of the pump and eventually lead to premature failure.

Pumps

In early hydronic systems, the natural tendency for hot water to rise was the method to move water through the system. These systems located the heating source in the basement of the building and allowed the heated water to rise up through the piping system and once in the heating terminal reject its heat to the building. The cooler water would be denser and then settle back down toward the lower sections and become heated again. These systems were difficult to control and were limited in the height of building they could serve. To provide for bigger and more efficient systems, pumps were added to increase the flow of water.

Since most hydronic systems operate at low pressure, centrifugal pumps are the best choice. Centrifugal pumps are nonpositive displacement pumps that can generate a high amount of flow at low pressure. The pump relies on a curved impeller that directs the incoming water from the center of the pump to the outer shell of the main housing. As its name implies, centrifugal pumps use centrifugal force to sling the water outward. Centrifugal forces can be seen in something as simple as a weight on a string. If a weight is tied to a string and then spun in a circle, the weight would pull against the string and remain in a circular path. The faster the weight is spun, the more force pulling along the string. This is an important concept to understand because as the pump's speed is increased, its ability to produce flow will also increase. However, centrifugal pumps are nonpositive displacement pumps because they do not have internal mating surfaces that seal off between the low-pressure (suction) side and the high-pressure (discharge) side. This means that as the pressure increases, there will be a point at which the water will slip past the impeller instead of moving forward. At this point, the pump will maintain the pressure, but the flow will drop off to near zero. This condition is called "dead headed," which means the pump cannot overcome the resistance of the system. It will generate a great deal of heat and become noisy and could eventually damage itself. This same condition can also occur if the pump is run at too low of an rpm. At low speeds, there will not be enough centrifugal force to move the water and the pump will simply churn the water and overheat. See Figure 8-20 to see a cutaway of a centrifugal pump.

System Start Up

When a new hydronic system is first installed, the system will need to be pressure tested and flushed. Before pressure testing, make sure to valve off or remove any components that are not rated for high pressure. Examples of such components would be expansion tanks and pressure reducing valves. Once the system is flushed, it will need to be filled. Since some hydronic systems use some form of antifreeze, a small pump may be needed to fill the system. Figure 8-21 shows an example of filling a solar water heater with a food grade glycol, which is a form of antifreeze. Other additives such as rust inhibitors can be added; always check with the system manufacturer or the glycol supplier to know what is recommended or compatible.

Air Separators and Air Vent

As the system is being filled with the heat transfer medium, it is important to remove all trapped air. Air in the system will come from two sources. The first is air

FIGURE 8-20 Cutaway of a centrifugal circulator pump.

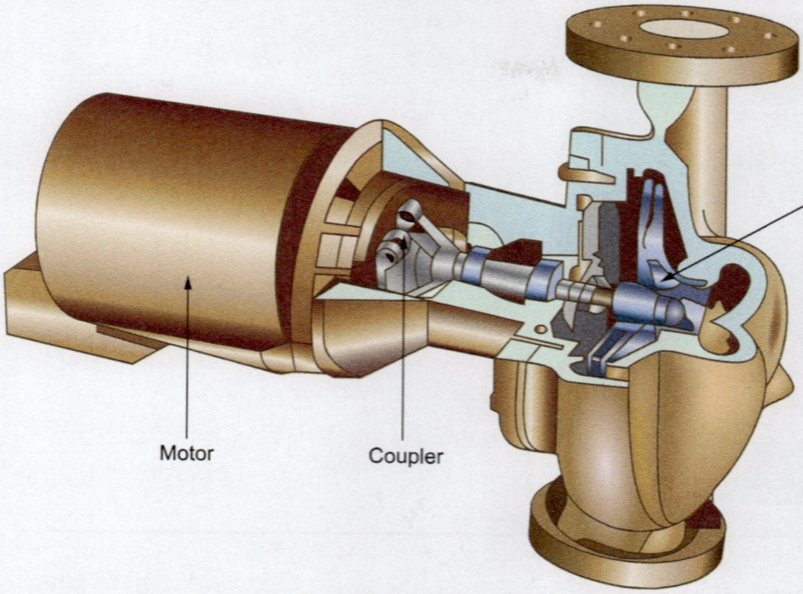

FIGURE 8-21 Hydronic heating systems must be filled and primed before start up. This system uses a food grade glycol mix to prevent freezing.

FIGURE 8-22 Automatic air vent.

Courtesy of Bell & Gossett

that is in the pipes prior to filling. This air can usually be vented out as the system is being filled. This is accomplished by placing air vents at high points along the system. The second source of air in the system is air that is dissolved in the cold water used to fill the system. As the water is heated, the air will separate out into micro bubbles. These are tiny bubbles of air that can cause all sorts of problems, such as noise and system inefficiency. Additionally, as described in the previous section on pumps, air trapped in the system can cause rapid wear within the pump through either cavitation or corrosion. At the elevated temperatures in system, the mixture of air and water will greatly increase the likelihood of corrosion and rapid wear of components. Air trapped in the system decreases system efficiency because it does not transfer heat as well as a liquid and it can prevent proper circulation. The trapped air will also increase the noise level of the system. To remove air from the system, it is advised to install air vents and air separators. Figure 8-22 shows a manually operated air vent that can be installed within the system.

Since air is lighter than water, it will tend to accumulate in the highest points throughout the system. Additionally, when water velocities are reduced to 2 ft./sec or less the water is incapable of carrying air bubbles with it. With this is mind, one of the most effective locations to vent air from the system is in the top of storage tanks (or boiler) or large manifolds located near the top of the system. A commercially available air separator that can be added to the system uses the principle of coalescence to separate the air from the water. *Coalescence* is a term that means the ability for smaller particles (or in this case bubbles) to combine into one larger bubble. A coalescence type air separator uses a stainless steel wire mesh to attract the smallest (micro) bubbles on

FIGURE 8-23 Air separator.

the mesh surface. These micro bubbles combine, making larger bubbles, which are much easier to separate from the water. This type of separator is especially suitable for radiant panel and other applications, where even small amounts of air are troublesome. Air separators of this type are usually made in small sizes, such as 1-inch to 2-inch pipe and require relatively low flow rates. This is typical of most HVAC applications. Figure 8-23 shows a typical air separator that has a built-in air vent to release the accumulated air.

Pressure Relief Valve

The American Society of Mechanical Engineers (ASME) Boiler and Pressure Vessel Code requires that all water heating appliances have a pressure relief valve. This pressure relief valve is required to be located within the top 6 inches of the water heater or boiler and is designed to protect the tank from rupturing due to overheating and expansion. Most temperature-and-pressure relief valves for standard residential and commercial water heaters are rated to open at a temperature of 210°F and 150 psi. These limits are well above the normal operating temperatures and pressures of a hydronic system. This means that an additional pressure relief valve is located somewhere within the hydronic system itself.

Figure 8-24 shows a typical pressure relief valve that is designed for low-pressure systems. This particular valve is rated to open if the system pressure reaches 30 psi. Do not confuse this valve with the pressure-reducing valve that was discussed earlier. A pressure-reducing valve only limits incoming water to a set level. Once it closes due to higher pressure, it will not reopen to relieve that pressure. This is important to remember when placing the pressure relief valve in the system.

FIGURE 8-24 Safety relief valve.

Courtesy of Bell & Gossett

REVIEW QUESTIONS

Short Answer

1. What two system requirements are necessary to size a tankless water heater?
2. What size (Btu/hr requirement) tankless water heater would be needed to supply 5.0 gpm at a temperature rise of 70°F? Assume 85 percent efficiency.
3. What size electric tankless water heater would be needed to supply a lavatory with a 0.75 gpm faucet? Assume a 70°F temperature rise and 95 percent efficiency.
4. What does the "first hour rating" of a water heater mean?
5. According to Table 8-3, what size of electric water heater is required for a 4-bedroom house with 2.5 baths?

True or False

6. __________ The anode rod is a sacrificial device that helps protect the tank.
7. __________ Water heaters made after 2004 can be installed directly on the floor, no matter what the code says.
8. __________ Energy cutoff devices are used on typical water heaters to stop the energy input in case of over-temperature of the water.
9. __________ The dip tube helps to get the maximum amount of hot water from the water heater before dilution occurs with incoming cold water.
10. __________ Automatic controls on gas-fired water heaters prevent main burner gas flow if a pilot flame is not present and proved.
11. __________ Generally, hot water is water with a temperature of 110°F or higher.
12. __________ Direct water heaters use heat from another source to direct heat to the potable water.
13. __________ Code always requires a vacuum breaker to be installed on a water heater.
14. __________ Scale can build up on the inside of a water heater and reduce the amount of heat transfer.
15. __________ Circulating hot water in a loop helps to minimize the waste of water.
16. __________ If the boiler in a hydronic system has a pressure relief valve, one is not needed for the hydronic system itself.
17. __________ Air separators are designed to remove air that was dissolved in the cold fill water.

Chapter 9

LEARNING OBJECTIVES

The student will:

- Describe types of backflow and how they occur.
- Describe the degree of backflow hazard.
- Describe types of mechanical equipment for cross-connection protection.
- Summarize methods of eliminating the risk of backflow.
- Explain the importance of identifying piping as an aid in guarding against backflow.

Cross-Connection Protection and Pipe Identification

BACKFLOW PREVENTION

The most important requirement of plumbing is to protect the potable water supply system. Under normal conditions, potable water will flow from the water source to the point of use through a network of pipes known as the water service. Once the water reaches its point of use, it is either consumed, returned to the environment (uncontaminated), or it may flow into a sewage system where it can be properly disposed of or recycled. Normally, the water service is pressurized so there is little chance for contaminants to enter; however under some conditions the system can experience fluctuations in pressure due to sudden changes in flow rates, system failures, or pressure from external systems improperly connected to it. Since we use water for a variety of reasons, it makes sense that the potable water system could be connected to other systems. For example, the water that fills the tank on a water closet is indirectly connected to the sewage system through the water closet itself. There are also numerous instances where water is used in systems, such as swimming pools, clothes washing machines, industrial processes, and food preparation. All of these instances are called cross-connections.

Cross-connection is the general term applied to any arrangement that has the potential to violate the safety of the potable water system. A cross-connection, according to the 2012 National Standard Plumbing Code, is

> *Any connection or arrangement between two otherwise separate piping systems, one that contains potable water and the other either water of questionable safety, steam, gas, or chemical, whereby there may be a flow from one system to the other, the direction of flow depending on the pressure differential between the two systems.*

There are three recognized methods of producing a pressure differential that can result in contamination entering the potable system. These methods are described below.

Gravity Backflow

Gravity backflow occurs when water is at a higher elevation. Remember that water is affected by gravity so it will always want to go downstream.

Back-Pressure Backflow

Back-pressure backflow is a reversal of normal flow in a system due to downstream pressure being greater than the supply pressure. In the case with back-pressure backflow, both systems are at a pressure that is above atmospheric pressure. For example, if a check valve on a boiler feedwater system fails in the open position and the water supply pressure were to drop to a lower pressure than the boiler, then treated boiler water would overpressure the potable water supply and push its way into the potable water supply.

Back-Siphonage Backflow

Back-siphonage backflow is a reversal of the normal flow in a system caused by a negative pressure (vacuum) in the potable supply. A vacuum can develop at elevated fixtures whenever there are very large flows in the street mains, such as firefighting in the area or a water main rupture. Back-siphonage can also occur when the water supply inlet is submerged in a container. For example, if a garden hose end was submerged into a bucket of water, a vacuum in the supply main would siphon the water out of the bucket.

Sometimes back-siphonage can occur due to the piping arrangement. Whenever a large pipe transitions to a smaller diameter pipe and then back to a larger pipe, there will be an increase in water velocity and a dramatic decrease in static pressure. This reduction in pipe diameter could also be a kink or some other restriction in the piping. If there is an opening in this section of pipe, fluids can be drawn into the pipe. This is actually the operating principle that allows a venturi to work.

In practice, the three causes of backflow are really one—backflow due to differences in pressure. We choose to think of these three causes separately because of the different equipment types available to guard against backflow and the many piping arrangements that can lead to problems. Consider these examples:

- Codes require that the water service and building sewer be separated by a suitable distance. Assume that a situation occurs where both the water supply and the building sewer were damaged and leaking. Water from the water service would flood the soil, but when the pressure is removed to make the repairs, the contaminants in the flooded soil could backflow into the now depressurized water service pipe.
- Any pressurized devices (boilers, pressure tanks, stills, refrigeration systems, etc.) with potable water connections can produce back-pressure backflow under abnormal conditions.
- A nonapproved ballcock could siphon the contents of the water closet tank into the potable system if a subatmospheric pressure occurs in the supply.
- Back-siphonage can also occur if a hose is left lying in a puddle, pond, swimming pool, or bucket containing chemicals or undesirable solutions. When the faucet is opened, a subatmospheric pressure exists in the potable supply.
- Lawn sprinkler systems offer another possibility. Lawn chemicals could be drawn back into the potable system when system pressure is off.

DEGREE OF HAZARD

Before describing ways to guard against the hazards of backflow, we must consider the idea of degree of hazard. Consider these definitions:

Toxic Substance

A toxic substance is a fluid, solid, or gas that, when added to the potable supply, is detrimental to health.

Nontoxic Substance

A nontoxic substance is any product that, when added to the potable system, is found to create a moderate hazard, nuisance, or an unpleasant condition.

Polluted

A polluted supply system contains a nontoxic substance. The polluted system is impaired in taste, odor, or color but is not deemed a health hazard. A polluted system is referred to as a low-hazard risk.

Contaminated

A contaminated supply system contains a dangerous/toxic substance that is a health hazard.

The level of backflow protection is usually influenced by our judgment about the degree of hazard.

CROSS-CONNECTION PROTECTION

Five methods or devices used to prevent or protect against cross-connections are listed next. See Table 9-1 for a summary.

Air Gap

An air gap is the unobstructed vertical distance through the free atmosphere between the flood level rim of the receptor and the lowest opening from any pipe or faucet supplying water to a tank, plumbing fixture, or other device. With this arrangement, the contents of the tank or fixture cannot be drawn or forced back into the potable water system (see Figure 9-1).

The air gap is the simplest and most effective protection against cross-connection. The air gap can be used on the supply connection to receptors containing any toxic substance.

The vertical distance between the supply pipe and the flood level rim should be at least two times the diameter of the effective opening of the supply pipe outlet and never less than 1 inch. If the supply outlet is within three diameters of a wall or similar vertical surface, the air gap must be three times the diameter of the effective opening. Table 10.5.2 of the 2012 National Standard Plumbing Code gives several examples of this requirement and other backflow preventers.

TABLE 9-1 Backflow Protection*

Cross-Connection Piping Configuration	Air Gap	Reduced Pressure Zone Backflow	Pressure Type Vacuum Breaker	Atmospheric Vacuum Breaker	Double Check Valve Assembly
Low inlet to receptors containing toxic substances (vats, storage containers, plumbing fixtures)	X	X	X	X	
Low inlet to receptors containing nontoxic substances (steam, air, food, beverages, etc.)	X	X	X	X	X
Outlets with hose attachments that may constitute a cross-connection	X	X	X	X	X
Coils or jackets used as heat exchangers in compressors, degreasers, and other such equipment involving toxic substances	X	X			
Direct connections subject to back pressure:					
Toxic	X	X			
Nontoxic	X	X			
Sewage & lethal substances	X				

*Acceptable under National Standard Plumbing Code. Consult your local code for acceptance.

FIGURE 9-1 Air gaps are the most dependable of all backflow protection methods. Notice that spouts cannot have threaded ends.

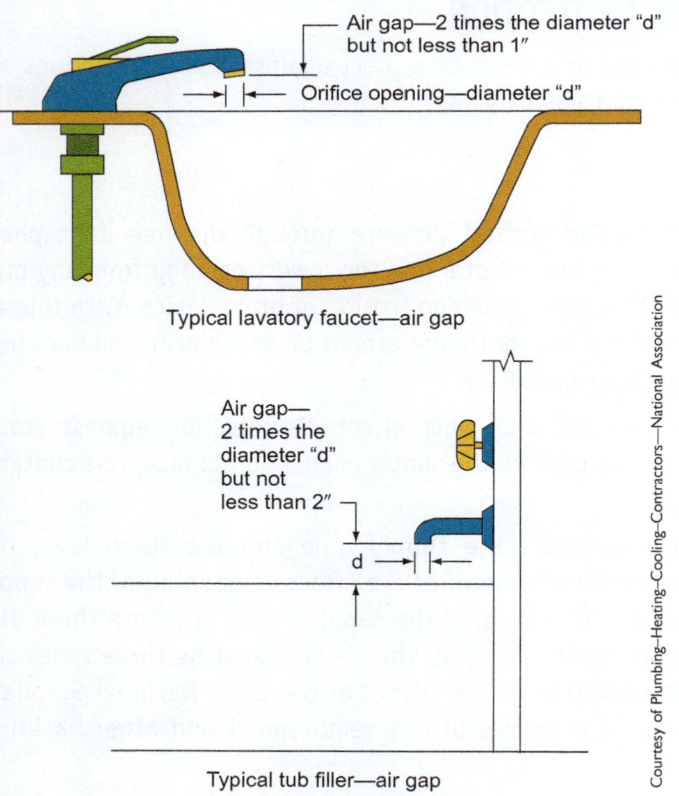

Some appliances, such as most dishwashers and automatic clothes washers, are fitted with internal air gap fittings to protect the potable water supply.

Reduced Pressure Zone Backflow Preventer

The reduced pressure zone (RPZ) backflow preventer is a single device that incorporates a check valve in the inlet, a vented chamber in the middle, and a check valve in the outlet. The check valves are spring-loaded to the closed position. RPZ devices include test ports to determine the operating condition of all the component parts. Figure 9-2 shows a common RPZ.

These devices protect against back-pressure or back-siphonage backflow, but they are costly, complicated, and require servicing. They must also be tested frequently. Most codes require a minimum of an annual test.

A strainer is recommended on the inlet to prevent fouling of the check valves.

RPZ devices may be used on all direct connections that might be subject to back-pressure or back-siphonage and where there is the possibility of contamination by a material that constitutes a potential health hazard.

An important fact to keep in mind with RPZs is that, because of the spring-loaded check valves and the need for a pressure drop across these products at all times in normal operation, they produce a 9–24 psi drop in pressure in the water line. In some conditions, where the line pressure is close to the minimum for satisfactory performance, the RPZ-type backflow preventer may require the use of a booster pump. RPZs cannot be installed in pits or other areas subject to flooding, and they must be installed 12 inches to 60 inches above the finished floor, finished grade, or flood level zone.

Double Check Valve Assembly

A double check valve assembly is a single device that incorporates two check valves, spring-loaded to a closed position, in a series flow arrangement. The product is equipped with test cocks and upstream and downstream valves for testing. Figure 9-3 shows an example.

FIGURE 9-2 This reduced pressure zone (RPZ) backflow preventer is located on the incoming water supply line for a public building.

Photo by Ed Moore

FIGURE 9-3 For non–health hazard conditions, a double check valve assembly can be used.

Photo by Ed Moore

Double check valve assemblies are only used on low-hazard risks. An inlet strainer may also be used, and inlet and outlet valves are required. Double check valve assemblies may be used as protection for all direct connections through which foreign material, such as air, steam, food, or other material that does not constitute a health hazard might enter the potable system in concentrations that would constitute a nuisance or be aesthetically objectionable.

It should also be pointed out that some areas have adopted the use of smaller residential dual check valves that operate in the same manner as the above double check valve, but are made without the test ports and are more economical. They are typically installed directly after the water meter (see Figure 9-4).

Pressure-Type Vacuum Breaker

A pressure-type vacuum breaker contains a spring-loaded check device in the flow path and a chamber vented to atmosphere. During normal operation, the check valve is open or closed (depending on flow state) and the vented chamber is closed. Figure 9-5 shows a typical pressure-type vacuum breaker. In this diagram, water enters the bottom ball valve and leaves via the horizontal ball valve.

Pressure-type vacuum breakers may be used as protection for connections to all types of nonpotable systems where the vacuum breakers are not subject to backpressure. These devices are suitable for applications that involve continuous supply pressure. They must be installed at least 12 inches above the highest downstream piping, opening, or usage. Traditional designs were normally used outdoors, or where spillage from the vent would not cause damage. However, new spill-resistant pressure-type vacuum breakers are available for indoor applications.

FIGURE 9-4 In some locations, residential dual check valves are becoming a standard requirement when connecting to the public water system.

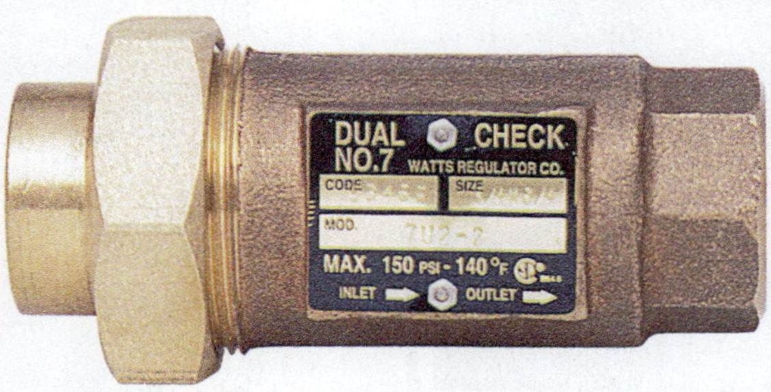

Photo courtesy of Watts Regulator Company

FIGURE 9-5 Pressure-type vacuum breakers are rated for continuous pressure.

Photo courtesy of Watts Regulator Company

Backflow preventers with intermediate atmospheric vents may be used as acceptable alternates for $\frac{1}{2}$ inch and $\frac{3}{4}$ inch pressure-type vacuum breakers and will also provide protection against back pressure. Always check your local code to determine whether a particular backflow preventer is acceptable for use in the application you are installing or servicing.

Atmospheric-Type Vacuum Breaker

An atmospheric-type vacuum breaker uses a float valve, which means that it can only be used where the contained water is basically at atmospheric pressure. Therefore, it must be installed after the last control valve to the fixture or system to be isolated. Figure 9-6 shows a typical example.

Atmospheric-type vacuum breakers may be used only on connections to a nonpotable system, where the vacuum breaker is never subjected to back-pressure and is installed on the discharge side of the last control valve. The breaker is normally 6 inches above the highest point of the fixture or system being isolated but can be as low as 1 inch for products tested to that lower range. It cannot be used under continuous pressure. The pressure should not be "on" to the valve more than 12 hours per day. Prolonged, continuous exposure to pressure could cause the checking member to remain attached to the vent port, which negates the function of the valve.

Hose connection vacuum breakers may be used on sill cocks and service sinks (see Figure 9-7). They are not required on hose-threaded outlets such as water heater tank drains and washing machine valves. A hose-type atmospheric vacuum breaker may not allow proper drainage of exterior faucets. This may cause freezing of the faucet and piping.

Atmospheric-type vacuum breakers should never be installed in pits or similar locations that could become submerged. Devices that vent to atmosphere should not be installed in exhaust hoods or areas where the air quality is questionable. All devices should be readily accessible for servicing and testing.

Vented devices should not be installed where spillage could cause structural or equipment damage. Thus, suitable surfaces and floor drains or other disposal facilities should be close by.

Devices should not be subjected to freezing.

FIGURE 9-6 Atmospheric-type vacuum breakers are not rated for continuous pressure applications.

Courtesy of Cash Acme

FIGURE 9-7 Atmospheric hose vacuum breakers will discharge water when there is a pressure imbalance, so make sure the water discharge will not cause a mess or damage to the structure.

Photo courtesy of Watts Regulator Company

TABLE 9-2 Size of Color Bands and Letter Heights

Outside Diameter (OD) of Pipe or Pipe Covering in Inches	Width of Colored Bands in Inches	Height of Letters in Inches
$\frac{3}{4}$–$1\frac{1}{4}$	8	$\frac{1}{2}$
$1\frac{1}{2}$–2	8	$\frac{3}{4}$
$2\frac{1}{2}$–6	12	$1\frac{1}{4}$
8–10	24	$2\frac{1}{2}$
Over 10	32	$3\frac{1}{2}$

Note: check most recent ANSI A13.1 for current data. Local code requirements supersede these data.

PIPE IDENTIFICATION

To minimize the chance for error in connecting two pipelines or taking branches from mains, it is recommended that all extensive piping systems be marked according to the ANSI Standard No. A13.1-2007, *Scheme for the Identification of Piping Systems*. This standard calls for piping to be identified as to content by four broad categories, using lettering and color-coded markers attached to the piping. The categories and codes are as follows:

- F category is for fire-protection material and equipment, such as hydrants, alarm boxes, fire doors, piping, conduit, and so on. Paint is red with white lettering.
- D category is for dangerous materials. Flammable, toxic, and corrosive materials, such as acids, alcohol, ammonia, blow-off water, feedwater, carbon dioxide and monoxide, chlorine, fuels, and steam are in this category. Paint is yellow with black lettering.
- S category is for safe materials, such as air, carbonated water, domestic water, plumbing vents, most refrigerants, and many gases. Paint is green, black, aluminum, or gray with black lettering (white if background is black).
- P category is for protective materials. These materials are used to minimize the danger of toxic or lethal materials. Examples are antidote gas, calcium chloride, or filtered water. Paint is blue with white lettering.

See Table 9-2 for a summary and for the sizes of bands and lettering for various pipe sizes.

It is important that piping in any large building be marked at frequent intervals. In case of emergency, such markings could save time and minimize the chance of injury or property loss.

REVIEW QUESTIONS

Short Answer

1. Name the three commonly recognized causes of backflow.
2. Which backflow prevention method is considered the simplest and most effective of all the methods?
3. If a faucet with a $\frac{1}{2}$-inch opening is located 4 inches from an adjacent wall, what is the required air gap distance?
4. Give one location in which an RPZ should not be installed.
5. According to the ANSI code, what color should a fire sprinkler line be colored?

Chapter 10

LEARNING OBJECTIVES

The student will:

- Contrast the different types of sewage disposal and aeration systems.
- Explain the basic processes for large systems and private sewage disposal systems.
- List the allowable materials for building sewers.
- Describe typical installation practices for building sewers.

Sewers & Sewage Disposal Methods

SEWAGE DISPOSAL

Sewage is defined in the ASSE dictionary as *the liquid waste conducted away from residences, business buildings, or institutions, together with those from industrial establishments, and with such ground, surface, and storm water present. Also included is any liquid waste containing animal or vegetable matter in suspension or solution, which may include liquids containing chemicals in solution.*

Thus, sewage is liquid-borne undesirable or questionable material that commonly contains two broad types of materials—organic and inorganic. Organic materials include human waste products and food products. Inorganic materials found in sewage include dirt or grit that is introduced into the system in random ways or from precipitates from the water chemicals. The organic materials are subject to change and give rise to gases and other breakdown chemicals. Organic chemicals are based on carbon compounds in one form or another.

Sewage treatment is required to remove or neutralize all of the noxious and hazardous materials that are found in sewage. The treatment converts these materials into harmless products or removes and disposes of the materials that cannot be converted (principally the inorganic components).

Organic materials are broken down by bacterial action. Two categories of this action are known as aerobic and anaerobic.

Aerobic bacteria are active only in the presence of oxygen. The process is fast and does not produce odors but requires oxygenation of the sewage. This process takes place in the sewers as well as in the treatment plant, which is a benefit to the entire system and is why ventilation is so important.

If the sewage mixture becomes oxygen-deficient, breakdown of the organic material continues, but the action becomes anaerobic. The anaerobic bacteria process is slower than the aerobic process and usually results in offensive odors.

Three stages of sewage breakdown actions are usually found:

1. Fresh sewage with high oxygen content that is undergoing aerobic processes.
2. Stale sewage with very little oxygen present, but where putrefaction has not yet occurred.
3. Septic sewage that is undergoing total anaerobic activity.

Putrefaction, according to the ASSE dictionary, *is the decomposition of organic matter by the agency of bacteria and fungi with the formation of foul smelling, incompletely oxidized products; decay*. In public sewage disposal systems, the sewage mixture is kept oxygenated, so this stage does not develop. However, smaller systems, and especially private systems, operate extensively within this third step.

Many of the bacteria that promote the breakdown of sewage components are harmful to human beings. Therefore, we must foster conditions that encourage the growth of these bacteria in appropriate locations that will prevent these products from gaining entry into the potable water system.

It is usually necessary to purify the final effluent from the treatment plant before that effluent is released to natural waterways.

PUBLIC SYSTEMS

Public or municipal systems are made up of extensive systems of mains and branches, properly sloped to the treatment plant. By the time the sewage travels to the plant, the decomposition process is well underway. As the flow gets closer to the treatment plant, the pipe sizes increase in this layout to accommodate the increasing load.

At the treatment plant, the chemical processes are completed and the effluent is treated before being discharged into the waterway. The products in the incoming sewage that will not become liquids—materials that are inactive solids—end up as sludge. Think of this as a kind of mud. This sludge is dried and then disposed of in landfills. Some treatment plants sell the dried sludge as a fertilizer or compost material, thus raising revenue for the treatment plant and reducing the volume of dried material to be discarded. In some areas, the inert sludge is spread over pastures or crop fields as fertilizer. This does require special equipment and posting a public notice.

Complete public sewage treatment processes include these steps:

1. Primary
2. Settling
3. Aeration
4. Secondary settling
5. Filtration
6. Chlorination or other form of final treatment before discharge into a waterway

The treatment plant operators are responsible for reporting to government offices routine chemistry levels at several points on a prescribed cycle of observation. The operators are also required to report at once if any of several critical materials exceeds certain limits.

These government regulations are subject to change as the scientific community discovers new limits that are appropriate for safety of people, wildlife, and vegetation in the vicinity of the treatment plant.

It is difficult for the layman to get a proper perspective for the precision and sensitivity of these measurements. The limitations involve small numbers of parts per million—in some cases, parts per billion. This is compounded by the fact that, in some cases, millions of gallons of sewage are being treated, all of varying composition.

PRIVATE SYSTEMS

Private systems must be used for any building that cannot be connected to a public system. These private systems are sized to accommodate the building served. They are usually based on the septic tank—a watertight tank where sewage decomposes—and a leaching field or leaching pit to dispose of the effluent. The sludge that accumulates in the tank must be removed periodically.

An alternate arrangement uses a cesspool to receive the sewage and hold it for breakdown. The walls of the cesspool are not watertight, so the contents of the tank slowly seep into the surrounding earth. These are unsatisfactory for general use because there is no assurance that the sewage is broken down before seepage into the earth occurs.

A new variation on the septic tank uses a blower system to aerate the sewage, producing aerobic action. The only difficulty with this system is the chance of equipment failure or power interruption. If these conditions are guarded against, these make excellent systems.

Private systems require proper installation; this means that separation distances between all components of the private system and items such as potable water sources and property lines must be maintained (see Table 10-1). Particular attention must be paid to the following points:

- All pipe joints (except in the leaching field) must be watertight.
- Cleanouts should be no more than 75 feet apart.
- All tanks should be marked with some permanent item in the ground.
- The septic tank should be accessible for periodic removal of contents by pumping.
- The inlet pipe should have a uniform slope of about $\frac{1}{4}$ inch per foot to minimize churning of the tank contents.
- The upper half of effluent distribution line joints must be covered with tar paper or plastic to prevent dirt from entering these pipes.
- All piping must be supported properly.
- Only clean backfill should be used to cover the piping. Tamp the backfill in layers as it is being placed.
- Leaching fields must not be placed in a location that can permit effluent to enter the water table.

TABLE 10-1 Minimum Distance Between Components of an Individual Sewage Disposal System (in ft.)*

	Shallow Well	Deep Well	Single Suction Line	Septic Tank	Distribution Box	Disposal Field	Seepage Pit	Dry Well	Property Line	Building
Bldg. Sewer, Other than Cast Iron	50	50	50							
Bldg. Sewer, Cast Iron	10	10	10							
Septic Tank	100	50	50		5	10	10	10	10	10
Distribution Box	100	50	50	5			5	5	10	20
Disposal Field	100	50	50	10	5				10	20
Seepage Pit	100	50	50	10	5				10	20
Dry Well	100	50	50	10	5				10	20
Shallow Well				100	100	100	100	100		
Deep Well				50	50	50	50	50		
Suction Line				50	50	50	50	50		

*May be closer to building when permission is given by the local health department or other authority having jurisdiction.

Source: Table 16.4.3 2012 National Standard Plumbing Code.

Septic Tank Detail

The septic tank must hold the waste long enough for the conversion process to take place. This usually means that the tank must be large enough to hold about 24 hours of effluent from the system.

Most modern tanks are prefabricated from concrete, steel, or plastic. However, since the average life expectancy of a steel tank is only 7 years, most jurisdictions no longer allow them to be installed. Site-built tanks can also be used, especially for larger jobs. Figure 10-1 shows the cross-sectional view of a typical septic tank.

Since the conversion process taking place in a septic tank is anaerobic, the tank must have a means to vent the gases. The building plumbing vent system is also venting the tank, so these odors are released above the roof line.

The tanks usually contain a baffle to divide the tank into two compartments. The first compartment is a settling area for solids. In the second compartment, most of the bacterial conversion process takes place.

The inlet and outlet of the tank may be fitted with a sanitary tee, installed vertically or with baffles, so that the inlet fluids are discharged below the surface and effluent is drawn from below the surface. This entry method does not churn up the scum layer floating at the top of the tank contents and assures that only liquid flows out of the tank, rather than any of the floating scum. The open top of the tee vents the lines so that they will not become air-bound. The outlet is always 2 inches to 4 inches below the inlet elevation.

For residential applications, the tank size is selected based on the number of bedrooms or apartment units in the building. For other applications, the tank is sized to be equal to or greater than the total volume of sewage discharged in a 24-hour period. In any case, no septic tank should be smaller than 1,000 gallons.

FIGURE 10-1 Interior of a typical septic tank. The water level should not reach the top. If so, there is a problem with the discharge line or the leach field.

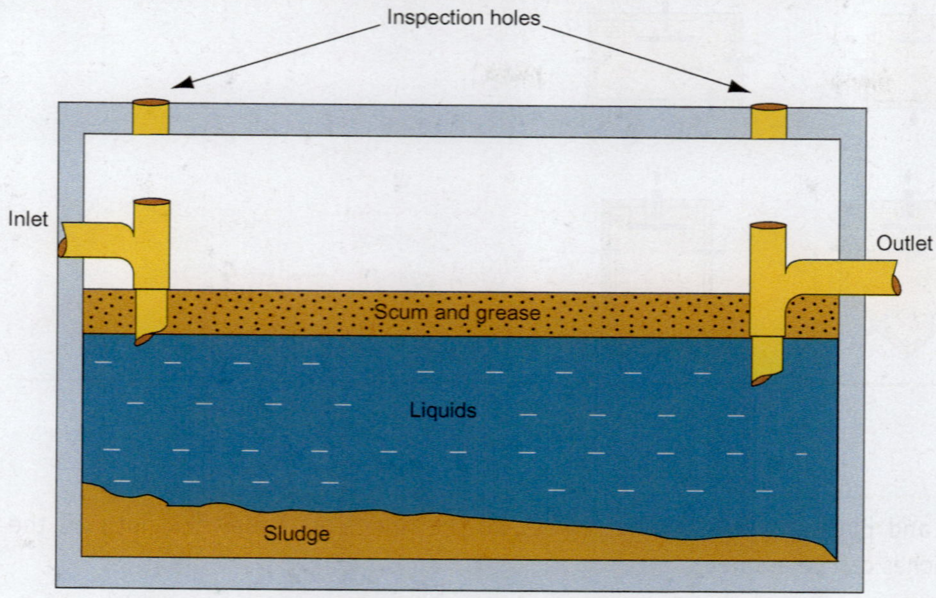

As pointed out in Table 10-1, the tank should also be a minimum of 10 feet from the building. The tank and leaching field should be positioned in an area that will receive very little heavy traffic and slopes slightly away from the building. For buildings that have both private water supply and sewage systems, local codes require a minimum separation distance between the two. The soil type and rock formations in your area may result in greater distance requirements, but if there is no regulation, maintain at least 50 feet between any well and the septic tank or any part of the disposal field. There are many specific clearances for these items, so check local codes for exact values. Remember that this separation applies to well and septic systems within the same property lines as well as to those systems in adjoining properties.

Septic tanks should be installed as shallow as possible to assure satisfactory performance. The actual depth required will be determined by the elevation of the building drain where it exits the building and the slope of the building sewer, but the depth of cover over the tank should not exceed 36 inches.

Distribution Box

The second component of the private system is the distribution box. Typical layouts for distribution boxes are shown in Figure 10-2. The effluent from the septic tank flows through piping with watertight joints to the distribution box, which is a box set in the ground to which the individual lines of the leaching system are connected. Small systems may not require these boxes, but consider a usual problem: if multiple disposal paths are used, it is difficult to have each path share equally in the effluent coming from the septic tank. By connecting all the parallel paths to the distribution

FIGURE 10-2 Distribution box designs.

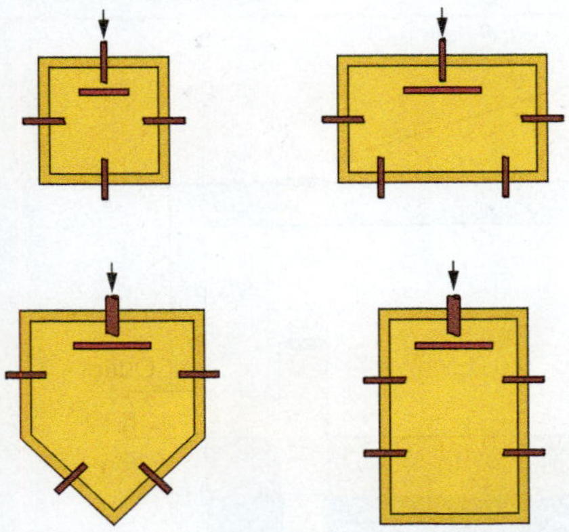

box and placing a level barrier across all the openings, equal flow into all the branches can be assured.

A further improvement entails the use of an alternating valve and two full-capacity disposal fields and distribution boxes. This arrangement permits use of a full-capacity field for a year, then alternating the field in use for the next year. This installation provides as near a trouble-free system as can be developed.

Disposal Fields

The third part of the private system is the disposal field, leaching field, or absorption field. The field is made with a 4 inch or larger open-jointed or perforated pipe, placed in trenches 18 inches to 36 inches wide. The trenches are filled with gravel around the pipe. The trenches should be installed a minimum of 7 feet 6 inches on centers and a maximum of 4 feet deep. Note that the 7 feet 6 inches center-line dimension is required for satisfactory liquid disposal. If the field is dug with a conventional backhoe, however, the trenches will have to be about 10-foot centers because of the need for solid support of the backhoe during the trenching operation.

The total length of trench required is determined by the effluent quantity and the ability of the soil to absorb water from the drain pipes. A properly operating system will drain the effluent from the septic tank into the ground at the same rate as the septic tank receives waste material.

The capability of the earth to accept the effluent is determined by a sampling procedure known as the percolation test, or perc test. The test involves digging several holes (a minimum of five) of approximately 12" × 12" cross-section and depth of 12 inches to 24 inches in the area of the proposed disposal field. Place at least 7 inches of water in the holes and allow the water to soak away. Later the same day or the next day, again place 6 inches to 7 inches of water in the hole. After the water has dropped about 1 inch, record the time for the next 1 inch of water to percolate away. The pre-wetting and waiting for the first inch to dissipate assures that the absorption has reached equilibrium.

The septic system is sized as follows:

1. The amount of sewage flow is determined by Table 10-2.
2. The size of the septic tank is selected using Table 10-3.
3. The drainage lines in the sewage disposal fields are sized using Table 10-4.
4. If a seepage pit is used, apply Table 10-5.

If the percolation time is excessive (greater than 30 minutes), consideration should be given to dry wells (or leaching pits) to leach the water into the earth.

TABLE 10-2 Sewage Flows According to Type of Establishment*

Type of Establishment	Sewage Flow
Schools (toilets and lavatories only)	15 gallons per day per person
Schools (with above plus cafeteria)	25 gallons per day per person
Schools (with above plus cafeteria and showers)	35 gallons per day per person
Day workers at schools and offices	15 gallons per day per person
Day camps	25 gallons per day per person
Trailer parks or tourist camps (with built-in bath)	50 gallons per day per person
Trailer parks or tourist camps (with central bathhouse)	35 gallons per day per person
Work or construction camps	50 gallons per day per person
Public picnic parks (toilet wastes only)	5 gallons per day per person
Public picnic parks (bathhouse, showers, and flush toilets)	10 gallons per day per person
Swimming pools and beaches	10 gallons per day per person
Country clubs	25 gallons per day per person
Luxury residences and estates	150 gallons per day per person
Rooming houses	40 gallons per day per person
Boarding houses	50 gallons per day per person
Hotels (with connecting baths)	50 gallons per day per person
Hotels (with private baths—2 persons per room)	100 gallons per day per person
Boarding schools	100 gallons per day per person
Factories (gallons per person per shift—exclusive of industrial wastes)	25 gallons per day per person
Nursing homes	75 gallons per day per person
General hospitals	150 gallons per day per person
Public institutions (other than hospitals)	100 gallons per day per person
Restaurants (toilet and kitchen wastes per unit of serving capacity)	25 gallons per day per person
Kitchen wastes from hotels, camps, boarding houses, etc. serving three meals per day	10 gallons per day per person
Motels	50 gallons per day per bed space
Motels with bath, toilet, and kitchen wastes	60 gallons per day per bed space
Drive-in theaters	5 gallons per day per car space
Stores	400 gallons per day per toilet room
Service stations	10 gallons per day per vehicle served

(continued)

TABLE 10-2 Sewage Flows According to Type of Establishment* (Continued)

Type of Establishment	Sewage Flow
Airport terminals	3–5 gallons per day per passenger
Assembly halls	2 gallons per day per seat
Bowling alleys	75 gallons per day per lane
Churches (small)	3–5 gallons per day per sanctuary seat
Churches (large with kitchen)	5–7 gallons per day per sanctuary seat
Dance halls	2 gallons per day per person
Laundries (coin operated)	400 gallons per day per machine
Service stations	1,000 gallons (first bay) per day 500 gallons (each additional bay)
Subdivisions or individual homes	75 gallons per day per person
Marinas—flush toilets	36 gallons per fixture per hr
Marinas—urinals	10 gallons per fixture per hr
Marinas—wash basins	15 gallons per fixture per hr
Marinas—showers	150 gallons per fixture per hr

*Source: Table 16.3.7 2012 National Standard Plumbing Code.

TABLE 10-3 Capacity of Septic Tanks*

Single Family Dwelling—Number of Bedrooms	Multiple Dwelling Units or Apartments—One Bedroom Each	Other Uses—Maximum Fixture Units Served	Minimum Septic Tank Capacity in Gallons
1–3		20	1,000
4	2 units	25	1,200
5 or 6	3 units	33	1,500
7 or 8	4 units	45	2,000
	5 units	55	2,250
	6 units	60	2,500
	7 units	70	2,750
	8 units	80	3,000
	9 units	90	3,250
	10 units	100	3,500

Extra bedroom, 150 gallons each.
Extra dwelling units over 10, 250 gallons each.
Extra fixture units over 100, 25 gallons per fixture unit.

Note: Septic tank sizes in this table include sludge storage capacity and the connection of domestic food waste disposer units without further volume increase.

*Source: Table 16.6.1 2012 National Standard Plumbing Code.

These units usually require special permission of the authority having jurisdiction (building inspector, health department, or other governmental agency responsible), but in some soil types, they are the only device that works.

TABLE 10-4 Required Length of Trenches (in ft.) in Sewage Disposal Fields per 100 Gallons of Sewage per Day

	Trench Width		
Time in Minutes for 1 Inch Drop	1 ft.	2 ft.	3 ft.
1	25	13	9
2	30	15	10
3	35	18	12
5	42	21	14
10	59	30	20
15	74	37	25
20	91	46	31
25	105	53	35
30	125	63	42

Source: Table 16.5.4 2012 National Standard Plumbing Code.

TABLE 10-5 Required Absorption Area in Seepage Pits per 100 Gallons of Sewage per Day

Time in Minutes for 1 Inch Drop	Effective Absorption Area in Square Feet
1	32
2	40
3	45
5	56
10	75
15	96
20	108
25	139
30	167

Source: Table16.5.6 2012 National Standard Plumbing Code.

EXAMPLE 1

How much sewage flow (in gallons) would be expected for an elementary school that is planned to have 400 students in attendance and a support staff of 25 adults? Assume the school has a cafeteria, toilets, and lavatories.

Table 10-2:

$$\text{Gallons of sewage flow} = \frac{25 \text{ gallons}}{\text{person}}(425 \text{ persons}) = 10{,}625 \text{ gallons}$$

EXAMPLE 2

What is the minimum size of the septic tank that can be used on a 4-bedroom single family dwelling?

Table 10-3: 1,200 gallons

EXAMPLE 3

A leaching field is being installed with 2-foot-wide trenches on a piece of land that had a percolation test result of 2 minutes per 1 inch drop. If the leach field is expected to handle 300 gallons of sewage flow per day, what length runs would be required?

Table 10-4: 2-foot-wide, 2-minute test: therefore 15 ft. per 100 gallons

$$\text{Length} = 300 \text{ gallons}\left(\frac{15 \text{ ft.}}{100 \text{ gallons}}\right) = 45 \text{ ft.}$$

Dosing Siphon and Dosing Tank

When leaching fields exceed 500 feet in length, a popular method of evenly distributing the effluent is the use of a dosing siphon and dosing tank. Under normal conditions, only the first section of the leaching field would be used because of the slow but continuous flow from the septic tank. The dosing tank arrangement acts to hold the effluent in the septic tank (increasing the time for bacterial action to take place) until the liquid runs over in the dosing siphon. The liquid flow then continues until the siphon is broken in the dosing unit, flooding the disposal field and assuring that the entire field is actively percolating effluent into the earth. By supplying the leaching field in a shotlike manner, there will be periods of time when there is no flow. This will give the effluent time to seep into the ground. Figure 10-3 shows a dosing siphon device and dosing tank installed as part of a septic tank.

FIGURE 10-3 Dosing siphons and tanks are used on large fields to saturate them more evenly.

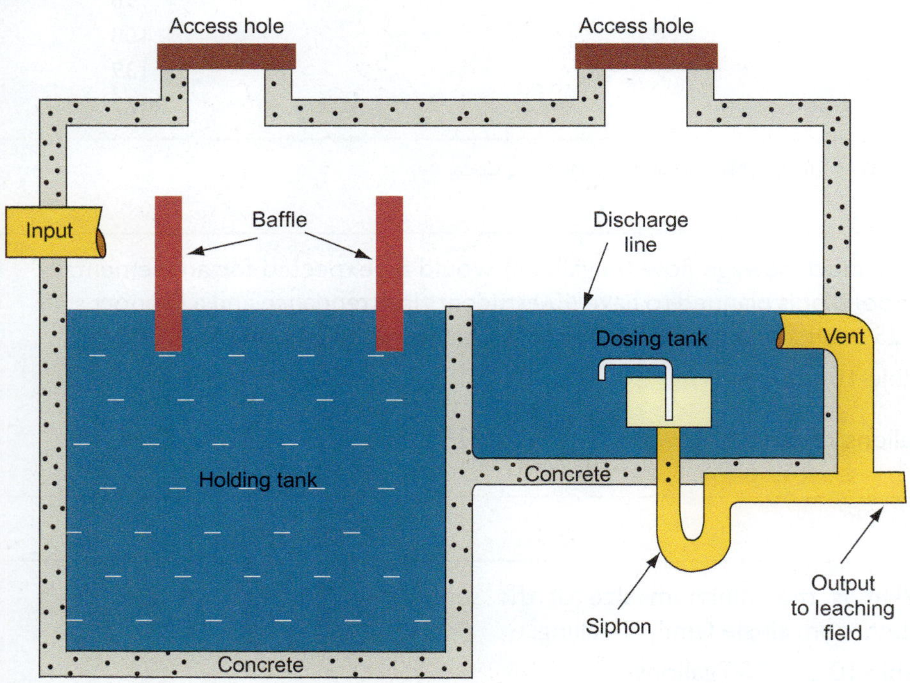

FIGURE 10-4 The top figure shows how the serial layout can work across the slope of the land. The bottom figure shows a cross section of the pipe rows.

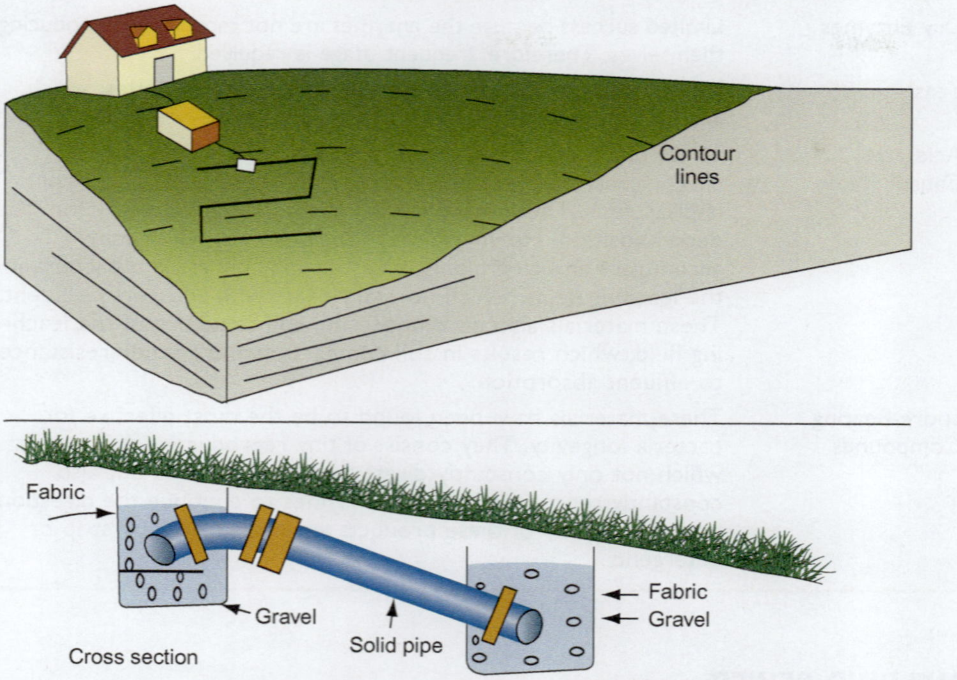

It is important to note that according to National Standard Plumbing Code, the length of lateral lines shall not exceed 100 feet. If more than 100 feet were required, two or more laterals must be provided.

Most large systems on sloped land can use a serial layout for the leaching fields. The field is sectioned into different levels on a series of elevations. Effluent will not flow to a lower level unless the previous level is saturated. Figure 10-4 shows an illustration of one particular type of serial distribution.

PRIVATE SYSTEM MAINTENANCE

A septic tank system requires the people in the building to minimize the use of water as well as avoid or restrict the use of many chemicals, such as soaps, detergents, acids, lotions, bleaches, polishes, cleaning compounds, antiseptics, caustic soda drain cleaners, toilet cleaners, disinfectants, sink and tub cleaners, chlorine products, and medicines. Even clear-water wastes, which dilute the septic tank contents and add load to the disposal system, must be kept to a minimum.

Many additive agents have been suggested to increase septic tank activity. Table 10-6 shows the suggested action and the likely result.

TABLE 10-6 Septic Tank Additives

Material	Consequences
Dry Enzymes	Limited success because the enzymes are not capable of reproducing themselves. Therefore, frequent usage is required.
Yeast	Limited success because yeast only digests carbohydrates, which are only 5 percent of the solids in a septic tank.
Acid, Lye, Caustic Yeast	These materials may produce a temporary success in that clogging materials are burned away. However, the total result is disastrous. These materials kill the bacteria necessary for decomposition. However, they loosen solids, which may accumulate and clog piping. Caustics may kill ground bacteria in the leaching fields, which normally react with and purify effluent. These materials also decompose the soil structure of the leaching field, which results in soil compaction and greater resistance to effluent absorption.
Spore-Bearing Compounds	These materials have been found to be the most effective for bacteria longevity. They consist of tiny reproductive spore cells, which not only constantly digest the waste material but also constantly reproduce cells and enzymes to continue the digestion processes. Many of these products are not affected by soap or detergents.

BUILDING SEWER

The building sewer is an integral part of the sewage disposal system. The building sewer is the drainage pipe that extends from about 3 feet (codes vary on the exact length) outside the building wall to the sewer main, the septic system, or some other approved point of disposal.

IN THE FIELD
Planning and scheduling with safety in mind can greatly reduce the possibilities of an accident.

The installation of the building sewer requires special attention to safety, as sewer construction usually requires excavation, trenching, and shoring. Each year there are hundreds of excavation-related injuries or fatalities. Always make sure that an OSHA-certified "competent person" has reviewed the work site and conditions daily, before entering a trench.

MATERIALS

Table 3.5 of the 2012 National Standard Plumbing Code indicates which materials are suitable for use in building sewers. The 2012 International Plumbing Code lists all approved building sewer pipe in Table 702.3 and lists all suitable water service pipes in Table 605.3. While both agree on the majority of suitable materials, there are some differences; always check for local approval before you begin a project. Table 10-7 shows the materials approved by the National Standard Plumbing Code.

TABLE 10-7 Materials for Sanitary Waste and Drain (3)

		Sewers Outside of Buildings	Underground Within Buildings	Above Ground Within Buildings	Pipe	Fittings
1	ABS Plastic Pipe and Fittings	X	X	X	ASTM D2661 (DWV, sched 40 IPS)	ASTM D2661 (socket, threaded)
2	ABS Cellular Core Plastic Pipe	X	X	X	ASTM F628 (DWV, sched 40 IPS)	ASTM F628, ASTM D2661 (socket, threaded)
3	ABS Sewer Pipe and Fittings	X			ASTM D2751 (SDR 23.5)	ASTM D2751 (SDR 23.5, socket)
4	ABS and PVC Composite Sewer Pipe	X			ASTM D2680	ASTM D2680 (socket), PVC (gasket)
5	Cast Iron Soil Pipe and Fittings (bell & spigot)	X	X	X	ASTM 74	ASTM A74
6	Cast Iron Soil Pipe and Fittings (no-hub)	X	X	X	CISPI 301, ASTM A888	CISPI 301, ASTM A888
7	Concrete Sewer Pipe, Non-reinforced	X			ASTM C14	ASTM C14
8	Concrete Sewer Pipe, Reinforced	X			ASTM C76	ASTM C76
9	Copper Drainage Tube, DWV (7)	X	X	X	ASTM B306	ASME B16.23 (cast copper solder joint)
10	Copper Water Tube, K, L, M (7)	X	X	X	ASTM B88	ASME B16.23 (cast copper solder joint)
11	Fiberglass Sewer Pipe	X			ASTM D3754	ASME D3840
12	Galvanized Steel Pipe			X	ASTM A53	ASME B16.12 (cast iron drainage)
13	PE Plastic Pipe, SDR-PR	X			ASTM F714 (6)	
14	PVC Plastic Pipe and Fittings, DWV	X	X	X	ASTM D2665	ASTM D2665 (socket), ASTM F1866 (socket)
15	PVC Plastic Pipe and Fittings, 3.25″ OD		X	X	ASTM D2949	ASTM D2949
16	PVC Sewer and Drain Pipe (cellular core)	X			ASTM F891	ASTM D2665 (socket), ASTM F1866 (socket)
17	PVC Sewer and Drain Pipe (cellular core)	X	X		ASTM F891 (4)	ASTM D2665 (socket), ASTM F1866 (socket)
18	PVC IPS Schedule 40 (cellular core)	X	X	X	ASTM F891	ASTM D2665 (socket), ASTM F1866 (socket)
19	PVC Plastic Sewer Pipe (PSM) and Fittings	X			ASTM D3034 (1)	ASTM D3034 (1)

(continued)

TABLE 10-7 Materials for Sanitary Waste and Drain (3) (Continued)

		Sewers Outside of Buildings	Underground within Buildings	Above Ground within Buildings	Pipe	Fittings
20	Stainless Steel DWV Systems - Type 316L (5)	X	X	X	ASTM A112.3.1	ASTMA112.3.1
21	Stainless Steel DWV Systems - Type 304 (5)			X	ASTM A112.3.1	ASTMA112.3.1
22	Vitrified Clay Pipe, Standard Strength	X			ASTM C700	ASTM C700
23	Vitrified Clay Pipe, Extra Strength	X	X		ASTM C700	ASTM C700
24	Ductile Iron Gravitiy Sewer Pipe	X			ASTMA746	AWWA C110, AWWA 153 (push-on, mechanical)
25	Ductile Iron Culvert Pipe	X			ASTM A716	AWWA C110, AWWA 153 (push-on, mechanical)
26	CPVC Chemical Waste Systems	X	X	X	ASTM F2618 (schd 40 IPS)	ASTM F2618 socket
27	PVC Non-Pressure, Reprocessed-Recycled Content	X	X		ASTM F1760, Sewer-Drain	ASTM D3034, ASTM F1336
28	PVC Non-Pressure, Reprocessed-Recycled Content	X	X		ASTM F1760, IPS, Schedule 40	ASTM D2466, ASTM D2665

Notes:

1. Plastic waste and drain piping classifed by standard dimension ratio shall be SDR 26 or heavier (lower SDR number).
2. Plastic sewer pipe classifed by pipe stiffness shall be PS-46 or stiffer (higher PS number).
3. Piping shall be applied within the limits of its listed standard and the manufacturer's recommendations.
4. PS-100 pipe or stiffer (higher PS number) is required for underground cellular core PVC sewer and drain pipe underground within buildings.
5. The alloy shall be marked on stainless steel DWV piping systems.
6. Minimum DR-17 is required for trenchless sewer replacement systems.
7. Copper tube shall not be used for drainage from non-water urinals to the point where it connects to piping containing drainage from one or more water closets.

FIGURE 10-9 Special fittings can help direct drain-cleaning equipment in various directions.

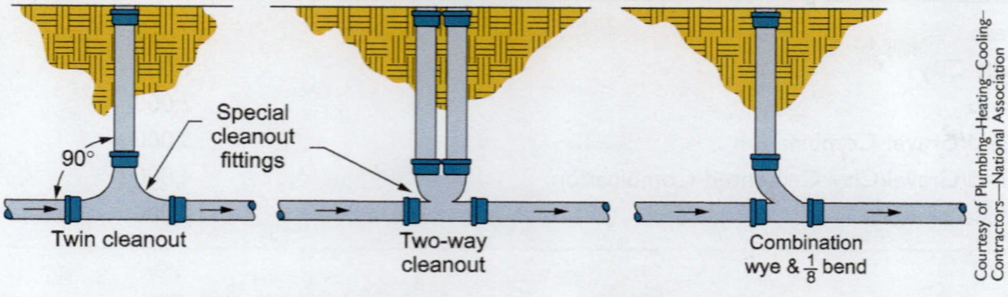

To provide easy access for rodding, the size of a cleanout should be equal to the size of the line it serves up to 4 inches and at least 4 inches for larger sewer lines. The 2012 National Standard Plumbing Code requires sewer pipe sizes 12 inches and over to be provided with manholes to serve as cleanouts. The 2012 International Plumbing Code has a similar requirement but for pipes 8 inches and larger. Most codes also allow manholes to be provided in place of cleanouts for larger sewer lines. It is always necessary to consult your local code for specific requirements and allowances.

Thrust Blocks

Under certain conditions, large pipe (such as pipe used in sewers) can experience significant forces from internal pressures. These forces can be great enough to separate joints. One way to control or counter such forces is to use thrust blocks, which are poured concrete masses that prevent movement of a fitting by utilizing the resistance of the earth over the area of the thrust block. See Figure 10-10. The size of the block is determined by the possible force to be restrained and the

FIGURE 10-10 Thrust blocks help prevent pipe from shifting out of position.

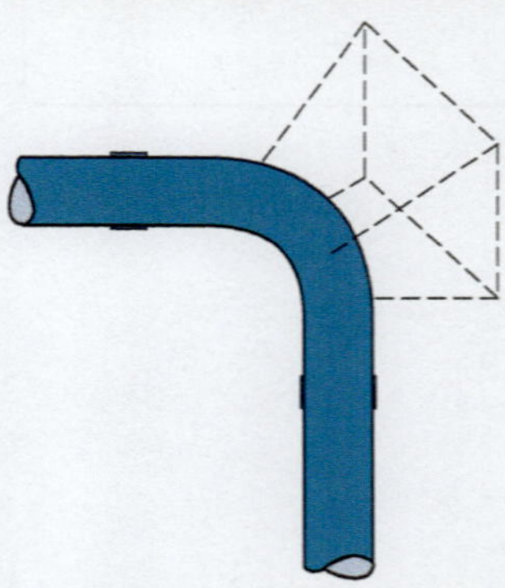

TABLE 10-9 Soil Load Capacities

Soil Type	Load Capacity (lbs/sq ft.)
Peat	0
Soft Clay	1,000
Sand	2,000
Sand/Gravel Combination	3,000
Sand/Gravel/Clay Cemented Combination	4,000
Shale or Rock	10,000

ability of the soil type to resist that load. Table 10-9 lists the load capacity of several soil types.

EXAMPLE 4

A 90-degree ell must be restrained in soft clay soil. The total load is 6,000 lbs. What area must the thrust block bear against?

$$\text{Area} \times \text{soil resistance} = 6{,}000 \text{ lbs total force}$$

$$\text{Area}\left(\frac{1{,}000 \text{ lbs.}}{\text{ft.}^2}\right) = 6{,}000 \text{ lbs}$$

$$\text{Area} = 6 \text{ sq. ft.}$$

EXAMPLE 5

A force-main installed in sandy soil is subjected to 20 psi pressure. If the main is 8 inches in diameter, what force must be resisted at fittings?

The force is the cross-sectional area of the pipe times the pressure.

$$\begin{aligned}\text{Area} &= \pi r^2\\ &= 3.14 \times (4 \text{ in.})^2\\ &= 50.3 \text{ in.}^2\end{aligned}$$

$$\text{Force} = 50.3(20)$$

$$\text{Force} = 1{,}000 \text{ lbs (approximately)}$$

The soil is sandy, so the thrust block must bear on the following area:

$$\text{Area} = \frac{1{,}000}{2{,}000} = 0.5 \text{ sq ft.}$$

REVIEW QUESTIONS

True or False

1. ________ The building sewer begins where the sewer line leaves the property.
2. ________ The slope of the building sewer (sizes through 6 inches) should not be less than $\frac{1}{8}$ inch per foot.
3. ________ A septic tank is a watertight vessel meant to receive sewage from a building.
4. ________ The outlet of a septic tank should be 2 inches to 4 inches below the inlet elevation.
5. ________ The minimum septic tank size should be equal to the volume of sewage discharged from the building in 24 hours.
6. ________ The distribution field is usually made of 4-inch perforated or open-jointed pipes.
7. ________ Aerobic action means that oxygen is present.

Chapter 11

LEARNING OBJECTIVE

The student will:

- Describe the basic concepts and application of drainage fixture units.

Drainage Fixture Units

DRAINAGE FIXTURE UNITS

The sizing of drainage piping has been studied for many years. The following items affect the individual fixture drain sizing details:

- Diameter of the waste outlet
- Trap size
- Scouring action of the fixture
- Desired velocity of waste material in the drain

An individual drain line must never be smaller than the trap to which it connects because a constriction will lead to fouling and stoppages. An exception to this literal provision is that a 4″ × 3″ closet bend, stub, or flange is not considered a restriction by most codes.

The method used to size drain pipes serving several fixtures is similar to the method used for water pipe sizing. It involves assigning a drainage fixture unit (DFU) value to each fixture type and its use. This concept was first used in the 1920s to normalize the loads that different fixtures would have on a plumbing system. The idea was to have a rating that would take into account the rate of waste flow, total quantity of waste, and frequency of use of the fixture.

In the early 1990s, a rational basis for evaluating low-water-volume fixtures was developed at Stevens Institute under the leadership of Professor Tom Konen. As a result of this research, it became widely recognized that a given plumbing fixture puts different loads on the drainage system depending on its frequency of use. The frequency of use varies depending on the type of installation, such as a single residence or apartment, office or store, or an industrial application. With the reduced flows of newer fixtures, even lower drainage fixture unit values should be applied.

The drainage fixture unit value is an index number that permits the combination of different fixture types to determine the probable maximum load on a drainage system.

For any fixture layout, the DFU values are added and pipe sizes are selected from appropriate tables. Table 11-1 shows the DFU values for various fixture types.

Because the orientation (vertical or horizontal) and location of the pipe can affect how the water flows in the pipe, there are special tables that need to be used for each section. Table 11-2 shows the DFU loadings permitted on horizontal branches and stacks, while Table 11-3 shows the loadings allowed on building drains and sewers. In this table, "gpf" stands for "gallons per flush."

TABLE 11-1 Drainage Fixture Unit (DFU) Values

Type of Fixtures	Individual Dwelling Units	Serving Three or More Dwelling Units	Other than Dwelling Units	Heavy Use Assembly
Bathroom Groups Having 1.6 gpf Gravity-Tank Water Closets				
Half-Bath or Powder Room	3	2		
1 Bathroom Group	5	3		
$1\frac{1}{2}$ Bathrooms	6	3.5		
2 Bathrooms	7	4.5		
$2\frac{1}{2}$ Bathrooms	8	5		
3 Bathrooms	9	5.5		
Each Additional Half-Bath	0.5	0.5		
Each Additional Bathroom Group	1	1		
Bathroom Groups Having 1.6 gpf Pressure-Tank Water Closets				
Half-Bath or Powder Room	3.5	2.5		
1 Bathroom Group	5.5	3.5		
$1\frac{1}{2}$ Bathrooms	6.5	4		
2 Bathrooms	7.5	5		
$2\frac{1}{2}$ Bathrooms	8.5	5.5		
3 Bathrooms	9.5	6		
Each Additional Half-Bath	0.5	0.5		
Each Additional Bathroom Group	1	1		
Bathroom Groups Having 3.5 gpf (or higher) Gravity Tank Water Closets				
Half-Bath or Powder Room	3	2		
1 Bathroom Group	6	4		
$1\frac{1}{2}$ Bathrooms	8	5.5		
2 Bathrooms	10	6.5		
$2\frac{1}{2}$ Bathrooms	11	7.5		
3 Bathrooms	12	8		
Each Additional Half-Bath	0.5	0.5		
Each Additional Bathroom Group	1	1		
Bathroom Group With 1.6 gpf Flushometer Valve	5	3		
Bathroom Group with 3.5 gpf (or Higher) Flushometer Valve	6	4		
Individual Fixtures				
Bathtub or Combination Bath/Shower, $1\frac{1}{2}''$ Trap	2	2		
Bidet, $1\frac{1}{4}''$ Trap	1	1		
Clothes Washer, Domestic, 2″ Standpipe	3	3	3	
Dishwasher, Domestic, with Independent Drain, $1\frac{1}{2}''$ Minimum Trap	2	2	2	
Drinking Fountain or Watercooler			0.5	
Food-Waste-Grinder, Commercial, 2″ Min. Trap			3	

(continued)

TABLE 11-1 Drainage Fixture Unit (DFU) Values (Continued)

Type of Fixtures	Individual Dwelling Units	Serving Three or More Dwelling Units	Other than Dwelling Units	Heavy Use Assembly
Floor Drain, Auxiliary			0	
Kitchen Sink, Domestic, with One $1\frac{1}{2}''$ Trap	2	2	2	
Kitchen Sink, Domestic, with Food-Waste-Grinder	2	2	2	
Kitchen Sink, Domestic, with Dishwasher	3	3	3	
Kitchen Sink, Domestic, with Grinder and Dishwasher	3	3	3	
Laundry Sink, One or Two Compartments, $1\frac{1}{2}''$ Waste	2	2	2	
Laundry Sink, with Discharge from Clothes Washer	2	2	2	
Lavatory, $1\frac{1}{4}''$ Waste	1	1	1	1
Mop Basin, 3″ Trap			3	
Service Sink, 3″ Trap			3	
Shower Stall, $1\frac{1}{2}''$ Trap	2	2	2	
Shower Stall, 2″ Trap	2	2	2	
Showers, Group, per Head (Continuous Use)			5	
Sink, $1\frac{1}{2}''$ Trap	2	2	2	
Sink, 2″ Trap	3	3	3	
Sink, 3″ Trap			5	
Trap Size, $1\frac{1}{4}''$ (Other)	1	1	1	
Trap Size, $1\frac{1}{2}''$ (Other)	2	2	2	
Trap Size, 2″ (Other)	3	3	3	
Trap Size, 3″ (Other)			5	
Trap Size, 4″ (Other)			6	
Urinal, 1.0 GPF			4	5
Urinal, Greater Than 1.0 GPF			5	6
Wash Fountain, $1\frac{1}{2}''$ Trap			2	
Wash Fountain, 2″ Trap			3	
Wash Sink, Each Set of Faucets			2	
Water Closet, 1.6 GPF Gravity or Pressure Tank	3	3	4	6
Water Closet, 1.6 GPF Flushometer Valve	3	3	4	6
Water Closet, 3.5 GPF Gravity Tank	4	4	6	8
Water Closet, 3.5 GPF Flushometer Valve	4	4	6	8
Whirlpool Bath or Combination Bath/Shower, $1\frac{1}{2}''$ Trap	2	2		

Notes:

1. A Bathroom Group, for the purposes of this Table, consists of not more than one water closet, up to two lavatories, and either one bathtub, one bath/shower combination, or one shower stall. Other fixtures within the bathing facility shall be counted separately to determine the total drainage fxture unit load.
2. A Half-Bath or Powder Room, for the purposes of this Table, consists of one water closet and one lavatory.
3. For unlisted fixtures, refer to a listed fixture having a similar fow and frequency of use.
4. When drainage fixture unit (DFU) values are added to determine the load on the drainage system or portions thereof, round the sum to the nearest whole number before referring to Tables 11.5.1A, 11.5.1B, or 12.16.6A for sizing the drainage and vent piping. Values of 0.5 or more should be rounded up to the next higher whole number (9.5 = 10 DFU). Values of 0.4 or less should be rounded down to the next lower whole number (9.4 = 9 DFU).
5. "Other Than Dwelling Units" applies to business, commercial, industrial, and assembly occupancies other than those defined under "Heavy-Use Assembly." Included are the public and common areas in hotels, motels, and multi-dwelling buildings.
6. "Heavy-Use Assembly" applies to toilet facilities in occupancies that place heavy, but intermittent, time-based loads on the drainage system, such as schools, auditoriums, stadiums, race courses, transportation terminals, theaters, and similar occupancies where queuing is likely to occur during periods of peak use.
7. Where other than water-supplied fixtures discharge into the drainage system, allow 2 DFU for each gallon per minute (gpm) of flow. (See Section 11.4.2.)

Source: Table 11.4.1 2012 National Standard Plumbing Code.

TABLE 11-2 Horizontal Fixture Branches and Stacks

Maximum Number of Drainage Fixture Units (DFU) That May Be Connected to Any Horizontal Fixture Branch, a Stack of Three Branch Intervals or Less, or Stacks of more than Three Branch Intervals

			Stacks with more than Three Branch Intervals	
Diameter of Pipe (in.)	Any Horizontal Fixture Branch[1] (DFU)	Stacks with Three Branch Intervals or Less (DFU)	Total for Stack (DFU)	Discharged into One Branch Interval (DFU)
$1\frac{1}{4}$	1	1	1	1
$1\frac{1}{2}$	3	4	8	2
2	6	10	24	6
3	20[2]	48[3]	72[3]	20[3]
4	160	240	500	90
5	360	540	1,100	200
6	620	960	1,900	350
8	1,400	2,200	3,600	600
10	2,500	3,800	5,600	1,000
12	3,900	6,000	8,400	1,500
15	7,000			

Notes:

1. Does not include branches of the building drain.
2. See Section 11.5.6.a & d for the restrictions on the number of water closets installed on 3″ horizontal fxture branches.
3. See Section 11.5.6.b & d for the restrictions on the number of water closets on 3″ stacks and discharged into any branch interval of a 3″ stack.

Source: Table 11.5.1B 2012 National Standard Plumbing Code.

TABLE 11-3 Building Drains and Sewers[1]

Maximum Number of Drainage Fixture Units (DFU) That May Be Connected to Any Portion of the Building Drain or the Building Sewer

Diameter of Pipe (in.)	Slope per Foot $\frac{1}{16}$-inch	Slope per Foot $\frac{1}{8}$-inch	Slope per Foot $\frac{1}{4}$-inch	Slope per Foot $\frac{1}{2}$-inch
2			21	26
3		36[2]	42[2]	50[2]
4		180	216	250
5		390	480	575
6		700	840	1,000
8	1,400	1,600	1,920	2,300
10	2,500	2,900	3,500	4,200
12	3,900	4,600	5,600	6,700
15	7,000	8,300	10,000	12,000

Notes:

1. On-site sewers that serve more than one building may be sized according to the current standards and specifications of the Authority Having Jurisdiction for public sewers.
2. See Section 11.5.6.c & d for the restrictions on the number of water closets installed on 3″ building drains and sewers.

Source: Table 11.5.1A 2012 National Standard Plumbing Code.

Note that the revised information shown in Table 11-1 has eliminated some former fixture types, such as dental units and floor drains. It should also be noted that since most floor drains are installed for emergency use, they are shown as 0 DFU. On the other hand, floor drains that are intended for regular, planned use should be given a DFU that is consistent with that planned use.

EXAMPLE 1

Compute the drainage fixture unit values for the following fixtures in a new commercial building using low-consumption (1.6 gpf) water closets (see Table 11-4).

The drain pipe to serve this total group must be sized to serve 25 DFU. Table 11-2 or Table 11-3 would be used depending upon whether the drain pipe is a horizontal branch, stack, or building drain.

Note that the total would be different if you were calculating the load in an existing building with older fixtures.

EXAMPLE 2

Size the drain stack for a bathroom group on the upper floor and powder room on the lower floor for a new single family residence as shown in Figure 11-1. The fixtures combine for the following calculations (see Table 11-5):

The stack has a total of 8 DFU, which is less than the 48 DFU maximum allowed on a 3-inch stack; therefore, the stack should be 3 inches. A smaller stack size is not permitted because 3 inches is the smallest size that is permitted to serve a water closet.

TABLE 11-4 Drainage Fixture Unit Values for New Commercial Building

Quantity	Fixture	DFU	Total
2	Bathroom Groups (WCs, Lavs, Showers)		
	2 WCs, Gravity Tank	2 × 4 =	8
	2 Lavs	2 × 1 =	2
	2 Showers	2 × 2 =	4
3	Dental Lavatories (Assume Same as Lav)	3 × 1 =	3
3	Dental Units (Assume Same as Lav)	3 × 1 =	3
1	Floor Drain (Assume for Emergency Use)		0
2	Wash Sinks	2 × 2 =	4
2	Drinking Fountains	2 × 0.5 =	1
		Total	25 DFU

FIGURE 11-1 Single family residence.

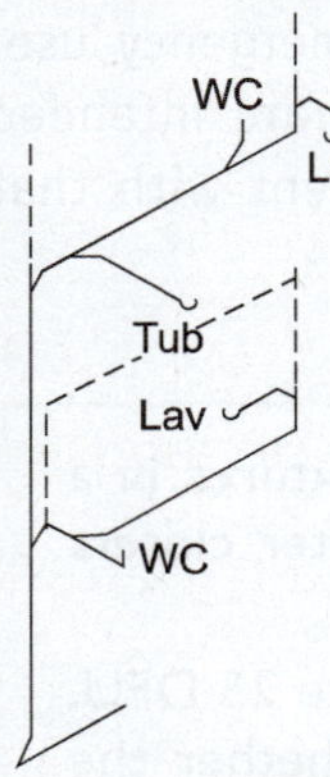

TABLE 11-5 Drain Stack Sizing for Residence

Fixture	Fixture DFU	Total DFU	Branch Size (in.)
	Second Floor		
Bathtub	3		$1\frac{1}{2}$
Lavatory	1		$1\frac{1}{2}$
Water Closet	3		3
Second Floor Bathroom Group:		5	3
	First Floor		
Water Closet	3		3
Lavatory	1		$1\frac{1}{2}$
First Floor Bathroom Group:	3	3	3
Total		8	3

Table 11-3 shows the sizing for building drains and sewers with DFU values for four different values of pipe slope. As discussed in Chapter 10 the preferred slope is $\frac{1}{4}$ inch per foot, with $\frac{1}{8}$ inch and $\frac{1}{2}$ inch per foot being acceptable. A slope of $\frac{1}{16}$ inch per foot is permitted for sizes 8 inches and larger. Maintain trap arm distances to vent connections as required by code.

Offsets

Offsets in stacks can be sized with the tables found in this chapter. If an offset is made with an angle change of 45° or greater from the horizontal, the pipe is sized as a vertical line (Table 11-2). If the angle is less than 45° from the horizontal, the offset must be sized as a building drain (Table 11-3).

FIGURE 11-2 Offset patterns.

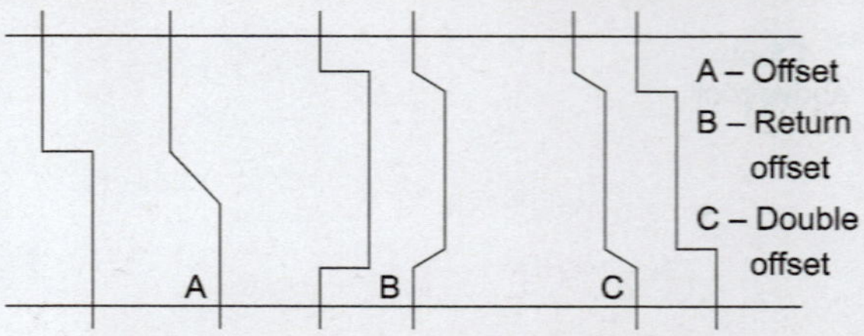

Figure 11-2 shows three typical offset patterns using either ells or 45° angles. Note that while return offsets and double offsets look different, they are hydraulically identical.

Summary of Sizing Methods

Sizing of drainage piping is generally performed as indicated earlier. However, some special circumstances must be explored.

First, most codes have minimum sizes for certain drainage elements, either to provide some reserve for future changes or because stoppages are less likely to occur if the mandatory sizes are used. For example, most codes require the minimum drain size below a slab to be 2 inches unless it is a condensate line, which may be as small as $1\frac{1}{4}$ inches.

Second, continuous or semi-continuous flow into a drainage system is accounted for by allowing 2 DFU for each gpm of steady flow, so 3 gpm would be calculated as 6 DFU.

Third, for large systems (over 1,000 DFU) and where the ratio of plumbing fixtures to occupants is more than the minimum required, a diversity factor may be used for sizing stacks, branches, building drains, and sewers. Final approval must be obtained from the authority having jurisdiction (building inspector, plumbing inspector, or other governmental authority).

EXAMPLE 3

Figure 11-3 represents a bathroom on the upper floor, powder room, kitchen, and laundry for a new single family residence. Use Tables 11-1, 11-6, 11-7, and 11-8 to size this system.

This kitchen load calls for $1\frac{1}{2}$ inch pipe, but since the drain line is placed below the slab floor, it must be sized at 2 inches minimum.

The total stack load is 13 DFU, so a 3-inch stack and 3-inch building drain (42 DFU maximum) would be selected. Note that many codes require a minimum size of 4 inches for the building drain and sewer, although a 3-inch line will provide acceptable flow, especially with the low-consumption water closets presently in use.

FIGURE 11-3 New single family layout. Notice how the branch interval is measured.

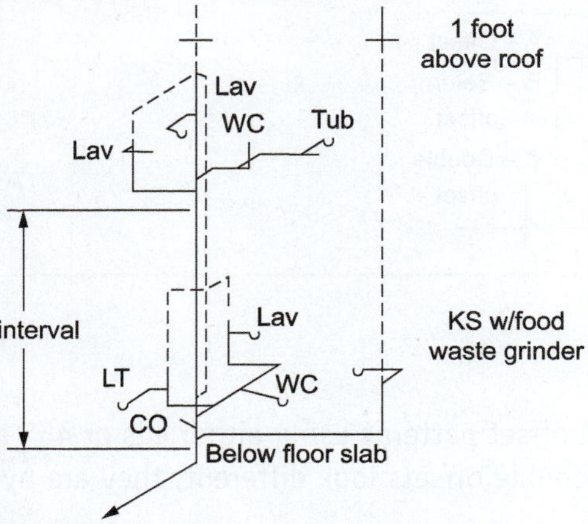

TABLE 11-6 Sizing of Drainage System

Fixture	DFU	Size (in.)
	Second Floor	
Water closet (private use)	3	3
Lavatory	1	$1\frac{1}{2}$
Bathtub	3	$1\frac{1}{2}$
Bathroom group	5 (less than the above sum, when looked at as a group)	3
The 3″ stack continues to the lower floor:		
	First Floor	
Water closet half bath	3	3
Total on the stack to this point:	8	3
First Floor		
Laundry tray with clothes washer	2	$1\frac{1}{2}$
The automatic washer standpipe size is from Table 11-1.		
Total load connected at this point is:	10	
The 3″ stack is still satisfactory (rated 48 DFU):	3	
Kitchen sink, food waste grinder, and dishwasher combination	3	$1\frac{1}{2}$

TABLE 11-7 Minimum Size of Non-integral Traps

Plumbing Fixture	Trap Size (in.)
Bathtub (with or without overhead shower)	$1\frac{1}{2}$
Bidet	$1\frac{1}{4}$
Clothes washing machine standpipe	2
Combination sink and wash (laundry) sink with food waste grinder unit	$1\frac{1}{2}$*

(continued)

TABLE 11-7 Minimum Size of Non-integral Traps (Continued)

Plumbing Fixture	Trap Size (in.)
Combination kitchen sink, domestic dishwasher, and food waste grinder	$1\frac{1}{2}$
Dental unit or cuspidor	$1\frac{1}{4}$
Dental lavatory	$1\frac{1}{4}$
Drinking fountain	$1\frac{1}{4}$
Dishwasher, commercial	2
Dishwasher, domestic (nonintegral trap)	$1\frac{1}{2}$
Floor drain	2
Food waste grinder—commercial use	2
Food waste grinder—domestic use	$1\frac{1}{2}$
Kitchen sink, domestic, with food waste grinder unit	$1\frac{1}{2}$
Kitchen sink, domestic	$1\frac{1}{2}$
Lavatory, common (private and public)	$1\frac{1}{4}$
Lavatory (barber shop, beauty parlor, or surgeon's)	$1\frac{1}{2}$
Lavatory, multiple type (wash fountain or wash sink)	$1\frac{1}{2}$
Laundry tray (1 or 2 compartments)	$1\frac{1}{2}$
Shower stall or shower drain (single shower head)	$1\frac{1}{2}$
Shower stall or shower drain (multiple shower heads)	2
Sink (surgeon's)	$1\frac{1}{2}$
Sink (flushing rim type, flushometer valve supplied)	3
Sink (service type with floor outlet trap standard)	3
Sink (service type with P-trap)	2
Sink, commercial (pot, scullery, or similar type)	2
Sink, commercial (with food grinder unit)	2

*Separate trap required for wash tray and and separate trap required for sink compartment with food waste grinder.

Source: Table 5.2 2012 National Standard Plumbing Code.

TABLE 11-8 Horizontal Fixture Branches and Stacks

Maximum Number of Drainage Fixture Units (DFU) That May Be Connected to Any Horizontal Fixture Branch, a Stack of Three Branch Intervals or Less, or Stacks of more than Three Branch Intervals

			Stacks with more than Three Branch Intervals	
Diameter of Pipe (in.)	Any Horizontal Fixture Branch[1] (DFU)	Stacks with Three Branch Intervals or Less (DFU)	Total for Stack (DFU)	Discharged into One Branch Interval (DFU)
$1\frac{1}{4}$	1	1	1	1
$1\frac{1}{2}$	3	4	8	2
2	6	10	24	6
3	20[2]	48[3]	72[3]	20[3]
4	160	240	500	90
5	360	540	1,100	200
6	620	960	1,900	350

(continued)

TABLE 11-8 Horizontal Fixture Branches and Stacks (Continued)

Maximum Number of Drainage Fixture Units (DFU) That May Be Connected to Any Horizontal Fixture Branch, a Stack of Three Branch Intervals or Less, or Stacks of more than Three Branch Intervals

			Stacks with more than Three Branch Intervals	
Diameter of Pipe (in.)	Any Horizontal Fixture Branch[1] (DFU)	Stacks with Three Branch Intervals or Less (DFU)	Total for Stack (DFU)	Discharged into One Branch Interval (DFU)
8	1,400	2,200	3,600	600
10	2,500	3,800	5,600	1,000
12	3,900	6,000	8,400	1,500
15	7,000			

Notes:

1. Does not include branches of the building drain.
2. See Section 11.5.6.a & d for the restrictions on the number of water closets installed on 3″ horizontal fxture branches.
3. See Section 11.5.6.b & d for the restrictions on the number of water closets on 3″ stacks and discharged into any branch interval of a 3″ stack.

Source: Table 11.5.1B 2012 National Standard Plumbing Code.

EXAMPLE 4

In Figure 11-4, an apartment layout includes service sink, bathroom, kitchen sink–food waste grinder–dishwasher, clothes washer, and laundry tray for each of three floors. The DFU count for each stack for each floor should be noted first. (See Table 11-9.)

The stacks are as follows (see Tables 11-10 and 11-11):

The building drain size is selected from Table 11-3.

Practice sizing different systems to improve your technical skills.

FIGURE 11-4 Apartment layout with multiple stacks.

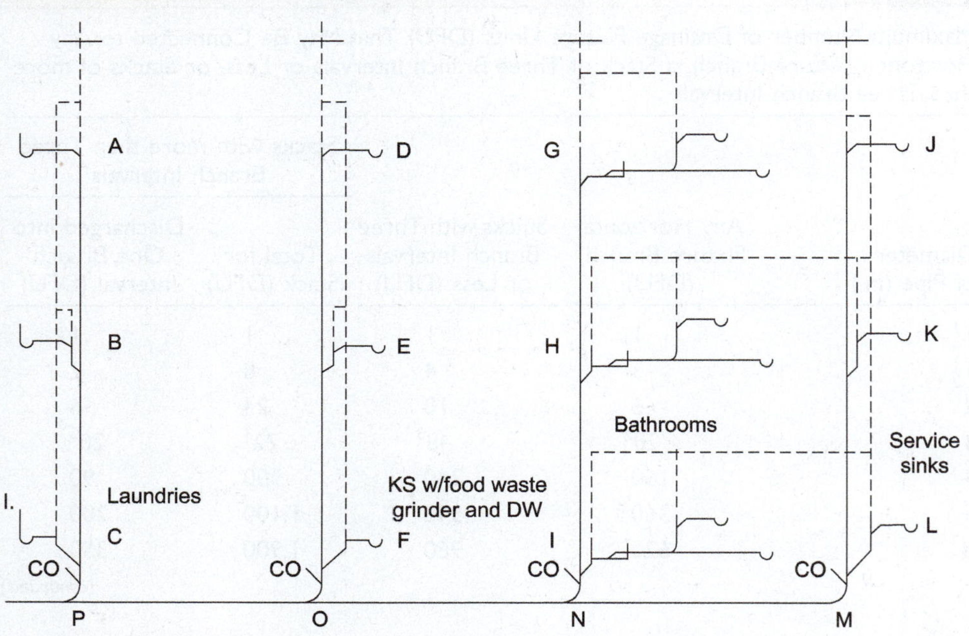

TABLE 11-9 DFU per Floor

Fixture	DFU per Floor
Laundry—Washer Plus Laundry Tray	2
Kitchen—Sink–Food Waste Grinder–Dishwasher	3
Baths—Bathroom Group	3
Service Sinks	3

TABLE 11-10 Stack Sizing

Stack		Load (DFU)	Branch Size (in.)	Stack Size (in.)
1	A	2	2	2
	B	4	2	2
	C	6	2	2
2	D	3	$1\frac{1}{2}$	$1\frac{1}{2}$
	E	6	$1\frac{1}{2}$	2
	F	9	$1\frac{1}{2}$	2
3	G	3	3	3
	H	6	3	3
	I	9	3	3
4	J	3	2	2
	K	6	2	2
	L	9	2	2

TABLE 11-11 Total Load Calculation for Building Drain

Location	Drainage Fixture Units	Building Drain (in.)
M	9	2
N	18	3
O	27	3
P	33	4

EXAMPLE 5

A two-story school has adjacent boys and girls restrooms on each floor. Calculate the fixture unit load on the branch drains and the stack for the system as shown in Figure 11-5.

The water closets are 1.6 gpf pressure tank; the urinals are 1.0 gpf. The loadings should be selected from the heavy use column since school schedules tend to produce very heavy usage during short periods of time (see Table 11-12).

FIGURE 11-5 Two-story school bathroom layout.

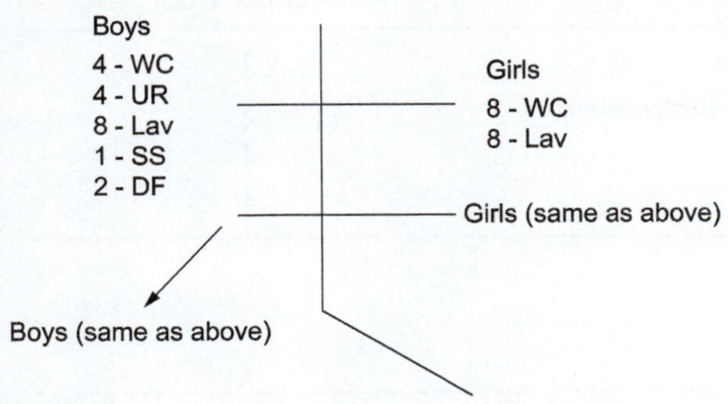

TABLE 11-12 School Restroom Fixture Unit Load

Boys' Branch DFUs	Girls' Branch DFUs	Total DFUs
4 WC × 8 = 32	8 WC × 8 = 64	96
4 UR × 5 = 20 (1 gpf type)		20
8 Lav × 1 = 8	8 Lav × 1 = 8	16
1 SS × 3 = 3		3
2 DF × 0.5 = 1		1
Stack to First Floor = 64	Stack to First Floor = 72	136
First Floor Boys = 64	First Floor Girls = 72	136
Total Load of Four Restrooms:		272

REVIEW QUESTIONS

True or False

1. ________ Studies indicate that the drainage fixture unit for a fixture is influenced by the type of building in which the fixture is installed.
2. ________ Fixture unit values are different for 3.5 gpf and 1.6 gpf water closets.
3. ________ Offsets are sized as stacks if the offset angle is 45° or more from the horizontal.
4. ________ One bathroom group serving an individual dwelling is equal to 5 DFUs.
5. ________ A domestic clothes washer (2-inch standpipe) has 5 DFUs.

Fill in the Blank

6. Figure 11-6 shows a stack serving four water closets (1.6 gpf) and tubs in an existing two-story apartment building (the lavatories are served by another stack). Size the branches to each fixture, the stack, and the building drain on both sides of the connection to the stack if the building drain is serving 170 DFU upstream of this stack connection, and the slope of the building drain is $\frac{1}{8}$ inch per foot.
   WC branches ________ Tub branches ________
   Stack first to second ________ Stack drain to first ________
   Building drain upstream ________ Building drain downstream ________

FIGURE 11-6 Stack in two-story apartment building.

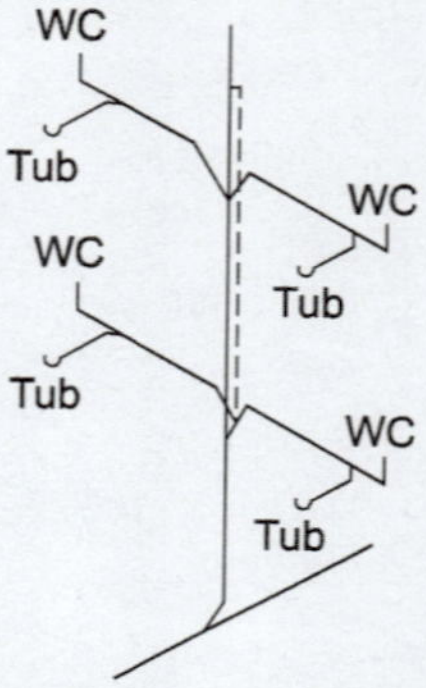

Chapter 12

LEARNING OBJECTIVES

The student will:

- Identify allowable materials for a building drain.
- Describe the construction details of the building drain.
- Discuss how floor drains contribute to DFU calculations.
- Calculate correct building drain sizes.

Building Drains

DEFINITION OF BUILDING DRAIN

The building drain is defined as *that part of the lowest piping of a drainage system which receives the discharge from soil, waste, and other drainage pipes inside the walls of a building and conveys it to the building sewer beginning three feet outside the building wall.*

This lengthy definition is usually considered to mean that the building drain accepts the discharge of stacks and has no fixture drain connected directly to it. This distinction is important. Building drains are allowed to be more heavily loaded than horizontal branch drains (drains that receive fixture discharges directly). The reason the heavier load is acceptable is that the stacks will develop some entrance energy into the building drain. This extra energy will help move materials through the building drain.

Building drains and building sewers are sized from the same table. The reason for the distinction between building drain and sewer is that most codes permit a wider variety of materials to be used for sewers than for building drains. Further, some jurisdictions permit persons other than licensed plumbers to install sewers, whereas only licensed plumbers may do installation work within the building.

ELEVATION

The building drain is usually the first piping installed on the job. In many cases, the elevation of the building drain must be carefully planned so that all the stacks in the building can be connected to the drain with acceptable slope for the drain and so that the drain, at the connection to the sewer, is high enough to permit acceptable slope to the sewer main connection.

EXAMPLE 1

What would the elevation difference have to be between the sewer main and the building drain if the building sewer length is 100 feet with a $\frac{1}{4}$ inch per foot slope?

$$\text{Rise} = (100 \text{ ft.}) \times \left(\frac{1}{4} \text{ in./ft.}\right)$$
$$= 25 \text{ inches}$$

MATERIALS

According to the 2012 National Standard Plumbing Code, all plastic sanitary drainage pipe and fittings must have an SDR 35 or lower when used underground outside of the building. If used within the building, all pipe and fittings must have an SDR 26 or lower. Remember that as the SDR number gets smaller the wall thickness of the pipe increases. All materials approved for sanitary drainage can be found in Table 3-5 of the 2012 National Standard Plumbing Code, Section 702 of the 2012 International Plumbing Code, and Section 701.1 of the 2012 Uniform Plumbing Code. Make certain you consult the proper code for your area because in a few instances they differ on approved materials. For example, vitrified clay pipe is allowed for underground building drains by the National Standard Plumbing Code and Uniform Plumbing Code but not by the International Plumbing Code. Remember, by definition, the building sewer is located outside the building and the building drain is within.

FITTINGS

Connections from the stacks to the building drain should be made using wyes and $\frac{1}{8}$ bends or equivalent fittings. The fittings should be arranged so that the wye opening is placed above the horizontal centerline of the building drain. Figure 12-1 shows two recommended and two not-recommended connection details. The important idea is that the stack should enter the building drain above the center line to minimize the possibility of a submerged inlet to the drain and to provide flexibility to minimize breakage of a rigid connection without any swing joints.

FIGURE 12-1 Connections to horizontal piping.

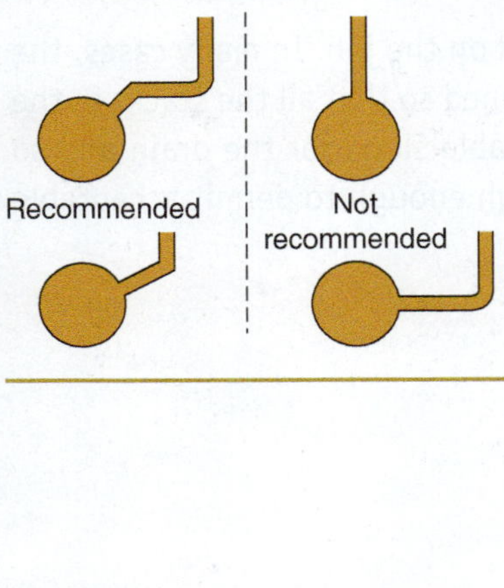

BACKFLOW PROTECTION

If a building drain is subjected to sewage backflow conditions, the fixtures that are subject to backflow must be protected with a backwater valve or with a sewage receiver sump and sewage pump equipped with a check valve. A good example of this situation is when the flood level rim of fixtures and/or drain inlets are below the overflow level of the first upstream manhole in the public sewer. As shown in Figure 12-2, a blockage in the main sewer line will cause sewage to back up into the building before it overflows into the nearest upstream manhole.

Whatever method of protection is selected, it should only be applied to the drain lines serving fixtures subject to backflow. Any fixtures that can drain by gravity to the building drain should be arranged to do so.

Backwater valves should be readily accessible for maintenance and repair. Figure 12-3 shows one type of backwater valve.

FIGURE 12-2 A backwater valve may be required when the flood level rim of fixtures are below the overflow level of the first upstream manhole in the public sewer.

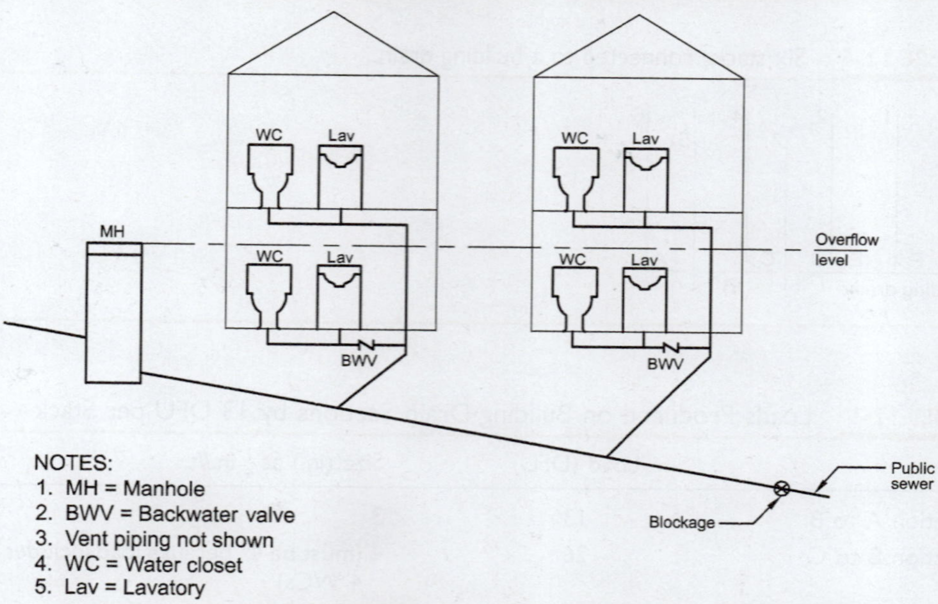

Courtesy of Plumbing–Heating–Cooling–Contractors—National Association

FIGURE 12-3 Backwater valves are large swinging check valves that prevent backflow of sewage.

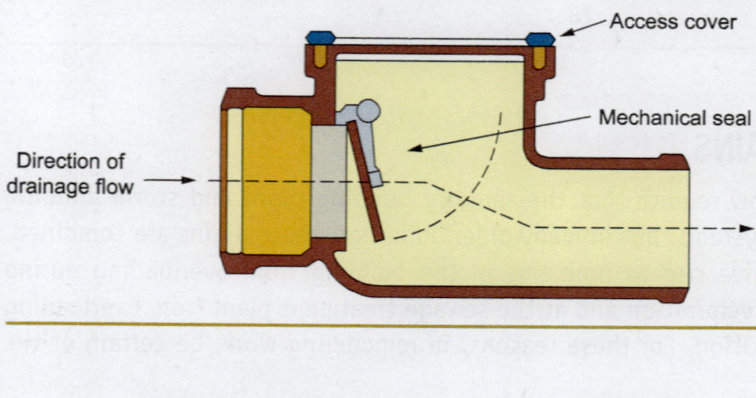

Courtesy of Plumbing–Heating–Cooling–Contractors—National Association

SIZING

As described in Chapter 11, the building drain is sized based on the drainage fixture unit load connected to the drain. As a review, Example 2 shows how the load increases as you move downstream.

EXAMPLE 2

Figure 12-4 shows six stacks connected to a building drain.

Each stack serves identical loading, which produces 13 DFU per stack on these sections of the building drain. See Table 12-1 for tabulation of loads. Use Table 12-2 to size the pipe sections in Figure 12-4.

EXAMPLE 3

Figure 12-5 presents another sizing example. For the stack loadings as marked in the illustration, the building drain sizes are selected from Table 12-3. Note that the main branch (A to E) of the building drain has to be sloped at $\frac{1}{8}$ in./ft. because of building constraints.

FIGURE 12-4 Six stacks connected to a building drain.

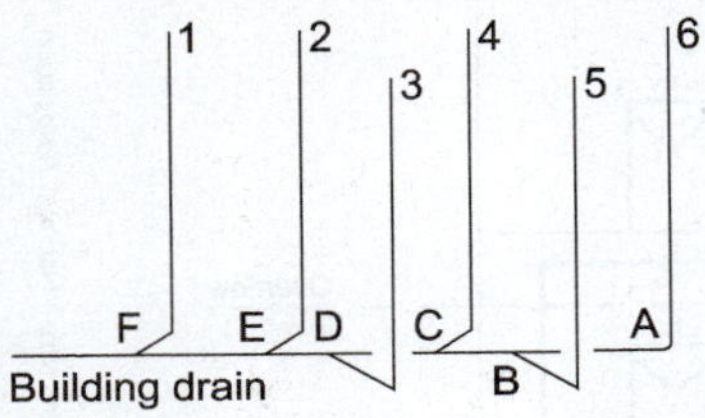

TABLE 12-1 Loads Produced on Building Drain Sections by 13 DFU per Stack

	Load (DFU)	Size (in.) at $\frac{1}{4}$ in./ft.
Section A to B	13	3
Section B to C	26	4 (must be 4″ because load includes 4 WCs)
Section C to D	39	4
Section D to E	52	4
Section E to F	65	4
Beyond F	78	4

COMBINED DRAINS

Nearly all codes now require that the sanitary building drain and storm building drain be separate systems, but in many older buildings, these drains are combined. Combined drains give rise to problems in the building from overloading during periods of heavy precipitation and at the sewage treatment plant from overloading and clear water dilution. For these reasons, in remodeling work, be certain of the

TABLE 12-2 Building Drains and Sewers[1]

Maximum Number of Drainage Fixture Units (DFU) That May Be Connected to Any Portion of the Building Drain or the Building Sewer

Diameter of Pipe (in.)	Slope Per Foot $\frac{1}{16}$-Inch	Slope Per Foot $\frac{1}{8}$-Inch	Slope Per Foot $\frac{1}{4}$-Inch	Slope Per Foot $\frac{1}{2}$-Inch
2			21	26
3		36[2]	42[2]	50[2]
4		180	216	250
5		390	480	575
6		700	840	1,000
8	1,400	1,600	1,920	2,300
10	2,500	2,900	3,500	4,200
12	3,900	4,600	5,600	6,700
15	7,000	8,300	10,000	12,000

Notes:

1. On-site sewers that serve more than one building may be sized according to the current standards and specifications of the Authority Having Jurisdiction for public sewers.
2. See Section 11.5.6.c & d for the restrictions on the number of water closets installed on 3″ building drains and sewers.

Source: Table 11.5.1A 2012 National Standard Plumbing Code.

FIGURE 12-5 Multiple stacks connected to a building drain.

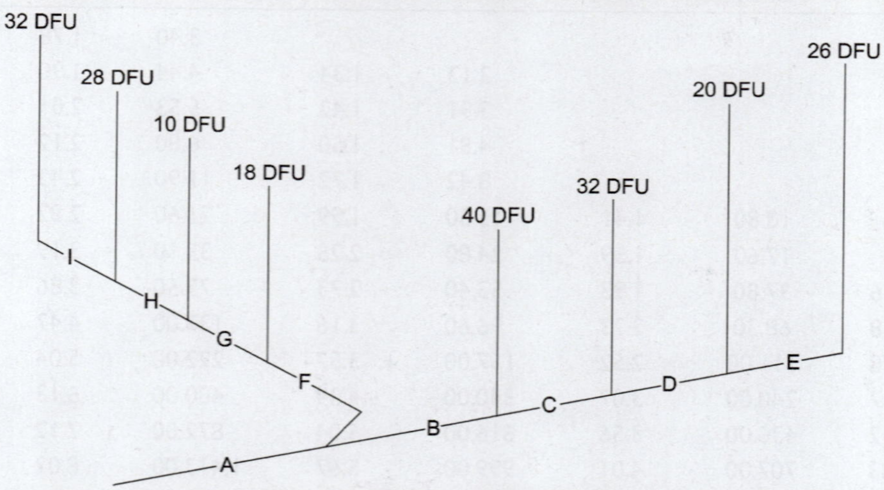

function of a pipe before you cut into it so that you do not cause sanitary wastes to enter the storm sewer or storm wastes to enter the sanitary system.

FLOW RATE

Flow rate (velocity) is important in any drain line. Many industry professionals consider the minimum satisfactory velocity to be 2 fps. Drains $2\frac{1}{2}$ inches and larger will develop this flow rate at a slope of $\frac{1}{4}$ in./ft., but only when flowing at least half full. Smaller drain lines do not meet this standard, even at $\frac{1}{2}$ inch per foot slope.

TABLE 12-3 Building Drain Sizes at Various Points in Figure 12-4

	Load (DFU)	Size (in.) $\frac{1}{4}$ in./ft.	Size (in.) $\frac{1}{8}$ in./ft.
Point I	32	3	
Point H	60	4	
Point G	70	4	
Point F	88	4	
Point E	26		4
Point D	46		4
Point C	78		4
Point B	118		4
Point A	206		5

TABLE 12-4 Approximate Discharge Rates and Velocities in Sloping Drains

	Flowing Half Full* Discharge Rate and Velocity**							
	$\frac{1}{16}$ in./ft. Slope		$\frac{1}{8}$ in./ft. Slope		$\frac{1}{4}$ in./ft. Slope		$\frac{1}{2}$ in./ft. Slope	
Actual Inside Diameter of Pipe (in.)	Discharge (gpm)	Velocity (fps)	Discharge (gpm)	Velocity (fps)	Discharge (gpm)	Velocity (fps)	Discharge (gpm)	Velocity (fps)
$1\frac{1}{4}$							3.40	1.78
$1\frac{3}{8}$					3.13	1.34	4.44	1.90
$1\frac{1}{2}$					3.91	1.42	5.53	2.01
$1\frac{5}{8}$					4.81	1.50	6.80	2.12
2					8.42	1.72	11.90	2.43
$2\frac{1}{2}$			10.80	1.41	15.30	1.99	21.60	2.82
3			17.60	1.59	24.80	2.25	35.10	3.19
4	26.70	1.36	37.80	1.93	53.40	2.73	75.50	3.86
5	48.30	1.58	68.30	2.23	96.60	3.16	137.00	4.47
6	78.50	1.78	111.00	2.52	157.00	3.57	222.00	5.04
8	180.00	2.17	240.00	3.07	340.00	4.34	480.00	6.13
10	308.00	2.52	436.00	3.56	616.00	5.04	872.00	7.12
12	500.00	2.83	707.00	4.01	999.00	5.67	1,413.00	8.02

*Half full means filled to a depth equal to one-half of the inside diameter.

**Computed from the Manning Formula for half-full pipe, $n = 0.015$.

For $\frac{1}{4}$ full, multiply discharge by 0.274 and multiply velocity by 0.701.

For $\frac{3}{4}$ full, multiply discharge by 1.82 and multiply velocity by 1.13.

For full, multiply discharge by 2.00 and multiply velocity by 1.00.

For smoother pipe, multiply discharge and velocity by 0.015 and divide by "n" value of smoother pipe.

Table 12-4 shows flow capacities (in gpm) and a list of velocities for the usual pipe sizes and slopes from $\frac{1}{16}$ inch per foot to $\frac{1}{2}$ inch per foot when flowing half full. Drains smaller than 2 inches are usually short and serve a single fixture, where at least occasional full discharges will occur.

Many drains probably operate for years and never see "design" flow rate, yet they perform satisfactorily over their lifetime. Therefore, it seems simplistic to say that a drain line must operate at 2 fps flow rate. The right idea is that the loading tables produce correct sizing; do not get concerned about a "rule-of-thumb" flow rate number.

GREASY WASTES

Lines conveying greasy wastes are special problems. Slope these lines at $\frac{1}{4}$ inch/foot and encourage users to wash the grease down with cold water. Cold water will cause the grease to ball up and move like any other solid. Hot water causes the grease to disperse and collect wherever the pipe wall is cooler, eventually producing a stoppage. In commercial installations, cold water should be piped to solenoid valve assemblies, which are connected electrically to food waste grinders. This can be a tremendous energy saver.

Commercial, industrial, or institutional fixtures that receive a significant amount of greasy waste should be drained through a grease trap or grease recovery device. Grease traps intercept and retain greasy and oily substances. Grease recovery devices are designed to heat the greasy/oily substances and skim them from the device. These skimmed waste solutions are safely removed for recycling. Remember to install the grease trap or grease recovery device so that it is readily accessible for cleaning.

Fixture branches that receive occasional greasy waste should be washed by an upstream fixture or group. Figure 12-6 shows an arrangement where a kitchen sink stack is washed with a lavatory and the building drain is washed with a bathroom group upstream of the grease waste point.

Grease Traps

Grease traps are used in the drains from commercial kitchens or other grease generators to minimize the grease load in a septic system. Grease traps must be accessible for cleaning, be watertight, and be sized for the flow rate of the kitchen or other load being served. Information concerning sizing of grease traps will be covered in future training. The materials used for construction should be nonabsorbent and suitable to contain the wastes conveyed. A typical grease trap is

FIGURE 12-6 Preferred method of drainage when dealing with greasy waste.

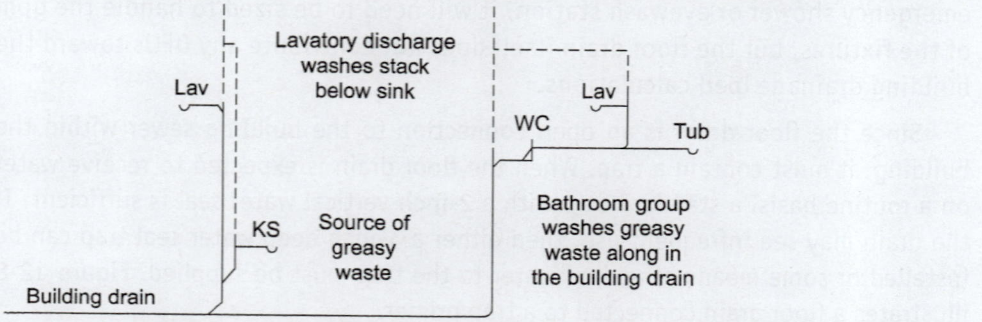

FIGURE 12-7 Example of a grease interceptor.

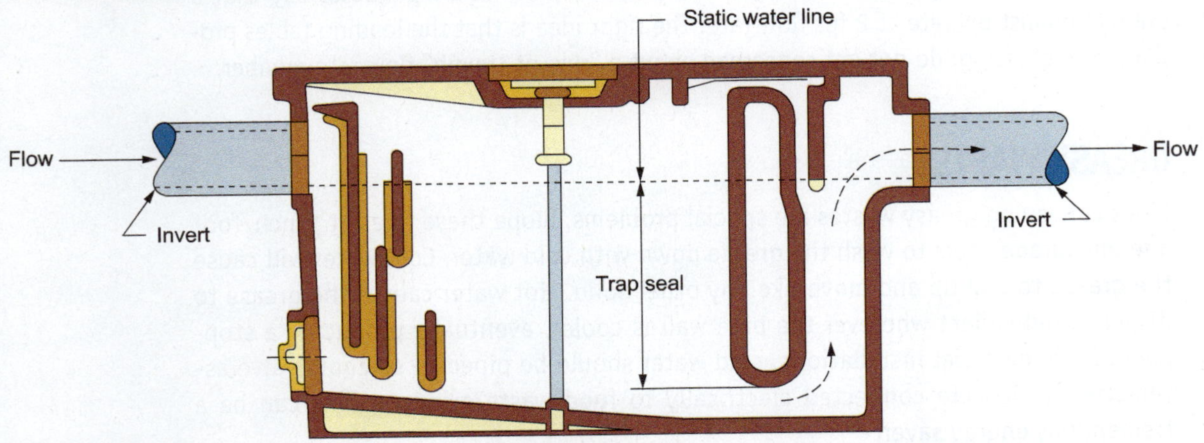

Courtesy of Plumbing-Heating-Cooling-Contractors—National Association

shown in Figure 12-7. Automatic skimming grease recovery units are now becoming quite common. These designs skim the grease from the unit for safe and easy disposal.

Installation of grease interceptors is highly regulated in each local area. Check local codes and ordinances before installing a grease interceptor.

FLOOR DRAINS

As their names imply, floor drains are drainage connections to the building drain or sewer that are located at the floor level. Depending on the nature of the building, they can be used for a wide variety of reasons, such as washing down a commercial kitchen in a restaurant or hospital, serving as a drain for an emergency shower in an industrial building, or as means to wash down a public restroom. Section 7.16.4 of the 2012 National Standard Plumbing Code lists where they are required by code.

Floor drains are typically sized based on their normal, expected flow rate, but cannot be less than 2 inches in nominal size. The expected flow could vary anywhere from a continuous low flow, such as the drain from a commercial ice maker or be a high-volume intermittent flow from a washdown of a food processing plant. In cases where the floor drain serves emergency fixtures, such as an emergency shower or eyewash station, it will need to be sized to handle the gpm of the fixtures, but the floor drain itself does not contribute any DFUs toward the building drainage load calculations.

Since the floor drain is an open connection to the building sewer within the building, it must contain a trap. When the floor drain is expected to receive water on a routine basis, a standard trap with a 2-inch vertical water seal is sufficient. If the drain may see infrequent use, then either a 4-inch deep water seal trap can be installed or some means of adding water to the trap must be supplied. Figure 12-8 illustrates a floor drain connected to a trap primer.

FIGURE 12-8 A trap primer is connected to a floor drain that is used intermittently to ensure that the trap seal is maintained.

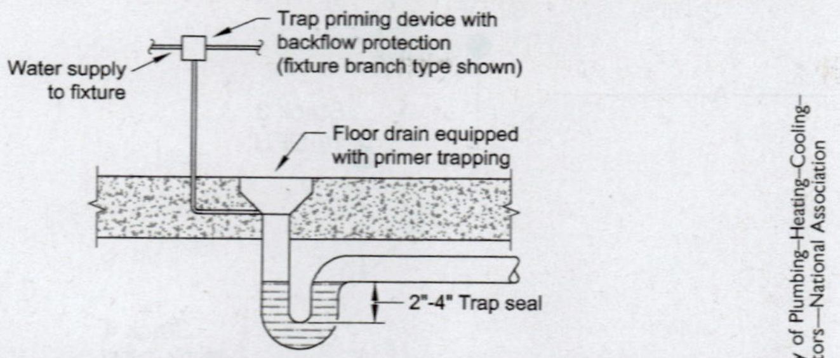

NOTE:
1. A fixture branch type primer can prime more than one trap if it is fitted with a multi-outlet manifold.

In recent green construction, floor drains are starting to be required whenever the crawl space of a residential building is enclosed and encapsulated. In this situation, the floor drain is connected to the waterproof membrane that seals the crawl space. It serves as a means to drain any water that should enter the crawl space either by flooding or by an accidental plumbing leak. This drain should be either connected to the building foundation drain or sump or drain to the outside.

PLANNING THE BUILDING DRAIN INSTALLATION

Usually starting with the plumbing floor plan, the designer (in many cases the plumber) must make a sketch of the stacks and the building drain. The stack loads (DFUs) are noted, and the stacks and building drain are sized as discussed in Chapters 11 and 12.

Next, general considerations must be reviewed, such as the elevation of the connection to the building sewer, total length of the building drain, maximum slope possible (and minimum slope required) for the building drain, fixtures located below the building drain, and possibility of backflow from the sewer (whether it needs a backwater valve or pump system).

Further, the designer must also consider the best locations for cleanouts, best routing for the building drain to serve stack locations, how to avoid obstructions or other utility piping, what piping material and joining method is best, and what soil conditions are present.

The condition or status of each of these topics will have an influence on the details of the installation.

EXAMPLE 4

Figure 12-9 shows a plan view of a two-story building. The building drain is properly sleeved where it passes through the building wall. The sleeve should be a smooth metallic pipe two pipe sizes larger than the building drain. The space between the drain and the sleeve is packed with a pliable waterproof material.

EXAMPLE 4
continued

FIGURE 12-9 Plan view of residential building drain with stacks.

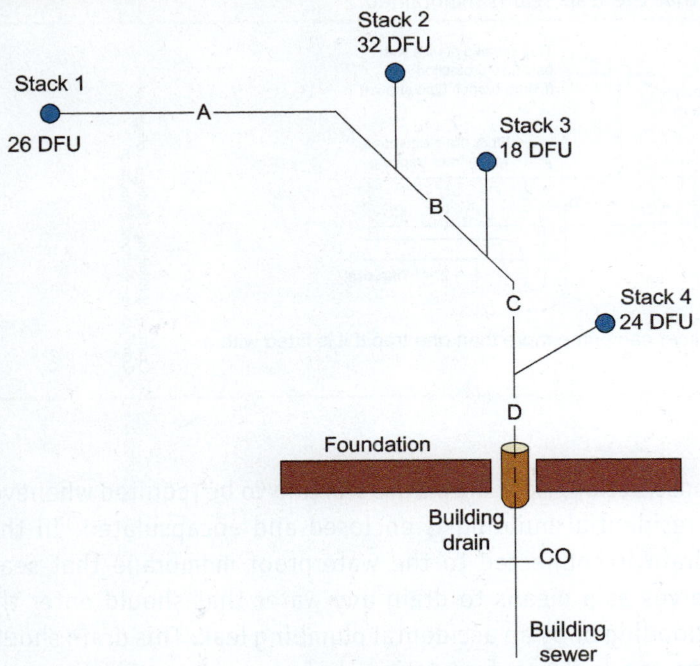

Assume the following loads on the stacks (see Table 12-5):

TABLE 12-5 Load on Stacks

	Load (DFU)	Size (in.)
Stack 1	26	3
Stack 2	32	3
Stack 3	18	3
Stack 4	24	3

The building drain loads are as follows (see Table 12-6):

TABLE 12-6 Building Drain Load

	Load (DFU)	Size (in.) $\frac{1}{4}$ in./ft. Slope
Stack 1	26	3
Stack 2	58	4
Stack 3	76	4
Stack 4	100	4

Thus, the sleeve size must be 6 inches, since it must be two sizes larger than the pipe passing through it.

FINAL CONSIDERATIONS

If the street sewer is a combination sewer and if there are fixtures installed in the basement, consideration should be given to placing a backwater valve in this branch.

Cleanouts must be located along the building drain unless the cleanouts at the base of stacks provide equivalent access. In this case, the stack cleanouts serve the drain requirements. Usually the routing chosen results in minimizing the length of the building drain, but if obstructions or other utilities require it, the alignment can be changed.

The piping material and joining method can be chosen from any selections permitted by the local code. The choice is usually made based on material cost, availability, and experience of the installers.

If the building drain is to be buried below the building floor, the problems associated with trenching must also be recognized. Such problems include avoiding undermining footings, timing of trenching to minimize delays to other trades, placement of excavated material, and all trenching safety considerations.

Adverse soil conditions can reduce the range of acceptable piping materials. If the soil within the building is not suitable for supporting the pipe, it will not be suitable for supporting the building, either. In such a case, the unsatisfactory soil will be removed by the site preparation contractor and replaced with suitable material. Hence, we seldom encounter unsatisfactory supporting material within a building.

Double tees or double tee-wyes should not be used in horizontal lines. Double fittings can only be laid with the side branches level, whereas we desire to enter the building drain above the centerline. Also, tees or crosses permit waste throw from one branch to backflow into the branch opposite it.

When roughing-in the below-slab drainage work, try to arrange fittings rising above the floor so that joints will be just above the finished floor elevation. This will facilitate future changes, as well as guard against having a pipe stub covered with concrete. See Figure 12-10.

HYDRAULIC JUMP

When the flow down a stack changes from vertical to horizontal at the base, the velocity of the fluids (which are a turbulent mixture of solids, liquid, and air) must

FIGURE 12-10 After the concrete has been poured, the water closet stubout can be cut so a closet flange can be installed.

FIGURE 12-11 Hydraulic jump is caused when fast-moving waste from the stack meets with slower-moving waste in horizontal piping.

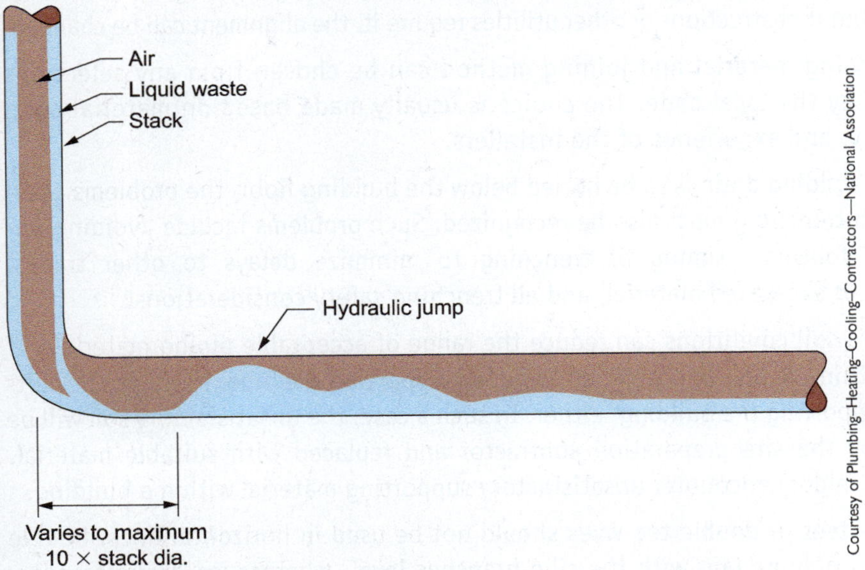

change quickly to the much lower rate of flow in the horizontal line. If there is sufficient material flowing, this rapid reduction in velocity results in a corresponding increase in the cross section of the waste in the pipe, sufficient to close off the pipe. This close-off happens about ten pipe diameters from the base of the stack. This phenomenon, called hydraulic jump, results in positive pressures and pressure variations in the vicinity of the jump. Downstream from the jump, the air and waste materials separate so that the air moves along the top of the liquid/solid mixture flowing along the lower part of the drain pipe. Hydraulic jump is shown in Figure 12-11. To avoid this turbulent, high-pressure region, do not make a connection to a horizontal drain within ten pipe diameters of the base of a stack unless additional venting is provided.

Similar circumstances are present in any vertical-to-horizontal change in flow where the vertical dimension is at least 15 feet and the change of direction is 45° or greater.

While it is rare to have hydraulic jump in buildings of only one or two stories, plumbing professionals should be aware of this condition and how to minimize it. Ways to minimize hydraulic jump include increasing the size of the horizontal drain line to be larger than that of the vertical stack, increasing the slope of the horizontal line, or increasing the radius of the turn from vertical to horizontal. This last method is most easily done by using multiple fittings to make the turn. See Figure 12-12 for a 90° turn using two 45° turns and Figure 12-13 for a change made using four $\frac{1}{16}$ bends ($22\frac{1}{2}$ ells).

FIGURE 12-12 Gradual transition from vertical to horizontal using two 45° fittings.

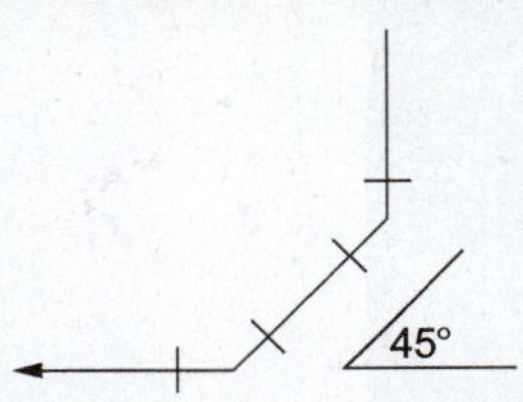

FIGURE 12-13 Gradual transition from vertical to horizontal using four $22\frac{1}{2}°$ fittings.

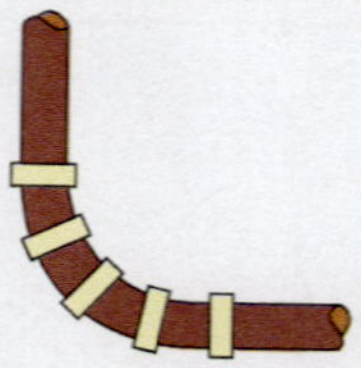

CLEANOUTS

When locating cleanouts, consider how drain-cleaning equipment can be brought to the location and used in the space. Avoid areas that will have expensive finishes or furnishings. Whenever possible, place cleanouts in utility rooms, storage spaces, or other rough areas.

Cleanout locations also must be selected so that access is provided to all parts of the building drain. The cleanout covers should be heavy enough for the traffic over them and they should be made of corrosion-resisting materials.

The spacing between cleanouts is based on the limitations of drain-cleaning equipment. The spacing in Table 12-7 is generally considered adequate.

INSTALLATION OF BUILDING DRAINS

As in other areas of plumbing, there are specific procedures for installing building drains. A uniform slope is maintained by preparing the trench properly, checking each section of pipe when installed, and backfilling properly.

1. Use large-radius turns, by using either long-sweep fittings or multiple fittings.
2. Run lines serving stacks as directly as possible.
3. Test all piping and call for inspection before backfilling.

The typical hydrostatic test is done by closing all lower openings, adding a 10-foot standpipe, and filling the piping with water so that a minimum of 10 feet of water is present throughout the system to be tested. The water must be retained as described for 15 minutes without water loss. Test time begins after air has been purged. A pneumatic test requires pressurizing the section to be tested to 5 psi with compressed air, carbon dioxide, or nitrogen. The pressure must be held without loss for 15 minutes.

Plastic pipe is not permitted to be air-tested per requirements of the plastic pipe industry.

Support

In installations where the building is built on a slab, the building drain installation involves trenching, continuous bedding, and covering with a concrete slab after

TABLE 12-7 Cleanout Spacing

Pipe Size (in.)	Maximum Spacing (ft.)	Minimum Cleanout Size
3 to 4	75	Line size
5 to 10	100	4″
Greater than 12	300	Manhole

FIGURE 12-14 J hangers are quick and easy to use when hanging pipe to a wooden structure.

backfilling. However, when the building is built on a crawl space or a basement with no plumbing, the exposed building drain must be supported on hangers or supported on blocks or piers. In either case, the slope of the building drain must be carefully calculated and maintained to meet the requirements discussed earlier. Special care must be observed when such supported pipe must be backfilled later. The backfill must be firmly tamped to provide continuous support for the pipe. Otherwise, problems will develop at the rigid support points. In a crawl space, special hangers can be used, such as those shown in Figure 12-14.

Where a buried pipe passes through a wall, the opening should be sleeved with a pipe two sizes larger than the line, and the annular space packed with a waterproof material that is flexible enough to let the drain line move somewhat. (See Figure 12-15.)

If the pipe line is supported with hangers, use a string line or laser to align the hangers. Be sure the hangers are substantially attached to the structure. Since a stoppage could cause the drain to fill with water, the hangers must be adequate to support the pipe and its contents. Additional support will be required at the base of the stack and at all changes of direction. In some cases, thrust blocks may

FIGURE 12-15 Piping must be sleeved when passing through masonry structural members.

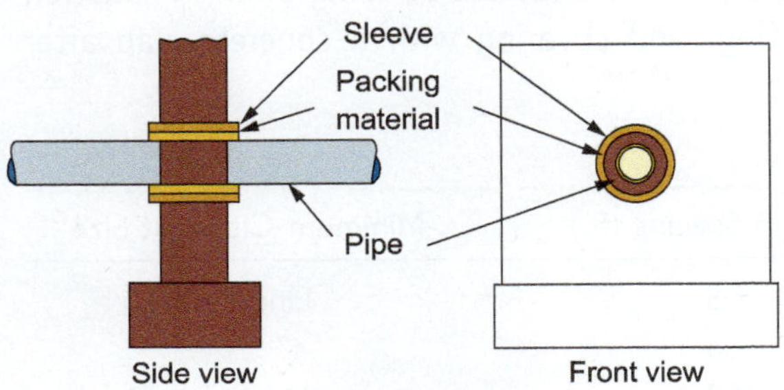

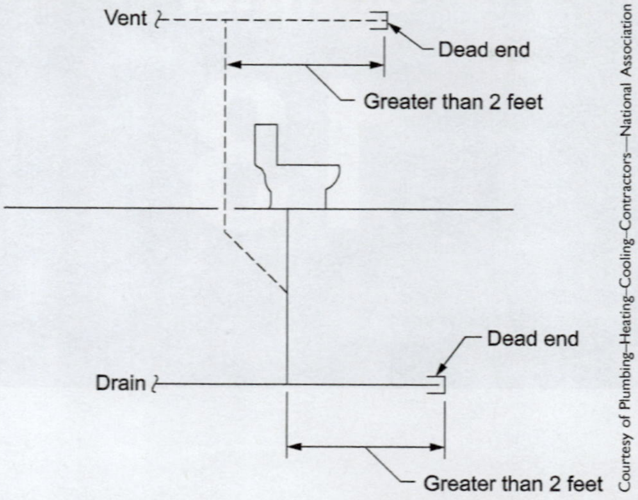

FIGURE 12-16 Dead ends can create collection areas for waste and must be avoided.

be required. In areas that are prone to earthquakes, additional seismic supports must be added to suspended pipe to prevent unwanted swaying of the pipe. Seismic supports must be calculated by an engineer.

Avoid dead-end installations, since they can harbor various disease-supporting colonies, or even be a home for vermin. Cleanout piping or extensions and stub ends for future expansion are, in effect, dead ends. Use them only where required. Examples of dead ends are shown in Figure 12-16.

PROTECTION

While the project is under construction, protect the work every night to protect passersby from injury. Vandalism after hours can introduce heavy expenses, and no expense is equal to someone being injured at a jobsite.

REVIEW QUESTIONS

Fill in the Blank

1. Fixtures that could be flooded when the sewer is overloaded should be protected by a ________ ________ or ________ ________.
2. The building drain is sized from the same table as for the building ________.
3. A flow rate of ________ is necessary to move solids in a drain line.

True or False

4. ________ Cleanouts or cleanout equivalents must be located for access to all of the drainage pipe.
5. ________ A dead end is an extension of more than 2 feet of a drain or vent beyond the last active connection.

Chapter 13

LEARNING OBJECTIVES

The student will:

- Identify proper location and techniques for notching, boring, and altering the structure for pipe penetrations.
- Describe the construction details of stacks, including supports, hangers, branch intervals, and branch wastes.

Stacks

STACKS

The waste stack is the vertical pipe that receives the discharge from horizontal branches that are used to convey waste away from fixtures. A stack is sized according to the drainage fixture unit (DFU) load that is connected to it. See Chapter 11 and Table 11-2 for details on DFU loads.

Stacks that receive fecal matter, human or animal wastes, or similar sewage from water closets, urinals, or similar fixtures are called soil stacks. Vent stacks are special stacks that do not receive any of the above-mentioned wastes; they convey air so that the pressures within the drainage, waste, and vent (DWV) system remain equalized. Most codes require at least one vent stack in a building to extend from the building drain to a point at least 6 inches above the roof. Storm water stacks are used to convey storm water that is collected from roofs and other elevated areas and discharge it into either a combined or storm water sewer. A stack that collects any other unwanted liquid is called a waste stack.

Thus, *stack* is a general term for any vertical line, including offsets of soil, waste, vent, or inside conductor piping. This term does not include vertical fixture and vent branches that do not extend through the roof or that pass through only one or two stories before being reconnected to the vent stack or stack vent.

The following are some important limitations of stacks:

- They must be vertical (an angle of 45° or more from the horizontal).
- They must extend through the roof or pass through at least two stories before connecting to another stack.

MATERIALS

Most codes allow any of the same materials used for building drains to be used for stacks. See Chapter 12. However, the support distances and loading will be different.

LOADING AND CAPACITY TABLE

Stacks are designed to be only about 30 percent full at design load. This seemingly light loading is necessary to minimize pressure changes and surges when the stack is at rated load. It was determined through testing by the National Institute of Standards and Technology that at very light loads, water will flow along the wall of the stack as it descends. As flow increases, air friction will cause the water to diaphragm across the pipe, forming a short slug of water, which moves down the pipe like a piston, increasing air pressure ahead of it and reducing air pressure behind it. This pressure difference will break up the slug, which will reform again in a short distance, break up, and reform, and so on.

Table 13-1 presents the capacity of stacks in terms of DFUs. The table also indicates the capacity of fixture branches. Notice that the ratings of fixture branches are less than the values for the same pipe sizes when they are used in the building drain. The reason for this difference is that building drains do not receive waste directly from fixtures, so the building drain has the capability to have more aggressive inputs from stacks.

Notice that stacks with three or fewer branch intervals have lower ratings than stacks with more intervals. This is because, when three or fewer branches are discharging into the stack, the limitation is the capacity of the branches, not the capacity of the stack.

TABLE 13-1 Horizontal Fixture Branches and Stacks

Maximum Number of Drainage Fixture Units (DFU) That May Be Connected to Any Horizontal Fixture Branch, a Stack of Three Branch Intervals or Less, or Stacks of more than Three Branch Intervals

			Stacks with more than Three Branch Intervals	
Diameter of Pipe (in.)	Any Horizontal Fixture Branch[1] (DFU)	Stacks with Three Branch Intervals or Less (DFU)	Total for Stack (DFU)	Discharged into One Branch Interval (DFU)
$1\frac{1}{4}$	1	1	1	1
$1\frac{1}{2}$	3	4	8	2
2	6	10	24	6
3	20[2]	48[3]	72[3]	20[3]
4	160	240	500	90
5	360	540	1,100	200
6	620	960	1,900	350
8	1,400	2,200	3,600	600
10	2,500	3,800	5,600	1,000
12	3,900	6,000	8,400	1,500
15	7,000			

Notes:

1. Does not include branches of the building drain.
2. See Section 11.5.6.a & d for the restrictions on the number of water closets installed on 3″ horizontal fixture branches.
3. See Section 11.5.6.b & d for the restrictions on the number of water closets on 3″ stacks and discharged into any branch interval of a 3″ stack.

Source: Table 11.5.1B 2012 National Standard Plumbing Code.

TERMINAL VELOCITY

Because of the actions of waste traveling down the stack, a maximum velocity of 10 to 15 fps is reached in 10 to 15 feet of drop. This means that after about 15 feet of vertical travel, the waste material in the pipe will be moving at a constant speed. This velocity is called the *terminal velocity*, and the vertical distance required for the falling waste to reach that velocity is called the *terminal length*.

VENTING

The venting system connected to the stack provides the path(s) for excess pressures to be dissipated or for reduced pressures to be equalized. A properly sized and located venting system will limit the pressure variation at any trap outlet to ±1-inch water column, thus keeping the trap seal from being blown into the room or being drawn into the stack. (Recall that the standard trap seal is 2 inches.)

STACK SUPPORT

One consequence of the terminal velocity concept is that the base of a stack in a high building does not have any more hydraulic load on it than a stack base in a two- or three-story building. The base of the stack needs to be supported, but the main function of this support is to keep the base fixed during the construction process and until stack supports can be installed at upper floors. The hydraulic forces encountered in use will be less than the structural loads during construction, at least for most metallic pipes.

The stack support may be concrete, brick set in mortar, metal brackets attached to the building, or other suitable devices. See Figure 13-1.

The following support intervals are recommended:

- Cast iron soil pipe—at the base and each story (at every joint when horizontally)
- Threaded pipe—at every other story
- Copper tubing—at each story and not more than 10 feet apart
- Plastic pipe—at each story and not more than 10 feet apart

FIGURE 13-1 One method of stack support.

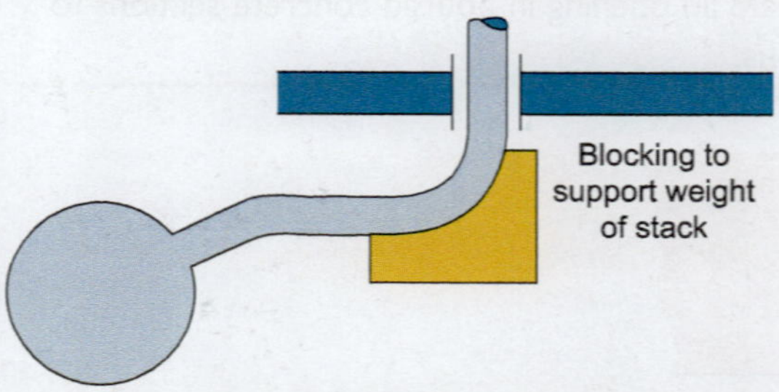

Support all offsets with proper hangers. Do not use any material or device that will compress, distort, or abrade the pipe surface. All hangers must be able to support the fully loaded pipe. Hangers should be compatible with the pipe, especially where moist conditions exist.

SLEEVES AND THIMBLES

For new work, lay out the stack openings as the building is being erected, especially if it is of poured-concrete construction. Not only is it more economical than core-drilling the openings later, it may not be possible to get permission to cut the openings later. The structural engineer can accommodate the openings if they are formed in the slab before the pour, but the structure may be unacceptably weakened by drilling unplanned holes. If metal reinforcing bars in prestressed concrete are cut, the structural member will be severely weakened.

Sleeves should be made of well-reamed metallic pipe, two sizes larger than the pipe line passing through the sleeve. The sleeve should extend at least 1 inch above the floor surface so that spillage on the floor will not pass through the opening. The annular space between the pipe and sleeve should be filled with elastic fireproofing sealant.

Sometimes thimbles are used in lieu of sleeves. Thimbles form the opening in the concrete and are then removed from the hole. Either sleeves or thimbles must be securely attached to the forms so that they are not moved during the pouring operation.

For either thimbles or sleeves, precise placement is very important. If the openings are out of line, the plumbing installation will be much more difficult. A thimble and a sleeve are shown in Figure 13-2.

For stacks running through wooden floors, the best work is performed by cutting the floor with hole saws slightly larger than the pipe to be installed. Close the clearance space remaining to provide a fire-stop and to prevent vermin from using the gap to move from floor to floor. Never wedge the pipe in place, as pipe movement will cause building noises.

Install stacks carefully, keep them straight and plumb, and use enough supports to carry the weight and maintain alignment. Fire-stop at each floor penetration according to local code.

FIGURE 13-2 Thimbles and sleeves create an opening in poured concrete sections to allow for pipe penetrations.

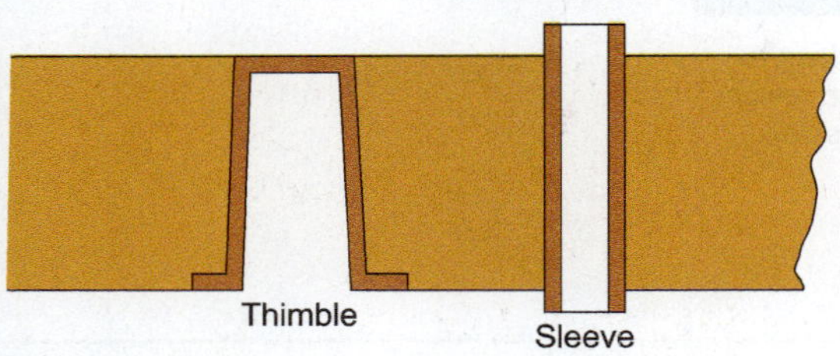

If freezing temperatures occur at the site, protect any stack exposed to freezing conditions by placing insulating material between the stack and the exposed wall.

FITTINGS AND SUPPORT

Connections from stacks to building drains should be made with wye and $\frac{1}{8}$ bends or similar fittings developing equivalent sweep. The fittings should be arranged so that the flow enters the building drain above the centerline of the horizontal line. Stack support is also important, as shown in Figure 13-1.

Double tees or tee-wyes should not be used in horizontal lines. If double fittings are installed in such lines, they can be laid only with the side branches level on the horizontal centerline. Tees or crosses also permit waste to cross over from one branch to the other.

CLEANOUTS

When locating cleanouts, consider how drain-cleaning equipment can be brought to the location and used in the space. Avoid areas that will have expensive finishes or furnishings. Whenever possible, place cleanouts in utility rooms, storage spaces, or other rough areas.

IN THE FIELD
While the project is under construction, protect the work every night and protect passersby from injury. Vandalism after hours can introduce heavy expenses and no expense is equal to someone being injured at a jobsite.

Cleanout locations also must be selected so that access is provided to all parts of the building drain. The cleanout covers should be heavy enough for the traffic over them and they should be made of corrosion-resisting materials.

Different piping materials should be joined with proper adapters. Stack bases and changes of direction should be supported with thrust blocks or other substantial methods.

FULL-SIZE STACKS

Some codes require at least one stack to be installed full-sized, from the base up through the roof. The reason for this requirement is that it is believed the building sewer and building drain will be better ventilated. When dealing with vent stacks, most codes do not allow them to be any smaller than one-half the diameter of the drain they serve.

FIGURE 13-3 Branch intervals.

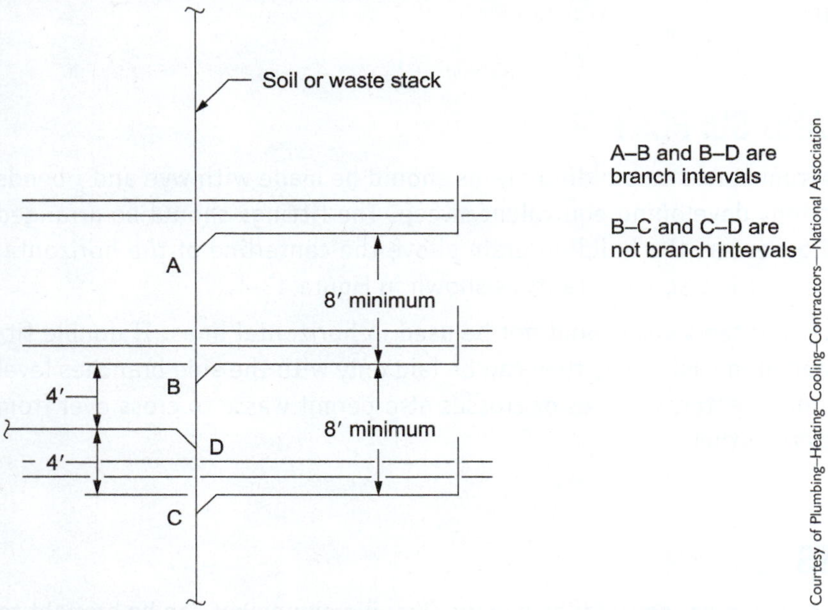

Other stacks in the building—or all stacks, where codes permit—are sized in each section according to the load connected and flow produced at each section.

Stacks are differentiated by the number of branch intervals connected. A branch interval is the distance along a soil or waste stack corresponding in general to a story height, but in no case less than 8 feet, within which the horizontal branches from one floor or story of a building are connected to the stack. See Figure 13-3.

Typical Branches

In many jobs, the fixture load connected at each branch interval is the same for several intervals. The fixture layout for a single interval can be referred to as "typical" on elevation drawings. This means that the pattern occurs more than once. The use of such recurring patterns aids in design and estimating, but is even more valuable in prefabricating the work and facilitating installation.

Frequently, it is convenient to test the work one branch interval at a time. Since this interval represents a height of 8 feet or more, the test pressure is in the range of the normal required value of 10 feet head of water.

Branch Loads

The allowable load that may be connected to the top interval of a stack is greater than the loading permitted on any horizontal branch below the top. The reason for this provision is that, with flow coming down the stack, a lighter load must be installed on the lower horizontal branches so that the back-pressure needed to punch through the flow in the stack does not produce problems in the branch.

At the present time, most codes do not recognize the many possible variations of stack size, branch size, and load above the branch to fine-tune the branch load for all the possible conditions. However, it would be possible for an engineered system to take advantage of these considerations to obtain maximum usefulness for a given installation. Computer sizing models have been developed to assist the professional in the design of a complete DWV system, based upon the discharge characteristics of individual fixtures. These systems will be more extensively used in the future.

STACK FITTINGS

For purposes of this discussion, you should realize that a horizontal branch load can best be introduced into the stack with a downward component of motion. Thus, a 45° wye or 60° wye, a combination wye and $\frac{1}{8}$ bend, or a tee-wye produces this downward component. This would be similar to how an off-ramp on a highway allows a person to gradually slow down and then make a 90° turn. If it were a sharp turn, like in a city or town, you would have to come to a complete stop and then turn, losing all of your momentum. Sanitary tees or tapped tees provide a satisfactory connection, but the horizontal flow will see a larger resisting pressure than the first types listed. See Figure 13-4.

Double wyes or tee-wyes will produce satisfactory connections, but crosses or tapped crosses should be avoided unless the side openings are at least two sizes smaller than the vertical run. Otherwise, crossover from branch to branch may occur.

Fittings selected for the base of the stack should have as large a radius as possible to minimize problems with hydraulic jump. See Chapter 12.

FIGURE 13-4 The flow pattern on the left allows for material to maintain its flow. The flow pattern on the right does not.

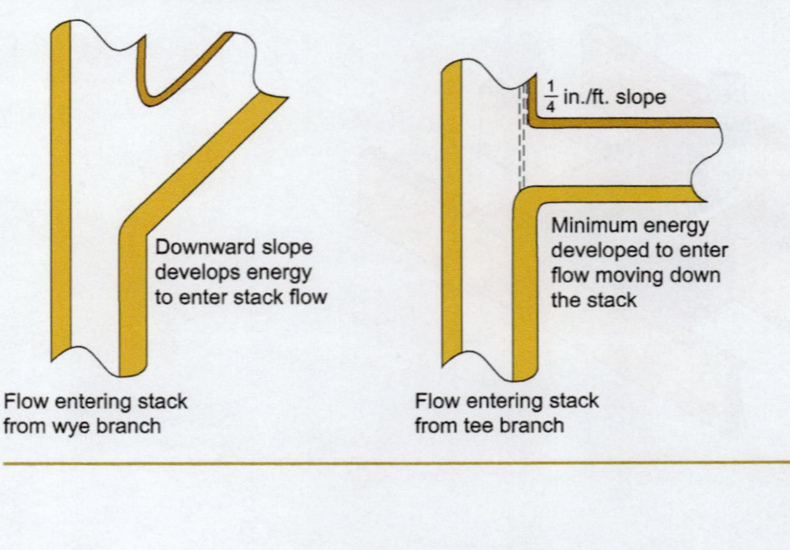

STACK ROUTE

Careful consideration must be given to the route of the stack through the building. If any structural members must be cut to provide space for the stack, you must get permission and instructions from the architect or general contractor before proceeding. Cut openings of sufficient size, but do not make the openings oversized. Tight or forced fits must also be avoided, as expansion and contraction noises will be present.

It is generally preferred to drill out the center of studs or joists to permit piping to pass, rather than notching the member, but again, obtain permission and instructions from the architect or general contractor.

If a joist or stud has to be notched, cover the opening with a protective metal plate to prevent sheetrock, lathing, or flooring nails from piercing the installed pipe. Figures 13-5 through 13-10 demonstrate preferred structural protection, cutting, and notching procedures.

The final building structure must remain in accordance with the building code. The amount of added support required is based on the load-bearing requirements of the wall or floor as well as the depth and location of cut.

Note 1 (see Figure 13-6)

Floor and ceiling joists may be notched in top or bottom surface. Notches should not exceed one-sixth ($\frac{1}{16}$) the depth of the joist. See Table 13-2.

Note 2 (see Figure 13-6)

The middle one-third ($\frac{1}{3}$) of the span can never be notched.

Note 3 (see Figure 13-6)

A notch no more than one-third ($\frac{1}{3}$) of the joist depth is acceptable in the top of the joist when located near a support. See Table 13-3.

FIGURE 13-5 Headers allow the two adjacent members to support the ends of the one that was cut.

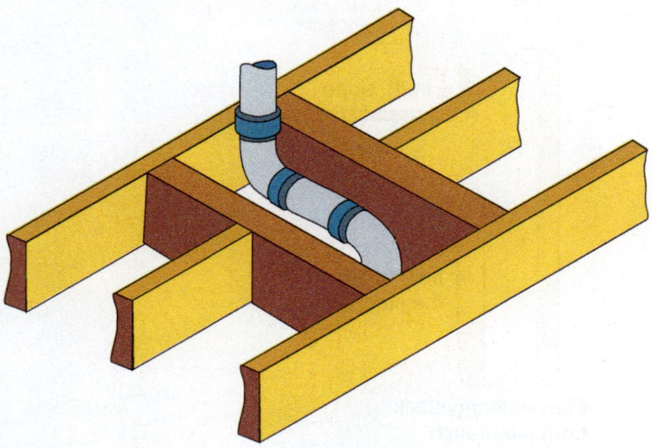

FIGURE 13-6 Cutting and notching a joist based on its location.

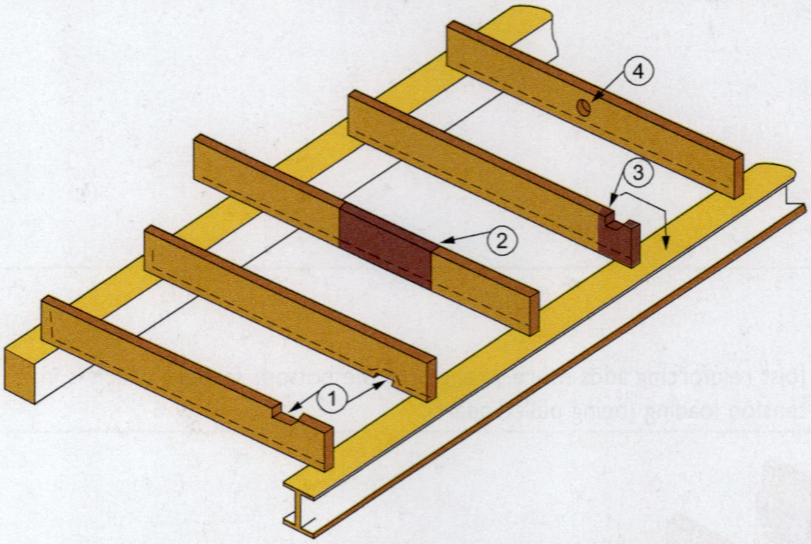

Note 4 (see Figure 13-6)

Holes drilled in joists cannot be within two (2) inches of the top or bottom edge of the joist, and their diameter cannot exceed one-third ($\frac{1}{3}$) the joist depth.

Note 5 (see Figure 13-7)

The maximum hole to be drilled in a load-bearing 2″ × 4″ stud is $1\frac{7}{8}$ inches in diameter.

FIGURE 13-7 Cutting and notching studs and top/bottom plates.

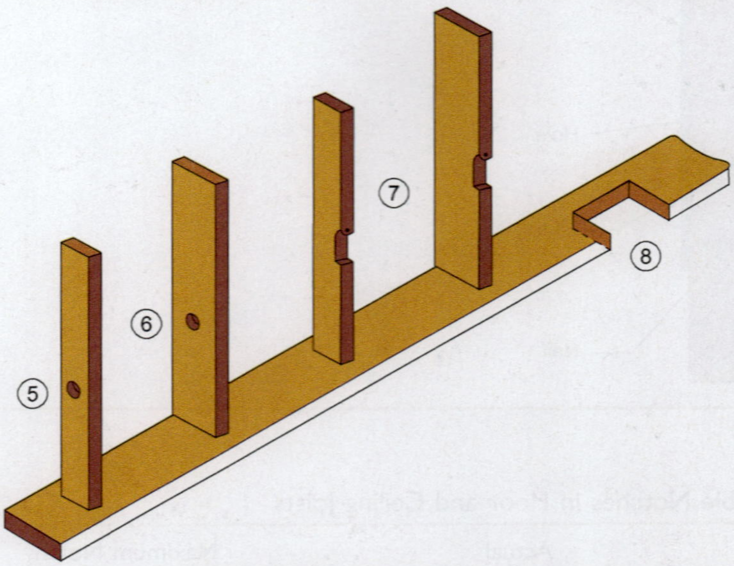

FIGURE 13-8 Stud reinforcing and strike protection.

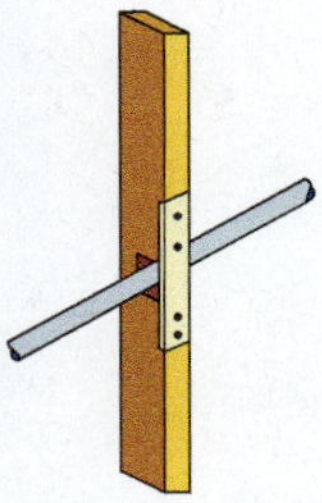

FIGURE 13-9 Joist reinforcing adds more strength to the bottom section, which is in tension loading (being pulled apart).

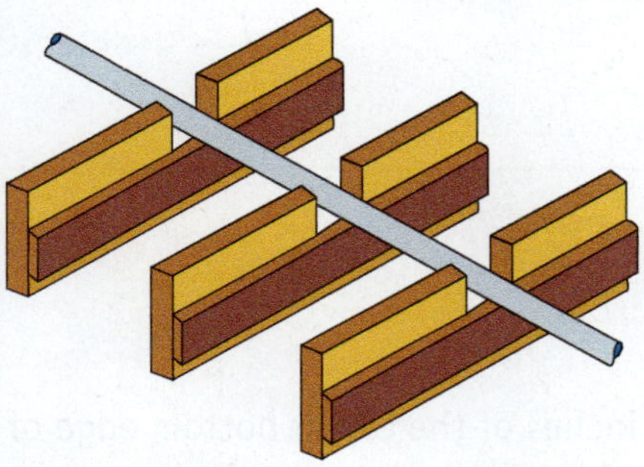

FIGURE 13-10 Nail plate protection.

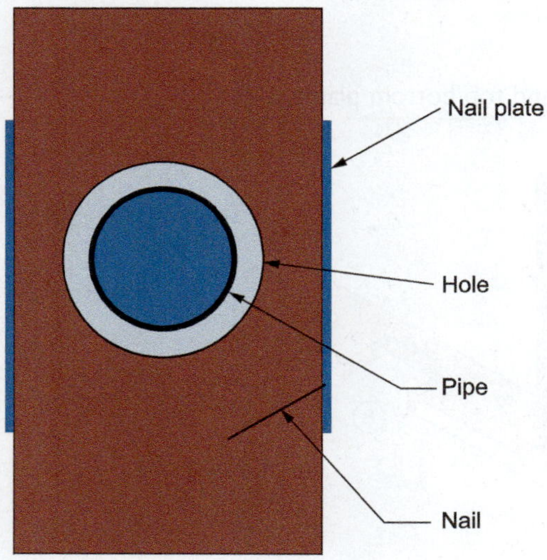

TABLE 13-2 Allowable Notches in Floor and Ceiling Joists

Size of Joist	Actual	Maximum Notch
2″ × 12″	$1\frac{1}{2}$″ × $11\frac{1}{2}$″	$1\frac{7}{8}$″
2″ × 10″	$1\frac{1}{2}$″ × $9\frac{1}{2}$″	$1\frac{1}{2}$″
2″ × 8″	$1\frac{1}{2}$″ × $7\frac{1}{2}$″	$1\frac{1}{4}$″

TABLE 13-3 Allowable Notches in Floor and Ceiling Joists, Near Supports

Size of Joist	Actual	Maximum Notch
2″ × 12″	$1\frac{1}{2}$″ × $11\frac{1}{2}$″	$3\frac{3}{4}$″
2″ × 10″	$1\frac{1}{2}$″ × $9\frac{1}{2}$″	3″
2″ × 8″	$1\frac{1}{2}$″ × $7\frac{1}{2}$″	$2\frac{1}{2}$″

TABLE 13-4 Allowable Notches in Studs

Stud Size	Actual	Maximum Notch
2″ × 6″	$1\frac{1}{2}$″ × $5\frac{1}{2}$″	$2\frac{1}{2}$″
2″ × 4″	$1\frac{1}{2}$″ × $3\frac{1}{2}$″	1″

Note 6 (see Figure 13-7)

The maximum hole to be drilled in a load-bearing 2″ × 6″ stud is $2\frac{1}{2}$ inches in diameter.

Note 7 (see Figure 13-7)

Studs may be notched as shown in Table 13-4.

Note 8 (see Figure 13-7)

When unsecured top plates are cut more than one-half ($\frac{1}{2}$) their width for piping, reinforce with 18-gauge steel straps, 24-gauge steel angle, or suitable blocking.

Note 9

The final building structure must remain in accordance with the building code or appropriate standard. Verify all cutting and notching requirements with the code or standard applicable to the specific job.

ROOF PENETRATIONS

Roof penetrations must be executed carefully to avoid water leakage. First, cut the opening with minimum clearance and align it properly for the stack. Support the stack so that it is held in place. Use a quality flashing made of long-lasting material.

Most modern flashings use a neoprene collar to make the joint at the pipe waterproof. Earlier flashings made of lead are folded over into the pipe to keep rain or snow passing down the pipe, rather than around it. See Figures 13–11 and 13–12.

The flashing must be set in asphalt, tar, or cement to seal the flashing to the roof. Don't be stingy with the sealant material. On sloping roofs, install the flashing beneath upper shingles and over lower shingles.

Be especially careful when working on older roofs. Some tend to be brittle and are more easily damaged than other roofs. Even walking on the roof may do damage.

Some codes governing colder areas require minimum pipe sizes passing through the roof. If the stack has to be increased, the size change should take place at least 1 foot below the roof in the insulated spacing. The fitting(s) used should assure flow down the stack. See Figure 13-13.

FIGURE 13-11 Flat roof penetration.

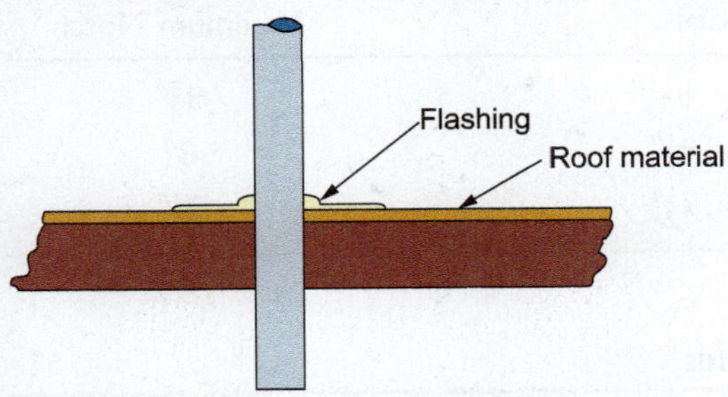

FIGURE 13-12 Waterproofing a vent terminal.

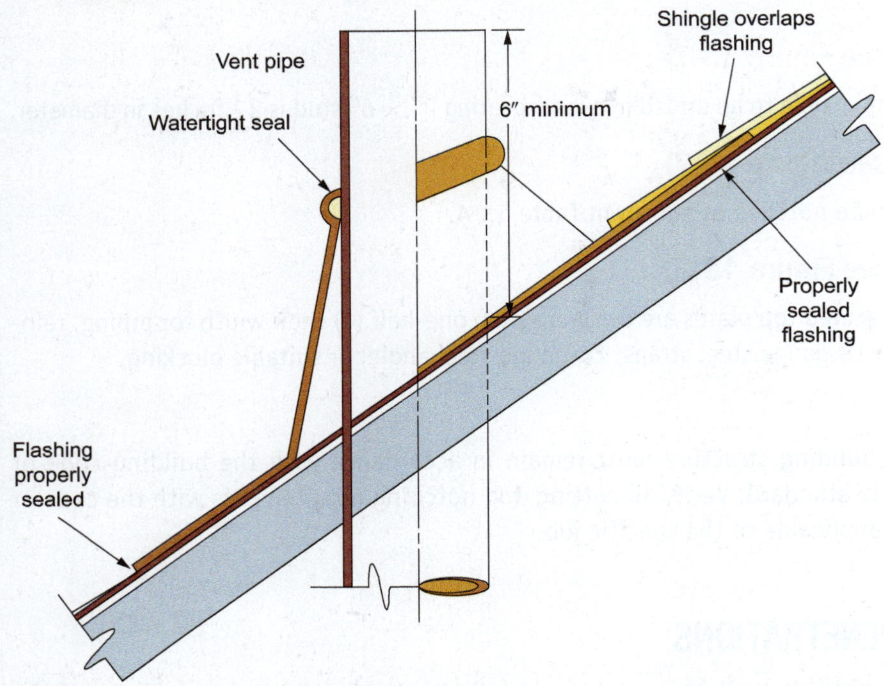

FIGURE 13-13 Protective method to prevent frost closure of a vent in colder climates.

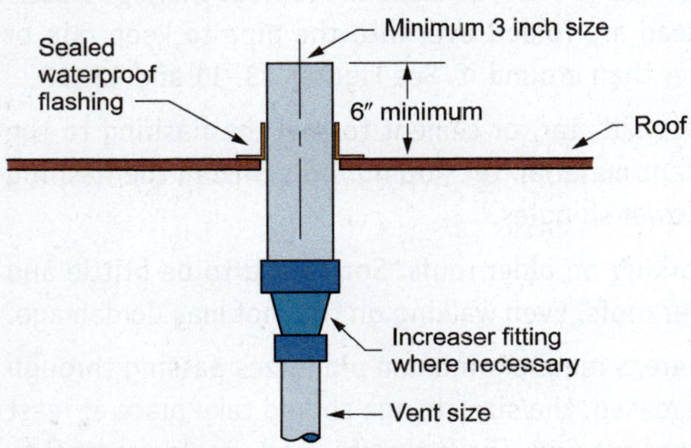

Courtesy of Plumbing–Heating–Cooling–Contractors—National Association

The vent should extend at least 6 inches above the roof—higher in heavy snow areas. The stack opening must not be located below a door, window, ventilating opening, or roof overhang, nor within 10 feet horizontally from any such opening.

Stacks must never be used for any structural purpose such as supporting aerials, satellite dishes, flagpoles, railings, and so on.

If a roof is used for any occupancy purpose (terrace, sun deck, etc.), then any vent in that area should be at least 7 feet above the finished deck surface.

Since plastic pipe above the roof will be exposed to sunlight, it must contain pigment protection from ultraviolet radiation or be otherwise so protected.

TESTING STACKS

Rough testing of stacks is usually accomplished with water by testing 10 foot to 12 foot vertical sections as construction progresses or, on nonplastic piping, by using a 5 psig air test on completed stacks. In some areas, codes require that the finished installation be tested with smoke or peppermint oil. Since the rough plumbing has already been tested, this final test is to verify trap seals and final connections.

If smoke is used, it is introduced into the stack, which is capped off when smoke appears at the terminal. The fixture traps and connections are then checked for any sign of leakage.

If peppermint oil is used, it must be placed in the stack by someone outside the building. The stack is then capped. The person who puts the peppermint oil in the stack must not come indoors as residual odors from the peppermint oil may confuse the inspection process. The person inside the building checks traps and connections for any sign of mint odor.

Both the rough and final tests may have to be conducted in the presence of the plumbing inspector.

Always use an extension hose to release air pressure from the test plug following the final test. Never reach into the fitting to release the air. There have been cases where the technician's hand was lodged in the fitting by the force of the water against the plug. In one incident, the technician was in a ditch and nearly drowned.

REVIEW QUESTIONS

Fill in the Blank

1. A stack is a vertical waste, soil, vent, or storm pipe that extends either through the __________, or for at least __________ stories.
2. At rated full load, a stack is about __________ full.
3. Some codes require at least one stack to extend __________ from the base through the roof.
4. A __________ is a distance along a stack corresponding to a story height.
5. A joist may not be notched in the middle __________ of the span.

Chapter 14

LEARNING OBJECTIVES

The student will:

- Apply the concepts of fractions as they relate to length calculations.
- Apply the concepts of decimals as they relate to length calculations.
- Describe the concept of squares.
- Describe the concept of cubes.
- Describe the concepts of roots of numbers.
- Demonstrate how to use a square/square root table and cube/cube root table.

Applied Plumbing Mathematics

Several mathematical operations and principles were presented in *Plumbing 101*. Some pertinent definitions and methods are given in this chapter to refresh your memory about these mathematical operations.

FRACTIONS

A fraction is the ratio of two whole numbers using the form "*a*" divided by "*b*," or (a/b). The value a is the numerator and b is the denominator. Fractions should normally be expressed in simplest terms, meaning that all common factors have been divided out. For example, if a pie were cut into four equal pieces and two of them were removed, you could say that two-fourths ($\frac{2}{4}$) remains. But it is customary to reduce the fraction to its simplest terms, which is to say that one-half ($\frac{1}{2}$) of the pie remains.

Improper Fractions

An improper fraction is a fraction in which the numerator is larger than the denominator. Sometimes, when adding fractions, the sum will be an improper fraction. Usually, an improper fraction is reduced to a mixed number, which is a whole number plus a fraction.

EXAMPLE 1

$$\frac{5}{8} + \frac{3}{8} + \frac{5}{8} + \frac{7}{8} + \frac{5}{8} = \frac{25}{8}$$

If this were a length that needed to be measured and then cut, the $\frac{25}{8}$ would have to be converted into a whole number and a proper fraction. The simplest method of doing this would be to divide the denominator into the numerator and place the remainder over the denominator. For example,

$$\frac{25}{8} \Rightarrow 8\overline{)25}^{\,3} \text{ with a remainder of 1}$$

Therefore, $\frac{25}{8} = 3\frac{1}{8}$

Adding or Subtracting Fractions

As you may have noticed in the example above, all of the fractions being added had a common denominator, which means the numerators can simply be added or subtracted. The resulting numerator is then set over the common denominator. If all of the fractions did not have the same denominator, a common denominator would need to be found first.

Multiplying Fractions

When fractions are multiplied together, the process consists of multiplying numerator by numerator and denominator by denominator to find the product. Of course the fraction is then reduced to its lowest terms.

EXAMPLE 2

$\frac{3}{4} \times \frac{1}{2} = \frac{3}{8}$

IN THE FIELD
Whenever you need to divide a fraction in half, such as to find the center of something, simply double the denominator and leave the numerator.

When multiplying or dividing by a whole number, it is useful to express the whole number in fraction form by setting it over the denominator one (1).

When multiplying fractions, it can simplify the math to cancel out all common factors in the numerators and denominators before carrying out the multiplication. For example,

EXAMPLE 3

$\frac{3}{1} \times \frac{2}{3} \times \frac{1}{4} \Rightarrow \frac{3 \times 2 \times 1}{1 \times 3 \times 4} \Rightarrow$, which can be rearranged as $\frac{3}{3} \times \frac{2}{4} \times \frac{1}{1}$.

Since any number divided by itself equals 1, the first and last fractions are reduced to 1.

The expression can now be reduced to $\frac{2}{4}$, which is $\frac{1}{2}$.

Dividing Fractions

To divide fractions, simply invert the second fraction and multiply it by the first fraction. For example,

EXAMPLE 4

$\frac{3}{4} \div \frac{1}{2}$ can be rearranged to $\frac{3}{4} \times \frac{2}{1} = \frac{6}{4} = \frac{3}{2} = 1\frac{1}{2}$.

At first this may seem confusing, but you have to remember when you divide by a small number (such as $\frac{1}{2}$), you are looking for how many times a small number will go into a bigger number.

Mixed Numbers

A whole number and fraction (mixed number) such as $3\frac{1}{4}$ must be expressed as an improper fraction before multiplying or dividing by the methods just described. Remember, this is done by multiplying the denominator by the whole number and then adding it to the numerator. The denominator stays the same.

EXAMPLE 5

$$3\frac{1}{4} = \frac{3 \times 4 + 1}{4} = \frac{13}{4}$$

DECIMALS

When adding or subtracting decimals, keep the decimal points in a straight vertical line to assure accuracy.

When multiplying decimals, the number of places after the decimal point in the answer is equal to the sum of places in both numbers being multiplied.

When dividing decimals, move the decimal point to make the second number (the divisor or the number you are dividing by) a whole number. Move the decimal point the same number of places in the first number (the dividend or the number being divided) and then locate the decimal point in the answer directly above the new location in the dividend.

Decimal to Fraction

A decimal is simply the numbers to the right of the decimal divided by 10, 100, 1,000, and so on. For example, 0.5 is the same as 5 divided by 10, or $\frac{1}{2}$, and 0.375 is 375 divided by 1,000, or $\frac{3}{8}$.

Fraction to Decimal

To convert a fraction to a decimal, divide the numerator by the denominator. In some cases, this will not come out evenly, so you may have to round off your answer.

SYSTEM OF MEASUREMENT

The system of units used in the United States is called the Imperial (formerly British) system of measurement. Lengths are measured in feet and inches, and weights are measured in pounds and ounces.

To convert back and forth from feet (F) to inches (I), use the following formula:

$$I = F \times 12$$

or

$$F = \frac{I}{12}$$

EXAMPLE 6

What length of pipe is necessary to connect two fittings that are 20 inches apart center-to-center (c–c)? The fitting allowance is $\frac{1}{2}$ inch.

$$\text{Length} = 20'' - 2 \times (\text{fitting allowance})$$

$$\text{Length} = 20'' - 2\left(\frac{1''}{2}\right)$$

$$\text{Length} = 20'' - 1''$$

$$\text{Length} = 19'' \text{ end-to-end (e–e)}$$

EXAMPLE 7

If an 18-inch piece of pipe is needed in each of 40 homes, how many feet of pipe must be ordered?

$$\text{Total} = \frac{(40 \times 18'')}{12}$$

$$\text{Total} = \frac{720''}{12} \text{ (divide by 12 to convert inches to feet)}$$

$$\text{Total} = 60'$$

EXAMPLE 8

How many 30-inch pieces of pipe can be cut from a 20-foot length of pipe? First, convert 20 feet into inches so there is a common set of units.

$$\text{Number of pieces} = \frac{(20 \times 12)}{30}$$

(Note that 10 can be factored out of the numerator and denominator)

$$\text{Number of pieces} = \frac{(2 \times 12)}{3}$$

(Note that 3 can be factored out of the numerator and the denominator)

$$\text{Number of pieces} = \frac{(2 \times 4)}{1}$$

$$\text{Number of pieces} = 8$$

Note that it is easier to work this problem if it is set up completely rather than working it out in stages. We converted 20 feet to inches by multiplying by 12. In this way, both the numerator and denominator values are in inches.

LINEAR MEASURE

Combining measures can only be done if they are all expressed in the same units. For example, addition of feet and inches must be done by first adding all the inches, then adding the feet; by changing all the inch values to a fraction of a foot and adding the fractions, then adding the feet; or by converting all the foot values to inches, then adding all the inches together.

EXAMPLE 9

What is the total length of the following pieces: 6 inches, 18 inches, 3 feet 6 inches, 42 inches, 8 feet 3 inches, and 4 feet 3 inches?

Method 1: Convert to feet and inches, then add feet and add inches:

$$6'' = 0'6''$$
$$18'' = 1'6''$$
$$3'6'' = 3'6''$$
$$42'' = 3'6''$$
$$8'3'' = 8'3''$$
$$4'3'' = 4'3''$$
$$= 19'30''$$
$$= 21'6''$$

Method 2: Convert inches to fractions of feet, then add:

$$6'' = \frac{1'}{2} = \frac{2'}{4}$$
$$18'' = 1\frac{1'}{2} = 1\frac{2'}{4}$$
$$3'6'' = 3\frac{1'}{2} = 3\frac{2'}{4}$$
$$42'' = 3\frac{1'}{2} = 3\frac{2'}{4}$$
$$8'3'' = 8\frac{1'}{4} = 8\frac{1'}{4}$$
$$4'3'' = 4\frac{1'}{4} = 4\frac{1'}{4}$$
$$= 19\frac{10'}{4}$$
$$= 21\frac{2'}{4}$$
$$= 21\frac{1'}{2}$$

EXAMPLE 10

Combine the following lengths by converting everything to inches: $2\frac{1}{2}$ feet, $3\frac{1}{4}$ feet, $\frac{1}{4}$ feet, $1\frac{3}{4}$ feet, and 5 feet.

$$2\frac{1'}{2} = \frac{5'}{2} = \left(\frac{5}{2} \times 12\right)'' = 30''$$
$$3\frac{1'}{4} = \frac{13'}{4} = \left(\frac{13}{4} \times 12\right)'' = 39''$$
$$\frac{1'}{4} = \left(\frac{1}{4} \times 12\right)'' = 3''$$
$$1\frac{3'}{4} = \frac{7'}{4} = \left(\frac{7}{4} \times 12\right)'' = 21''$$
$$5' = (5 \times 12)'' = 60''$$
$$= 153''$$

EXAMPLE 10 *continued*

$$153'' = \frac{153'}{12}\ 12\frac{3'}{4} \text{ or } 12'9''$$

Subtraction also requires expressing all items in common terms.

EXAMPLE 11

Subtract 18 inches from 10 feet 4 inches.

$$10'14'' = (10 \times 12)'' + 4'' = 124''$$
$$124'' - 18'' = 106''$$
$$106'' = \frac{106'}{12} = 8\frac{10'}{12} \text{ or } 8'10''$$

EXAMPLE 12

How much pipe remains after 8 feet 6 inches is cut from a 10 foot 3 inch piece?

First method:

$$10'3'' - 8'6'' = (120'' + 3'') - (96'' + 6'') = 123'' - 102'' = 21'' = 1'9''$$

Second method:

$$10'3'' = 9' + 1'3'' = 9'15''$$
$$9'15'' - 8'6'' = 1'9''$$

$10'3'' = 9'15''$ (converting 1 ft. into 12 in. and adding it to the 3) $9'15'' - 8'6'' = 1'9''$

Third method:

$$10'3'' - 8'6'' = 10\frac{3''}{4} - 8\frac{1'}{2} = 1\frac{3'}{4}$$

Multiplication and division are performed as shown in the following examples.

EXAMPLE 13

How much pipe (in feet) is needed for 40 sections, each 27 inches long?

First method:

$$27'' = 2'3'' = 2\frac{1'}{4} = \frac{9'}{4} \text{ each piece}$$

$$\text{Total Length} = 40 \times \frac{9'}{4} = \frac{(40 \times 9)'}{4}$$

Total Length = (factor out 4 from the numerator and denominator)

Total Length = 90′

EXAMPLE 13
continued

Second method:

$$40 \times 27'' = 1080''$$

$$1080'' = \left(\frac{1080}{12}\right)'$$

$$= 90'$$

EXAMPLE 14

What is the total length of 20 pieces of pipe, each 8 feet 6 inches long?

$$20 \times 8'6'' = (20 \times 8)' + (20 \times 6)''$$
$$= 160' + 120''$$
$$= 160' + 10'$$
$$= 170'$$

EXAMPLE 15

How many sections of drain cleaning cable 7 feet 6 inches long will be needed to reach an obstruction 60 feet into the line?

$$7'6'' = 7\frac{1}{2}'$$

$$\text{Number of pieces} = 60 \div 7\frac{1}{2}$$
$$= 60 \div \frac{15}{2}$$
$$= \frac{60}{1} \div \frac{2}{15}$$
$$= \frac{120}{15}$$
$$= 8$$

EXAMPLE 16

How many $7\frac{1}{2}$ inches nipples can be cut from 20 feet of pipe? How much is left over?

$$20' = (20 \times 12)'' = 240''$$

$$\text{Number of pieces} = 240 \div 7\frac{1}{2}''$$
$$= 240 \div \frac{15}{2}$$
$$= \frac{240}{2} \times \frac{2}{15}$$
$$= \frac{480}{15}$$
$$= 32 \text{ pieces, with nothing left over}$$

In many situations, using decimals is easier than using fractions.

EXAMPLE 17

What is the total length in feet and inches of two pipes 6.5 feet long and 15.7 feet long?

$$\begin{array}{r} 6.5 \\ +15.7 \\ \hline 22.2 \end{array}$$

$$22.2' = 22' + (0.2 \times 12)'' = 22'2.4''$$
$$= 22'2\frac{2}{5}''$$

EXAMPLE 18

Find the total length of the following pieces: 2.5 feet, 13.25 feet, 18 inches, 12.5 feet, and 6 inches.

Put all dimensions in common units:

$2.5' = 2.5'$
$13.25' = 13.25'$
$18'' = 1.5'$
$12.5' = 12.5'$
$6'' = 0.5'$

$$\begin{array}{r} 2.5 \\ 13.25 \\ 1.5 \\ 12.5 \\ \hline +0.5 \end{array}$$

$= 30.25'$
$= 30'3''$

Subtraction of decimals is straightforward after all parts are in the same units.

EXAMPLE 19

Two pieces of pipe, each 0.75 foot long, were taken from a piece 3.25 feet long. How much remains of the original piece?

Two pieces $= 0.75 + 0.75 = 1.5$

Original piece $= 3.25$

Remainder $= 3.25 - 1.5$

$= 1.75'$(or $1'9''$)

EXAMPLE 20

A building floor has an elevation of 700.75 feet. The building sewer line is 250 feet long, starts 2 feet 6 inches below the floor, and is installed with a slope of $\frac{1}{4}$ inch per foot.

What is the elevation of the connection to the main sewer?

Elevation at the main connection = 700.75 feet − 2 feet 6 inches − fall of building sewer

EXAMPLE 20

continued

Fall of building sewer $= 250\left(\frac{1''}{4}\right) = \frac{250}{4} = 62\frac{1}{2} = 62.5''$

In decimal form it would appear as

$250 \times 0.25'' = 62.5''$

$(62.5 \div 12) = 5.21'$

Thus, the sewer connection elevation is calculated:

$$\begin{array}{r} 700.75 \\ 2.5 \\ -5.21 \\ \hline = 693.04' \end{array}$$

Multiplication and division are also simplified by using decimals.

EXAMPLE 21

Twenty-five pieces of pipe measuring 2.25 feet each are required for a project. How many feet of pipe must be ordered?

Total $= 25 \times 2.25'$

$= 56.25'$ or

$= 56'3''$

EXAMPLE 22

How much solder (in feet) is needed for (12) 2-inch 45° ells? Assume 2 inches of solder is required for each joint.

12 ells = 24 joints

24 joints $\times$ 2″ solder = 48″ solder

48″ $\div$ 12 = 4′ solder

EXAMPLE 23

How many $\frac{3}{4}$ inch tees can be soldered with a 50-foot roll of solder? Assume each $\frac{3}{4}$ inch joint takes $\frac{3}{4}$ inch of solder.

$3 \times \frac{3''}{4} = \frac{9''}{4} = 2\frac{1''}{4}$ (three joints per tee)

$2\frac{1''}{4} = 2.25''$ of solder per fitting

Then, calculate the number of fittings:

$50' \div 2.25'' = (50 \times 12)'' \div 2.25''$

$= 600'' \div 2.25''$

$= 266.7$

A 50 foot roll of solder will solder approximately 267 tees, each $\frac{3}{4}$ inch. (In this case, however, it might be better to round *down*, giving 266—you don't want to run out of solder before you run out of tees!)

Review

Be sure to put dimensions in similar units so that they can be combined. In some cases, decimals will be easier than working with feet and inches or fractional parts of feet. We have seen that all the methods will work, so use the one(s) with which you are most comfortable.

SQUARES AND SQUARE ROOTS

Many problems arise that involve finding the square, square root, cube, or cube root of a number. These problems occur often enough that tables of numbers have been developed so these values can be found quickly. Table 14-1 shows squares, square roots, cubes, and cube roots for all whole numbers from 1 to 100.

You will recall that the square of a given number is the result of multiplying the number by itself. For example, $12^2 = 12 \times 12 = 144$. The square root is the exact opposite of the square. The square root of a number is that value which, when multiplied by itself, equals the original number. Using the previous example, we can say that 12 is the square root of 144. Symbolically, this is written $\sqrt{144} = 12$.

Several examples will show how useful the square and square root table can be.

EXAMPLE 24

Use Table 14-1 to solve the following problem: If you need 21 lengths of pipe, each 21 feet long, what would the total length equal?

Find 21 in the first column and then find the number in the same row, but under the square column. That would be the same as:

21×21 feet $= 441$ feet (notice this is not square feet, since the first 21 is the number of lengths of pipe)

The next example shows how the table can also be used to solve problems that deal with square roots.

EXAMPLE 25

What size square opening would be required to have 625 square inches of area?

Using Table 14-1, move down the square column until you find 625. Now move across the same row until you reach the number column and read 25. Therefore, $\sqrt{625} = 25$.

Another example of using the square root function would be to solve for the side of a right triangle, using the Pythagorean theorem: $A^2 + B^2 = C^2$. Remember that the shorter sides, A and B, are called sides (or base and height), and the longest side, C, is called the hypotenuse. Also recall that a triangle is called isosceles if it has two out of three sides with equal length.

TABLE 14-1 Squares, Square Roots, Cubes, and Cube Roots

Number	Square	Cube	Square Root	Cube Root	Number	Square	Cube	Square Root	Cube Root
1	1	1	1.000	1.000	51	2,601	132,651	7.141	3.708
2	4	8	1.414	1.260	52	2,704	140,608	7.211	3.733
3	9	27	1.732	1.442	53	2,809	148,877	7.280	3.756
4	16	64	2.000	1.578	54	2,916	157,464	7.348	3.780
5	25	125	2.236	1.710	55	3,025	166,375	7.416	3.803
6	36	216	2.449	1.817	56	3,136	175,616	7.483	3.826
7	49	343	2.646	1.913	57	3,249	185,193	7.550	3.849
8	64	512	2.828	2.000	58	3,364	195,112	7.616	3.871
9	81	729	3.000	2.080	59	3,481	205,379	7.681	3.893
10	100	1,000	3.162	2.151	60	3,600	216,000	7.746	3.915
11	121	1,331	3.317	2.224	61	3,721	226,981	7.810	3.936
12	144	1,728	3.464	2.288	62	3,844	238,328	7.874	3.958
13	169	2,197	3.606	2.351	63	3,969	250,047	7.937	3.979
14	196	2,744	3.742	2.410	64	4,096	262,144	8.000	4.000
15	225	3,375	3.873	2.466	65	4,225	274,625	8.062	4.021
16	256	4,096	4.000	2.520	66	4,356	287,496	8.124	4.041
17	289	4,913	4.123	2.571	67	4,489	300,763	8.185	4.062
18	324	5,832	4.243	2.621	68	4,624	314,432	8.246	4.082
19	361	6,859	4.359	2.668	69	4,761	328,509	8.307	4.102
20	400	8,000	4.472	2.714	70	4,900	343,000	8.367	4.121
21	441	9,261	4.583	2.759	71	5,041	357,911	8.426	4.141
22	484	10,648	4.690	2.802	72	5,184	373,248	8.485	4.160
23	529	12,167	4.796	2.841	73	5,329	389,017	8.544	4.179
24	576	13,824	4.899	2.881	74	5,476	405,224	8.602	4.198
25	625	15,625	5.000	2.921	75	5,625	421,875	8.660	4.217
26	676	17,576	5.099	2.962	76	5,776	438,976	8.718	4.236
27	729	19,683	5.196	3.000	77	5,929	456,533	8.775	4.254
28	784	21,952	5.292	3.037	78	6,084	474,552	8.832	4.273
29	841	24,389	5.385	3.072	79	6,241	493,039	8.888	4.291
30	900	27,000	5.477	3.107	80	6,400	512,000	8.944	4.309
31	961	29,791	5.568	3.141	81	6,561	531,441	9.000	4.327
32	1,024	32,768	5.657	3.175	82	6,724	551,368	9.055	4.344
33	1,089	35,937	5.745	3.208	83	6,889	571,787	9.110	4.362
34	1,156	39,304	5.831	3.240	84	7,056	592,704	9.165	4.380
35	1,225	42,875	5.916	3.271	85	7,225	614,125	9.220	4.397
36	1,296	46,656	6.000	3.302	86	7,396	636,056	9.274	4.414
37	1,369	50,653	6.083	3.332	87	7,569	658,503	9.327	4.431
38	1,444	54,872	6.164	3.362	88	7,744	681,472	9.381	4.448
39	1,521	59,319	6.245	3.391	89	7,921	704,969	9.434	4.465
40	1,600	64,000	6.325	3.420	90	8,100	729,000	9.487	4.481
41	1,681	68,921	6.403	3.448	91	8,281	753,571	9.539	4.498
42	1,764	74,088	6.481	3.476	92	8,464	778,688	9.592	4.514
43	1,849	79,507	6.557	3.503	93	8,649	804,357	9.644	4.531
44	1,936	85,184	6.633	3.530	94	8,836	830,584	9.695	4.547

(continued)

TABLE 14-1 Squares, Square Roots, Cubes, and Cube Roots (Continued)

Number	Square	Cube	Square Root	Cube Root	Number	Square	Cube	Square Root	Cube Root
45	2,025	91,125	6.708	3.557	95	9,025	857,375	9.747	4.563
46	2,116	97,336	6.782	3.583	96	9,216	884,736	9.798	4.579
47	2,209	103,823	6.856	3.609	97	9,409	912,673	9.849	4.595
48	2,304	110,592	6.928	3.634	98	9,604	941,192	9.899	4.610
49	2,401	117,649	7.000	3.639	99	9,801	970,299	9.950	4.626
50	2,500	125,000	7.071	3.684	100	10,000	1,000,000	10.000	4.642

EXAMPLE 26

An isosceles right triangle has legs equal to 5′. Find the hypotenuse.

$C^2 = 5^2 + 5^2 = 50$

$C = \sqrt{50}$

Using the square and square root table, we find that the square root of 50 = 7.07.

Therefore, $C = 7.07'$.

EXAMPLE 27

How many square feet are in 2,304 square inches?

$1\text{ ft.}^2 = (12\text{ in.})^2 = 144\text{ in.}^2$

Therefore,

$$2{,}304\text{ in.}^2 = \frac{2{,}304}{144}\text{ ft.}^2 = 16\text{ ft.}^2$$

We can also use the square and square root table. Note that you can find 2,304 in the square column, so you can write:

$\sqrt{2{,}304}\text{ in.}^2 = (48\text{ in.})^2$

or, since 48 inches = 4 feet,

$(48\text{ in.})^2 = (4\text{ ft.})^2$

$(4\text{ ft.})^2 = 4' \times 4' = 16\text{ ft.}^2$

Therefore, $2{,}304\text{ in.}^2 = 16\text{ ft.}^2$

Sometimes a problem will be very close to a square of a number. In the following example, two numbers very close in value are to be multiplied. By changing one value into the sum of two numbers and then using the rules of multiplication, it is possible to reduce the problem into a square of one number and a multiple of two smaller numbers.

EXAMPLE 28

If 26 workers worked 24 hours on a job, how many total hours were worked?

Total hours = 26 workers × 24 hours (change 26 into 24 + 2)

Total hours = (24 + 2) × 24 (now the 24 hours can be multiplied into both values in the parentheses)

Total hours = (24 × 24) + (2 × 24) (now the square table can be used to find the square of 24)

Total hours = 576 + 48

Total hours = 624 hours

CUBES

Up until this point, we have only used the squares and square roots portion of Table 14-1. If we were to multiply a squared number by itself one more time, it would now be a cube. A cube is the result of using a number as a factor three times. For example, the cube of 4 is:

$$4 \times 4 \times 4 = 64.$$

The operation of cubing is often expressed in the following form: a^3. Thus, $11^3 = 1{,}331$.

If a number representing physical units is squared or cubed, the units (length, weight, time, etc.) are also squared or cubed.

$$(2\text{ ft.})^3 = 8\text{ ft.}^3$$

EXAMPLE 29

If 8 workers labored 8 hours for 8 days, how many worker-hours are used? Use Table 14-1 to find the cube of the number.

$$8\text{ workers} \times 8\frac{\text{hours}}{\text{day}} \times 8\text{ days} = 8^3\text{ worker-hours} = 512\text{ worker-hours}$$

In Table 14-1, find the 8 in the numbers column and move across that row until it intersects with the cube column. Notice that the days cancel out, but since workers and hours are not the same, they are not squared or cubed.

EXAMPLE 30

If 12 workers toil 12 hours per day for $12.00 per hour, for a total of 3 days, what is the direct labor cost?

$$\text{Total dollars} = 12\text{ workers} \times 12\frac{\text{hours}}{\text{day}} \times \frac{\$12}{\text{worker-hour}} \times 3\text{ days}$$

$$\text{Total dollars} = \frac{\$(12)^3}{\text{day}} \times 3\text{ days}$$

$$\text{Total dollars} = \frac{\$1{,}728}{\text{day}} \times 3\text{ days} = \$5{,}184$$

The following is another example of how numbers can be changed to allow the cube table to be used.

EXAMPLE 31

Twenty feet of pipe is needed for 20 units on each floor of a 21-story building. How many total feet of pipe are needed?

Total feet of pipe $= 20\frac{\text{ft.}}{\text{unit}} \times 20\frac{\text{units}}{\text{story}} \times 21$ stories

Total feet of pipe $= 20 \times 20 \times (20 + 1)$

Total feet of pipe $= (20 \times 20 \times 20) + (20 \times 20 \times 1)$

Total feet of pipe $= 20^3 + 20^2$

Total feet of pipe $= 8{,}000 + 400$

Total feet of pipe $= 8{,}400$ ft.

Total feet of pipe $= 8{,}400$ ft.

Notice that units cancel and stories cancel in the top equation, leaving only feet. This is why the feet are not cubed. (Another way to think about it is that feet measure length, while cubic feet measure volume. Here, you want a total length, so you need feet, not cubic feet.)

CUBE ROOTS

The cube root of a value is that number which, when cubed, is equal to the original value. Thus, the cube root of 64 feet is written $\sqrt[3]{64}$, which is equal to 4 feet. We can see that the cube root form of the radical symbol is $\sqrt[3]{\ }$, where the 3 indicates that the cube root is desired. The following is an example showing how the cube root is found.

EXAMPLE 32

What would be the length of the sides of a cube (a box with all sides equal) if it contained 27 cubic feet?

Volume of a cube = side × side × side = $(\text{side})^3$

$(\text{side})^3 = 27\ \text{ft.}^3$ (to solve, we must take the cube root of both sides)

$\sqrt[3]{(\text{side})^3} = \sqrt[3]{27\text{ft.}^3}$

side = 3 ft. (using Table 14-1, move down the cube column until you find 27, then move left across the row until you find the intersection of the number column, which gives 3)

REVIEW QUESTIONS

Short Answer

1. Convert 20 feet 3 inches to inches.
2. Convert 163 inches to feet and inches.
3. How much pipe is needed to cut 15 pieces, each 1 foot 9 inches long?
4. How many 45-inch pieces can be cut from a 100 foot roll of pipe?
5. An acre is 66 feet × 660 feet. How many square feet are in an acre?
6. A house contains 10 windows, each 5′ × 5′. What is the total window area of the house?
7. A septic tank discharges to a drainage field that must be a minimum of 300 sq. ft. What are the approximate field dimensions if it is square?
8. Find the squares of the following numbers: 63, 42, 96, 31.
9. Find the square roots of the following numbers: 48, 6, 25, 18.
10. Find the cube roots of the following numbers: 48, 6, 25, 18.
11. Three equal-length nipples, plus a union ($1\frac{1}{2}$ inches laying length), plus a tee (laying length 3 inches), plus two ells (laying length $1\frac{3}{4}$ inches each) are made up going from one ell to the other ell. If the c–c distance between the ells is 20 inches, what is the length of each nipple?

Chapter 15

LEARNING OBJECTIVES

The student will:

- Calculate offsets through the use of constants for 90° and 45° offsets.
- Calculate 30° and 60° offsets.
- Describe the concepts of geometric shapes and the formulas involving perimeter, area, and volume.
- Apply the concepts of geometric shapes to practical plumbing applications.

Applied Geometric Concepts

OFFSETS

In the fabrication and installation of piping systems, we usually measure or know the distance from the center of one pipe to the center of another pipe. This length is known as the center-to-center (c–c) distance. In order for the two pipes to be joined, a third pipe must be cut to an exact length, known as the end-to-end (e–e) measurement. To determine this end-to-end length, the amount of space that the fittings will take up must be subtracted from the center-to-center distance. See Figure 15-1.

As you can see from Figure 15-1, the length of the pipe will be less than the center-to-center value because the fittings attached to the pipe ends require a certain amount of space. This space is often referred to as the "fitting allowance" or the "take out."

Unfortunately, as the diameter of the pipe changes, so does the fitting allowance. The fitting allowance can also change based on the different materials being used. The safest procedure is to measure the actual fittings with which you are working. This can be done easily by measuring a short section of pipe and then installing a fitting at one end. After the fitting is installed, measure the pipe and fitting combination. The difference between the first and second measurements is the fitting allowance. Figure 15-2 shows how this is accomplished.

In the case of 90° offsets, which are made with two 90° ells (or elbows) and a connecting pipe, the end-to-end length of pipe is equal to the center-to-center distance between the pipe lines minus the fitting allowance (or laying length) of the two ells. Example 1 shows how to determine the end-to-end measurement required.

EXAMPLE 1

A 1-inch pipe passing through the ceiling is 2 feet 6 inches from the center of the pipe to which it must be connected in the room. What is the length of pipe required for the offset? (One-inch 90° ells have a fitting allowance of 1 inch.) See Figure 15-3.

End-to-end length $= 2'6'' - 2(1'') = 2'6'' - 2'' = 2'4''$

FIGURE 15-1 As pipes are joined, the space required for the fitting must be subtracted from the center-to-center distance to determine the cut length of the pipe segment.

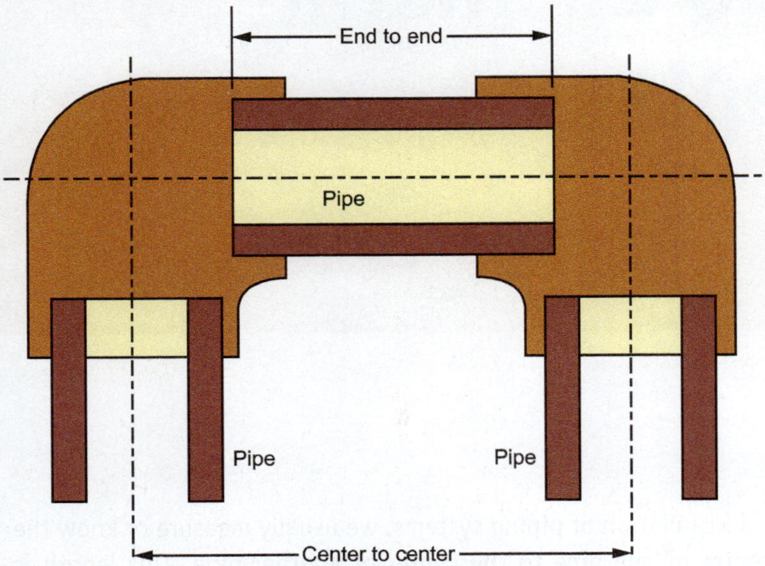

FIGURE 15-2 By installing a fitting on one end of a pipe with a known length, the allowance can be determined for that size of fitting. Center-to-end length minus the end-to-end length equals the allowance of the fitting.

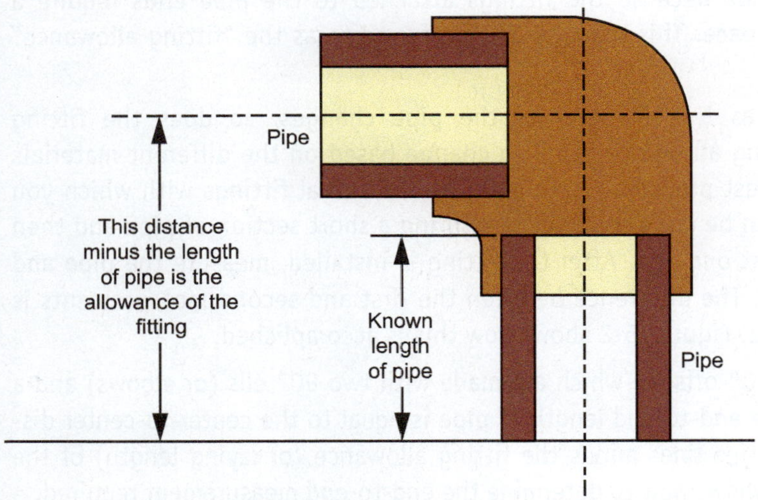

A 90° offset is the simplest to calculate, but it also has the highest pressure drop across the offset when compared to other offsets. If the offset connection were made along a 45° line to the original pipes, then the flow would be smoother and the pressure losses lower. The distance along this 45° line can be shown to be 1.414 times the distance between the pipes. You will recall that the 45° offset is the hypotenuse (C) of an isosceles right triangle with legs ($A = B$) equal to the distance between the pipes. Figure 15-4 shows that the distance that the pipes are offset is referred to as the *offset*, the distance that separates the pipe

FIGURE 15-3 Example of a 90-degree offset in a run of pipe.

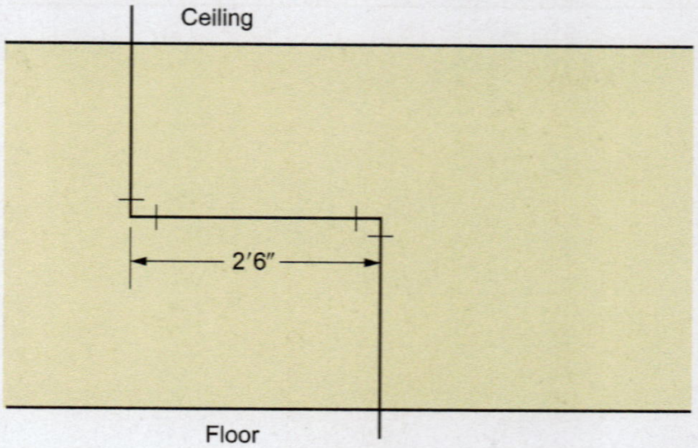

ends along their length is known as the *run*, and the diagonal length of the pipe segment that will join them is known as the *travel*.

The hypotenuse (travel) of that triangle is found from the formula $C^2 = A^2 + B^2$.

If $A = B$, then $C^2 = A^2 + A^2 = 2A^2$.

By taking the square root of both sides, the equation is reduced to

$$C = \sqrt{2A^2} = 1.414A$$

EXAMPLE 2

Using Figure 15-4, find the travel for a 2″ tubing to cut for a 45° offset where the pipes are $3\frac{1}{2}'$ apart. The laying length of 2″ 45° ells equals $\frac{9}{16}''$.

First, convert distance to inches: $3\frac{1}{2}' \times 12 \text{ in./ft.} = 42''$

Length along 45° offset $= 1.414(42'') = 59.4''$

Length of pipe required $= 59.4 - 2\left(\frac{9}{16}\right) = 59.4 - 1.13 = 58.27''$

OFFSETS WITH OTHER ANGLES

Besides the 45° offset, we can make 60° offsets with factory-made fittings and we can make 30° offsets with field-trimmed welding or hubless fittings. In this section, we will explore the relationships that exist between the parts of these offset patterns and also present a diagram that will show how the 30°, 45°, and 60° offsets compare with each other.

Consider an equilateral triangle (Figure 15-5a). The three sides are of equal length (c) and the angles are equal (60°). Next, we can bisect this triangle by drawing a vertical line AD (Figure 15-5b). This forms triangle ADC, which has interior angles of 30°, 60°, and 90°. The triangle has sides of lengths $c/2$, AD, and

FIGURE 15-4 When making offsets other than 90 degrees, the terms *run*, *offset*, and *travel* are used.

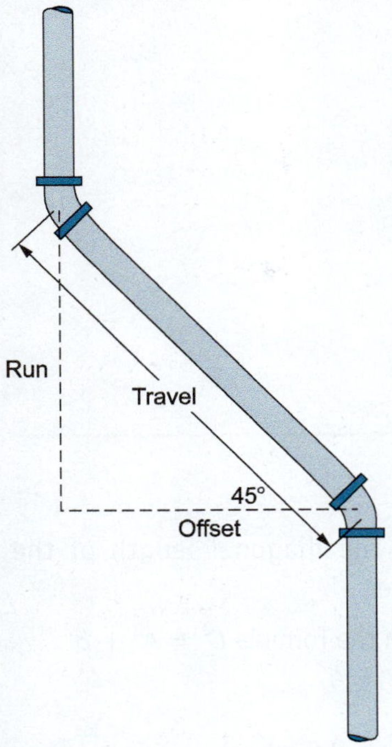

FIGURE 15-5a-b (a) Equilateral triangle. (b) Equilateral triangle bisected to form two right triangles.

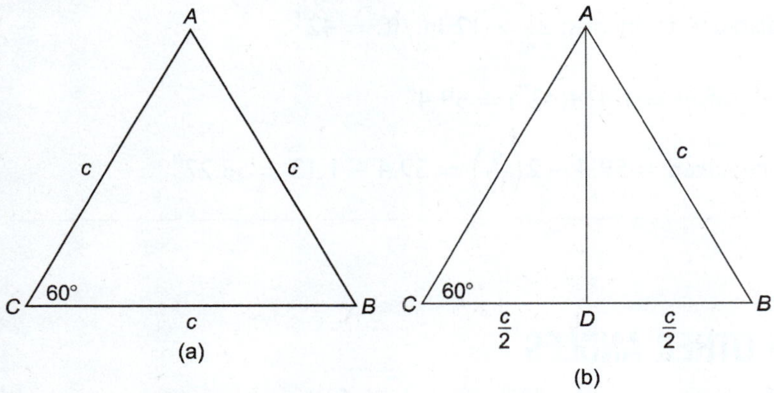

hypotenuse *c*. If we calculate *AD* in terms of *c*, we can express ratios of all the sides to each other.

$$c^2 = \left(\frac{c}{2}\right)^2 + AD^2$$

$$AD = \sqrt{\left(c^2 - \left(\frac{c}{2}\right)^2\right)} = \sqrt{c^2\left(1 - \frac{1}{4}\right)}$$

$$AD = \sqrt{0.75c^2} = 0.866c$$

If this is a unit triangle with sides equal to 1:

- The ratio of hypotenuse to length *AD* is $\frac{1}{0.866} = 1.154$
- The ratio of short side to length *AD* is $\frac{0.5}{0.866} = 0.577$
- The ratio of the hypotenuse to the short side is $\frac{1}{0.5} = 2$

See Figure 15-6 for diagrams of run, travel, and offset for 30°, 45°, and 60° offsets.

FIGURE 15-6 Notice that when the fitting angle changes, the lengths of the run and the travel also change.

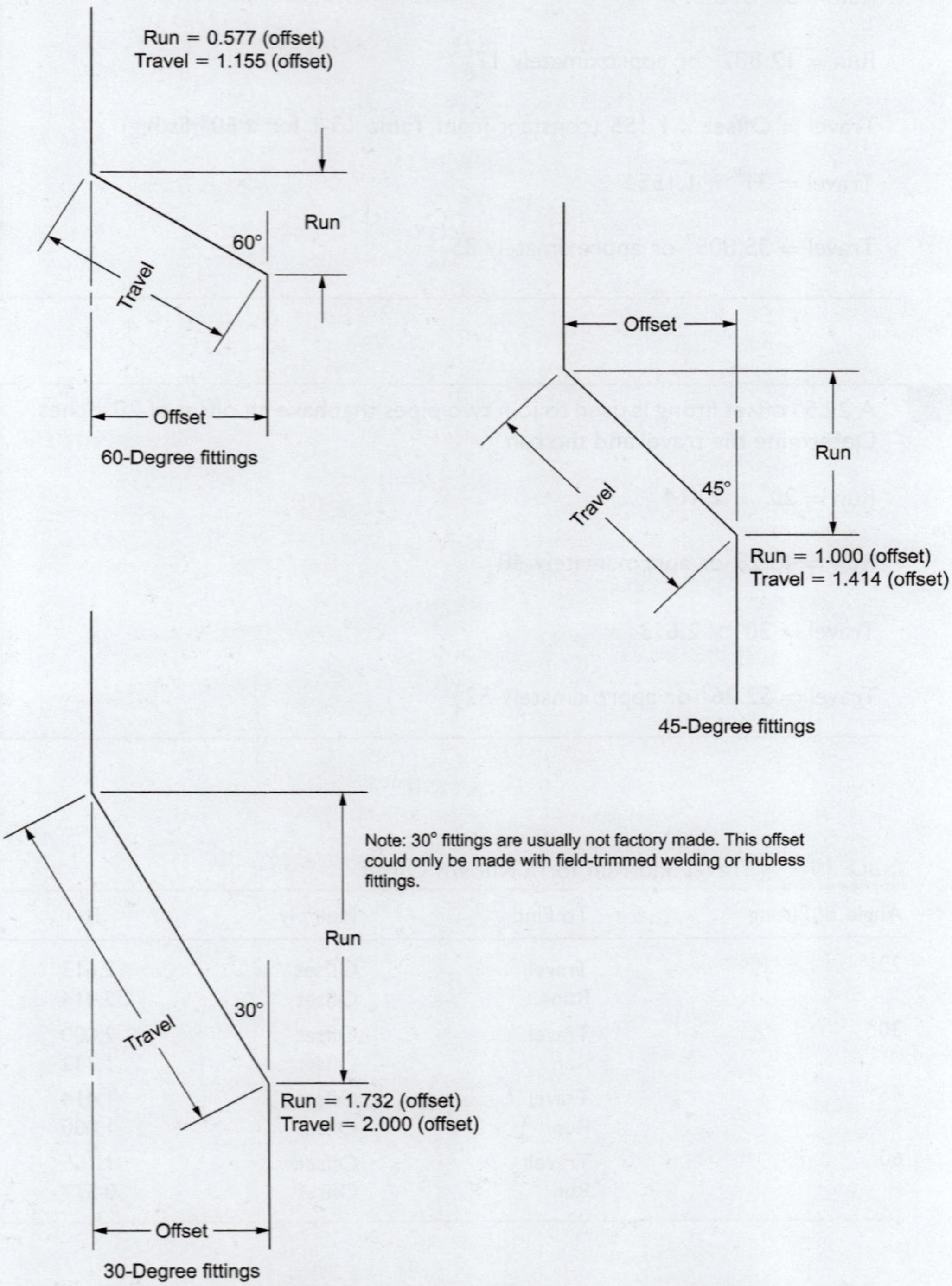

Because certain angled fittings are commonly used, tables have been developed to simplify the above calculations. Table 15-1 shows the constants used to find the travel and run of various fittings when the offset is known.

EXAMPLE 3

Two pipes are to be connected with a 60° offset. The offset is 31 inches. Find the lengths of the run and the travel.

Using the table:

Run = Offset × 0.577 (constant from Table 15-1 for a 60° fitting)

$\text{Run} = 31'' \times 0.577$

$\text{Run} = 17.887''$ or approximately $17\frac{7}{8}''$

Travel = Offset × 1.155 (constant from Table 15-1 for a 60° fitting)

$\text{Travel} = 31'' \times 1.155$

$\text{Travel} = 35.805''$ or approximately $35\frac{13}{16}''$

EXAMPLE 4

A 22.5° offset fitting is used to join two pipes that have an offset of 20 inches. Determine the travel and the run.

$\text{Run} = 20'' \times 2.414$

$\text{Run} = 48.28''$ or approximately $48\frac{1}{4}''$

$\text{Travel} = 20'' \times 2.613$

$\text{Travel} = 52.26''$ or approximately $52\frac{1}{4}''$

TABLE 15-1 Travel and Run for a Known Offset

Angle of Fitting	To Find	Multiply	By
$22\frac{1}{2}°$	Travel	Offset	2.613
	Run	Offset	2.414
30°	Travel	Offset	2.000
	Run	Offset	1.732
45°	Travel	Offset	1.414
	Run	Offset	1.000
60°	Travel	Offset	1.155
	Run	Offset	0.577

EXAMPLE 5

If a building sewer pipe is located 24 inches above the sewer connection on the sewer main (see Figure 15-7) and a 45° offset is required, what would be the length of the piece of pipe needed to make the connection? The 45° fitting has a 2-inch allowance and the coupling has a 0.5-inch allowance.

Travel = 24″ × 1.414

Travel = 33.936″

Length of pipe (e–e) = 33.936″ − 2″ − 0.5″

Pipe length = 31.436″ or approximately $31\frac{7}{16}''$

CIRCLE

A circle is a closed curved line whose points lie in a plane where all the points are equidistant from a single point called the center. Figure 15-8 shows a circle with the center, a radius, a diameter, a chord, and an arc indicated.

FIGURE 15-7 Offset used to connect to a sewer main.

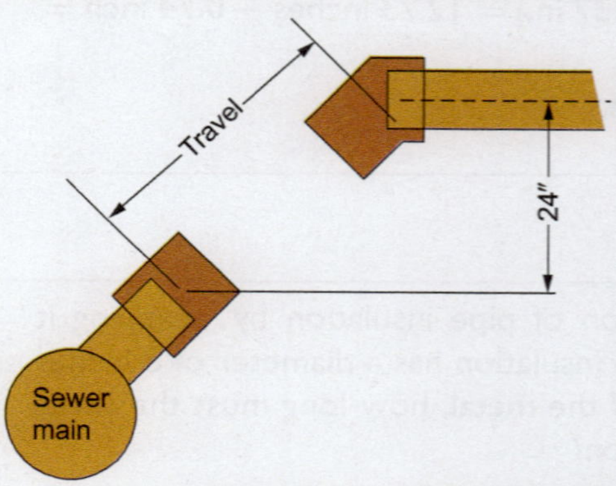

FIGURE 15-8 Components of a circle.

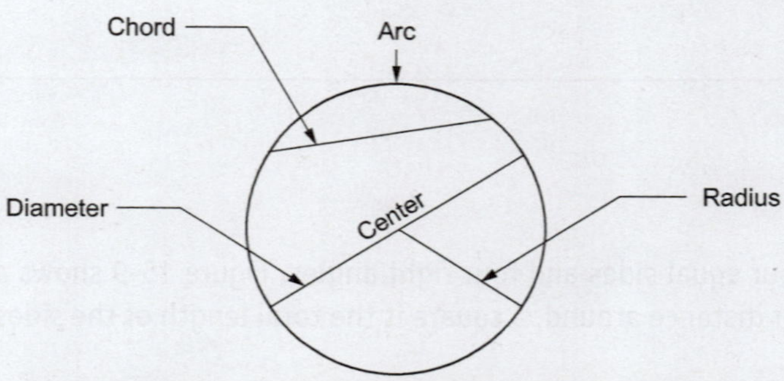

A radius (r) is any line from the center to the circle.

A diameter (d) is any line drawn from any point on the circle through the center to the other side of the circle.

An arc is any portion of the circle.

A chord is the straight line that connects the ends of an arc. The circumference (C) is the total distance around the circle. The relationships between these parts are expressed as follows:

$$C = \pi d$$
$$\text{Since } d = 2r$$
$$\text{Then } C = 2\pi r$$

The constant π is equal to approximately 3.1416.

EXAMPLE 6

A plumber has unearthed a cast iron soil pipe of unknown diameter. A string wrapped around the circumference of the pipe measures approximately 40 inches. Thus, the outside diameter must be equal to

$$C = \pi d$$
$$d = \frac{C}{\pi} = \frac{40^\circ}{3.1416} = 12.73''$$

Since 12-inch cast iron soil pipe has a wall thickness of 0.37 inch,
inside diameter = 12.73 inches − 2 (0.37 in.) = 12.73 inches − 0.74 inch = 12 inches approximately

The internal pipe diameter is 12 inches.

EXAMPLE 7

A plumber wants to protect a section of pipe insulation by wrapping it with a band of sheet metal. If the pipe insulation has a diameter of 8 inches and there is to be a 1-inch overlap of the metal, how long must the sheet metal be to surround the pipe insulation?

$$C = \pi d + 1'' \text{ (for overlap)}$$

$$C = (3.1416 \times 8'') + 1''$$

$$C = 25.13'' + 1'' = 26.13''$$

$$C = 26\tfrac{1}{8}''$$

SQUARE

A square is a figure with four equal sides and four right angles. Figure 15-9 shows a square. The perimeter of, or distance around, a square is the total length of the sides and is expressed as $P = 4s$.

FIGURE 15-9 A square is a rectangle in which all of the sides are equal in length.

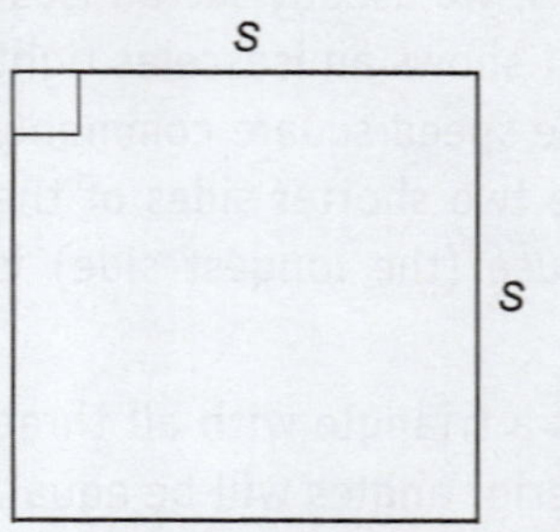

EXAMPLE 8

A square-shaped building has sides of 48 feet (inside dimension). A hot water recirculating line must be installed around the perimeter. What is the approximate length of pipe (without fittings) required for this loop?

$P = 4s$
$P = 4 \times 48'$
$P = 192'$

RECTANGLE

A rectangle is a figure similar to a square with all right angles, except that adjacent sides are not equal (opposite sides are equal). Figure 15-10 shows a rectangle with sides *l* (length) and *w* (width).

The formula for finding the perimeter is $P = 2(l + w)$.

EXAMPLE 9

A sprinkler line must be installed around the perimeter of a $26' \times 42'$ building. How much pipe (without fittings) must be obtained for this loop?

$P = 2(l + w)$
$P = 2(26' + 42')$
$P = 2(68')$
$P = 136'$

FIGURE 15-10 Like a square, the rectangle must have 90-degree corners.

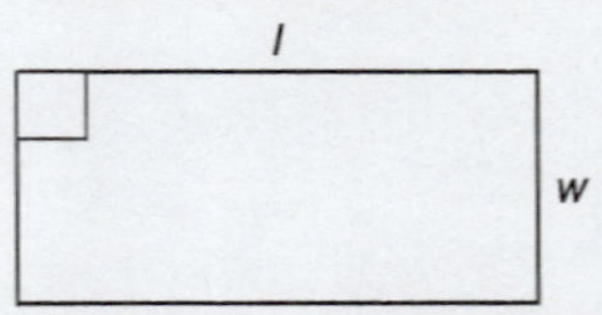

TRIANGLE

A triangle is a three-sided figure. In plumbing applications, we usually see an isosceles right triangle or an equilateral triangle. Figure 15-11 shows an isosceles right triangle. A common example of an isosceles triangle is the speed square commonly used by carpenters. Note for a 45°-45°-90° triangle, the two shorter sides of the right triangle are equal and the length of the hypotenuse (the longest side) is 1.414 times the length of each short side.

Figure 15-12 shows an equilateral triangle, which is a triangle with all three sides of equal length. This also means that all of the interior angles will be equal. Since the sum of all interior angles of a triangle is 180°, then each angle must be 60°.

If the equilateral triangle is divided straight down from its apex, then two smaller triangles would be formed. These would be the well-known 30°-60°-90° right triangle (see Figure 15-13). The parts of this triangle have the following relationships.

$$AB = BC = CA = 2(CD)$$
$$AD = (1.732/2)AC = 0.866(AC)$$
$$CD = AC/2 = 0.5(AC)$$

FIGURE 15-11 An isosceles triangle is one that has two equal sides. The angles opposite of these equal sides are also equal.

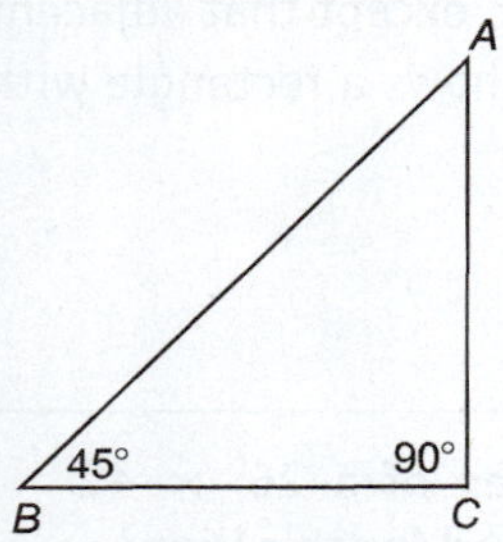

FIGURE 15-12 An equilateral triangle will have three 60° included angles and three sides of equal length.

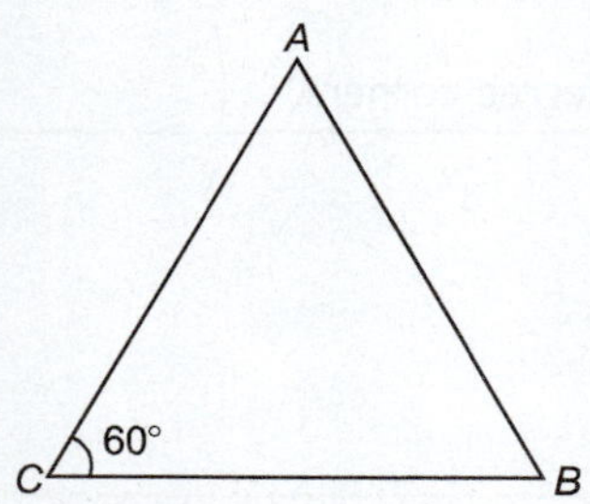

FIGURE 15-13 Dividing an equilateral triangle in half creates a 30°-60°-90° triangle.

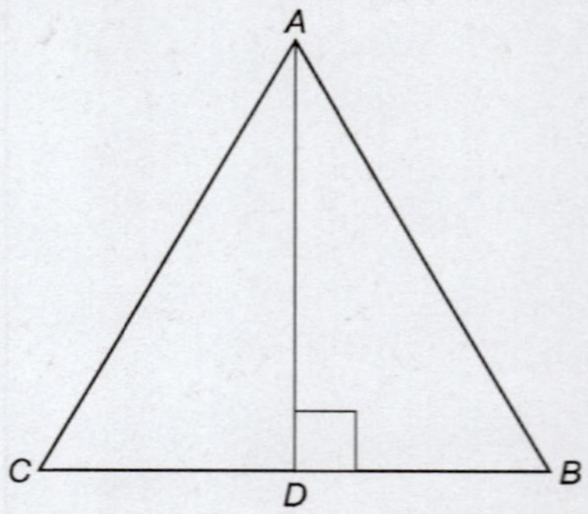

AREA CALCULATIONS

Extending the concepts just presented, we will now consider the calculation of areas of planar figures and then extend these ideas to volumes of three-dimensional shapes.

Area of a Circle

The area of a circle is given by the following expression (see Figure 15-14):

$$A = \pi r^2$$

where $A =$ area, $r =$ radius, and $\pi = 3.1416$

and since

$$d = 2r$$

$$A = \pi\left(\frac{d}{2}\right)^2 = \frac{\pi d^2}{4}$$

EXAMPLE 10

What is the area of a circle with a radius of 7 inches?

$A = \pi(7)^2 = 3.1416 \times 49 \text{ in.}^2 = 154 \text{ in.}^2$

FIGURE 15-14 Calculation of the area of a circle involves knowing or calculating the length of the radius (r).

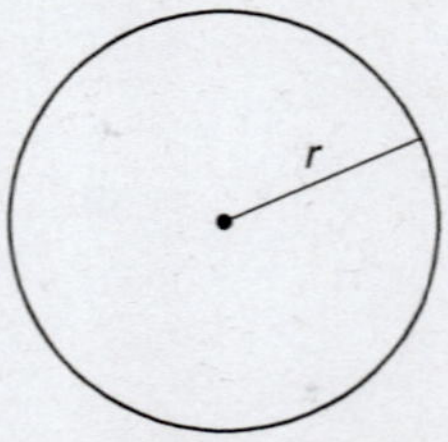

EXAMPLE 11

What is the diameter of a pipe with a cross-sectional area of 452.6 in.^{2}?

$$A = \pi\left(\frac{d^2}{4}\right)$$

$$452.6 \text{ in.}^2 = 3.1416\left(\frac{d^2}{4}\right) = 0.7854d^2$$

$$\frac{452.6}{0.7854} = d^2$$

$$\sqrt{\frac{452.6}{0.7854}} = \sqrt{d^2}$$

$$24 = d$$

Therefore the pipe diameter is 24″.

Area of a Square

The area of a square is the product of one side multiplied by the adjacent side (see Figure 15-15). Since the four sides of a square are equal, the formula is:

$$A = s^2$$

where A = area and s = side of square.

EXAMPLE 12

A square soak tub measures 48 inches on each side. How many square feet will this unit cover?

$$A = s^2$$
$$A = (48'')^2$$
$$A = (4')^2$$
$$A = 16 \text{ ft.}^2$$

FIGURE 15-15 The area of a square is calculated by multiplying one side by an adjacent side, which is the same as squaring one side of the square.

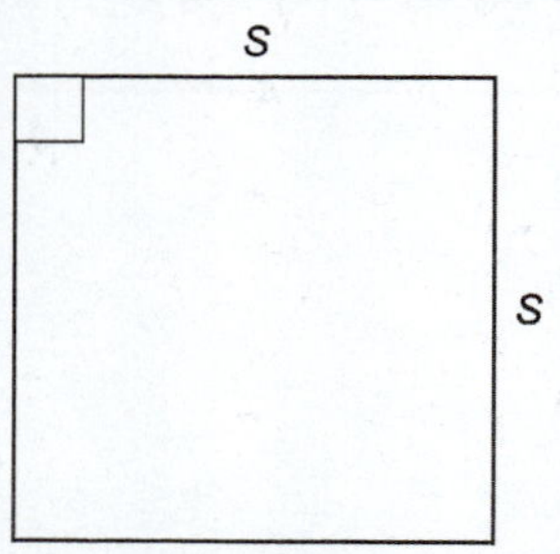

EXAMPLE 13

A shower stall base must be covered with a waterproof membrane. The stall is $6\frac{1}{2}'$ on a side. The pan must also have a 6-inch ledge added to all sides. What is the area of membrane required?

Side = base + 2 ledges = $6.5' + 2(6'') = 6.5' + 1' = 7.5'$

$A = (7.5)^2 = 56.25$ ft.^{2} of membrane

Area of a Rectangle

The area (*A*) of a rectangle is the product of the length (*l*) times the width (*w*). See Figure 15-16.

$$A = l \times w$$

where A = area, l = length, and w = width of the rectangle

EXAMPLE 14

What is the floor area covered by a standard $30'' \times 5'$ bathtub?

$A = \left(\frac{30}{12}\right)' \times 5'$

$A = 2.5' \times 5' = 12.5$ ft.^{2} floor area

EXAMPLE 15

The National Fuel Gas Code requires a 1 square inch ventilation opening for each 1,000 BTU/hr input of appliances in an enclosed room. If a 54,000 BTU/hr appliance is installed in an enclosed room, what is the length of opening required if the width is 6 inches?

Area $= \frac{54{,}000}{1{,}000} = 54$ in.^{2} required opening

$54 \text{ in.}^2 = l \times 6''$, so

$l = \frac{54 \text{ in.}^2}{6 \text{ in.}} = 9''$

FIGURE 15-16 The area of a rectangle is calculated by multiplying the length by the width.

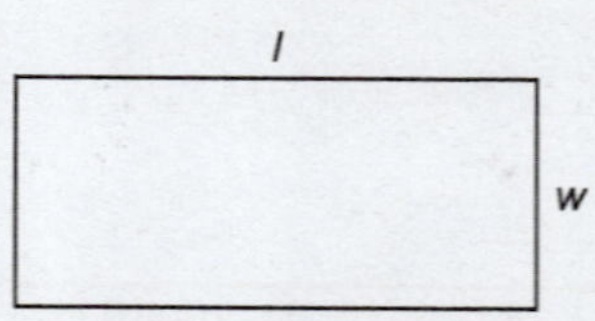

Area of a Triangle

The area of a triangle is the product of $\frac{1}{2}$ the base times the height of the triangle. See Figure 15-17.

$$A = \left(\frac{1}{2}\right)(b \times h)$$

where A = area, b = base, and h = height of triangle.

VOLUME CALCULATIONS

Volume calculations are used for such things as determining the capacity of tanks, the amount of fluid contained in piping, the amount of dirt to be removed from a trench or required to fill a hole, the amount of concrete to patch a floor, or the amount of air in a room. Volume computations are easy for regular shapes and approximations can be made for irregular shapes. In general, volume is the product of an area (length × width) and depth, so that the dimensions are

$$\text{feet} \times \text{feet} \times \text{feet} = \text{feet}^3$$

Volume of a Cylinder

A cylinder has two parallel, equal circular bases that are connected with a containing surface at right angles to the two bases. See Figure 15-18.

The volume formula for a cylinder is
volume = (area of base) × height = $\pi r^2 h$
where r = radius of base and h = height of cylinder.

FIGURE 15-17 To calculate the area of a triangle, multiply $\frac{1}{2}$ by the product of the base times height. h is drawn perpendicular to the base b.

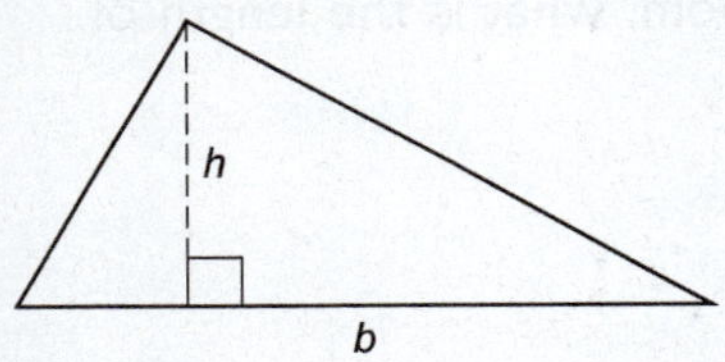

FIGURE 15-18 Cylinders are three-dimensional objects with a circular cross section.

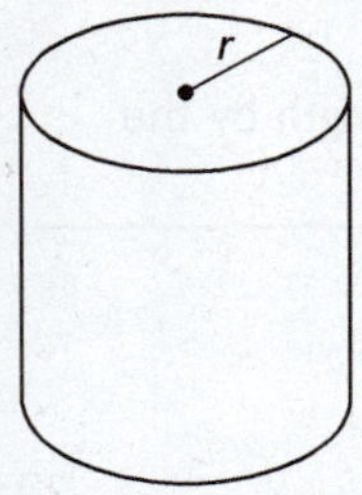

EXAMPLE 16

A 4 inch cast iron sewer is 250 feet long. Approximately how much water is needed to fill this line for testing? (One ft.^{3} equals 7.5 gallons.)

$$\text{radius} = \frac{4''}{2} = 2'' = \frac{1'}{6}$$

$$\text{Volume} = \pi r^2 \times h$$

$$\text{Volume} = 3.1416\left(\frac{1}{6}\right)^2 \times 250'$$

$$\text{Volume} = 21.82 \text{ ft.}^3 \text{ or } 21.82 \text{ ft.}^3 \times \frac{7.5 \text{ gallons}}{\text{ft.}^3} = 163.6 \text{ gallons}$$

EXAMPLE 17

What is the total weight of a cylindrical tank filled with water if the tank is 4′ in diameter and 10′ long? The empty tank weighs 850 lbs. (Water weighs 62.5 lbs per ft.^{3})

$$\text{Volume} = \pi(2)^2(10) = 125.6 \text{ ft.}^3$$
$$\text{Weight of water} = 125.6 \times (62.5) = 7{,}850 \text{ lbs}$$
$$\text{Total weight} = 7{,}850 + 850 = 8{,}700 \text{ lbs}$$

Volume of a Rectangular Prism

The volume of a rectangular prism is calculated using the formula length times width times height. See Figure 15-19.

$$V = l \times w \times h$$

where V = volume, l = length, w = width, and h = height of prism.

EXAMPLE 18

Workers must dig a ditch $2\frac{1}{2}$ feet wide, 4 feet deep, and 200 feet long. What is the volume of dirt that must be removed?

$$V = 2.5'(4')(200') = 2{,}000 \text{ ft.}^3$$

FIGURE 15-19 Rectangular prisms are shapes found in septic tanks and excavations.

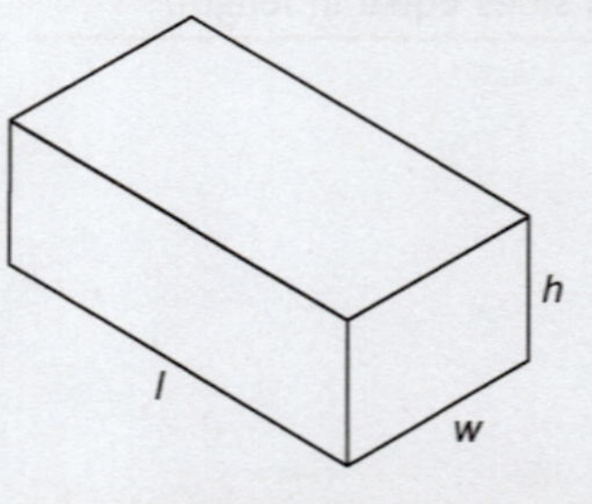

EXAMPLE 19

A septic tank is 4′ × 5′ × 8′. What is the total capacity in gallons?

$$V = \left(\frac{7.5 \text{ gallons}}{\text{ft.}^3}\right)(4')(5')(8') = \left(\frac{7.5 \text{ gallons}}{\text{ft.}^3}\right)(160 \text{ ft.}^3) = 1{,}200 \text{ gallons}$$

Volume of a Cube

A cube is a six-sided figure with all right angles and all edges equal. See Figure 15-20.

The volume is the product of the three perpendicular sides.

$$V = l \times w \times h$$

since $l = w = h$, $V = l^3$

or if the side is represented by s,

$$V = s^3$$

EXAMPLE 20

What is the volume of a cube-shaped tank with a length, width, and height of 5 feet 6 inches?

$V = (5'6'')^3 = (5.5)^3 = 166.4 \text{ ft.}^3$ or 1,248 gallons

EXAMPLE 21

A cube-shaped dry well is used to collect and hold water. The length of a side is 54 inches. What is the capacity in gallons?

$$V = \left(\frac{54}{12}\right)^3 = (4.5)^3 = 91.1 \text{ ft.}^3, \text{ or } (91.1 \text{ ft.}^3)\left(\frac{7.5 \text{ gallons}}{\text{ft.}^3}\right) = 683 \text{ gallons}$$

Volume of a Triangular Prism

The volume of a triangular prism equals one-half the area of the base (base × height) times the length. See Figure 15-21.

$$V = \left(\frac{1}{2}\right) b \times h \times d$$

where b = base, h = height, and d = length.

FIGURE 15-20 Cubes are special rectangular prisms with all sides equal in length.

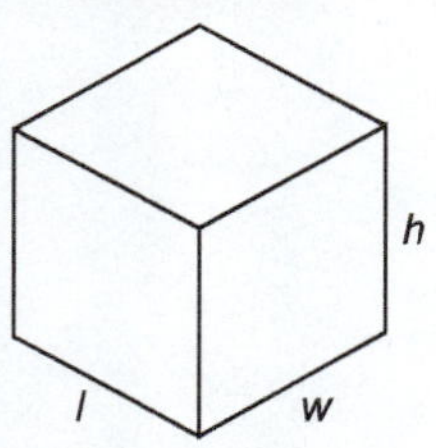

FIGURE 15-21 Triangular prisms are three-dimensional objects with a triangular cross section.

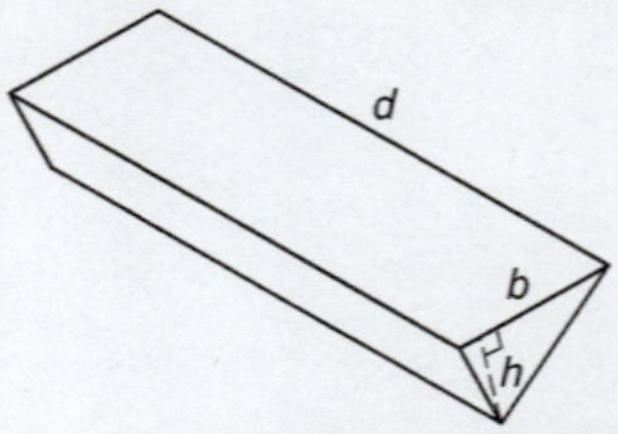

EXAMPLE 22

A ditch is to be dug to a depth of 5 feet, for a length of 100 feet. The soil type requires a one-to-one angle of repose of the sides. Figure 15-22 illustrates the ditch. How much dirt must be removed?

The dirt to be removed can be visualized as two 45° right triangular prisms with sides of 5 feet and length of 100 feet.

$$\text{Volume} = 2\left(\frac{1}{2}\right) \times 5' \times 5' \times 100' = 2{,}500 \text{ ft.}^3 \text{ of dirt to remove.}$$

Degree of Precision with Excavation Problems

Practical problems of excavating are going to produce only approximations of these shapes, but the calculations will give a reasonable first estimate of the amount of material that has to be handled.

LEAD AND OAKUM

If you are working on a caulked cast iron soil pipe job, it is helpful to know how much lead caulking and oakum are required to complete the work. Figure 15-23 shows a poured lead joint with a small section removed.

FIGURE 15-22 Sloped sides of a trench would be similar to a triangular prism.

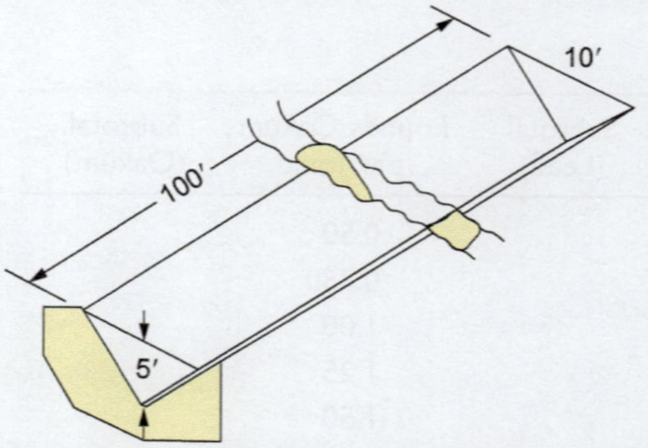

FIGURE 15-23 This cutaway shows a cross-sectional view of a poured lead and oakum joint.

Photo by Ed Moore

Experience has shown that, in general, one pound of lead and 0.25 pound of oakum are required for every inch of diameter of the pipe size. These values include allowances for some spillage or overpouring. Table 15-2 can be used to calculate the total amount of lead and oakum needed for a job of various pipe sizes. Always check with the pipe manufacturer for correct amounts of lead and oakum for joining their pipe.

EXAMPLE 23

How much lead and oakum are required for two 6-inch joints?

Lead = 2 joints × 6″ × 1 lb./inch = 12 lbs. of lead

Oakum = 2 joints × 6″ × 0.25 lb./inch = 3 lbs. of oakum

EXAMPLE 24

What is the lead and oakum required for the following joints?

(16) $1\frac{1}{2}''$, (24) 2″, (30) 3″, (42) 4″, (62) 5″, (76) 6″

The amounts are shown in Table 15-3.

TABLE 15-2 Lead and Oakum Takeoff

Pipe Size	Number of Joints	Pounds Lead per Joint	Subtotal (Lead)	Pounds Oakum per Joint	Subtotal (Oakum)
2		2		0.50	
3		3		0.750	
4		4		1.00	
5		5		1.25	
6		6		1.50	
		Total lead =		Total oakum =	

TABLE 15-3 Lead and Oakum Takeoff for Example 19

Pipe Size	Number of Joints	Pounds Lead per Joint	Subtotal (Lead)	Pounds Oakum per Joint	Subtotal (Oakum)
2	24	2	48	0.50	12.0
3	30	3	90	0.750	22.5
4	42	4	168	1.00	42.0
5	62	5	310	1.25	77.5
6	76	6	456	1.50	114.0
		Total lead =	1,096	Total oakum =	274.0

EXAMPLE 25

If lead is sold in 25 lb. ingots, and oakum in 10 lb. boxes, how many ingots of lead and how many boxes of oakum will be required in the last example? Round up to the next larger whole number.

No. of ingots of lead $= \frac{1{,}096}{25} = 43.84$, so 44 should be on hand.

No. of boxes of oakum $= \frac{274}{10} = 27.4$, so 28 should be on hand.

These amounts should be on hand as the job begins.

IN THE FIELD

When working with molten lead, most people recognize the danger of being burned by the hot lead, but many people are unaware of the dangers of lead poisoning from fume inhalation or absorption through the skin. While lead poisoning is more dangerous for children, adults can be harmed as well. Lead poisoning can happen either through continued exposure to low levels or by a single high-level exposure. Symptoms of lead poisoning may include:

- Irritability
- Aggressive behavior
- Low appetite and energy levels
- Difficulty sleeping
- Headaches
- Reduced sensations
- Very high levels may cause vomiting, muscle weakness, and seizures

To reduce exposure, always work in a well-ventilated area.

Another danger of working with molten lead is being splattered by lead as the pouring is done. If water is in the pipe joint, it will be turned to steam as it contacts the hot lead. The expansion of the steam will cause the lead to shoot out of the joint, and it could come into contact with workers. Always make sure that appropriate personal protective equipment is being used.

REVIEW QUESTIONS

Fill in the Blank

1. Fill in the missing values for Table 15-4 for the three different offsets (1, 2, and 3). Round values to the nearest $\frac{1}{8}''$.

TABLE 15-4 Offsets for $\alpha = 45°$

	1	2	3
Offset	20″	—	36″
Diagonal	—	16.5″	—
Fitting Allowance	1″	—	0.75″
Cut length (e–e)	—	15″	

2. Fill in the missing values for Table 15-5, for the three different offsets, each using different degree elbows (22.5°, 45°, 60°). Round values to the nearest $\frac{1}{8}''$.

TABLE 15-5 Offsets of Various Degrees

Fitting Angle (α)	22.5°	45°	60°
Offset	20″	15″	20″
Run	—	—	—
Travel	—	—	—
Fitting Allowance	$\frac{1}{2}''$	$1\frac{1}{2}''$	2″
Cut Length (e–e)	—	—	—

3. Calculate the circumference of a circle with a radius of 15 feet.
4. What is the perimeter of a square with sides of 100 inches?
5. Calculate the perimeter of a rectangle with sides of 15 inches and 10 inches.
6. Find the volume of a horizontal cylinder in gallons if the radius is 4 feet and the length is 16 feet.
7. If a rectangular tank is limited to 4′ of depth and 10′ of length, how wide does it have to be to hold 1,500 gallons?
8. How many cubic yards of dirt must be moved to open a 45° sloped trench 6 feet deep and 300 feet long?
9. Calculate the amount of lead and oakum required for the joints listed in Table 15-6.

TABLE 15-6 Lead and Oakum Takeoff for Question 9

Pipe Size	Number of Joints	Pounds Lead per Joint	Subtotal (Lead)	Pounds Oakum per Joint	Subtotal (Oakum)
2	10	2		0.50	
3	6	3		0.750	
4	20	4		1.00	
5	0	5		1.25	
6	10	6		1.50	
		Total Lead =		Total Oakum =	

Chapter 16

LEARNING OBJECTIVES

The student will:

- Describe the properties of water and the relationship between depth of water and pressure.
- Describe the relationship between pressure, force, and area in liquid systems.
- Describe the properties of air.
- Describe the methods and problems associated with measuring very small air pressures.
- Describe causes and remedies for noise in water systems.

Properties of Water and Air

WATER

Water, which is described chemically by the formula H_2O, is a compound of two atoms of hydrogen and one atom of oxygen. Water is the most abundant chemical on the surface of the Earth, and is essential to all life forms.

Water can exist in one of three *states* or forms:

- Solid
- Liquid
- Gas

Although almost all substances can exist in all three states, water is the only material that we may encounter in each state at more or less normal temperatures in our daily lives, and as plumbers we must learn how to handle these three states of water in our work.

Solid Form

Water changes from liquid to solid when cooled to 32°F (0°C). Note that water at 32°F must be cooled to form ice at 32°F. If ice and water are together in a container and mixed well, the temperature can only be 32°F. It takes almost as much heat removed (about 80 percent) from water at 32°F to form ice at 32°F as is required to cool water from 212°F to 32°F. No other material requires as much heat removed per pound to change from liquid to solid.

A most unusual characteristic of water is that the solid form is less dense (that means it expands) than the liquid form. This means that ice floats in liquid water, and if ice is formed in a closed container, very large forces will be developed in the ice. These forces will break nearly all containment structures (pipes or tubing, for instance) that are commonly employed to contain liquid water. The change in volume from liquid to ice can produce pressures as high as 30,000 psi. Such pressures will produce forces that no metal pipe can restrain.

In an open container (pan, tank, or lake), water freezes to ice when the air above the surface is colder than 32°F. Ice will form on the surface and eventually form a solid sheet. The water below the ice will gradually freeze (thickening the ice surface), but the process is slow because the ice sheet is a relatively poor conductor of heat. Remember that heat must be removed from liquid water to form solid water (ice), even though the temperature will remain 32°F.

Water containing any dissolved material (including gases) will freeze at a temperature below 32°F. The exact freezing point depends on the kind and amount of dissolved material. One combination that most of us are familiar with is antifreeze solutions in automobile cooling systems.

Even though antifreeze solutions can be used to maintain trap seals in empty buildings, they cannot be used for potable water piping protection. Such lines should be protected by keeping them above freezing temperatures or by draining them.

Liquid Form

Water is most abundant in nature in liquid form. Examples of the liquid state are oceans, rain, lakes, rivers, and streams. In liquid form, water will absorb gases and dissolve many solids. Some of the materials that are dissolved in the water that we use for potable supplies can create problems for us, such as staining fixtures, producing scale in piping and appliances, and making water hard (see Chapter 6 for more details) so that it is difficult to use for washing applications. The most serious problem of these dissolved materials is the effect they have on animal life, including our own.

Physically, the liquid state has some remarkable characteristics. It has the highest specific heat of any material, meaning that it takes more heat to change the temperature of one pound of water than the heat required for any other material. This feature makes water an ideal heat-transfer fluid.

When water is cooled, it contracts slightly in volume for every degree of cooling down to 39.2°F (4°C), but it expands slightly when cooled below that temperature. When water changes from liquid to solid at 32°F, an additional expansion of about 10 percent takes place. This means that the solid form of water (ice) is significantly less dense than the liquid form and, therefore, it floats.

Gas Form

Water exists in the third state (gas) at all temperatures, although we usually think of water changing to a gas at 212°F. This gas is called water vapor or steam. The temperature of 212°F is the boiling point of water at atmospheric pressure. There is a boiling point of water for each pressure over a very large range of pressures (up to 3,200 psi). The significance of this correlation is that if a volume of air contains water vapor at a certain pressure and the air is being cooled, there is a temperature at which liquid water will appear on the surfaces of the space. This temperature is called the dew point.

Heat must be added to a liquid to form a gas. Of all materials, water requires the greatest amount of heat to change 1 pound of liquid to 1 pound of gas. It takes over five times as much heat to change water at 212°F to a gas as is required to heat water from 32°F to 212°F. If liquid water and water vapor (steam) are present in the same container, the temperature of both must be the boiling point temperature for the vapor pressure that exists in the container. Likewise, if the water in an open pan is boiling on a stove at sea level, the water in the pan cannot be hotter than 212°F.

Steam is a clear gas. The white plume most of us think of as steam is steam with water droplets entrained in it. Water vapor exists in the atmosphere and is the source of clouds. Clouds are visible because of small droplets of liquid water that are suspended in the air.

When water in a partially filled, closed container is heated, more and more water will change to vapor state, and the pressure and temperature in the container will increase. It is essential to have a pressure relief valve in such a system to guard against pressure buildup that could rupture the vessel. Such a rupture (unlike a freeze-induced break) will release large amounts of energy and be a great hazard to life and property.

When all the water in a closed container is changed to a gas state, the pressure and temperature will rise if heating is continued, and the steam is said to be superheated. Superheated steam can contain huge amounts of energy and should be considered very dangerous.

Water as a Solvent

Water is often considered to be the universal solvent. To some degree, water will dissolve many of the substances found in nature. Water can also be used to dilute, or reduce the strength of, many products used in industry.

Potable Supplies

Basic Principle No. 1 of the National Standard Plumbing Code states: *All premises intended for human habitation, occupancy, or use shall be provided with a supply of potable water. Such a water supply shall not be connected with unsafe water sources, nor shall it be subject to the hazards of backflow*. This Basic Principle requires that we develop potable supplies and that we maintain their integrity so the supplies are free from inorganic or organic contamination.

Compressibility

One characteristic of most liquids, including water, is that they are essentially incompressible. Water, for example, will not change volume significantly until the applied pressure reaches several hundred thousand pounds per square inch. This incompressibility leads to freeze breaks of piping and water hammer as the result of the operation of quick-closing valves. Gases, on the other hand, are readily compressible with modest applied pressure.

Mass and Force

A property of all matter is mass. Mass is measured in pounds; the weight of a mass on Earth is the mass (in pounds) times the acceleration due to gravity, which is 32 ft./sec.

The result of this calculation is pounds-force. Throughout this book, when pounds are mentioned, pounds-force is the concept intended.

Water Pressure

In these discussions, references are made to water pressure. Pressure is defined as a force acting on a unit area. In our system of units, pressure is expressed in pounds per square inch, with the symbol psi. This definition gives rise to some surprising ideas. For example, a woman's high-heeled shoe develops more pressure on the ground than does the tread of a bulldozer or military tank. This is because all of the woman's weight is focused on a small area, while the bulldozer has large wide treads to spread out the load.

The typical pressure gauge is arranged to indicate the pressure in psi of the line to which it is connected. The pressure read on a gauge is referred to as pounds per square inch gauge (psig). The total pressure in a pipeline is gauge pressure corrected for atmospheric pressure.

Consider a cube 1 foot on each side (length, width, and height) full of water. See Figure 16-1. The water weighs 62.4 lbs. The total force attempting to push out the bottom is 62.4 lbs. Since there are 144 square inches in the bottom, the pressure on the bottom is:

$$\text{Pressure} = \frac{\text{Force}}{\text{Area}} = \frac{62.4 \text{ lbs}}{144 \text{ in.}^2} = 0.433 \text{ psi}$$

Note that the pressure on the bottom is based only on the distance from the bottom to the free surface (1-foot deep) of the water in the container. No matter what configuration of shapes, reductions, bends, offsets, and so on is used, the pressure at the bottom is a function only of the depth from the free surface of the water.

Thus, the pressure at the base of a 4-inch stack 30-feet high is the same as the pressure at the base of a $\frac{1}{2}$ inch pipe 30 feet high if they are filled to the same depth.

FIGURE 16-1 Column of water with a base of one square foot.

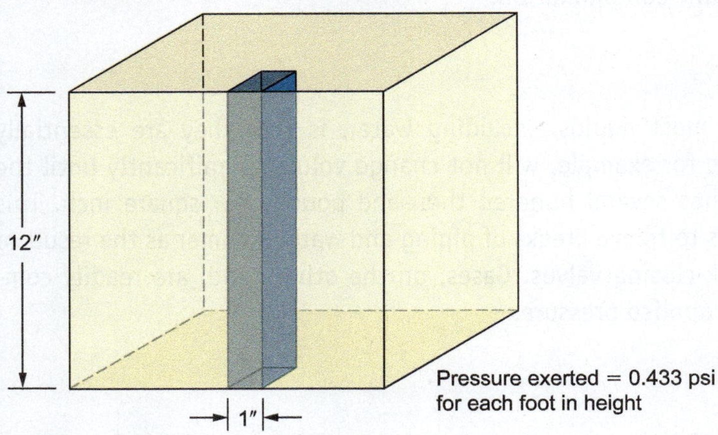

If both these pipes are full, the pressure at the base of each equals: $P = 30$ ft. $\times$ 0.433 psi/ft. = 12.99 psi

In practice, it takes 2.31 feet of water to exert 1 psi (1 lb. of pressure) at the bottom of the stack. In the equation above, we calculated the amount of pressure at the bottom of a column of water 1 foot in height as 0.433 psi. To calculate how many feet of water must be in a stack to exert 1 pound of pressure, we simply use the following ratio:

$$\frac{1\text{ ft.}}{0.433\text{ psi}} = \frac{X}{1\text{ psi}}$$

$$X = \frac{1\text{ psi} \times \text{ft.}}{0.433\text{ psi}} = 2.31\text{ ft.}$$

The following are some examples that will help develop your skill with these ideas.

EXAMPLE 1

Calculate the weight of water in a cylinder 2 inches in diameter and 3 feet high. Remember that the formula for calculating weight of water per cubic foot is:

$\text{Weight} = \text{Volume} \times 62.4\text{ lbs./ft.}^3$

In order to proceed, we must first know the volume of the cylinder. You may recall (see Chapter 15) that the equation for calculating a cylinder's volume is: $\text{Volume} = \pi r^2 h$

A note of caution: watch your units! Remember that we are working in feet, pounds, and gallons, and the information you have been given includes inches.

$r = 1''$ and $h = 3'$

$V = \pi\left(\frac{1}{12}\right)^2(3)$

$V = (3.14)(0.0069)(3)$

$V = 0.065\text{ ft.}^3$

Now that we have the volume, we can calculate the weight:

$W = (0.065\text{ ft.}^3)(62.4\text{ lbs./ft.}^3)$

$W = 4.06\text{ lbs.}$

Another way to solve this problem is to multiply the pressure at the bottom of the pipes by the area of the bottom:

$W = \text{pressure} \times \text{area}$

$\text{Pressure} = 3\text{ feet} \times (0.433\text{ psi/feet}) = 1.299\text{ psi}$

$\text{Area} = (3.14)(1\text{ inch})^2 = 3.14\text{ inch}^2$

$\text{Weight} = (1.299\text{ psi}) \times (3.14\text{ inch}^2)$

$\text{Weight} = 4.07\text{ lbs}$

EXAMPLE 2

Which stack has more pressure at the base: a 200-foot, $1\frac{1}{2}$ inch vertical pipe or a 100-foot, 3-inch vertical pipe, if each pipe is full?

The pressure at the base of the $1\frac{1}{2}$ inch stack is:

$P_1 = 200 \text{ ft.} \times 0.433 \text{ psi/ft.}$

$P_1 = 86.6 \text{ psi}$

The pressure at the base of the 3-inch stack is:

$P_2 = 100 \text{ ft.} \times 0.433 \text{ psi/ft.}$

$P_2 = 43.3 \text{ psi}$

The pressure of the first stack is twice the pressure of the second stack even though one pipe size is twice the diameter of the other. Remember, the controlling feature is the vertical height.

However, be aware that the total force (pounds of pressure over the surface area) at the bases of these stacks will show a different relationship.

$\text{Total Force} = P_1 \times A_1$ $\quad$ $\text{Total Force} = P_2 \times A_2$

$\text{Total Force} = 86.6(\pi)\left(\frac{3}{4}\right)^2$ $\quad$ $\text{Total Force} = 43.3(\pi)\left(1\frac{1}{2}\right)^2$

$\text{Total Force} = 153.0 \text{ lbs}$ $\quad$ $\text{Total Force} = 306.1 \text{ lbs}$

EXAMPLE 3

A 100-foot tall, 10-story building riser is full of water. The pressure pattern is shown in Table 16-1.

Consider adding a 100 psi booster pump to this system. See Table 16-2. If the connection is made at the top, the pressure pattern is as shown in the center column of Table 16-2. If the booster pump is connected at the bottom, the pressure pattern is as shown in the right column. You can see that there is a great difference throughout the building depending upon where the booster pump is added.

TABLE 16-1 Pressures Versus Elevations

Height (ft.)	Pressure (psi)
100	0.00
90	4.33
80	8.66
70	12.99
60	17.32
50	21.65
40	25.98
30	30.31
20	34.64
10	38.97
Ground	43.30

EXAMPLE 3
continued

TABLE 16-2 Pressures Due to Pumps and Elevations

Height (ft.)	Pressure (psi) (Boost Added at Top)	Pressure (psi) (Boost Added at Bottom)
100	100.0	56.7
90	104.3	61.1
80	108.7	65.4
70	113.0	69.7
60	117.3	74.1
50	121.7	78.4
40	126.0	82.7
30	130.3	87.0
20	134.6	91.3
10	139.0	95.7
Ground	143.3	100.0

Many other considerations are involved in selecting a booster pump for a tall building. This example simply points out that the effect of the same boost is very different depending on where the boost is added.

Pressure Independent of Shape of Container

Figure 16-2 shows several pipes of different shapes and types connected to a header. The pressure in the header is the same for all risers, and the water level in all risers is the same, demonstrating that the shape of the riser does not matter when determining pressure.

Consider now any point in a closed container of water. Pascal's Law states that pressure exerted by the liquid at any point is distributed evenly throughout the liquid: the same in all directions and at right angles to the surfaces of the vessel. This means that the liquid is pushing out evenly on all sides of the container.

FIGURE 16-2 Pressure remains the same regardless of riser shape.

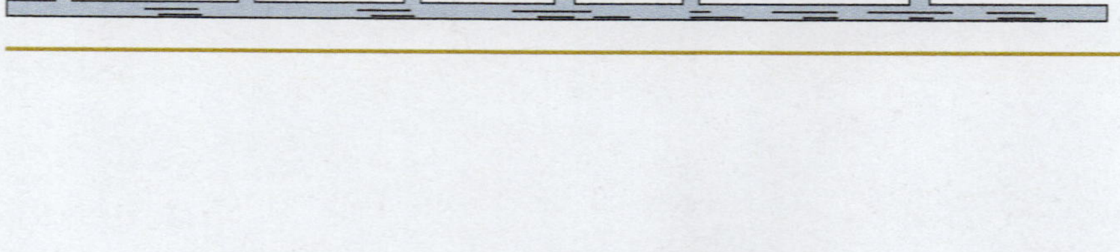

FIGURE 16-3 If a force is applied to the left-side piston and it lowers 1 foot, then the same force could be lifted the same height by the right-side piston.

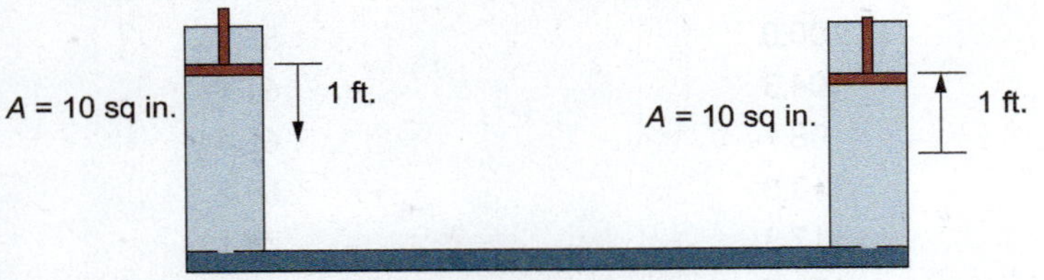

For example, if you have a long enclosed tube of water with pistons of equal size at both ends, when you push on one of the pistons, the pressure in the water will increase evenly throughout the container. If you had nothing holding the other piston down, the increased pressure of the water would move the other piston. Because the pressure is distributed equally throughout the water, the other piston would move exactly as far as the first piston moved. See Figure 16-3.

Carrying this concept even further, consider a container with a big open space in the middle and two tubes sticking out of opposite ends of the container with a piston in the end of each tube. In this case, one of the tubes is much larger than the other—the small piston has 1 square inch of surface area and the large piston has 100 square inches of surface area. This means that the large piston is 100 times larger than the small piston.

According to Pascal's Law, when you push on the small piston with a force of 1 pound, you increase the pressure in the entire container by 1 psi. This means that the large piston will be pushed out by the same pressure (1 psi), but since it has a surface area 100 times the small piston's surface area, the amount of force on the large piston will be 100 pounds.

This is useful because it means you can turn a small force into a large one without doing a lot of extra work. This is the main concept behind hydraulic pumps.

HYDRAULICS

Hydraulics is the study of liquids in motion and the work that can be done by transmitting pressures or by liquids under pressure.

There are circumstances where plumbing systems, or portions of systems, can be analyzed conveniently based on the principles of hydraulics. One such example is the consideration of the actions of water descending in a stack. Another is the study of pressure effects when water freezes or valves close quickly. These topics will be investigated in future training.

An example of the consequences of the magnitude of total force as a result of pressure is the relative strength between a $\frac{1}{2}$ inch copper tube and a 2-inch tube of the same wall thickness. The total force available to break the $\frac{1}{2}$ inch line is only $\frac{1}{16}$ inch (the ratio of [$\frac{1}{2}$ inch 2 to 2], squared) in the 2-inch line at the same pressure. Therefore, the $\frac{1}{2}$ inch line could be said to be 16 times as strong as the 2-inch line.

Another example of the importance of recognizing these situations is that large liquid storage tanks are typically limited to 5 psi pressure for testing because higher pressures could break the tank.

EXAMPLE 4

Examine Figure 16-4. A container has two pistons mounted on it, one of 10 in.^{2} area and the other of 100 in.^{2} area. If a force of 100 lbs. is applied to the piston with 10 in.^{2} area, a pressure is developed in the common casing as shown in the Figure.

FIGURE 16-4 If a small force is applied to piston A, a much larger force can be exerted by piston B, but over a smaller distance of travel.

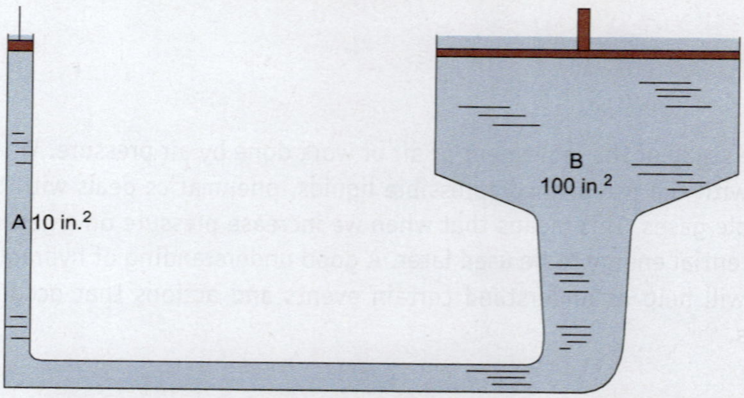

This pressure is also exerted on the 100 in.^{2} piston, which will thus have a total force of 1,000 lbs. applied. This is the basic principle of operation of a hydraulic jack. Be sure to always use a hydraulic jack rated for the type of job you are doing.

A force developed by a pressure acting on an area is the operating mode of many items, including flushometer valves, regulating valves, hydraulic backhoes, cranes, and presses.

AIR

We live in a sea of air. It surrounds us and extends upward for about 100 miles. Air is essential to all life that is seen on Earth. Air is a mixture of many gases—mainly nitrogen, oxygen, carbon dioxide, and water vapor. Air is colorless, odorless, and tasteless. Only pollutants produce other effects.

The amount of water vapor actually present in the air is influenced by many things. At any given air temperature, there is some maximum amount of water vapor that air can hold (defined by the vapor pressure at that boiling point). When air contains the maximum amount of moisture, it is said to be saturated. The term *relative humidity* is used to describe how much moisture is in the air compared to the maximum amount it could contain. For example, if the air contains the maximum amount of water vapor, then it is said to have a 100 percent relative humidity. When the air contains only half the water vapor it could normally hold, it is said to have 50 percent relative humidity.

Air can be used to transfer heat from one place to another. Like all gases, air is compressible. Air can be transported in pipes or ducts. Because air is a gas, it expands when heated and contracts when cooled.

Air can be beneficial in plumbing systems under controlled circumstances. For example, the air within an expansion tank allows pressure surges to be minimized. If not properly controlled, excess air pressure in the DWV system can cause a loss of trap seals.

Pneumatics

Pneumatics is the study of the movement of air or work done by air pressure. Where hydraulics deals with the use of incompressible liquids, pneumatics deals with the use of compressible gases. This means that when we increase pressure on a gas, we can store this potential energy to be used later. A good understanding of hydraulics and pneumatics will help us understand certain events and actions that occur in plumbing systems.

Drain Lines

To completely analyze the actions in a drain line, we must combine the studies of hydraulics and pneumatics. When water passes into a trap inlet, the trap seal changes from a passive condition to an active condition. Water enters the trap and the air is displaced. On the inlet side, the air comes up out of the trap inlet, helping to develop the scouring motion. On the discharge side, the air is pushed to the trap arm outlet, where it can escape out the vent (Figure 16-5). The water entering and falling down the stack again must displace air. The air also prevents the slug of water from remaining intact for more than a few seconds. Thus, the actions and motions within a trap are dynamic and complex. Remember that when a pipe is empty, it really contains air, and when water enters the pipe the air must exit. If the air does not leave, then the pressure will increase.

In Chapter 19, the trap seal is described as only a 2-inch vertical column of water. For this reason, the goal of any drainage and venting systems is to limit pressure variations at trap outlets to $\pm$1-inch water column pressure. A pressure of 1-inch water column is equal to $\frac{1}{12}$ of a 1-foot head (the pressure at

FIGURE 16-5 The swirling action in lavatory tailpiece helps water flow through trap.

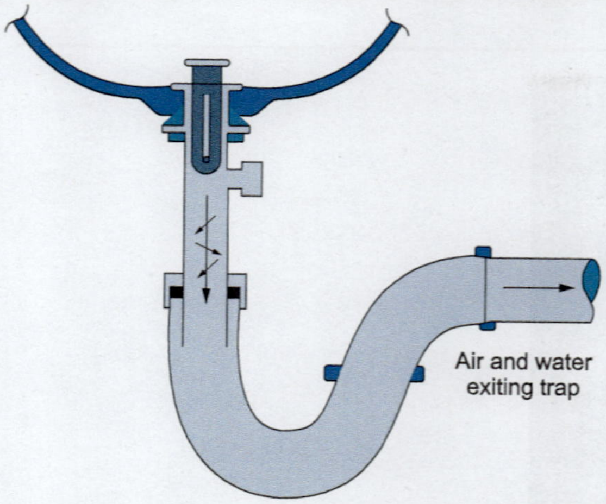

the bottom of a vertical column of water). Earlier in this chapter, it was explained that:

$$1' \text{ head} = 0.433 \text{ psi}$$

$$1'' \text{ water column} = \frac{1}{12} \times 0.433 \text{ psi} = 0.036 \text{ psi}$$

AIR IN HEATING LINES AND WATER LINES

Air in water lines or hot water heating lines can be a problem. In typical heating lines, air will act like a plug and prevent flow. Since air does not conduct heat as well as water, it reduces the amount of heat transfer. Water supply lines will vent the air without problems, but if a water line serves flushometers (see *Plumbing 101*), problems could arise. Since most heating system lines are closed circulation loops, small automatic or manual valves at high points are needed to bleed air from the lines.

To reduce problems associated with the presence of air in plumbing systems, install domestic water or space heating piping to minimize air traps. Use uniform slope upward in the direction of flow whenever possible. Install vent valves or fixture connections at any high points.

More details on air in heating and water lines will be covered in the fourth-year book (*Plumbing 401*) in this series, specifically as it relates to hydronic heating.

Compressed Air

Air is used in water hammer arrestors as the cheapest and most available cushioning medium. Compressed air is used to operate valves, dampers, air vises, impact tools, and similar pneumatic devices.

FIGURE 16-6 Air compressor tanks must be drained periodically to prevent water buildup in the tank. Here an automatic tank drain periodically discharges condensate.

Photo by Ed Moore

The typical compressed air source is an air compressor, as shown in Figure 16-6. Most small compressors are electric-motor-driven, reciprocating-piston types. They are usually controlled to turn on when the storage tank pressure is at the minimum setting and to turn off when pressure reaches a preset maximum.

When air is compressed, all the components of the air are compressed, including the water vapor. Condensed water in the compressed air storage tank and in the air piping can be a source of constant problems. Water in air lines tends to clog air mufflers and prevent valves from working properly. The compressed air tank is equipped with a drain so that water that collects in the tank can periodically be drained.

It is considered to be a good practice to install drip legs at the bottom of any long vertical runs. This allows moisture to collect in low points in the piping system. (See Figure 16-7.)

FIGURE 16-7 Drip legs allow for removal of condensate in the piping system. Here the pipe has been cut at an angle so no one can accidentally connect to the system.

Photo by Ed Moore

IN THE FIELD
The discharge from an air compressor tank drain will contain emulsified oil. Discharging large amounts of this can be an ecological problem and may also stain concrete or other surfaces.

SAFETY ISSUES WHEN TESTING WATER PIPING SYSTEMS

Sometimes air is used to test water piping systems. Never use oxygen as a substitute for compressed air. Pure oxygen makes many normally safe materials dangerously flammable. A suitable substitute for air is compressed nitrogen, which is available from welding supply companies and is an excellent compressed gas for pressurizing or testing. Be sure to use a regulator to control the pressure, as the initial pressure in nitrogen cylinders is in the range of 2,500 to 3,000 psi. See Figure 16-8.

Other chemicals, such as propane, acetylene, and carbon dioxide, should never be used to test water piping systems. Residues from these gases can chemically react with piping materials.

The best and safest method of testing water piping systems is with water.

IN THE FIELD
Testing plastic piping systems with compressed gas should not be done. Because there is a considerable amount of stored energy being released in a short period of time, explosive forces can be generated.

FIGURE 16-8 This nitrogen tank is equipped with a pressure regulator to safely pressurize piping systems.

Photo by Ed Moore

MANOMETERS

Pressures in the ranges encountered in plumbing drainage and fuel gas systems are usually expressed in terms of inches of water column. The logical way to measure these pressures is to connect a column of water to the point under consideration and measure the height obtained.

The instrument used for measuring pressure in a water system is a very simple device called a *manometer*. A traditional manometer is a U-shaped tube made of glass or clear plastic and filled with a liquid (see Figure 16-9). It is usually equipped with a measuring scale to make reading convenient. Make sure that the end not connected to the point being measured is left open to atmosphere.

Traditional manometers used either water or mercury as the liquid medium, but due to problems with accuracy, size of the manometers, and health hazards related to mercury exposure, these materials are no longer used. If a manometer needs to be refilled, make sure to use the correct fluid for that instrument.

Digital manometers are now available for use and reduce the margin of error in reading pressures. See Figure 16-10. Unlike U-tube manometers, digital manometers must be periodically checked for accuracy.

FIGURE 16-9 U-tube manometers are capable of reading very low pressures.

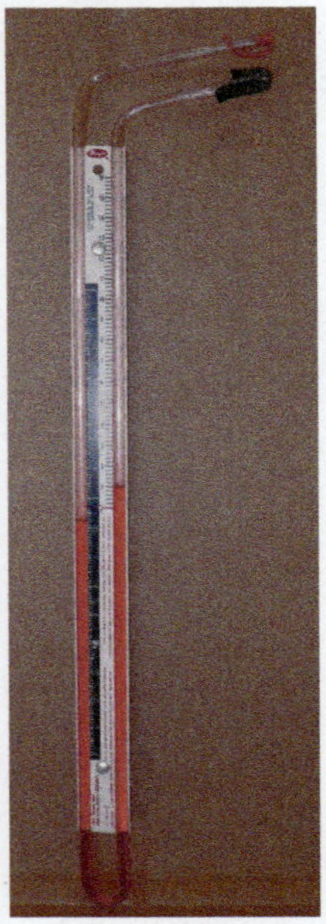

Photo by Ed Moore

FIGURE 16-10 Digital manometers are easier to read but must be periodically checked for calibration.

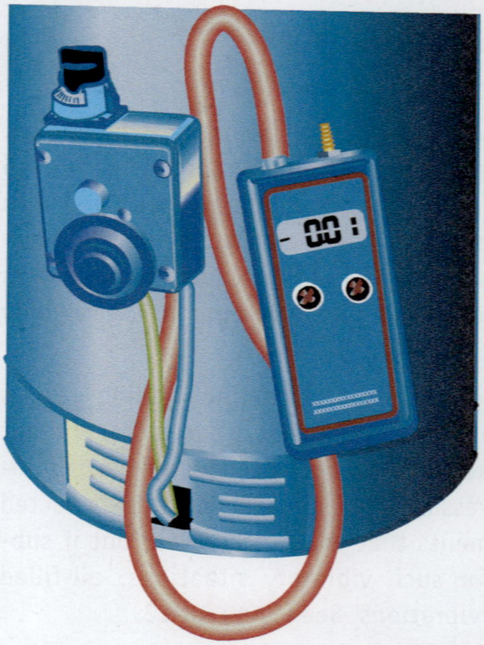

BOURDON TUBE

A manometer is the ideal pressure-measuring device for very low pressures, but it is not useful for measuring pressures above the range of a very few psi. A device for measuring higher pressures is the Bourdon tube, a gauge that uses a curved tube of brass or steel that distorts when pressure is applied. Most modern dial gauges that are used today contain a Bourdon tube. Figure 16-11 shows the working mechanism of a modern dial gauge.

The end of the curved tube is connected by linkages to an indicating pointer, which moves in proportion to the applied pressure. Gauges of this sort

FIGURE 16-11 The Bourdon tube is a primary part of any dial gauge.

Photo by Ed Moore

can be made small, portable, and inexpensive with accuracy levels of 95 percent to 98 percent. Large laboratory types with accuracy levels up to 99.9 percent are also available.

Gauge Selection and Use

Bourdon tubes can be used on any application where pressure must be known. There is a suitable gauge for almost any application. The gauge should be selected for the fluid to be measured, of size and accuracy suitable for the task, and with the proper scale range. The range selected should be such that the normal pressure to be read should be one-half to three-quarters of full scale. Therefore, you would not select a 0–300 psi gauge to read the pressure from a pump that can only develop 15 psi.

Many types of special purpose gauges are made, including welding gauges with blow-out backs, refrigeration gauges with special calibration faces, and hot water boiler gauges with feet of altitude as well as pounds per square inch calibrations. Gauges that are to be under pressure continuously must be selected carefully—many linkages have gear elements that will quickly wear out if subjected to constant pulsing pressure. For such vibratory situations, oil-filled gauges are available that damp out such vibrations. See Figure 16-12.

When installing gauges or other sensitive measure devices on steam lines, it is common practice to install a "pigtail siphon." A pigtail siphon is a device that acts as a trap. Condensate collects in the loop and prevents live steam from entering the Bourdon tube of the gauge. Figure 16-13 shows how one works.

FIGURE 16-12 A dial gauge is filled with an oil-like substance to damp vibration, which increases gauge life and makes it easier to read.

Photo by Ed Moore

FIGURE 16-13 A pigtail siphon prevents live steam from entering sensitive gauges.

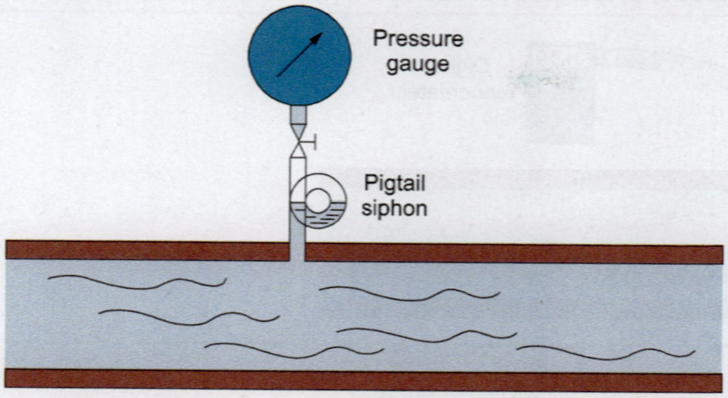

INSTALLATION TESTS

Installation tests can be made with a variety of devices. A static pressure tube (see Figure 16-14) is an open tube placed in a moving stream of air or liquid so that the moving fluid moves across the opening of the tube. With the tube connected to a measuring device, the pressure the fluid exerts on the walls of the pipe or duct is measured.

If the test tube is bent so that the fluid is directed into the tube, both the velocity pressure and the static pressure can be measured. While static pressure is present in any pressurized container, velocity pressure is only present when there is flow. Total pressure is the sum of both static and velocity pressure. Figure 16-15 shows a total pressure reading being taken.

Rough-In Pressure Test

An air test can be applied to a drainage rough-in (see Figure 16-16). All openings are closed, and compressed air is introduced into the test section. Figure 16-17 shows some typical pipe plugs used to seal pipe openings. Five psi is equivalent to about 12 feet water head, so 5 psi is the usual test pressure. If the gauge does not show a change after 15 minutes, the test is satisfactory.

FIGURE 16-14 When the test tube is positioned as illustrated, it reads static pressure much like a regular pressure gauge.

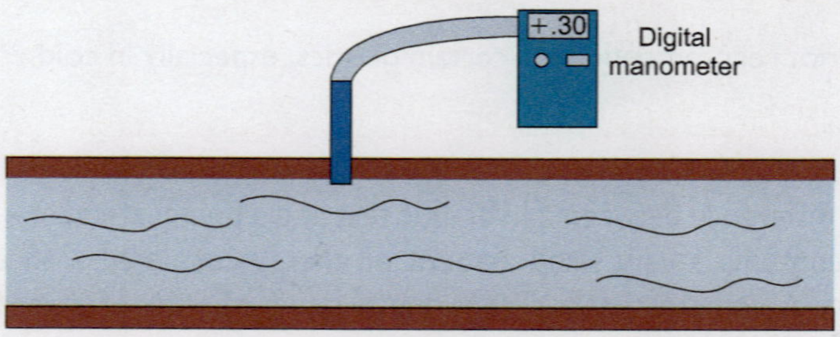

FIGURE 16-15 When the test tube is positioned as illustrated, it reads static pressure plus the velocity pressure. Both add up to be total pressure.

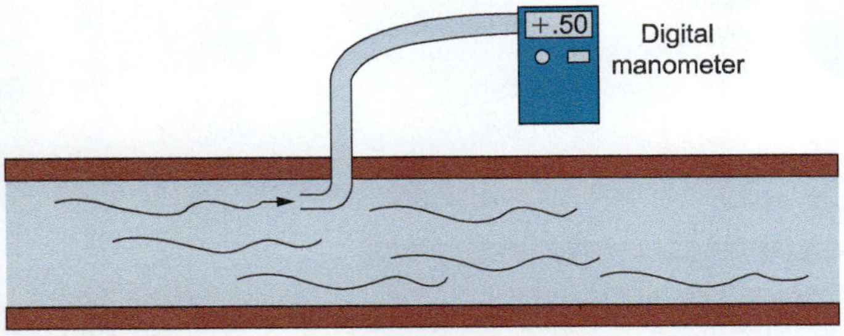

FIGURE 16-16 With all openings capped off, the system can be pressure tested with either air or water.

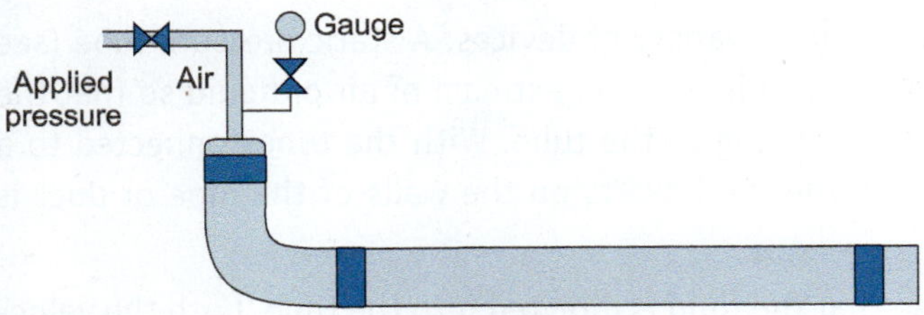

FIGURE 16-17 Three common internal pipe plugs that can be reused.

IN THE FIELD

Air pressure testing is not recommended for certain plastics, especially in cold climates.

There is a significant difference between a hydraulic test and a pneumatic test—the hydraulic test fluid contains a very small amount of energy compared with a pneumatic test situation. A compressed gas contains a large amount of stored energy, while a pressurized liquid does not.

Thus, pneumatic testing above 5 psi must be carefully evaluated. It is not unusual to test very small systems (small diameter, limited length) to 100 or 125 psi, but a large-diameter system of significant length could contain dangerous amounts of energy at as little as 10 or 15 psi.

A basic rule is never to remove test plugs or other closure devices unless you are sure that the pneumatic test pressure has been completely removed. Always release plugs slowly, and never expose your face to the plug as it is being removed.

NOISES IN WATER SYSTEMS

Three major causes of plumbing system noises are vibration, turbulence, and water hammer.

Vibration

Vibration noises can be the result of inadequate or insufficient hangers or inadequate hanger attachment to the building structure. In such cases, any event that initiates vibration will start the piping rattling in the hangers.

Possible corrections include adding or improving hangers or adding insulating or vibration-deadening material on the pipe where it contacts the hangers. Figure 16-18 shows a flexible pipe connector used to prevent vibration from the air compressor from rattling the pipes. Figure 16-19 shows special pipe hangers that are designed to stop vibration from being transmitted.

FIGURE 16-18 Braided flexible connectors help prevent the transfer of vibration throughout the system.

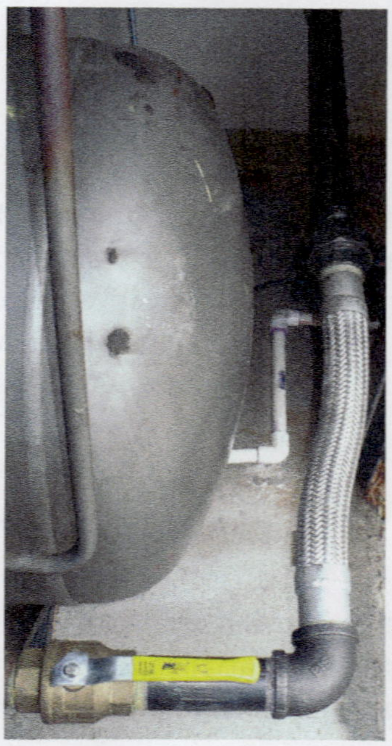

Photo by Ed Moore

FIGURE 16-19 Special pipe hangers help prevent the transfer of vibration throughout the system.

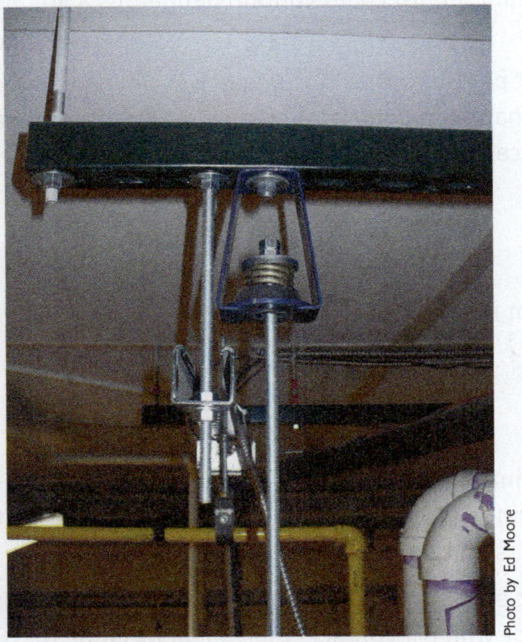

Turbulence

Turbulence from high flow velocities in the piping can develop objectionable noises under the following conditions:

- Thin wall tubing
- Lightweight material
- Short-turn fittings
- Loose parts in plumbing devices
- Undersized piping

In existing work, insulation on the pipe and maintenance of devices are about the only remedies. For new work, select heavy materials or insulate lines that could produce noise problems, and consider oversizing water lines in noise-sensitive areas. Oversizing is accomplished by selecting pipe or tubing sizes that will limit water flow velocities to not more than 4 fps.

Water Hammer

Water hammer is a nuisance for small lines in residential work, but it is a serious hazard in larger work. Valves, fittings, or joints can be ruined by water hammer.

Water hammer is caused by the rapid pressure rise in a moving column of water that is suddenly stopped by a quick-acting valve. Water is essentially incompressible, so when a moving column of water is suddenly stopped, inertia causes the water to try to keep moving. Since the water movement cannot continue, the water tries to compress. A very small volume change represents a very large pressure change, which is exerted on the valve closure and piping elements in the vicinity of the closure. The pressure rise front is then reflected back along the pipeline, which may cause heavy vibrations for many seconds. The effects can also be experienced far from the closing point, because the water hammer usually sets the pipe moving significantly in the hangers. This movement can cause problems far from the quick-closing valve, such as water heater relief valves opening due to water hammer pressure surges.

Water hammer may not make a perceptible sound. The noise most of us associate with water hammer is a pitch in the audible hearing range, and it is produced in small ($\frac{1}{2}''$ or $\frac{3}{4}''$) piping. Larger piping will have much lower resonance frequencies, so you may have water hammer experiences and not realize it.

Three common ways water hammer occurs are:

- A swing-check valve on pump discharge, especially sump or sewage pumps, will slam after pump shutoff when the water in the discharge starts to flow backward to the pump. This reverse flow closes the check, which stops the flow, and water hammer results.
- A butterfly valve with a manual handle on a medium-size line may be physically difficult to close off. If the operator pushes the handle vigorously to get it to seat, he has closed the line very quickly and water hammer results.
- A quick-closing valve, such as those found in dishwashers, washing machines, ice makers, and so on involves a solenoid valve. A true stop valve, the solenoid is electrically activated and shuts quickly. Its effect is like that of the swing-check valve.

Water hammer may be remedied through one of these methods:

- Limit water flow rate by sizing the water line to the quick-closing valve to 4 fps or less.
- Install shock arrestors or air chambers at fixtures and fixture branches.
- Install a spring-loaded check valve at pump discharge location.

Shock Arrestor

A shock or water hammer arrestor is made up of an outer container with air inside and a bellows that contains the water in the system. Figure 16-20 shows an illustration of commercially available water hammer arrestors.

When a pressure rise occurs in the system, the water within the bellows can move the bellows into the air volume. The air will compress with a large volume change and relatively small pressure change, thus absorbing the shock wave from the inertia of the water. Piston or flexible tube designs work in similar fashion.

These devices are somewhat expensive and limited in capacity, but they retain the air between the bellows and outer container indefinitely. They need to be

FIGURE 16-20 Sectional view of manufactured shock arrestors.

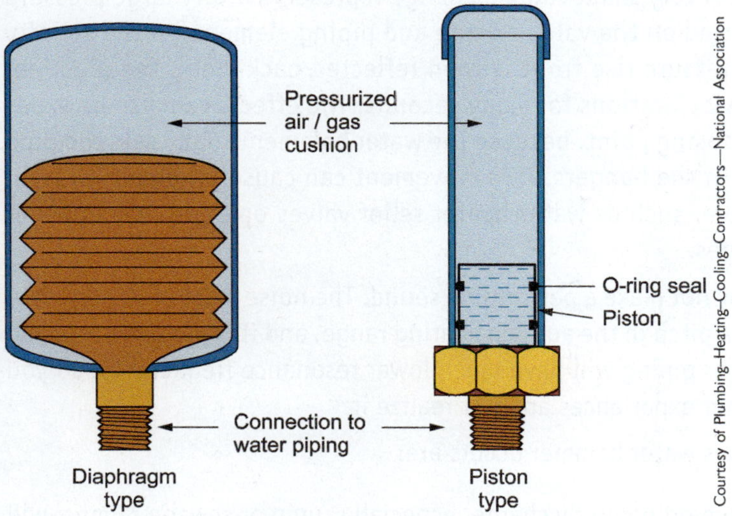

installed where they can be examined or replaced. If the water system cannot be shut down, install a valve below these units so that servicing or inspection will be made easier. There is an industry standard (ASSE 1010) for these devices, so be sure that any items purchased conform to those standards.

Air Chamber

Even though air chambers are no longer recognized in plumbing codes, many are still in use today. Air chambers produce the same effect as shock arrestors at much less cost. The drawback to an air chamber, however, is that the air in an air chamber is gradually dissolved into the water and it becomes waterlogged (filled with water), making it less effective.

A field-fabricated air chamber (Figure 16-21) is a vertical, capped pipe that is installed on the water line to provide an air cushion to absorb or arrest the water hammer pressure surge. Field-fabricated air chambers should be at least one pipe size larger than the branch line and should be at least 18 inches in length to be minimally effective. This is a rule-of-thumb for air chamber sizing.

Another problem with the air chamber is that if atmospheric air is the source of refilling the chamber, it will be compressed to a fifth or a sixth of the measured volume of the pipe that makes up the air chamber. Hence, the air cushion is not as big as you think it is.

Spring-Loaded Check Valve

Where the flow in a pump discharge is subject to reversal upon pump shutoff, water hammer results when the pump stops, forward flow stops, and flow reverses until swing-check valve closure stops the reverse flow. If a spring-loaded check

FIGURE 16-21 Field fabricated air chambers are similar to manufactured air chambers but lack a device to separate the air and water.

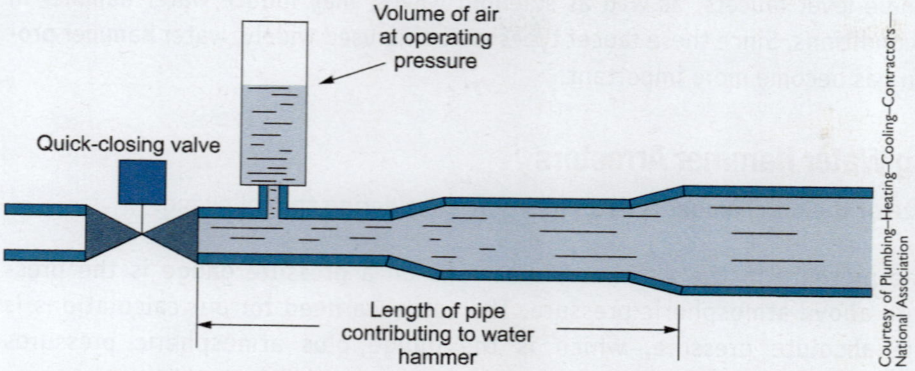

valve is used, the valve closes at the moment the forward flow stops and there is no hammer.

Pipe Material as Shock Absorber

Since the formation of the water hammer wave is caused by suddenly stopping a moving water stream, avoiding or preventing water hammer involves eliminating either the sudden or the moving part. Shock arrestors and air chambers give the moving water someplace to go (like pulling off on the shoulder of the road when you can't stop your car behind the car in front of you), and the spring-loaded check valve prevents the reverse flow from getting started. If the piping material itself is able to expand even a little bit, the shock wave buildup will be much less intense than it would be otherwise.

Locating Shock Arrestors

Practical applications of water hammer arrestors require some concept of the volume of air needed to hold the water hammer pressure rise to an acceptable maximum value. Unprotected lines can be seriously damaged if no water hammer protection is available.

Most codes require the installation of either air chambers or shock arrestors on any piping that contains quick-acting valves. Dishwasher or laundry machine solenoid valves are examples of quick-closing valves. The pressure shock of water hammer is at its maximum at the quick-closing valve, and it is dampened quickly in time and distance along the pipe. For this reason, the protection must be installed as close to the initiating valve as possible. It would be nice if a big air chamber installed at the water meter would dampen out water hammer, but it would be totally ineffective if the quick-acting valve is not close by. Protection must be installed at each and every quick-acting valve on the job if water hammer is to be suppressed.

The National Standard Plumbing Code specifically requires water hammer arresting devices to be installed as close as possible to the quick-acting valves, and also requires that they be accessible for repair, replacement, or replenishment of air. Consult your local code for specific requirements.

Single-lever faucets, as well as solenoid valves, may induce water hammer in some conditions. Since these faucet types are being used widely, water hammer protection has become more important.

Sizing Water Hammer Arrestors

The size of the air chamber is determined by considering the following:

- The pressure in the air chamber as read on a pressure gauge is the pressure above atmospheric pressure. The value we need for our calculations is the absolute pressure, which is the gauge plus atmospheric pressures (14.7 psia).
- The higher the working pressure in an air chamber, the larger the chamber must be to limit water hammer pressure rise to some maximum value (usually 150 psi).
- The longer the water line at constant diameter (or one pipe size larger), the larger the air chamber must be.
- The larger the quick-closing valve orifice, the larger the air chamber must be.
- If the initial pressure in the chamber can be higher than atmospheric, the chamber can be smaller.
- If the flow velocity can be reduced (i.e., increase line size) to 4 fps or less, water hammer can be reduced or eliminated.

The usual convention is that the length of water line contributing to the water hammer is the distance of water pipe back to a point where the pipe line is two sizes larger. Figure 16-22 illustrates the above ideas.

For sizing an air chamber, it is important to understand the concept of Boyle's Law. Boyle's Law states that the product of the volume and absolute pressure of

FIGURE 16-22 Water hammer can be affected by diameter changes in pipe.

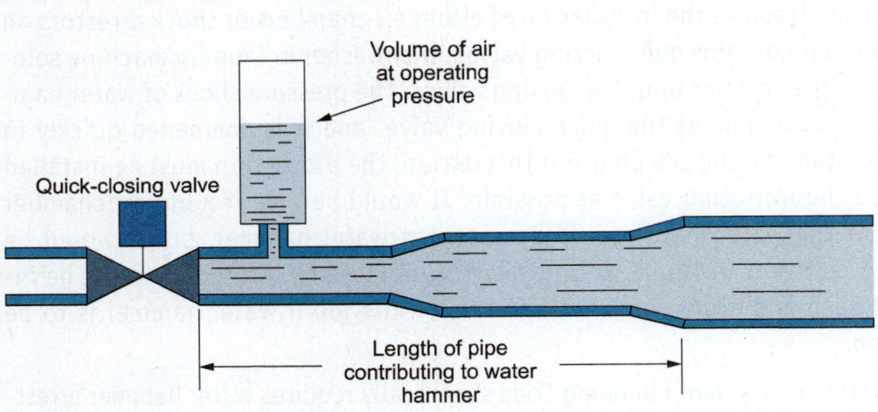

Courtesy of Plumbing–Heating–Cooling–Contractors—National Association

a gas at one condition is equal to the product of volume and absolute pressure at a second condition, as long as the temperature remains constant. This means that we can use the change in pressure in an air chamber to offset the change in pressure of the water system, and we can calculate how large the air chamber needs to be with a simple formula:

$$P_1V_1 = P_2V_2$$

Remember it is important to use psia in this formula. Psia is the pressure reading read from the gauge (psig) plus atmospheric pressure (14.7 psi). Also note that P means pressure and V means volume.

EXAMPLE 5

What is the volume of air in a cylindrical air chamber 2 inches in diameter and 24 inches long at a pressure of 80 psig if the chamber is initially filled with air at atmospheric pressure?

For this condition, if the air chamber is full of air at atmospheric pressure, $P_1 = 14.7$ psia, and V_1 is as follows:

$$V_1 = \pi r^2 h$$
$$V_1 = \pi(1)^2(24)$$
$$V_1 = (3.14)(1)^2(24)$$
$$V_1 = 75.4 \text{ in.}^3$$

V_2 (the volume of air at 80 psig) becomes:

$$V_2 = (V_1P_1)/P_2$$
$$V_2 = (75.4 \times 14.7)/(80 + 14.7)$$
$$V_2 = (75.4 \times 14.7)/94.7$$
$$V_2 = 11.7 \text{ in.}^3$$

Thus, for a 2-inch pipe, 11.7 in.^{3} is the volume of air in the 24-inch section of pipe when the pressure is increased to 80 psig. This small volume of air at the elevated pressure is all that is available for arresting water hammer.

It is clear that once the system has become pressurized, there is considerably less volume available to absorb shock. While this volume may be sufficient for many circumstances, you can see that this volume can be reduced quickly by being dissolved into the water during periods of turbulence.

The variable in the water hammer problem that is difficult to determine with accuracy is the volume of the water surge. Because of this uncertainty, we use a table to determine the air chamber sizes. Table 16-3 shows the minimum size of air chambers as a function of pipe size, pipe length, operating pressure, and maximum velocity. If you use this table, the pressure rise in the piping will be limited to 150 psi.

Always be sure that the air chamber is connected to the water line with a full-flow valve, and that the air chamber is accessible. Note that the air chamber is required if the total flow given is to be stopped suddenly. For example, a 1 inch line supplying a $\frac{1}{2}$ inch line would not see any water hammer at all as a result of an event on the $\frac{1}{2}$ inch line.

Be aware that quick-closing valves in large line sizes are rare, so it is unlikely you will see one.

TABLE 16-3 Capped-Pipe Air Chambers (Schedule 40 Pipe)

Nominal Pipe Diameter	Length of Pipe (ft.)	Flow Pressure (psig)	Velocity (fps)	Flow (gpm)	Required Air Chamber: Volume ($in.^3$)	Required Air Chamber: Physical Size (in.)
$\frac{1}{2}''$	25	30	10	9.5	8	$\frac{3}{4} \times 15$
$\frac{1}{2}''$	100	60	10	9.5	60	1×69.5
$\frac{3}{4}''$	50	60	5	8.4	13	1×15
$\frac{3}{4}''$	200	30	10	16.75	108	$1\frac{1}{4} \times 72.5$
1″	100	60	5	13.5	19	$1\frac{1}{4} \times 12.7$
1″	50	30	10	27	40	$1\frac{1}{4} \times 27$
$1\frac{1}{4}''$	50	60	10	46.5	110	$1\frac{1}{2} \times 54$
$1\frac{1}{2}''$	200	30	5	32	90	2×27
$1\frac{1}{2}''$	50	60	10	63.5	170	2×50.5
2″	100	30	10	104	329	3×44.5
2″	25	60	10	104	150	$2\frac{1}{2} \times 31$
2″	200	60	5	52	300	3×40.5

MECHANICAL SHOCK ARRESTOR SIZING

Sizing of mechanical shock arrestor devices must be performed in accordance with the manufacturer's recommendations.

REVIEW QUESTIONS

True or False

1. ________ Air is a single chemical constituent.
2. ________ To assure the integrity of trap seals, the pressure variations in a drainage system should not exceed 1 psi.
3. ________ Oxygen is not a safe substitute for compressed air.
4. ________ The precise amount of water in a U-tube is critical for accurate measurement.
5. ________ A U-tube manometer must be open at both ends in order to get an accurate reading.

Fill in the Blank

6. The three states of a substance are ________, ________, and ________.
7. The addition of ________ is required to change water into steam.
8. Water in ________ phase is not compressible.
9. Pressure is force acting on a unit ________.
10. As the height of a water column increases, the pressure at the bottom ________.

Chapter 17

LEARNING OBJECTIVES

The student will:

- Describe the forces that can produce pressure changes in a drain stack and explain why there is a need for venting those stacks.
- Differentiate between types of vents used in plumbing systems.
- Identify the types of hangers and supports available for vent systems.
- Describe the types of vent configurations used in vent piping.

Venting Plumbing Drainage Systems

PLUMBING VENT SYSTEMS

A plumbing vent system is *a pipe or pipes, installed to provide a flow of air to and from a drainage system or to provide a circulation of air within such system to protect trap seals from siphonage and back pressure* (National Standard Plumbing Code 2012).

The trap seal is provided by the water that remains in the trap after the fixture is cycled. This water seals off the pathway so that gases from the plumbing system cannot pass into the building. If the water in the trap is siphoned out or blown out by pressure differences in the drains, then offensive or hazardous gases from the sewer can enter the room.

The function of the vent system is to prevent pressure differences large enough to eliminate the trap seal. The maximum pressure difference permitted at any trap outlet is plus or minus 1 inch of water column. This is a very small pressure, equal to 0.036 psi. Recall that the standard trap seal depth is a minimum of 2 inches.

The vent piping conveys any buildup of air pressure to the free air (outdoors) or brings air in from outside to make up low-pressure conditions. Thus, we state that the principal function of the vent system is to provide air circulation throughout the entire system.

Allowing air to circulate also helps materials flow through the system. If air pressures are not equalized, pressure differences can cause problems. To illustrate this fact, consider the following example. Hold your finger on the end of a soda straw and place the other end in a glass of water. No water will enter the straw until you release your finger from the end. Water could not come into the straw because of the trapped air; once the air is allowed to flow out, the water can come in.

If you again close the top of the straw with your finger and withdraw the straw from the glass, the water will remain in the straw until the top is opened. The water will then flow out of the straw because air can flow in and replace the departing water.

Proper placement and sizing of vents guards against a significant reduction of the trap seal. In a plumbing system, fixtures that are not vented will drain erratically, with gurgling sounds and fluctuating water levels.

Figure 17-1 illustrates a vertical pipe with a slug of water descending inside the pipe. The air pressure below the slug will be increased while the air pressure above the slug will be reduced.

Figure 17-2 shows that similar effects happen when a mass of water moves down the pipe—increased pressure below the water mass and reduced pressure above. Further details on the science behind this example are presented in Chapters 16 and 19.

Figures 17-3 and 17-4 show that traps connected to a vertical pipe with water descending in it will have their seals affected by the pressure changes in the stack. These changes could be great enough to cause the loss of the trap seal.

FIGURE 17-1 A large slug of water can descend in a stack like a big plug or ball.

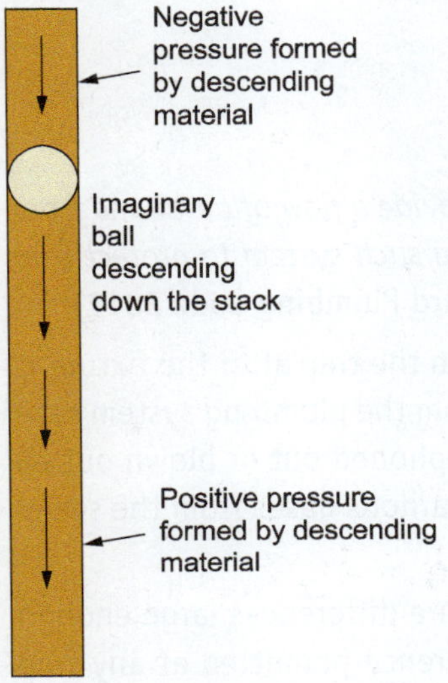

FIGURE 17-2 Effects of discharges from an upper level when the system is poorly vented.

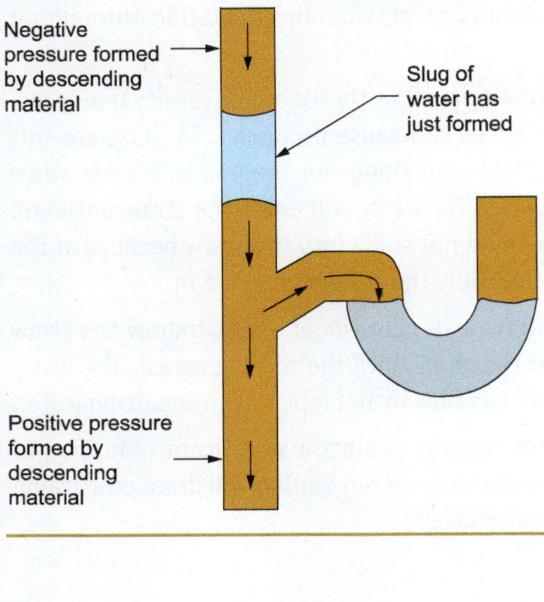

FIGURE 17-3 Back-pressure can push on the trap seal.

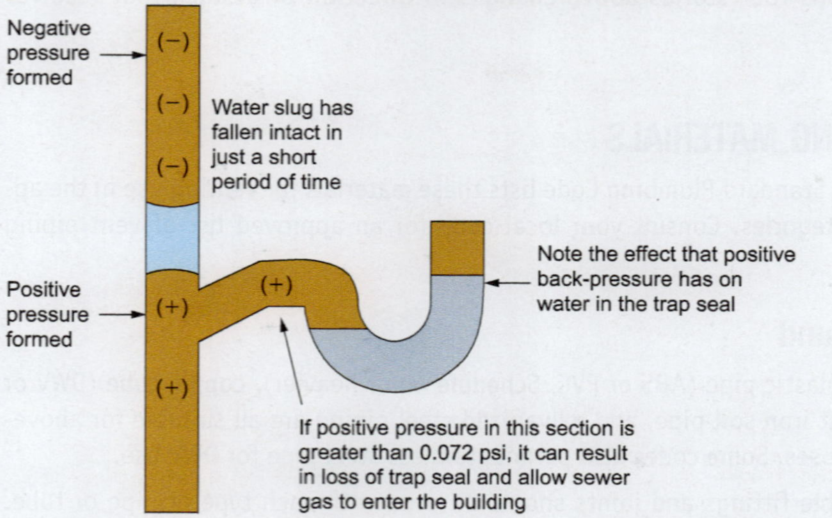

FIGURE 17-4 Siphonage and subsequent loss of seal can occur.

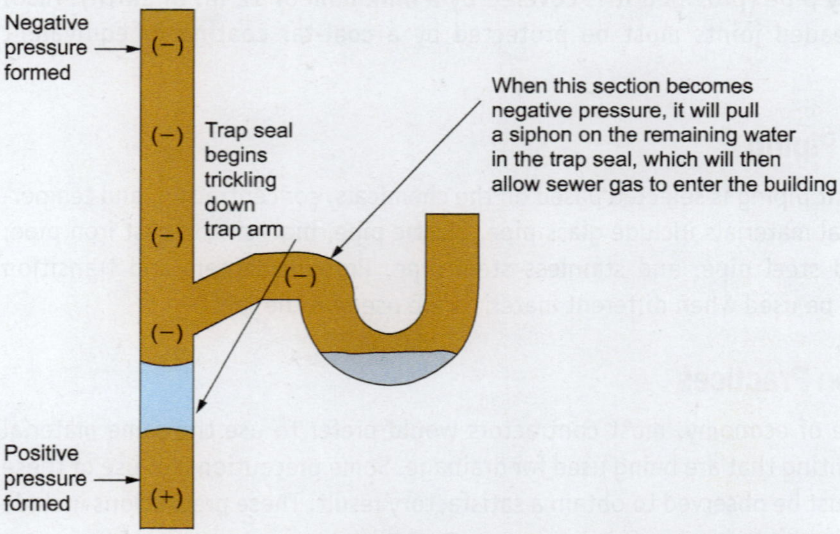

The free passage of air produces benefits beyond limiting pressure changes. Adequate air moving in the system encourages aerobic bacteria to thrive and continue their activity of breaking down the sewage. When air is depleted, the breakdown process becomes anaerobic, and odors become more prevalent. The vent piping also permits the removal of noxious or hazardous gases—such as sulfur dioxide, methane, and hydrogen sulfide—from the system. These gases and vapors can exit the building safely above the roof.

Venting is also useful to alleviate problems caused by soap suds. Soap suds can blow traps four stories above changes in direction of a stack that receives sudsy waste.

VENT PIPING MATERIALS

The National Standard Plumbing Code lists these materials for venting use in the appropriate categories. Consult your local code for an approved list of vent piping material.

Above Ground

Brass pipe, plastic pipe (ABS or PVC, Schedule 40 or heavier), copper tube (DWV or heavier), cast iron soil pipe, and galvanized steel piping are all suitable for above-ground purposes. Some codes also permit stainless steel pipe for DWV use.

Compatible fittings and joints should be used with each type of pipe or tube. Combinations of different pipe materials are acceptable if proper transition or adapter fittings are used.

In Ground

In-ground piping for venting includes the above materials plus extra-strength vitrified clay pipe (provided it is covered by a minimum of 12 in. of earth). Also, ferrous threaded joints must be protected by a coal-tar coating or equivalent protection.

Chemical Piping

Chemical vent piping is selected based on the chemicals, concentration, and temperature. Typical materials include glass pipe, plastic pipe, high-silicon cast iron pipe, plastic-lined steel pipe, and stainless steel pipe. Proper adapters and transition pieces must be used when different materials are used on the job.

Installation Practices

For the sake of economy, most contractors would prefer to use the same material types for venting that are being used for drainage. Some precautions for use of these materials must be observed to obtain a satisfactory result. These precautions include the following points:

- Install all horizontal vent piping so it slopes toward the drain. In this way, any moisture accumulation will be drained away and not form a blockage to the free flow of air.
- Piping should be supported to maintain its position.
- Offsets should be made with 45° fittings, rather than 90°, wherever possible. The flow characteristic for the 45° fittings is superior to other changes of direction, and if scaling occurs inside the fitting, 45° fittings will permit cleaning more readily.
- Ream and deburr the pipe before installation to obtain the full capacity. Remember that vent systems carry air as well as sewer gases and condensate, so full capacity is critical for optimal performance.

- Vent piping is normally installed within inside partitions. In warm climates, it may be permissible to install vent piping on the outside of buildings. If the climate is such that freezing is possible, the vents must be kept in a warm place or ice buildup can occur.
- Vent piping is connected above the center of horizontal lines. Figure 17-5 shows a vent takeoff from a horizontal line. Note that the vent invert is kept above the centerline of the drain. This detail helps to assure that debris does not collect in this branch opening. This horizontal vent would only be allowed below a fixture level if it is a wet vent (see definition below).
- The branch vent connection to a vent stack should be at least 6 inches above the flood level rim of the fixtures served (see Figure 17-6). When the branch vent is above the flood level rim, the vent cannot flood if the fixture drain stops up.

IN THE FIELD

The 2012 International Plumbing Code does not allow vitrified clay pipe to be used as building drain or vent pipe, but it does allow it to be used for building sewer pipe. Always be sure to check your local code!

FIGURE 17-5 Vent tie-ins should be above the centerline of the horizontal drain pipe.

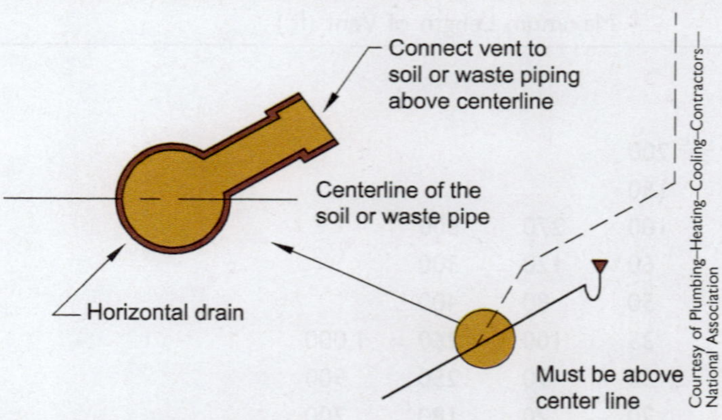

FIGURE 17-6 Vent connection below the flood level rim.

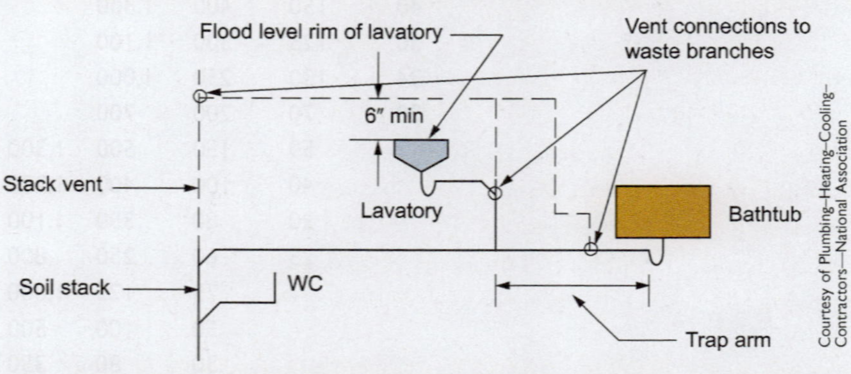

VENT SIZING

Sizing the vent system is determined by the drainage fixture unit (DFU) load to be served, the size of the stack to be vented, and the length of piping to the open air (i.e., outdoors). The correlation of these values is shown in Table 17-1. Consult your local code for specific standards in your area.

Two limitations for sizing the vent pipe according to most code provisions are that the vent pipe may not be less than half the diameter of the drain being vented, and the smallest size permitted is $1\frac{1}{4}$ inch pipe.

Some codes permit vent sizing of pipe less than $1\frac{1}{4}$ inches; however, these systems require engineered plans. Some residential codes also permit smaller sizes for engineered systems, but the risks of using smaller sizes may be greater than the saving in using such sizes. Consult your local code for specific requirements.

The actual vent sizing procedure is to:

1. Determine the DFU load on the stack, branch, or fixture.
2. Note the stack or drain size.

TABLE 17-1 Size and Length of Vents

Size of Drainage Stack or Fixture Drain (in.)	Drainage Fixture Units Connected	Diameter of Vent Required (in.)								
		$1\frac{1}{4}$	$1\frac{1}{2}$	2	$2\frac{1}{2}$	3	4	5	6	8
		Maximum Length of Vent (ft.)								
$1\frac{1}{4}$	1	1*								
$1\frac{1}{2}$	8	50	150							
2	12	30	75	200						
2	20	26	50	150						
3	10		30	100	270	600				
3	30			60	120	300				
3	60			50	80	400				
4	100			35	100	260	1,000			
4	200			30	90	250	900			
4	500			20	70	180	700			
5	200				35	80	350	1,000		
5	500				30	70	300	900		
5	1,100				0	10	50	150		
6	350					40	150	400	1,300	
6	620					30	125	300	1,100	
6	960					24	100	250	1,000	
6	1,900					20	70	200	700	
8	600						50	150	500	1,300
8	1,400						40	100	400	1,200
8	2,200						30	80	350	1,100
8	3,600						25	60	250	800
10	1,000							75	125	1,000
10	2,500							50	100	500
10	3,800							30	80	350
10	5,600							25	60	250

*The length of the vent is not limited.

3. Measure or compute the distance to open air along the developed length of the vent pipe.
4. Select a pipe size from Table 17-1 that has a capacity that equals or exceeds the requirements.

If the vent connects to a stack vent before continuing to the outside air, the total vent length includes the portion of the stack vent that conveys the vent to the outdoors.

Figures 17-7 and 17-8 show developed length and the concept of including both the vent pipe and the continuation of the stack vent.

Some practice will aid in understanding the information presented in Table 17-1. When sizing the vent piping, you must start at the farthest point and proceed toward the vent terminal. In other words, you must start at the lowest fixtures and work up toward the vent opening to outdoor air.

Figure 17-9 shows one example of wet vents, where the wet vent serves a drain for a bathroom group of fixtures.

FIGURE 17-7 Developed length (*x*).

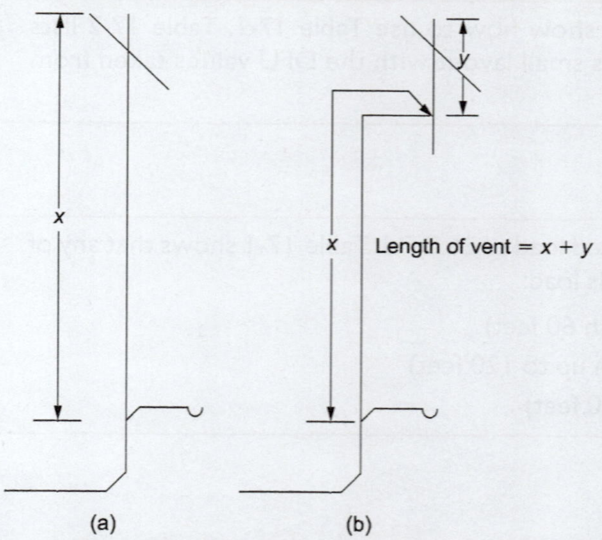

FIGURE 17-8 Stack vent developed length.

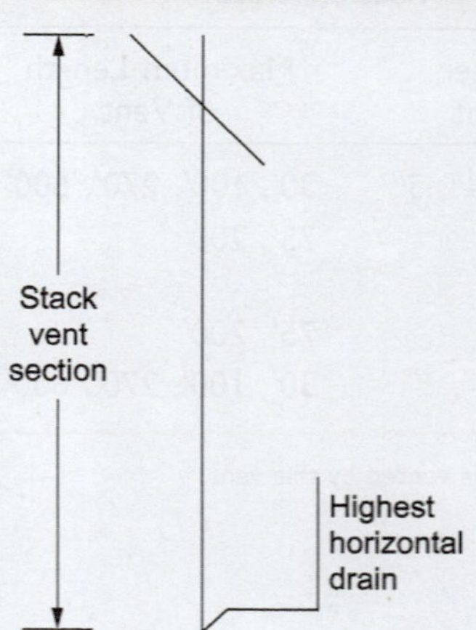

FIGURE 17-9 Vent layout for example.

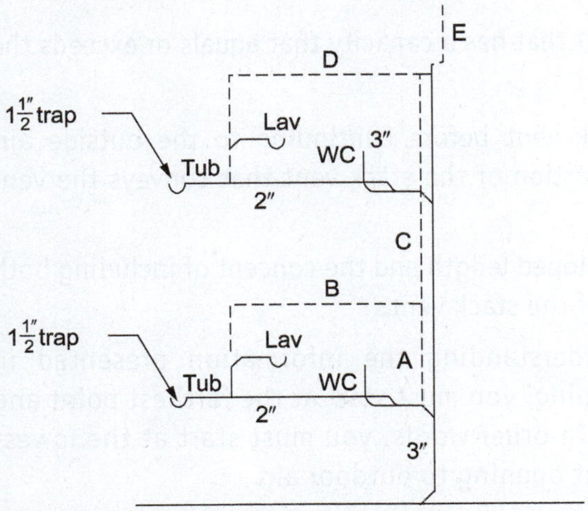

EXAMPLE 1

Figure 17-9 can be used to show how to use Table 17-1. Table 17-2 lists the options possible for this small layout with the DFU values taken from Table 17-3.

EXAMPLE 2

A 3-inch stack has a connected load of 26 DFU. Table 17-1 shows that any of the following could serve this load:

- 2 inches (maximum length 60 feet)
- $2\frac{1}{2}$ inch (developed length up to 120 feet)
- 3 inches (length up to 300 feet)

TABLE 17-2 Drainage Fixture Unit Loads for Figure 17-9

Vent (Section)	Size of Stack or Drain Served	Connect Load (DFU)	Possible Sizes: Diameter of Vent	Possible Sizes: Maximum Length of Vent
A	3″	3	$1\frac{1}{2}''$*, 2″, $2\frac{1}{2}''$, 3″	30′, 100′, 270′, 600′
B, C	2″	3 (Bathroom Group)	$1\frac{1}{2}''$*, 2″	75′, 200′
D	2″	3	$1\frac{1}{2}''$*, 2″	75′, 200′
E	3″	6	$1\frac{1}{2}''$, 2″, $2\frac{1}{2}''$, 3″	30′, 100′, 270′, 600′

*This vent must be a minimum of $1\frac{1}{2}''$ as drawn because the bathtub arm is vented by this vent.

TABLE 17-3 Drainage Fixture Unit (DFU) Values

Type of Fixtures	Individual Dwelling Units	Serving Three or More Dwelling Units	Other than Dwelling Units	Heavy Use Assembly
Bathroom Groups Having 1.6 gpf Gravity-Tank Water Closets				
Half-Bath or Powder Room	3	2		
1 Bathroom Group	5	3		
$1\frac{1}{2}$ Bathrooms	6	3.5		
2 Bathrooms	7	4.5		
$2\frac{1}{2}$ Bathrooms	8	5		
3 Bathrooms	9	5.5		
Each Additional Half-Bath	0.5	0.5		
Each Additional Bathroom Group	1	1		
Bathroom Groups Having 1.6 gpf Pressure-Tank Water Closets				
Half-Bath or Powder Room	3.5	2.5		
1 Bathroom Group	5.5	3.5		
$1\frac{1}{2}$ Bathrooms	6.5	4		
2 Bathrooms	7.5	5		
$2\frac{1}{2}$ Bathrooms	8.5	5.5		
3 Bathrooms	9.5	6		
Each Additional Half-Bath	0.5	0.5		
Each Additional Bathroom Group	1	1		
Bathroom Groups Having 3.5 gpf (or higher) GravityTtank Water Closets				
Half-Bath or Powder Room	3	2		
1 Bathroom Group	6	4		
$1\frac{1}{2}$ Bathrooms	8	5.5		
2 Bathrooms	10	6.5		
$2\frac{1}{2}$ Bathrooms	11	7.5		
3 Bathrooms	12	8		
Each Additional Half-Bath	0.5	0.5		
Each Additional Bathroom Group	1	1		
Bathroom Group With 1.6 gpf Flushometer Valve	5	3		
Bathroom Group with 3.5 gpf (or Higher) Flushometer Valve	6	4		
Individual Fixtures				
Bathtub or Combination Bath/Shower, $1\frac{1}{2}''$ Trap	2	2		
Bidet, $1\frac{1}{4}''$ Trap	1	1		
Clothes Washer, Domestic, 2″ Standpipe	3	3	3	
Dishwasher, Domestic, with Independent Drain, $1\frac{1}{2}''$ Minimum Trap	2	2	2	
Drinking Fountain or Watercooler			0.5	
Food-Waste-Grinder, Commercial, 2″ Min. Trap			3	
Floor Drain, Auxiliary			0	
Kitchen Sink, Domestic, with One $1\frac{1}{2}''$ Trap	2	2	2	
Kitchen Sink, Domestic, with Food-Waste-Grinder	2	2	2	
Kitchen Sink, Domestic, with Dishwasher	3	3	3	

(continued)

TABLE 17-3 Drainage Fixture Unit (DFU) Values (Continued)

Type of Fixtures	Individual Dwelling Units	Serving Three or More Dwelling Units	Other than Dwelling Units	Heavy Use Assembly
Kitchen Sink, Domestic, with Grinder and Dishwasher	3	3	3	
Laundry Sink, One or Two Compartments, $1\frac{1}{2}''$ Waste	2	2	2	
Laundry Sink, with Discharge from Clothes Washer	2	2	2	
Lavatory, $1\frac{1}{4}''$ Waste	1	1	1	1
Mop Basin, 3″ Trap			3	
Service Sink, 3″ Trap			3	
Shower Stall, $1\frac{1}{2}''$ Trap	2	2	2	
Shower Stall, 2″ Trap	2	2	2	
Showers, Group, per Head (Continuous Use)			5	
Sink, $1\frac{1}{2}''$ Trap	2	2	2	
Sink, 2″ Trap	3	3	3	
Sink, 3″ Trap			5	
Trap Size, $1\frac{1}{4}''$ (Other)	1	1	1	
Trap Size, $1\frac{1}{2}''$ (Other)	2	2	2	
Trap Size, 2″ (Other)	3	3	3	
Trap Size, 3″ (Other)			5	
Trap Size, 4″ (Other)			6	
Urinal, 1.0 GPF			4	5
Urinal, Greater Than 1.0 GPF			5	6
Wash Fountain, $1\frac{1}{2}''$ Trap			2	
Wash Fountain, 2″ Trap			3	
Wash Sink, Each Set of Faucets			2	
Water Closet, 1.6 GPF Gravity or Pressure Tank	3	3	4	6
Water Closet, 1.6 GPF Flushometer Valve	3	3	4	6
Water Closet, 3.5 GPF Gravity Tank	4	4	6	8
Water Closet, 3.5 GPF Flushometer Valve	4	4	6	8
Whirlpool Bath or Combination Bath/Shower, $1\frac{1}{2}''$ Trap	2	2		

Notes:

1. A Bathroom Group, for the purposes of this Table, consists of not more than one water closet, up to two lavatories, and either one bathtub, one bath/shower combination, or one shower stall. Other fixtures within the bathing facility shall be counted separately to determine the total drainage fxture unit load.
2. A Half-Bath or Powder Room, for the purposes of this Table, consists of one water closet and one lavatory.
3. For unlisted fixtures, refer to a listed fixture having a similar fow and frequency of use.
4. When drainage fixture unit (DFU) values are added to determine the load on the drainage system or portions thereof, round the sum to the nearest whole number before referring to Tables 11.5.1A, 11.5.1B, or 12.16.6A for sizing the drainage and vent piping. Values of 0.5 or more should be rounded up to the next higher whole number (9.5 = 10 DFU). Values of 0.4 or less should be rounded down to the next lower whole number (9.4 = 9 DFU).
5. "Other Than Dwelling Units" applies to business, commercial, industrial, and assembly occupancies other than those defined under "Heavy-Use Assembly." Included are the public and common areas in hotels, motels, and multi-dwelling buildings.
6. "Heavy-Use Assembly" applies to toilet facilities in occupancies that place heavy, but intermittent, time-based loads on the drainage system, such as schools, auditoriums, stadiums, race courses, transportation terminals, theaters, and similar occupancies where queuing is likely to occur during periods of peak use.
7. Where other than water-supplied fixtures discharge into the drainage system, allow 2 DFU for each gallon per minute (gpm) of flow. (See Section 11.4.2.)

Source: Table 11.4.1 2012 National Standard Plumbing Code.

EXAMPLE 3

A 5-inch stack has a load of 602 DFU. Table 17-1 shows that any of the following could serve this load:

- 3 inches (developed length up to 10 feet)
- 4 inches (length up to 50 feet)
- 5 inches (length up to 150 feet)

EXAMPLE 4

A laundry room is to be installed in a building. What are the vent options if the stack is 3 inches, the load is 16 DFU, and it is 90 feet (developed length) to the roof?

A 2-inch vent is only satisfactory for up to 60 feet developed length, so 2 inches is not sufficient. A $2\frac{1}{2}$ inch pipe will be suitable for a length up to 120 feet, and 3 inches would serve up to 300 feet length. Thus, $2\frac{1}{2}$ inch or 3-inch vent pipe would be acceptable.

EXAMPLE 5

An eight-story (80 feet) apartment building is provided with a 4-inch stack and a companion vent stack. The base of the vent stack contains a laundry room with a load of 6 DFU. Each story of the building contains a private bathroom load of 3 DFU.

Figure 17-10 shows the system. Table 17-4 shows the sizes of the elements with comments about the sections.

In addition, pipe J is the stack vent for the 3-inch stack, and it contributes capacity to the system. This section vents the top bathroom (3 DFU) only, so it would have to be $1\frac{1}{2}$ inch minimum (from Table 17-1, for $1\frac{1}{2}$ inch venting a 3-inch stack, the maximum developed length is 30 feet).

EXCEPTIONS AND SPECIAL CASES

Many exceptions or special cases are permitted in codes. One such exception is shown in Figure 17-11, as permitted in the National Standard Plumbing Code, Section 12.9.2. This section permits a single pipe to act as a drain for an upper fixture *and* the vent for a lower fixture if the two fixtures are on the same floor (connecting to the stack at different heights), and provided that the dual-purpose pipe is one size larger than the upper drain (but not smaller than the lower drain).

Another exception is the wet vent. A wet vent is both a drain line and a vent line. Wet vents may serve as the waste line for lavatories, drinking fountains, or bathtubs. The waste tends to wash the vent, helping to keep it clear. If the drain becomes clogged, the snaking of the drain also clears the vent.

You should take advantage of any such special arrangement that is permitted by the code used in your jurisdiction, but you must be careful to observe all the conditions required. Such exceptions are permitted because they provide satisfactory operation and cost savings in the installation.

FIGURE 17-10 High-rise DWV layout.

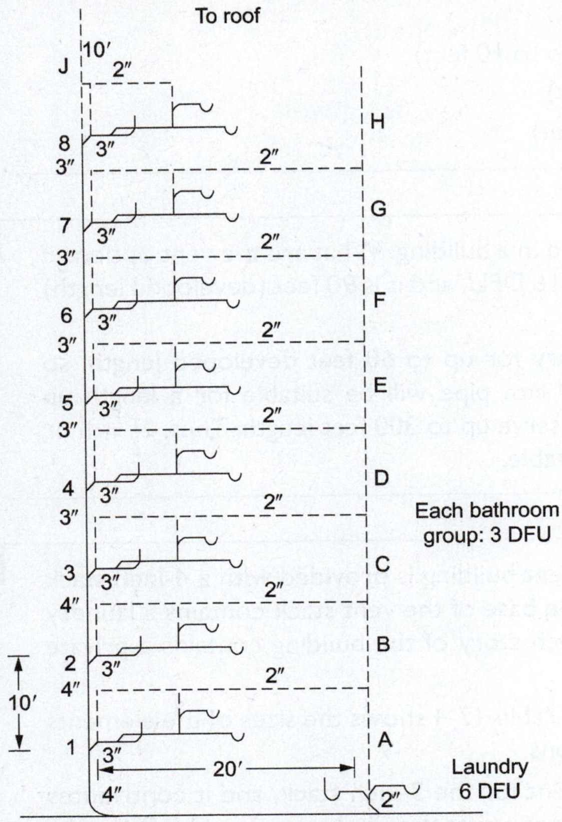

TABLE 17-4 Drainage Fixture Unit Loads for Figure 17-10

Vent (Section)	Size of Drainage Stack or Fixture Drain	Connected Load (DFU)	Minimum Vent Diameter (in.)	**Allowable Length to Open Air (ft.)	Actual Distance to Open Air (ft.)
A	2″	6	2*	200	100
B	3″	9	2*	100	100
C	3″	12	$2\frac{1}{2}$*	120	90
D	3″	15	$2\frac{1}{2}$*	120	80
E	3″	18	$2\frac{1}{2}$*	120	70
F	3″	21	$2\frac{1}{2}$*	120	60
G	3″	24	$2\frac{1}{2}$*	120	50
H	3″	27	$2\frac{1}{2}$*	120	40
J	3″	3	$1\frac{1}{2}$	30***	10

*Some codes require the vent stack to be full size to its base, which would require this section to be 3″.

**No entry in this column for the vent stack (items A through H) can be less than the greatest developed length (100′ for section A, and 80′ plus distance from lavatory [10′ allowed] for all sections except A in this example).

***Item J is not part of the common system, therefore it is sized based on its developed length to the outdoors (10′ in this example).

FIGURE 17-11 A common vent is a special venting exception. The vent portion must be one pipe size larger than the lower portion.

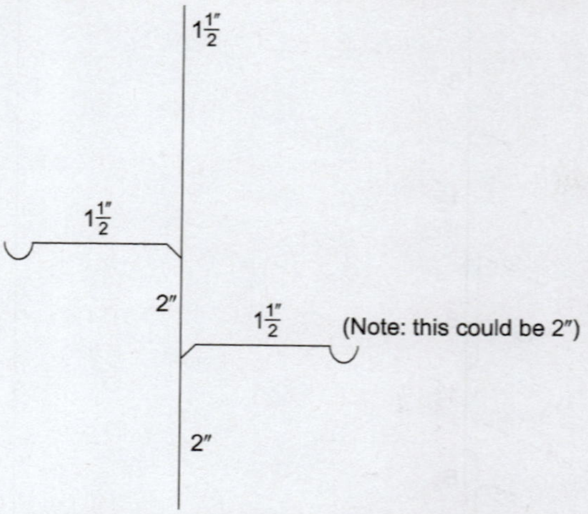

EXAMPLE 6

A somewhat more complicated plumbing system is shown in Figure 17-12. A portion of a four-story condominium is illustrated.

- Stack A serves a full bath and half bath on each floor.
- Stack B serves a sink, food waste grinder, and dishwasher on each floor.
- Stack C has a double standpipe for laundry machines in the basement.
- The rest of the stack is the vent stack for stacks A and B.
- The stacks are approximately 45 feet high, with 20 feet between stacks A and C and 10 feet between B and C.

In order to size the venting system, we must total the drainage fixture unit loads at all the points in the system and then use the information in Table 17-1. Note that the DFUs listed here are applicable in one code, but may not be applicable in all codes.

For three or more residential units:

- Each bathroom group (private use): 3 DFU
- Each half bath: 2 DFU
- Each kitchen sink, food waste grinder, and dishwasher: 3 DFU
- Each laundry standpipe: 3 DFU

To size the layout shown in Figure 17-12, use a table such as the one that follows, along with the notes and comments. Figures 17-13 and 17-14 show smaller sections of Figure 17-12 for clarity.

We have selected 3-inch size for the closet bends.

Table 17-1 provides the vent length/DFU load/stack size information:

- The branch vent from each kitchen is $1\frac{1}{2}$ inch (adequate for 8 DFU on a $1\frac{1}{2}$ inch stack up to 150 feet in length).
- The branch vent for each $1\frac{1}{2}$ inch bathroom group must serve 5 DFU. Since the length of the vent is 65 feet, a 2-inch vent will vent 10 DFU on a 3-inch stack for a length up to 100 feet (the length here is 65 feet). Thus, a 2-inch branch is acceptable.

EXAMPLE 6 *continued*

FIGURE 17-12 Condominium layout.

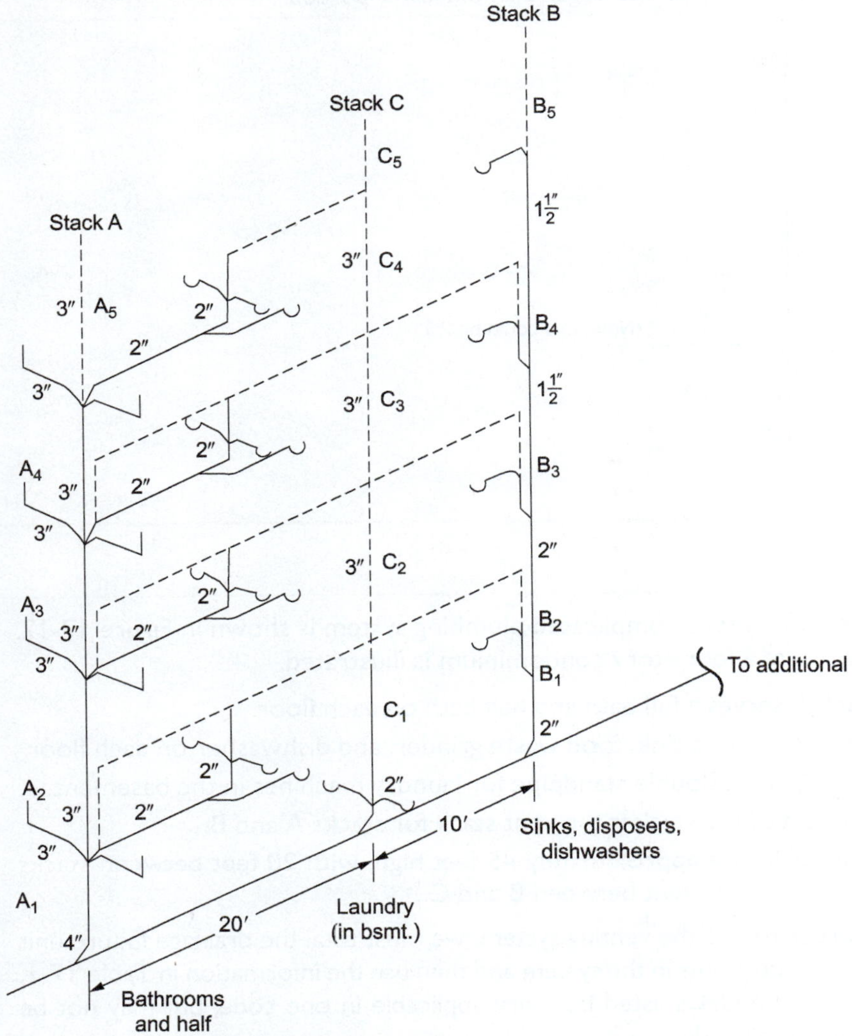

- Section C_2 serves $6 + 5 + 3 = 14$ DFU, $2\frac{1}{2}$ inch because the branch from A_1 is limited to 60 feet to open air if it is 2-inch size serving a 3-inch branch. Since the stack height is 45 feet, and the horizontal distance is 20 feet (total 65 feet), C_2 must be increased to a minimum of $2\frac{1}{2}$ inch.
- Section C_3 serves $6 + 10 + 6 = 22$ DFU (minimum $2\frac{1}{2}$ inch).
- Section C_4 serves $6 + 15 + 9 = 30$ DFU (minimum $2\frac{1}{2}$ inch).
- Section C_5 serves $30 + 5 = 35$ DFU (minimum $2\frac{1}{2}$ inch).
- Section A_5 serves 6 DFU ($1\frac{1}{2}$ inch up to 30 feet is satisfactory).

We can complete the answer using Table 17-5.

Tables 17-6 and 17-7 are copies of earlier tables in this book. They are repeated here for your convenience.

EXAMPLE 6
continued

FIGURE 17-13 Lower floor back-to-back layout.

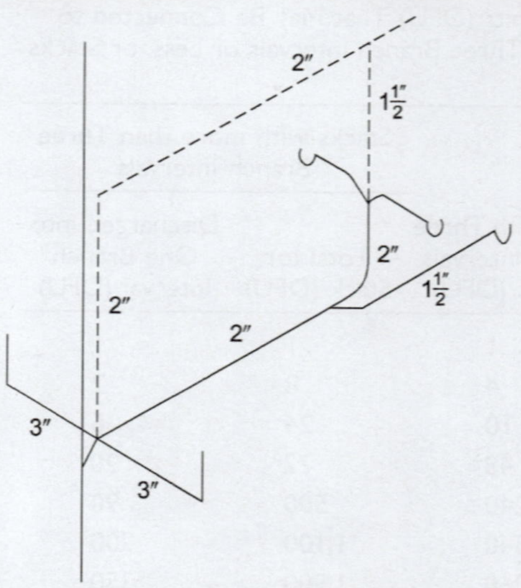

FIGURE 17-14 Kitchen sink and dishwasher layout.

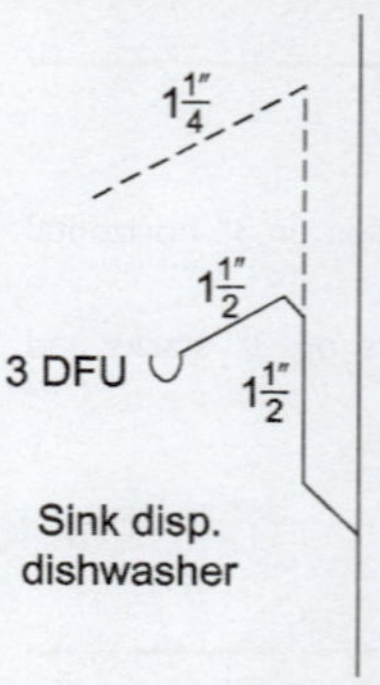

TABLE 17-5 Drainage Fixture Unit Loads for Figure 17-12

Drains				Vents		
Stack Element	Load (DFU)	Minimum Size (in.)	Branch (in.)	Stack Element	Load (DFU)	Minimum Size (in.)
A_1	20	4*	2	C_1	6	1½
A_2	15	3	2	C_2		2½
A_3	10	3	2	C_3		2½
A_4	5	3	1½	C_4		2½
B_1	12	2	1¼	C_5		2½
B_2	9	2	1¼	A_5	6**	1½
B_3	6	1½	1¼	B_5	3	1¼
B_4	3	1½				

*Note that pipe A_1 must be 4″ because it serves 8 water closets.

**Venting 2 water closets.

EXAMPLE 6
continued

TABLE 17-6 Horizontal Fixture Branches and Stacks

Maximum Number of Drainage Fixture Units (DFU) That May Be Connected to Any Horizontal Fixture Branch, a Stack of Three Branch Intervals or Less, or Stacks of more than Three Branch Intervals

Diameter of Pipe (in.)	Any Horizontal Fixture Branch[1] (DFU)	Stacks with Three Branch Intervals or Less (DFU)	Stacks with more than Three Branch Intervals	
			Total for Stack (DFU)	Discharged into One Branch Interval (DFU)
$1\frac{1}{4}$	1	1	1	1
$1\frac{1}{2}$	3	4	8	2
2	6	10	24	6
3	20[2]	48[3]	72[3]	20[3]
4	160	240	500	90
5	360	540	1,100	200
6	620	960	1,900	350
8	1,400	2,200	3,600	600
10	2,500	3,800	5,600	1,000
12	3,900	6,000	8,400	1,500
15	7,000			

Notes:
1. Does not include branches of the building drain.
2. See Section 11.5.6.a & d for the restrictions on the number of water closets installed on 3″ horizontal fxture branches.
3. See Section 11.5.6.b & d for the restrictions on the number of water closets on 3″ stacks and discharged into any branch interval of a 3″ stack.

Source: Table 11.5.1B 2012 National Standard Plumbing Code.

TABLE 17-7 Building Drains and Sewers[1]

Maximum Number of Drainage Fixture Units (DFU) That May Be Connected to Any Portion of the Building Drain or the Building Sewer

Diameter of Pipe (in.)	Slope per Foot $\frac{1}{16}$-inch	Slope per Foot $\frac{1}{8}$-inch	Slope per Foot $\frac{1}{4}$-inch	Slope per Foot $\frac{1}{2}$-inch
2			21	26
3		36[2]	42[2]	50[2]
4		180	216	250
5		390	480	575
6		700	840	1,000
8	1,400	1,600	1,920	2,300
10	2,500	2,900	3,500	4,200
12	3,900	4,600	5,600	6,700
15	7,000	8,300	10,000	12,000

Notes:
1. On-site sewers that serve more than one building may be sized according to the current standards and specifcations of the Authority Having Jurisdiction for public sewers.
2. See Section 11.5.6.c & d for the restrictions on the number of water closets installed on 3″ building drains and sewers.

Source: Table 11.5.1A 2012 National Standard Plumbing Code.

EXAMPLE 6
continued

The building drain and sewer size is based on the total load connected (20 + 12 + 6) or 38 DFU. Table 17-3 shows that a 4-inch line can be used, with a slope as small as $\frac{1}{8}$ in./ft. (rated 180 DFU). If other stacks are connected to the building drain, their loads would have to be added to be sure 4 inches is satisfactory.

TOTAL AREA OF VENT PIPES

The 2012 National Standard Plumbing Code requires the total area of vent pipes in a building to equal or exceed the required building drain size.

EXAMPLE 7

If a building has a required sewer pipe size of 4 inches, how many 2-inch vents are required to equal the area of the sewer?

For the 4-inch drain, $r = 2''$, area of sewer $= \pi r^2 = \pi(2)^2$

For a 2-inch vent, $r = 1''$, area of vent $= \pi r^2 = \pi(1)^2$

Ratio of areas equals $\frac{\pi(2)^2}{\pi(1)^2} = \frac{(2)^2}{(1)^2} = \frac{4}{1}$

Thus, four 2-inch vents are required.

Economical Sizing

Many designers and contractors will limit their vents to the principal pipe sizes in the interest of overall economy. Remember that the code calls for minimum sizes—you are free to increase them if you desire.

VENT HANGERS

Vent piping must be supported properly so that the pipe will retain its original alignment and so that low points will not form in the vents. Vertical lines must be kept at proper elevations or considerable damage can occur.

Piping must remain fixed in place or problems will occur with leakage at roof flashings, such as rain entering the building or sewer gas being released into the space. Inadequately supported pipe can come apart at joints as a result of expansion and contraction. Sections allowed to sag could fill with water and block the vent.

Depending on the type of pipe and the structure to which it will be secured, the following materials can be used to support vent piping:

- Split ring hangers
- Suspension brackets
- Riser (or friction) clamps
- Braces or short beams attached to structural elements
- Beam clamps
- Pipe hooks

Figures 17-15 through 17-20 show examples of some of these items.

FIGURE 17-15 Split ring hangers mount on a threaded rod. Be sure to secure with a locking nut.

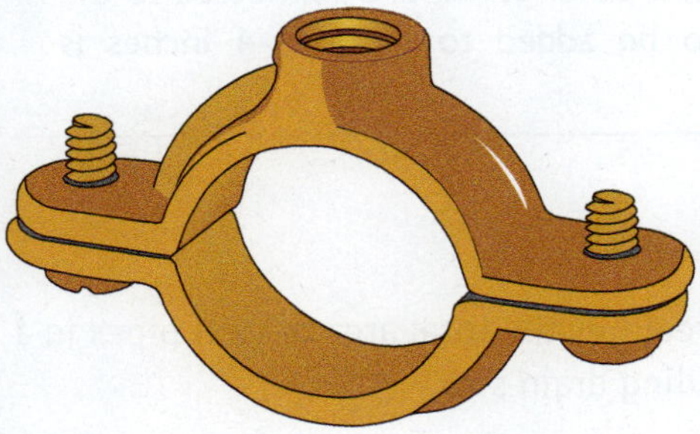

Courtesy of Sioux Chief Manufacturing Co., Inc.

FIGURE 17-16 Suspension brackets with a plastic collar to prevent corrosion and vibration.

Courtesy of Sioux Chief Manufacturing Co., Inc.

FIGURE 17-17 Riser clamps are used to secure vertical pipes.

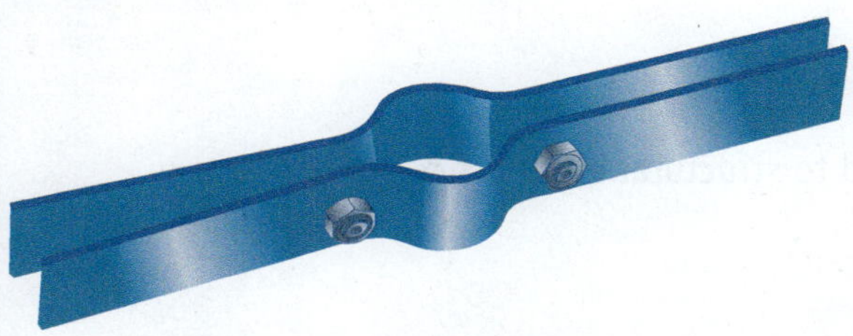

Courtesy of Sioux Chief Manufacturing Co., Inc.

FIGURE 17-18 Beam clamps are used to attach threaded rod to structural beams with a flange, such as I-beams.

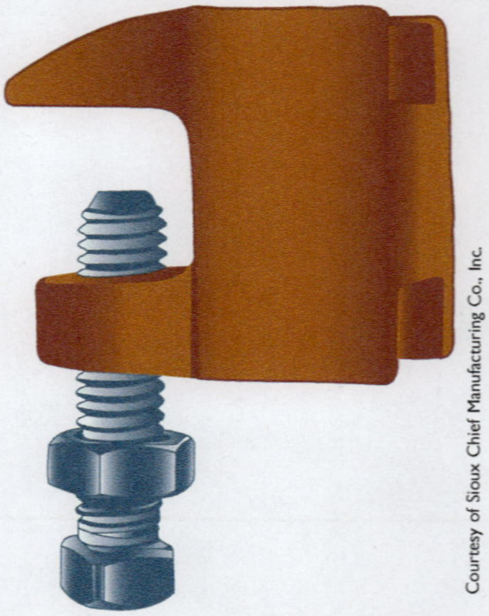

FIGURE 17-19 Pipe hooks are used when hanging metallic pipe on a wooden structure.

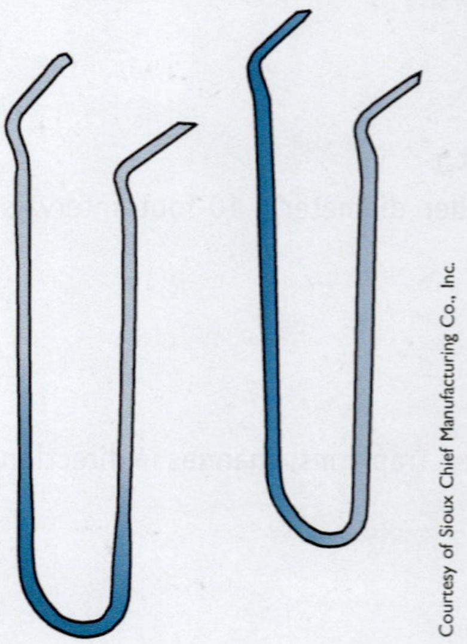

Recommended spacings for hangers and supports are listed below. Check your local code to confirm correct spacing in your area.

Cast Iron Soil Pipe

- Vertical: At base and each story
- Horizontal: 5-inch spacing or each joint (10 in. permitted with 10-ft. pipe lengths)

FIGURE 17-20 Clevis hangers allow the pipe to move due to expansion and contraction.

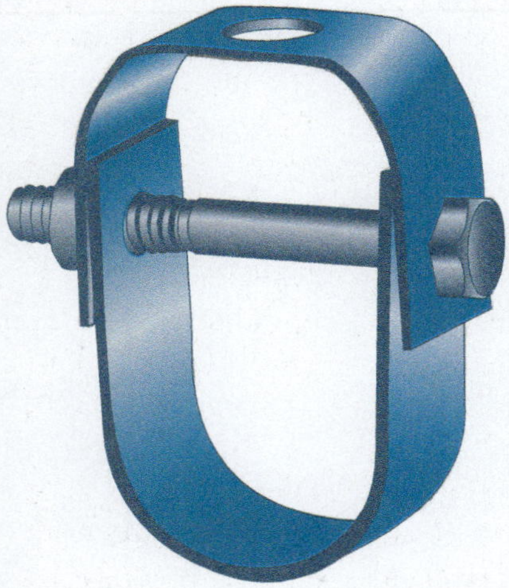

Courtesy of Sioux Chief Manufacturing Co., Inc.

Steel Threaded Pipe

- Vertical: Every other story
- Horizontal: 10-foot intervals ($\frac{3}{4}$ in. or smaller diameter), 12-foot intervals (1 in. and larger)

Copper Tube

- Vertical: Every story (not more than every 10 ft.)
- Horizontal: 6-foot intervals ($1\frac{1}{4}$ in. and smaller diameter), 10-foot intervals ($1\frac{1}{2}$ in. and larger diameter)

Plastic

- Vertical: 4-foot intervals
- Horizontal: 4-foot intervals, at ends of branches, trap arms, changes in direction or elevation

Stainless Steel

- Vertical: At each story
- Horizontal: 10-foot intervals, changes in direction and branch connections

General Considerations

Hangers should not compress or distort the material being supported. Be especially careful with plastic so that the hangers do not cut into the pipe.

Hangers for horizontal piping should allow for free movement of pipe. The hangers must be able to support the pipe when full of water (remember that some testing procedures require the pipe to be full of water).

Anchors and fasteners must be substantial and last throughout the life of the building. Expansion anchors should be used in concrete or masonry to develop adequate holding power. Wood construction permits the use of nails or screws for attachment of hanger devices. Nails are satisfactory if driven in such a way that vibration of the load will not cause them to pull out; do not drive nails into the end grain of lumber. Screws are preferable for long-term holding ability.

Use care when working with vertical lines to maintain them plumb. Use blocking or spacers to make the pipe secure when placed next to a rough wall or one that is not plumb.

VENT TYPES

In order to explain and improve the venting process, parts of venting systems and special venting methods have been given names. To be sure that the terms are used consistently, we will review venting definitions as defined by the 2012 National Standard Plumbing Code—Illustrated.

Back Vent

A back vent is used to vent a group of fixtures to the main vent or stack vent.

Branch Vent

A branch vent is a vent connecting one or more individual vents with a vent stack or stack vent. It should not transition horizontally until at least 6 inches above the flood level rim of the fixtures served. See Figure 17-21.

Circuit Vent

A circuit vent is a vent that connects to a horizontal drainage branch and vents from two to eight traps or trapped fixtures connected in a battery. See Figure 17-22.

The circuit vent is usually limited to these situations:

- The horizontal drain must be uniform in size.
- Not more than eight fixtures may be circuit vented. Additional fixtures are permitted, but an additional circuit vent connection is required.

FIGURE 17-21 Branch vent.

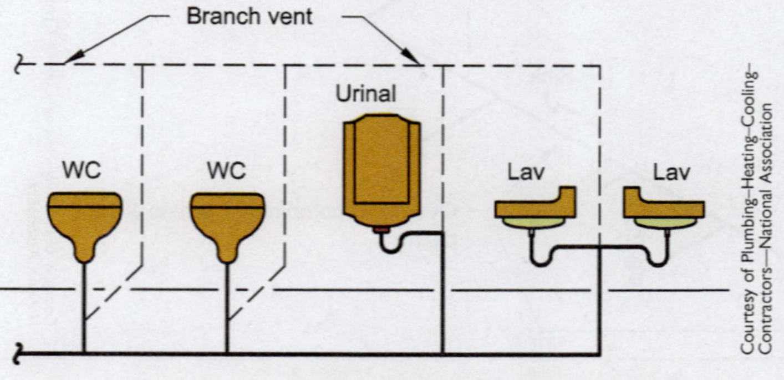

FIGURE 17-22 Circuit vent.

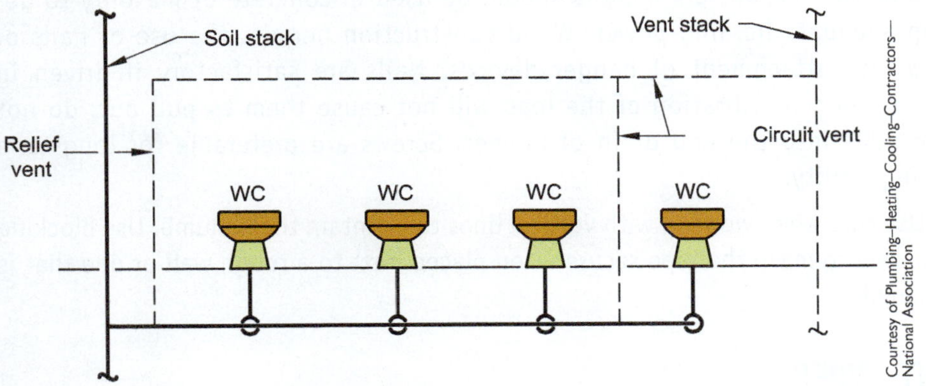

- No fixtures may drain into the circuit vent connection.
- The vent is installed between the last two (i.e., highest) fixture connections.
- A relief vent must be connected in front of the first (i.e., lowest) fixture connection when there are fixtures draining into the stack from above (no need for a relief vent on the top floor).

Combination Waste and Vent

A combination waste and vent is a horizontal wet-vented drain that serves as both waste and vent for sinks or floor drains. This arrangement is used only in special circumstances where structural conditions preclude the installation of conventional systems, such as in a large, open supermarket where there are no walls to run a vent off of each fixture. Combination waste and vents must be preapproved by the authority having jurisdiction. Larger traps are usually required by code in these situations. See Figure 17-23.

FIGURE 17-23 Combination waste and vent. Note that these are used in special circumstances only. Traps must be sized larger than in normal installations.

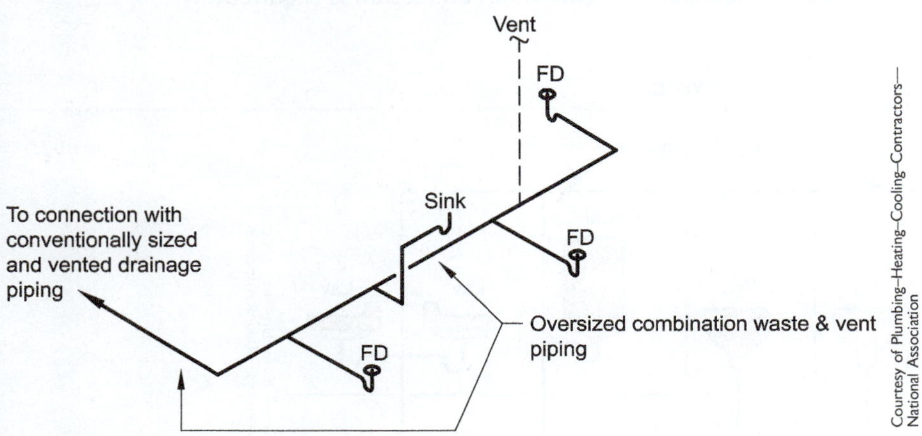

Common Vent or Dual Vent

A common vent is a vent connected at the common connection of two fixture drains and serving as a vent for both fixtures. Fittings for a common vent should be selected with care so that the waste from one fixture does not flow into the branch of the opposing fixture (throw over). It is best not to use a cross fitting for this connection. See Figure 17-24.

Continuous Vent

A continuous vent is a vertical vent that is a continuation of the drain to which it connects. See Figure 17-25.

Crown Vent (Prohibited)

A crown vent is one of the earliest vent configurations. It is a connection made at the top of a trap, configured like an S. To prevent crown venting, the vent may not connect within two pipe diameters of the trap. It has been found that this connection point is quickly fouled, so this vent arrangement is prohibited.

Individual Vent

An individual vent is a pipe installed to vent a single fixture drain. It connects with the vent system above the fixture served or terminates outside the building into the open air. See Figure 17-26.

FIGURE 17-24 Common vent.

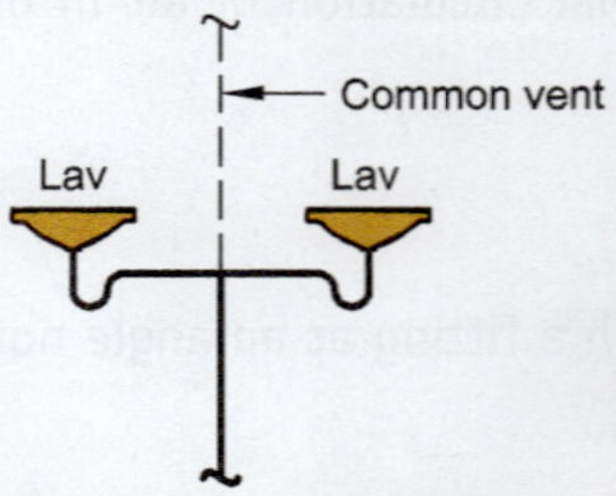

Courtesy of Plumbing–Heating–Cooling–Contractors—National Association

FIGURE 17-25 Continuous vent.

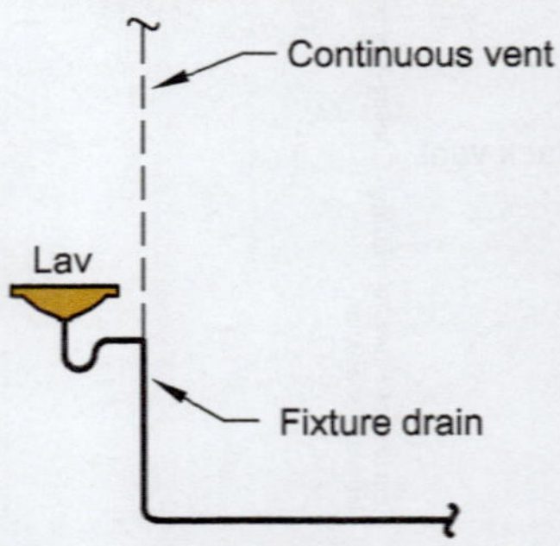

Courtesy of Plumbing–Heating–Cooling–Contractors—National Association

FIGURE 17-26 Individual vent.

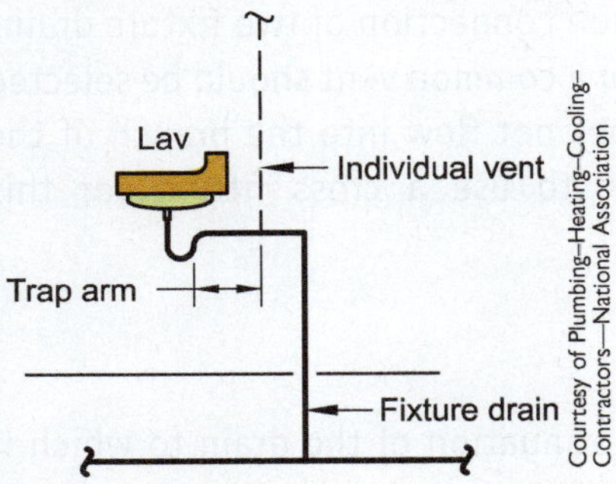

Local Vent

A local vent is an exhaust fan system or other room exhaust, such as a window.

Loop Vent

A loop vent is a circuit vent that loops back to connect with a stack vent instead of a vent stack. See Figure 17-27.

Main Vent

The main vent is the principal vent line in the system. It connects to the drainage system at the base of a drainage stack. See Figure 17-28.

Relief Vent

A relief vent is an auxiliary vent that permits additional circulation of air in or between drainage and vent systems. See Figure 17-29.

Side Vent

A side vent is a vent connecting to a drain pipe through a fitting at an angle not greater than 45° to the vertical. See Figure 17-30.

FIGURE 17-27 Loop vent.

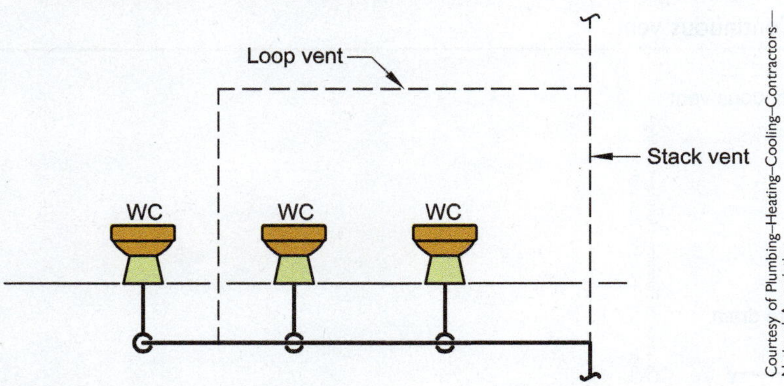

FIGURE 17-28 Main vent.

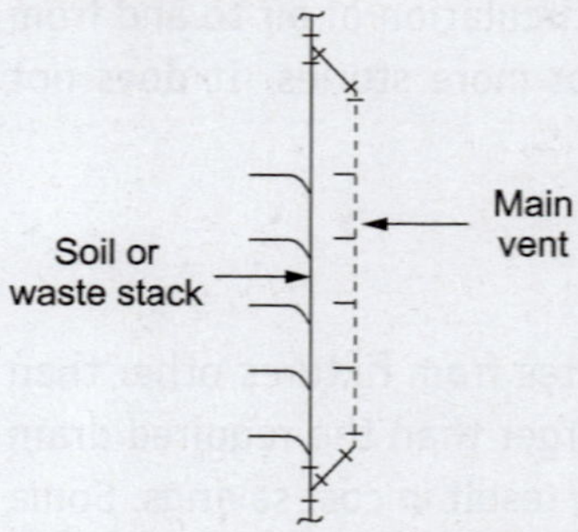

FIGURE 17-29 Relief vent.

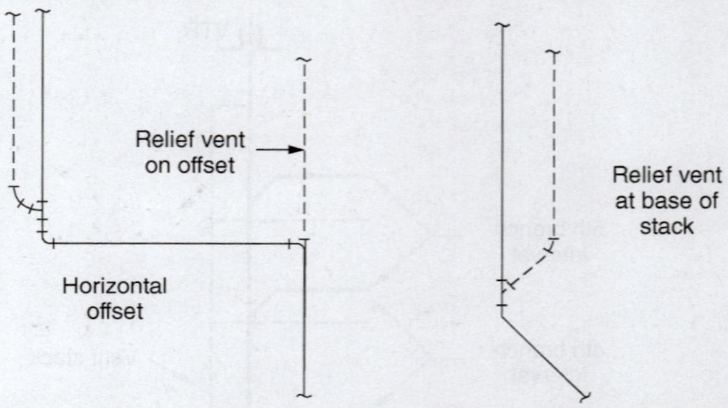

FIGURE 17-30 Side vent.

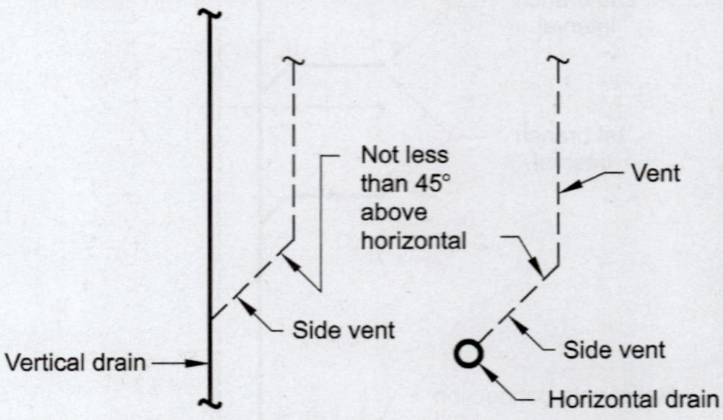

Courtesy of Plumbing–Heating–Cooling–Contractors—National Association

Soil Vent or Stack Vent

A soil or stack vent is the extension of a soil or waste stack above the highest horizontal drain connection to the stack.

Vent Stack

A vent stack is a vertical vent pipe installed to provide circulation of air to and from the drainage system and which extends through one or more stories. It does not receive discharge from any fixtures. See Figure 17-31.

Wet Vent

A wet vent is a vent that receives the discharge of wastes from fixtures other than water closets and sinks. Wet vents are sized one size larger than the required drain size to allow air circulation in the pipe. Wet vents usually result in cost savings. Some codes only allow fixtures that are located on the same floor to be wet vented. See Figure 17-32.

FIGURE 17-31 Vent stack with five or more branch intervals.

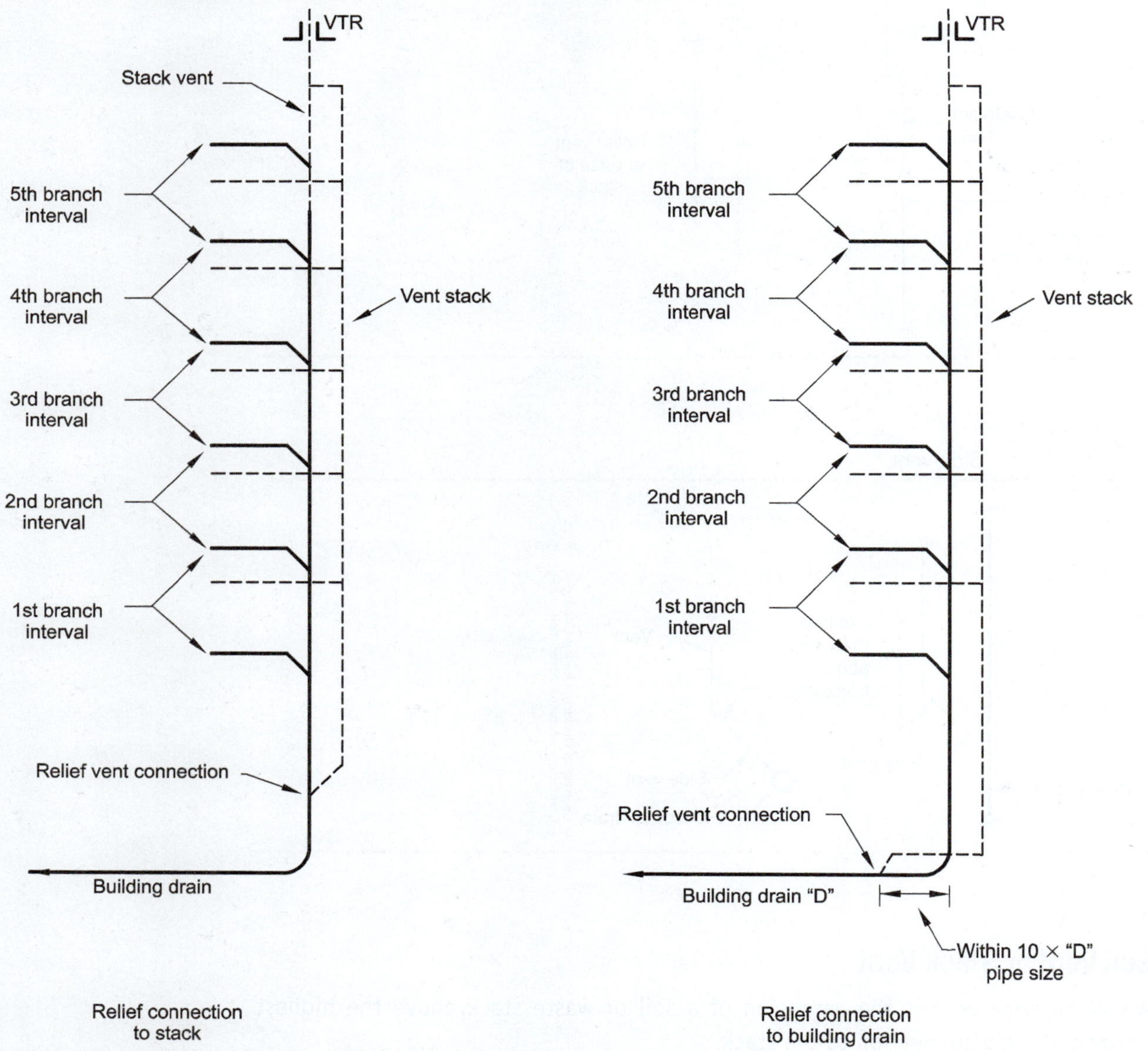

Courtesy of Plumbing–Heating–Cooling–Contractors—National Association

FIGURE 17-32 Wet vent.

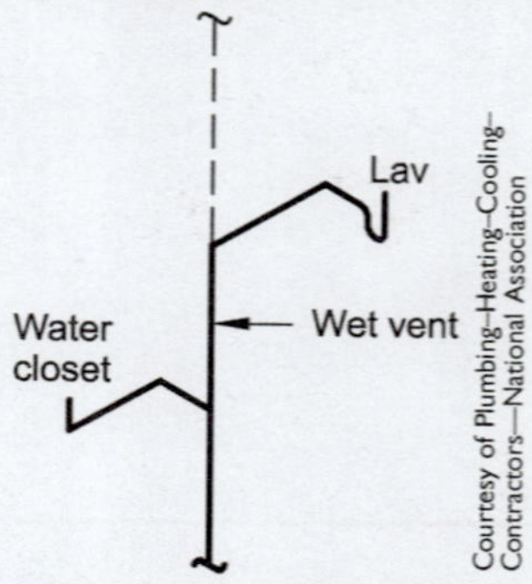

Yoke Vent

A yoke vent is a pipe connecting upward from a soil or waste stack to a vent stack for the purpose of equalizing the pressures in the stacks. A yoke vent is vertical. No portion of a yoke vent can be horizontal. See Figure 17-33.

TRAP ARM SLOPE

One limiting consideration for trap arm length and slope is illustrated in Figure 17-34. The top of the inside of the trap arm at the point of connection to the vented stack cannot be below the invert of the trap outlet. The theory is that there will always be space along the top of the waste flow in the pipe for air travel to equalize the pressure at the trap outlet. If the trap arm slopes downward steeply enough (or is long enough), the arm could fill with water and siphon the trap. A hydraulic gradient of $\frac{1}{4}$ inch per foot should never be exceeded on trap arms with fixtures that are not self-replenishing.

SEPARATE VENTS REQUIRED

Pneumatic sewage ejectors require vents separate from the other plumbing vents because they use high-pressure air or steam to empty the contents of the receiver.

FIGURE 17-33 Yoke vent.

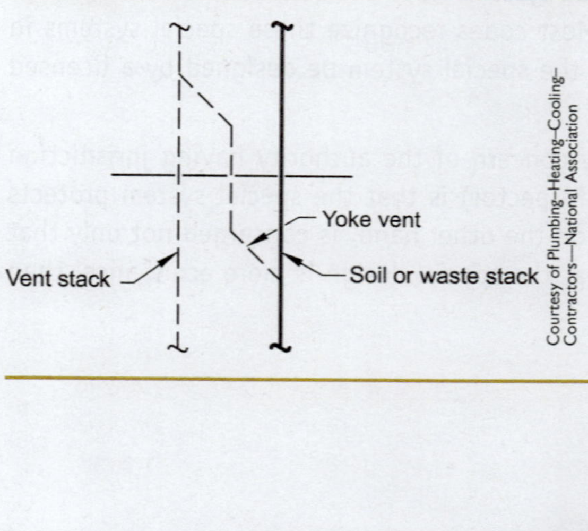

FIGURE 17-34 Point B should not be installed above Point A, or siphoning could occur.

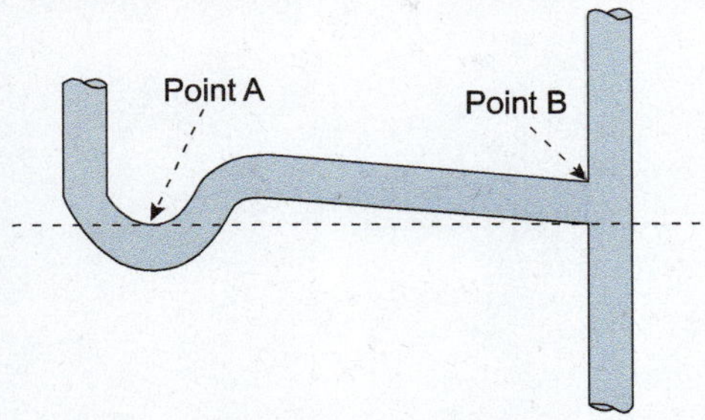

When this compressed air is vented from the receiver, very large pressure surges occur in the vent pipe. Therefore, this pipe is taken outside separately from any other vent system.

Laboratory and chemical sinks, sterilizers, dilution tanks, steam and hot water boiler blow-off lines, neutralizing tanks, and similar devices that could involve aggressive chemicals, high pressures, or high temperatures also require separate vents to the outdoors. Make sure that they vent into an area where pedestrians are not allowed.

SUDS

Suds can produce severe venting problems. Stacks that serve laundry rooms or sinks are vulnerable to stoppage from sudsy wastes. Remember that it is possible for suds to exert pressure on traps four stories above the base of a stack that receives soapy wastes.

Suds venting recognizes the problem of suds in high-rise buildings by requiring extra venting to relieve the suds in areas where they are generated (called suds pressure zones) and also requiring separate drainage stack sections in the vicinity of stack base or stack offsets where suds are a problem. See Figure 17-35.

ENGINEERED SYSTEMS

Special vent systems called engineered systems have been developed that have application in certain circumstances. Most codes recognize these special systems in some way, usually by requiring that the special system be designed by a licensed engineer.

You should realize that the only concern of the authority having jurisdiction (e.g., plumbing inspector, building inspector) is that the special system protects trap seals. The contractor or owner, on the other hand, is concerned not only that the special system works but also that the special design is more economical than other systems.

FIGURE 17-35 Venting suds pressure zones.

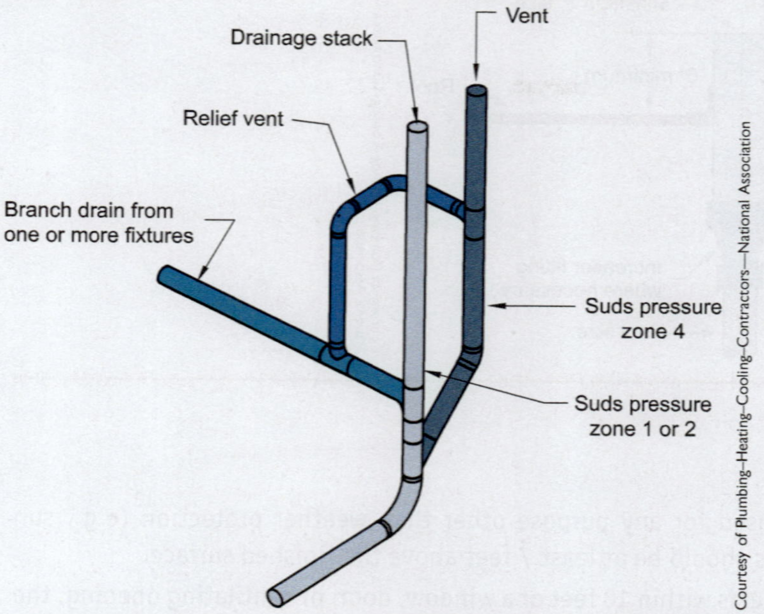

Roof Terminals

Unless permitted by local code, roof terminals must be installed per the following rules:

- The roof terminal of the vent must permit free flow of air into and out of the vent, and prevent rainwater and snow from entering the building where the vent pipe penetrates the roof. The terminal should be at least 6 inches above the roof (more if heavy snows occur at the site), and should be sealed tight to the pipe and roof. See Figure 17-36.

FIGURE 17-36 Vent extension through roof.

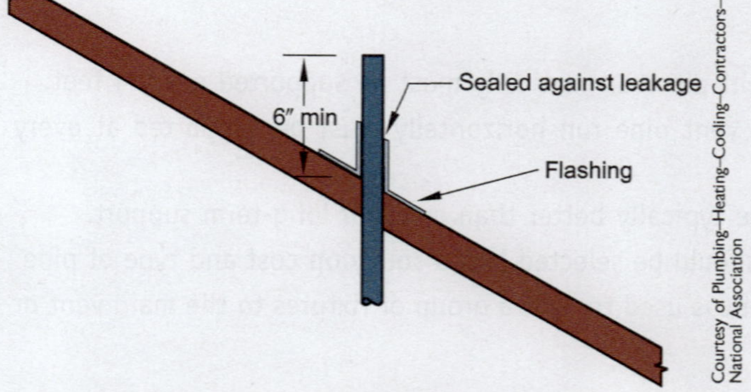

FIGURE 17-37 In cold climates, vent piping is increased to prevent frost closure.

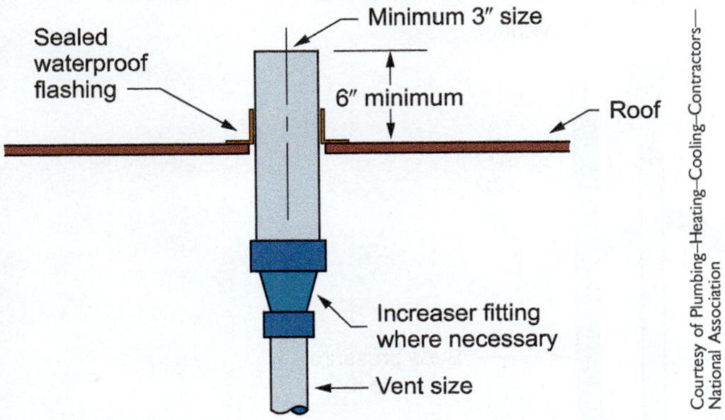

- If the roof is used for any purpose other than weather protection (e.g., sun deck), the vents should be at least 7 feet above the finished surface.
- If a vent terminal is within 10 feet of a window, door, or ventilating opening, the vent should extend 2 feet above the opening.
- Vent terminals should not be placed under a building or roof overhang.
- Vents should not be used to support antennas or flagpoles.

In mild climates, vent stacks may be installed on the outside of buildings. The stack must be solidly supported by the structure. If the weather in the area ever falls below freezing, the stacks should be inside, and the terminal through the roof should be one pipe size larger, up to 3 inches. This increase in diameter helps prevent frost and ice from building up around the opening and eventually closing it off. Note that if the vent must be increased to meet the 3-inch requirement, the increaser must be at least 12 inches below the roof in the insulated/conditioned space. See Figure 17-37.

REVIEW QUESTIONS

True or False

1. __________ Plastic vent pipe run vertically must be supported every 4 feet.
2. __________ Cast iron vent pipe run horizontally must be supported at every joint.
3. __________ Screws are typically better than nails for long-term support.
4. __________ Hangers should be selected based solely on cost and type of pipe.
5. __________ A back vent is used to vent a group of fixtures to the main vent or stack vent.

6. _________ The maximum pressure variation allowed in a waste line is ±1-inch water column.
7. _________ Fixtures that are not vented will not drain properly.
8. _________ Back vent and individual vent mean the same thing.
9. _________ A crown vent is an excellent way to vent a trap.
10. _________ Vent terminals above the roof may be used to support small antennas.

Chapter 18

LEARNING OBJECTIVES

The student will:

- Contrast the differences between sump pumps, sewage pumps, and sewage ejectors.
- Describe the design features and uses of pumps and ejectors.

Sump Pumps, Sewage Pumps, and Sewage Ejectors

BUILDING SUBDRAIN SYSTEMS

Whenever portions of a drainage system are below the elevation of the sewer connection, a building subdrain system must be used. The device selected to raise the waste products to the gravity sewer elevation depends upon the nature of wastes to be handled. There are three common devices used to remove wastes: sump pumps, sewage pumps, and sewage ejectors.

Sump pumps are used for liquid or clear wastes (lavatories, laundry machines, floor drains, etc.). Recent model code changes require sumps to be installed in receivers that are sealed airtight and thus have to be vented to the outdoors.

A sewage pump installation is shown in Figure 18-1, where a lavatory drains into a sump vent, making it a wet vent. The principal waste connections to this sump are not shown. The sump must be sealed and vented, and the discharge pipe should contain a check valve. The check valve prevents waste material from backflowing into the sump, once the pump has shut off. For convenience in servicing the check valve, a full port ball or gate valve may be installed after the check valve. Always install a union or other quick-disconnect device to allow removal of the pump for servicing.

The pump raises the waste in the sump just high enough to drain by gravity into the building drain or sewer.

SUMP PUMPS

According to the 2012 National Standard Plumbing Code, a sump pump is a *permanently installed mechanical device for removing clear water or liquid waste from a sump*.

Sump pumps are selected to transport wastewater that contains no human wastes. Thus, they are used for storm water, cooling water, condensate drainage, laundry water, cooled condensate from boilers, cooled boiler blowdown water, or any other wastewater that does not contain significant foreign material.

FIGURE 18-1 Sewage pump installation with lavatory discharge washing the sump vent (wet vent).

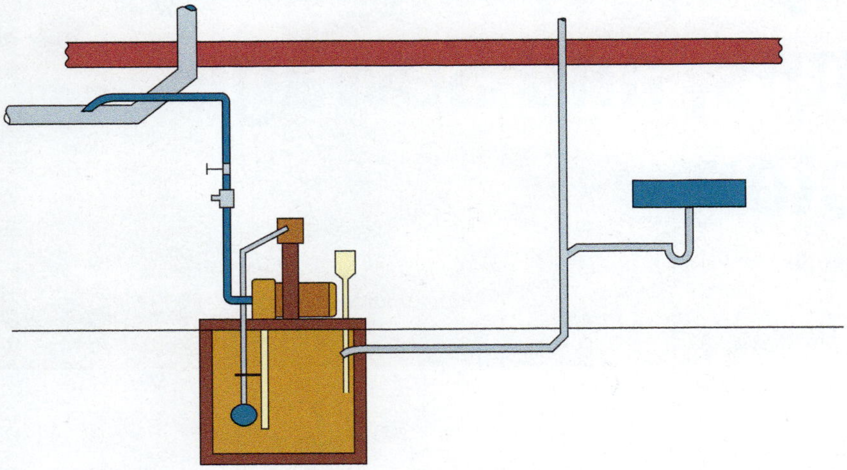

The sump or receiver for either clear water waste or sewage should be sized based on load circumstances. The discharge line should be sized based on the manufacturer's recommendation, but not less than $1\frac{1}{4}$-inch diameter.

Until the mid-1990s, clear water waste sumps did not need to be sealed and vented unless required for radon abatement. Most plumbing codes now require sumps to be sealed and vented to the outdoors. An exception is made for sumps that collect groundwater around building foundations. These sumps may be installed without being sealed. Either type should have a substantial cover to protect against accidental falls into the sump. The sumps and pumps should be accessible and arranged for convenient servicing as required.

Sump construction should be substantial and of nonabsorbent materials. The sump size should be large enough to hold the pump but not so large that there will be a large volume of waste remaining at the end of a pump cycle.

A sump pump is the simplest and least expensive of the three products. Clear water waste types have a minimum of accompanying solids (lint from laundry would be the worst example), so the pumps do not have to be able to pump water with significant solid content. Also, these waters do not produce significant odors, so the sumps that receive them have not typically had to be sealed or vented to the outside unless required locally for radon abatement. The last characteristic of these units is that most of them serve small loads—a single fixture, single laundry machine, foundation drain from a residence, and so on—so most of them are quite small with a small sump.

Thus, most sump pumps are small, inexpensive, for limited flows and limited lifts (20 to 40 gpm, 10 in. to 20 in. lifts), and placed in fairly small sumps (15 in. to 18 in. in diameter, 2 ft. to 3 ft. deep, or even smaller for small condensate systems). They are usually controlled by some type of float switch or pressure switch that detects water in the sump and activates the pump accordingly.

IN THE FIELD

According to the 2012 National Standard Plumbing code, sump pits cannot be less than 15 inches in diameter and 18 inches deep.

Components common to all sump pumps are a motor that drives the impeller, a rotating impeller in a housing, a screened inlet that keeps out solids that could block the impeller, and a switch. Simplified sketches of three impeller types are shown in Figure 18-2.

The most common pumps that use these types of impellers are known as centrifugal pumps. Water enters through the center of the impeller and rides along the outside curve of the fins. The rotation forces the water to the outside of the impeller, and the housing shape causes the water to flow out the discharge port, which is connected to the discharge pipe. Because these pumps have relatively large clearances between the impeller and the pump housing, they do not build up large pressures. The clearance does, however, allow the pump to handle fluids with suspended particles rather well. See Figure 18-3.

Sump Pump Configurations

Four types of sump pump configuration are in use; the first two are far more common than the others.

Submersible

The pump and motor are in a compact assembly with the motor and electrical portions sealed off so that the entire unit can be placed within the sump and submerged. These pumps are usually considered to be superior to the pedestal type. They are available in flow and head ratings similar to the pedestal type. See Figure 18-4.

Pedestal

The pump impeller and housing are located at the bottom of the sump, and the motor is mounted on top of a tube that is long enough to place the motor above the floor level. A light rod attached to a control switch has a float on it that rides up and down with variations in water level. With this arrangement, a standard open-frame motor (the most economical style) is sufficient to power the pump. See Figure 18-5.

FIGURE 18-2 Impeller designs.

FIGURE 18-3 The impeller rotates so fluid is moved from the center of impeller to outer diameter. The centrifugal force pushes the fluid out the discharge pipe.

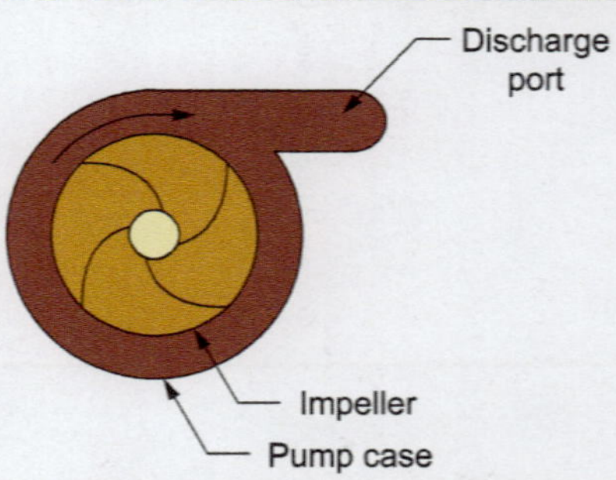

FIGURE 18-4 Submersible sump pumps are popular because they are self-contained and usually maintenance-free.

FIGURE 18-5 Pedestal installations are typical on larger pumping stations, especially where alarm monitoring is required.

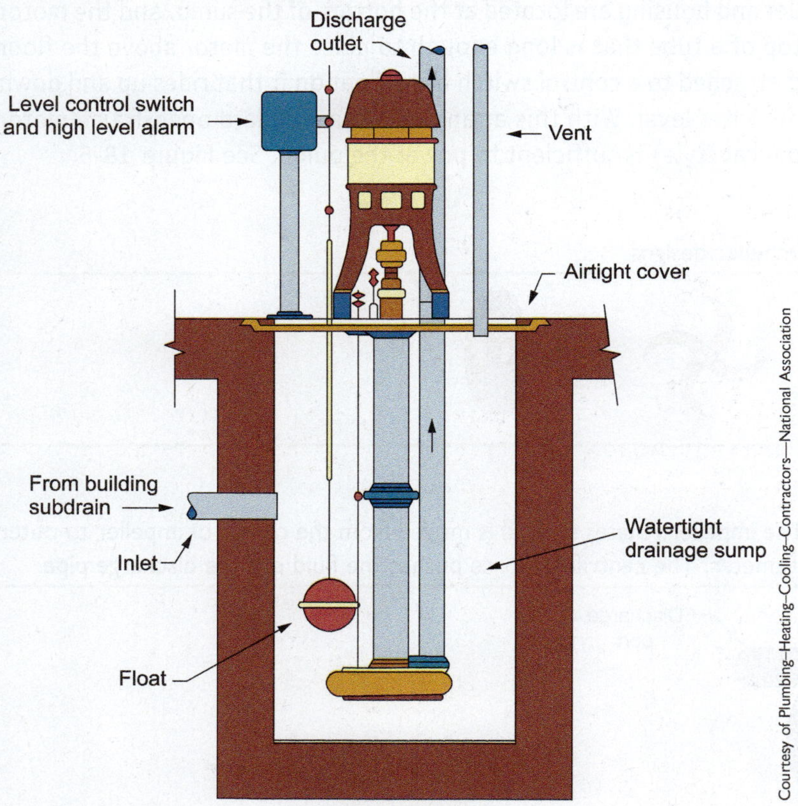

Top-Mounted

The pump and motor are located above the sump rim so that inspection and service are more easily accomplished. Only the suction line and control device sensor are within the sump. These pumps are usually larger than submersible or pedestal sump pumps. See Figure 18-6.

Dry Pit

The pump and motor are located in a pit adjacent to the receiver sump (Figure 18-7). The pump is connected to the wet sump with a pipe. This arrangement eliminates the problems associated with a suction lift and with having the pump directly submerged in the sump. The greater expense of the double pit and more elaborate pump limit this type to larger systems, where continuity of service is very important.

FIGURE 18-6 Top-mounted installations use a suction line to lift contents up to pump and motor combination.

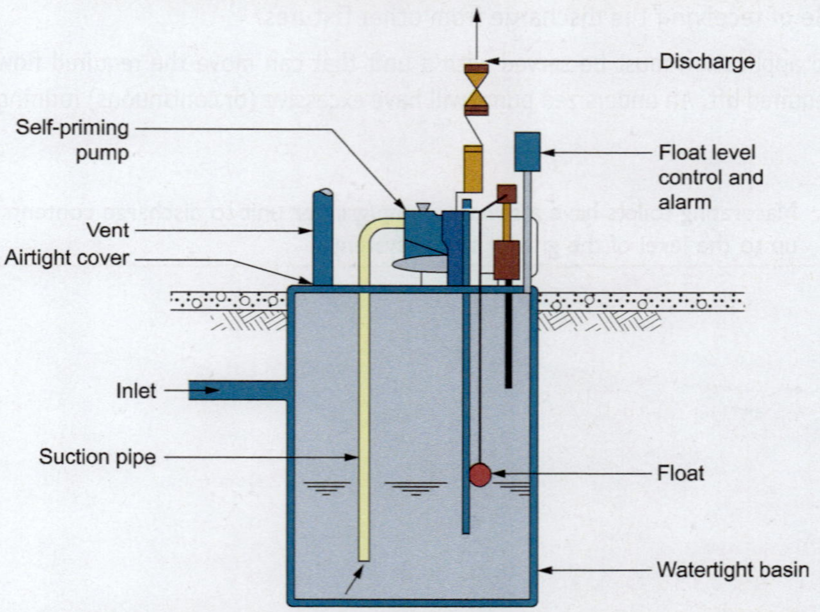

FIGURE 18-7 Dry pit installations are common on large-scale jobs. Notice a second sump pump is needed to keep the pit dry in case of a leak or failure of larger pump.

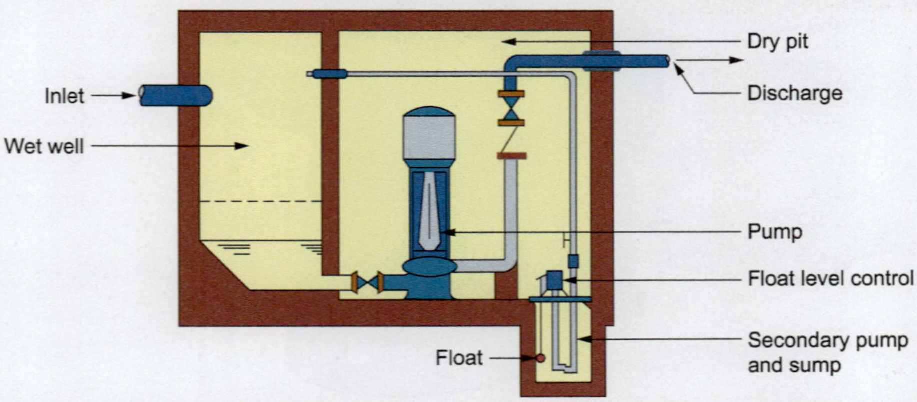

SEWAGE PUMPS

A sewage pump is a *permanently installed mechanical device other than an ejector for removing sewage or liquid waste from a sump* (National Standard Plumbing Code 2012). These devices are used to remove sewage from a receiver.

Sewage pumps are used to transport wastes that contain animal or vegetable matter in suspension. These pumps must be capable of handling material with considerable solids without fouling the passageway through the pump. Some sewage pumps, called grinder pumps, shred the wastes as part of the discharge cycle. Single bath or single fixture macerating systems are now available to facilitate bathroom or fixture placement in virtually any location. Such systems include a sump, a grinder or macerating pump, and all connectors to enable you to install a bathroom without the need to tear up the building structure. The wastes that enter the sump from the fixtures are ground to a fine slurry and discharged through a small diameter pipe ($\frac{3}{4}$ in. is typical) to a stack. Such systems are becoming common where expensive structural modifications would otherwise be required. Figure 18-8 shows a macerating toilet that is capable of receiving the discharge from other fixtures.

Any pump application must be served with a unit that can move the required flow through the required lift. An undersized pump will have excessive (or continuous) running

FIGURE 18-8 Macerating toilets have a sewage pump/grinder unit to discharge contents up to the level of the gravity sewer system.

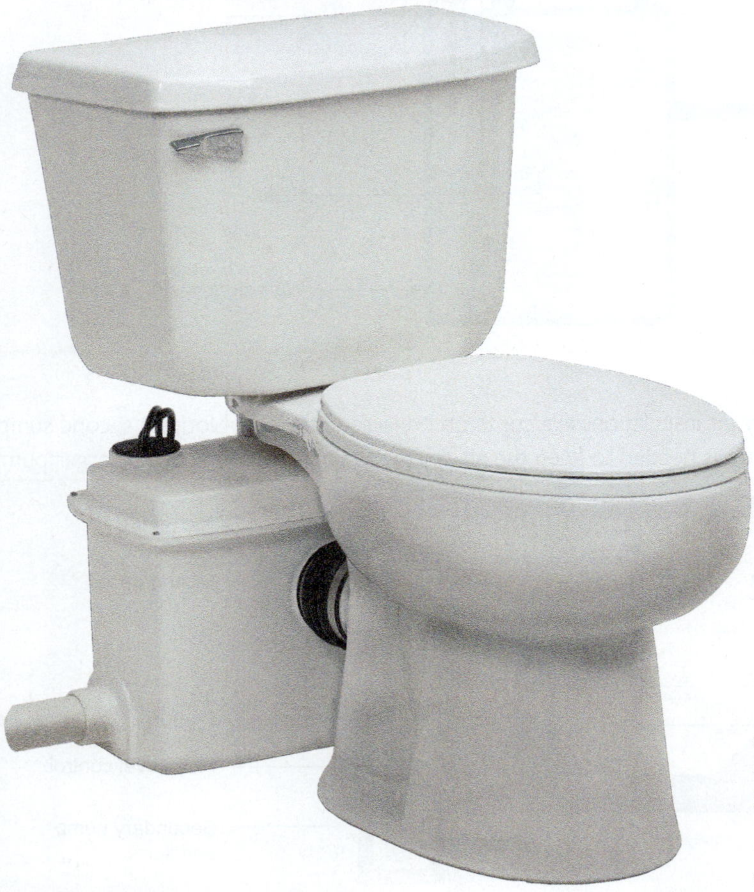

Courtesy of Zoeller Company

time, undue noise, and high operating costs. An oversized pump represents an unnecessary investment cost and can have problems with cycling on and off under light loads.

Duplex systems are installed where reliable performance is desired. Two pumps (activated either alternately or on an as-needed basis) ensure complete removal of wastes from the fixtures to the building drainage system.

Sewage pumps are similar in general concept to sump pumps, with these important differences:

- Sewage pumps are specially designed centrifugal pumps that can move solids. The internal features of these pumps permit $1\frac{1}{4}$ inch, $1\frac{1}{2}$ inch, and in some models even up to 3-inch solids to be pumped through the unit. Since the ability to pass such solids implies large passageways and hence large flow rates, sewage pumps would usually be rated at 100 gpm or more, with 20-foot to 30-foot lift. Pump configurations are the same as for sump pumps (similar to details shown in Figures 18-4 through 18-7).
- Because sewage pumps have higher pumping rates, the sewage receiver must be larger than the typical small sump or the pump "on-off" cycles will be too short. The receiver should be sized so that the pump operates for a minimum of 5 minutes for each "on" cycle.
- One variation on the sewage pump is the grinder pump or macerator. As mentioned above, rather than discharging large solids, the wastes are ground into small particles as part of the pumping discharge cycle. This pump design is available to serve one or multiple fixtures.

SEWAGE EJECTORS

A sewage ejector *is a device for lifting sewage by air pressure* (National Standard Plumbing Code 2012).

An ejector requires that the sump be a pressure vessel capable of operating with ejector pressures within the receiver. These pressures could be as high as 100 to 125 psig, but pressures in the range of 10 to 15 psig are more often sufficient to accomplish what has to be done. At the completion of an ejection cycle, the pressure within the vessel must be vented away—this requirement means that a separate vent to the outdoors must be provided for the sump.

For handling sewage of random-sized solids, where limited electric power is available, or where electrical hazards must be minimized, a sewage ejector is selected. The receiver is a pressure vessel equipped with electric solenoid valves, gate or ball valves, and sewage level sensors. The contents of the receiver are expelled from the vessel by pressurizing it with compressed air, which lifts the sewage to the point of connection with a gravity drain. Upon completion of this emptying cycle, the air pressure is vented outdoors and the receiver is again available to receive sewage. The ejector has no close-fitting parts in the sewage discharge piping, so the chance of fouling is minimized.

IN THE FIELD

Since each of these systems uses electrical energy as a part of its operation, installations must be done carefully to protect individuals from electric shock. An electrical potential of 120 volts can be lethal to a human body. A current of 100 milliamperes (0.1 amp) is deadly.

Various control methods can be used, but a popular system for these ejectors is to use the conductance of the sewage as an indicator of sewage depth. Simply put, the sewage, because it is mostly water, will conduct electricity. An electrode mounted high in the receiver will turn the pump on when the sewage level is high enough to submerge it in fluid. The sewage now conducts electricity between the top and bottom electrodes and initiates the pump operation. When the sewage level drops below the lower electrode, the action stops, the receiver is vented, and the receiver is available to accept sewage.

Each of the three basic waste lift devices has its own best applications. Review the discussions in this chapter and the installation methods described below for familiarity with these products.

DIFFERENCES AND DISTINCTIONS

The terms *sump pump*, *sewage pump*, and *sewage ejector* are sometimes used almost interchangeably in our industry. For example, see the National Standard Plumbing Code definitions for sump pump and sewage pump above. Despite this similarity of terminology, the three devices are different and the differences are quite important because these three product types are very different in cost.

Sump Pump Installation

The installation of sump pumps and sewage units begins with the installation or fabrication of the sump or receiver. Whenever possible, use a prefabricated unit because it will be a more economical choice. Prefabricated units are made of coated cast iron, reinforced concrete, clay tile, or plastic.

Field-erected sumps are made of concrete, brick, or large concrete or tile pipe sections. The least cost of the finished, installed sump is the basis for deciding the details of construction. Large prefabricated units may require considerable rigging and handling for installation, whereas field-erected units require much on-site labor. Different local conditions and job requirements may change the choice from one type to another.

If the sump is made of a large-diameter pipe, it should be set on gravel or concrete with the top flush with the floor. The bottom should be concrete so that the contents will not seep out.

Connections and Controls

The submersible pump unit is placed on the bottom of the sump and connected to electric power and the discharge pipe. The discharge pipe should contain a readily removable coupling or union (located above the floor), a check valve (so that the contents of the riser pipe will not flow back to the sump on the off cycle), and a full port gate or ball valve (so the check valve can be serviced). The discharge pipe should rise high enough so that the final horizontal pipe to the gravity drain will be above the gravity drain. In this way, the pump discharge will flow by gravity to the drain line. Electric power to the pump should have a switch close to the pump, located at least 4 inches above the floor.

A high-water alarm should be installed on any system where service interruption could produce damage. Greater reliability is obtained with duplex installations.

Many codes permit a pump handling only storm water or foundation water to discharge onto a splash block outside or into the storm drainage system. In many cases, sump pumps handling such drainage are not allowed to be connected to the sanitary sewer. Check local sewage plant rules before you make such connections to the sewer system.

Observe the pump through a few cycles to be sure the water level control is working properly. If a float rod is operating a switch above the sump cover, be sure that the rod is not binding where it passes through the cover.

For a cord-and-plug connected pump, be sure to use a grounded receptacle for the source of power, protected with a ground fault circuit interrupter. Fasten the cord to the riser pipe to provide a neat, safe job.

Sewage Pump Installation

Sewage pumps are installed using the same type of valved connections and electrical precautions. The sumps used must be tightly sealed and vented to the outdoors. Float rods pass through the cover with a sleeve filled with grease to provide free movement with a gastight seal. If a sump is serving six or more water closets, it should be provided with a duplex pump. Duplex pumps may be required by code on all installations that have "public use" occupancy. Check your local code for requirements specific to your area.

Sewage Ejector Installation

Ejectors are placed in pressure vessel receivers and all the special piping and equipment are connected. Large units must be anchored to the building as recommended by the manufacturer. Vibration and sound-deadening controls are also used to improve the installation.

Each of these larger units requires a custom installation; consult the manufacturer's representative or literature and the owner before deciding on job details.

REVIEW QUESTIONS

True or False

1. __________ A sump pump and sewage pump are the same thing.
2. __________ A sewage ejector requires a pressure-related sump.
3. __________ Installation of duplex pumps should be considered if the interruption of service cannot be tolerated.
4. __________ Sewage pumps must be made airtight and vented outside.
5. __________ A sewage ejector vent is not permitted to connect to the general plumbing vent system.

Chapter 19

LEARNING OBJECTIVES

The student will:

- Describe the operation of plumbing traps.
- Identify the types of plumbing traps.

Plumbing Traps

PLUMBING TRAPS

A plumbing trap is defined in the National Standard Plumbing Code as *a fitting or device which provides a liquid seal to prevent the emission of sewer gases without materially affecting the flow of sewage or waste water through it.* Figure 19-1 shows a conventional P-trap. The parts of the trap and the features that determine the depth of the seal are indicated in the illustration.

The first part of this definition, a liquid seal, is accomplished by the trap seal. Figure 19-1 shows how a vertical column of water blocks the flow path between the top of the trap dip and the crown weir. This seal is the barrier that keeps offensive or dangerous gases in the drainage piping and out of the occupied space. The usual trap seal is 2 inches deep with some variations, but because of certain conditions this may vary. Most codes limit the seal to a maximum of 4 inches, except for specialty traps. These specialty traps include grease interceptors (which stop grease, fat, oil, wax, and other debris from entering the DWV system) and plaster traps (which prevent solids from entering the DWV system). For more detailed information concerning specialty traps and interceptors, see *Plumbing 301*.

Figure 19-2 shows specialty traps or interceptors.

Since the second requirement of a trap is that it shall not materially affect the flow of waste, the interior of the trap should be smooth and the diameter uniform throughout the flow path (for one exception to this goal, see the discussion of S-traps later in this chapter). The trap should never be smaller than the discharge opening of the fixture.

Every fixture must be equipped with a trap before it discharges into the drainage system. Some fixtures, such as water closets, clinic sinks, and certain urinals have traps built into them, so there is no need for the plumbing technician to install one. It is customary not to show a trap on a line drawing if the fixture or appliance has a built-in, or integral, trap. Figure 19-3 shows a single-family dwelling unit in which all of the fixtures except the water closet, which has an integral trap, are shown with a trap.

FIGURE 19-1 Components of a typical P-trap.

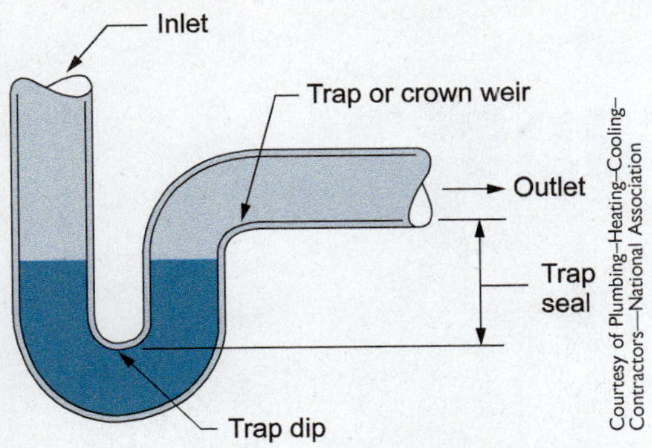

FIGURE 19-2 Specialty traps or interceptors, each particularly well suited for a certain task.

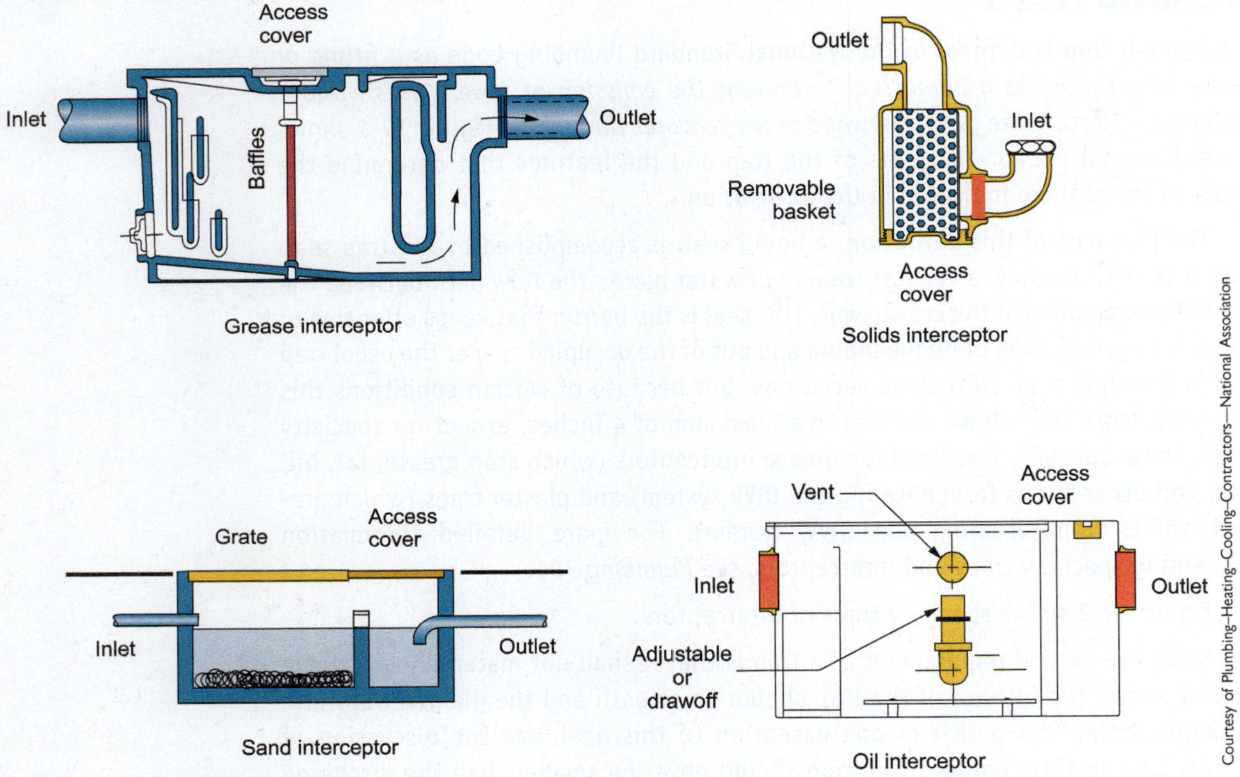

Trap Requirements

The particular requirements for traps include the following items:

The Trap Must Be Self-Scouring

Every time water passes through the trap, the trap will clean itself of any debris but still leave enough water to form an effective seal.

FIGURE 19-3 Typical line drawing of a residential building. Notice that no trap is shown for fixtures with an integral trap.

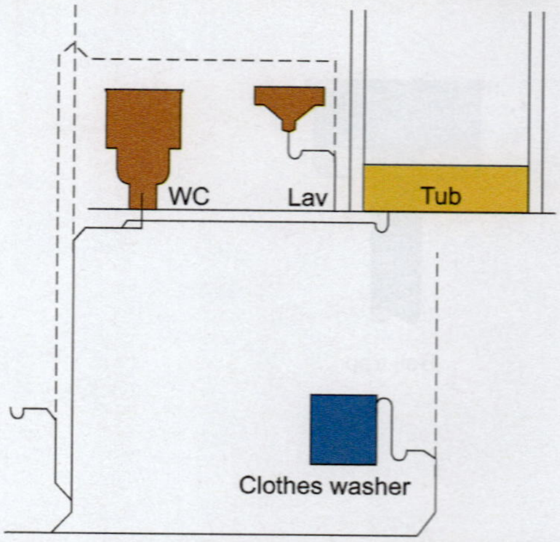

Least Amount of Fouling Surface Possible

Fouling means an accumulation or buildup of material on the surface of the pipe. The less fouling surface available, the less likely the trap will accumulate debris and become a health hazard.

No Interior Partitions Permitted

Unless the trap material is corrosion-resisting, no interior partitions are permitted. Interior partitions can fail and show no evidence of the failure, thus defeating the intended function of the trap. Even though most bottle traps have internal parts made of plastic or other corrosion-resisting material, many states do not allow them to be used.

No Internal Moving Parts to Maintain the Trap Seal

Internal parts can fail and the trap seal can be lost without any evidence that it has been lost.

The Trap Seal Should Be a Minimum of 2 Inches and a Maximum of 4 Inches

Except for interceptors, the trap seal should be 2 inches or greater, but not more than 4 inches. The self-scouring characteristic is less effective with greater seal depth.

Permitted and Prohibited Traps

Figure 19-4 shows a ball trap, bell trap, crown-vented trap, trap with partition, S-trap, and drum trap. Generally, these traps are prohibited for the following reasons:

- **Ball trap:** moving parts are required to maintain the seal and the parts could foul and not operate, thus losing the seal.

FIGURE 19-4 Prohibited traps.

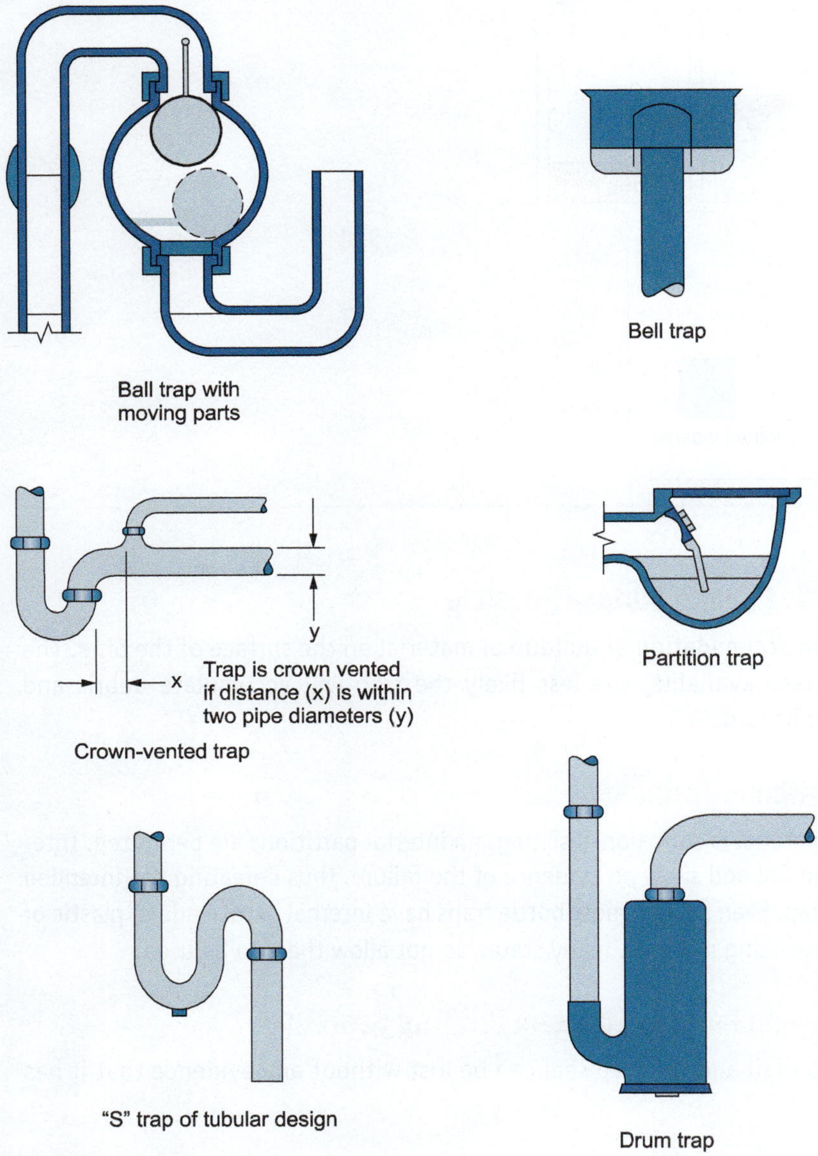

Courtesy of Plumbing–Heating–Cooling–Contractors—National Association

- **Bell trap:** not self-scouring and has an insufficient trap seal.
- **Crown-vented trap:** vent connection may be quickly fouled because of turbulence in the area of the connection. Notice there is a separation distance between the vent and crown. If the distance is greater than two pipe diameters, it is not classified as a crown vent.
- **Partition trap:** relies on interior partitions resulting in many fouling problems and the possibility of unrecognized leakage around partition.
- **S-trap:** tends to self-siphon.
- **Drum trap:** not self-scouring.

Some special chemical waste drainage piping systems allow the use of bottle traps. As expected, a bottle trap looks like a bottle, with a center drop tube inlet and a side tapped outlet from the "bottle." As part of their unique design, some waterless urinals use a form of partition trap to function. Such designs are critical to the performance of these fixtures.

Building Traps

Building traps (usually running traps) are prohibited by most codes. However, sometimes local municipalities require them if the sewer gas in the municipal sewage system is particularly corrosive or aggressive and could damage building DWV systems. Check with your local jurisdiction for ordinances affecting these codes. Building traps were discontinued because they prevented the building vent system from helping to vent the building sewer and also provided a settling place for solids and other wastes. This led to stoppages and reduced the amount of aerobic decomposition that occurred. Adequate drainage system venting has reduced the need for these traps. When they are required, they must be installed according to local code requirements.

Figure 19-5 shows a building trap installed outside the building, with cleanout risers to grade level.

P-Trap

The manufactured P-trap combines all of the desirable features of a trap. Manufactured P-traps are made of smooth, nonabsorbent materials, are self-scouring, and minimize the tendency to self-siphon. Figure 19-6 shows both a P-trap that uses slip joints, which allow for easy removal, and a P-trap with solid connections.

FIGURE 19-5 Building traps are rarely used in new construction.

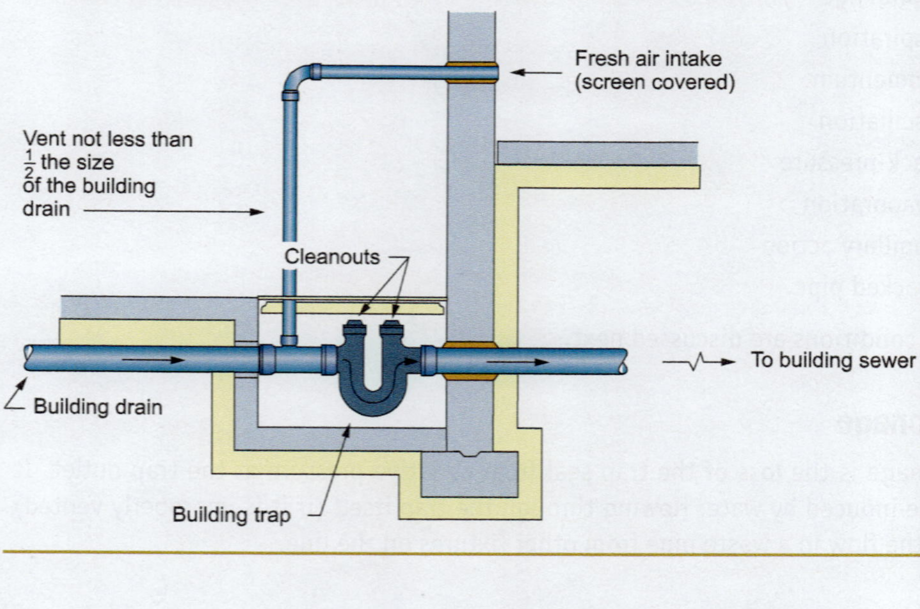

Courtesy of Plumbing–Heating–Cooling–Contractors—National Association

FIGURE 19-6 P-traps are available in both slip joint (top) and solvent joint (bottom) construction.

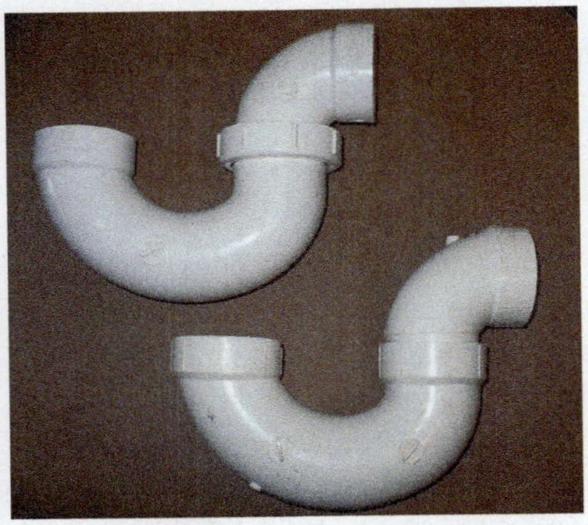

Photo by Ed Moore

Deep seal traps (4-in. seal) should be used where evaporation or marginal venting are problems. Examples are floor drains, catch basins, and separators. Since they are physically larger, they require more space for their rough-in and they are less likely to clean themselves in normal use. Deep seal traps offer more resistance to flow, so use them only where needed.

IN THE FIELD
Slip joint connections must be in accessible locations.

LOSS OF TRAP SEAL

Trap seals can be lost in many ways. The acronym SAMOBECC has been coined to aid in remembering them:

Siphonage
Aspiration
Momentum
Oscillation
Back-pressure
Evaporation
Capillary action
Cracked pipe

These conditions are discussed next.

Siphonage

Siphonage is the loss of the trap seal from negative pressure at the trap outlet. It may be induced by water flowing through the trap itself (if it is improperly vented) or by the flow in a waste pipe from other fixtures on the line.

Figure 19-7 shows an unvented trap connected to a stack. If the stack is not vented above, water flowing down the stack will develop a positive pressure at the trap outlet. This positive pressure will make the water in the trap rise. When the water in the stack has passed the opening for the trap, the positive pressure will be reduced (possibly even become a vacuum) and cause the trap water to drop back into the trap and then spill over the crown weir. This process is illustrated in Figure 19-7.

Self-Siphoning

Self-siphoning of a trap occurs when a negative pressure is induced in the outlet of a trap as a result of the flow through the trap. A partial vacuum can be developed easily as a result of such flow, especially if venting is inadequate or nonexistent. Figure 19-8 shows an established siphon. The difference in length of the two legs is the motivating head pressure that causes the fluid to flow from the higher level to the lower level.

A tubular S-trap of uniform diameter is the type of trap that can self-siphon. Once flow is developed in the trap, the flow will continue until the siphon is broken on the fixture side. Siphons are broken when air is introduced into

FIGURE 19-7 Water levels in a trap can change due to dynamic conditions in the system.

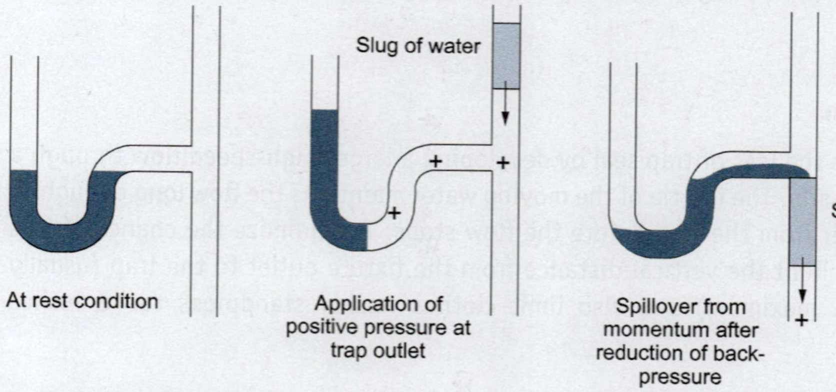

FIGURE 19-8 Basic concept of how a siphon is formed. The siphon will flow until air is introduced.

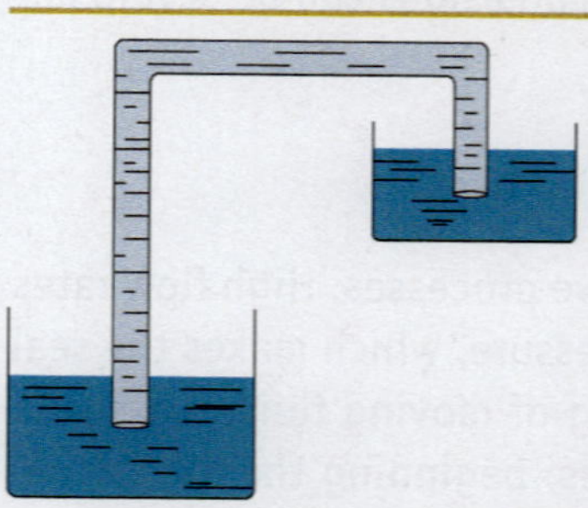

the fluid stream. In this case, the air will be introduced when the water level in the trap drops below the trap dip. Whether or not most or all of the trap seal will be lost depends on the type of fixture being served; a fixture that drains rapidly to the drain opening will not have enough water left in the fixture to reseal the trap.

The essential characteristic of an S-trap is that it has a uniform diameter throughout the length of the trap. If the drop piece has a larger diameter, and the change to the vertical at the crown of the trap has any horizontal extent, then even though the configuration still looks like an S, the action is changed. These modifications mean that the velocity is not uniform throughout the extent of the trap and the result is that the siphon will be broken by air coming up the discharge side before the trap seal is lost. This fact is the basis for the combination waste-and-vent (with the trap/drain configuration that looks like an S) system permitted in some codes (see Sections 12.17 and 12.8.2 of the 2012 National Standard Plumbing Code).

Aspiration

Aspiration is also the loss of the seal from negative pressure at the trap outlet, but the source of the negative pressure is high velocity flow through a fitting, and is due to the geometry of the fitting. A vent on the trap outlet will prevent this occurrence. Reducing the flow through the fitting or increasing the size of the fitting will prevent aspiration.

Momentum

Momentum is the loss of trap seal by developing a large, high-speed flow through a trap and trap arm. The inertia of the moving water maintains the flow long enough to pull the water from the trap before the flow stops. To minimize the chance of this event, codes limit the vertical distance from the fixture outlet to the trap (usually about 24 in. maximum) and also limit clothes washer standpipes to 48 inches maximum.

Oscillation

Oscillation is the loss of trap seal by the water in the trap moving back and forth rapidly enough to cause the seal water to run over the crown weir. The process starts as a result of cycling high and low pressures at the trap outlet due to flow from other fixtures or from wind blowing over vent outlets. Some conditions can be severe enough that trap primers are required to maintain seals.

Back-Pressure

Back-pressure leads to trap seal loss by aiding the above processes. High flow rates approaching the trap outlet produce an increase in pressure, which makes the seal rise on the inlet side of the trap. Then, when the slug of moving fluids passes, a negative pressure develops and the seal water reverses, beginning the oscillation (see Figure 19-7). If the back-pressure is high enough, stack gases will be forced into the room through the seal, splashing seal water into the space or causing a bubbling noise.

Evaporation

Evaporation is the loss of the seal because the water in the trap changes to vapor state and is lost from the trap. Chapter 16 discusses the gas state of water. Any traps that are infrequently used are subject to loss by evaporation, such as floor drains. Trap primers must be used to maintain the seal in these traps.

Capillary Action

Capillary action is the loss of the water in the trap due to debris lying in the trap over the crown weir. Water is wicked over the weir where it can gradually be dripped away. Figure 19-9 shows a mop string located so as to be a problem. This situation is difficult to avoid and more difficult to recognize. Vigorous rinsing of fixtures that receive this type of waste is the best defense.

Cracked Pipes

Cracked pipes permit the loss of seal through a broken or failed pipe or fitting. This problem will show up as a leak below the fixture and can only be remedied by repairing the leak. Use quality materials and proper installation methods to minimize this risk. Encourage the building occupants to use only proper drain cleaners and to avoid the use of strong or aggressive chemicals. Do not install traps in areas where freezing temperatures can occur.

TRAP PRIMER

A trap primer is a device that periodically adds water to a trap (see Figure 19-10). It is piped to the trap body, and, if connected to the potable water supply, must be fitted with proper backflow protection at the connection to potable water. Some potable water types use an electric valve and timing device and others are connected to a nearby fixture drain with fittings that divert a small amount of wastewater to the primer every time the fixture is used. In residential applications, infrequently used traps are usually refilled with a pail of water.

FIGURE 19-9 Fibrous materials can suck the water out of a trap, much like a wick in an oil lamp or kerosene heater.

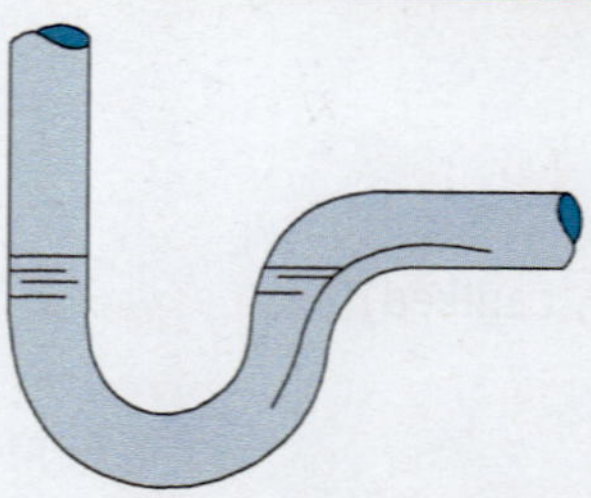

FIGURE 19-10 Trap primers are used to replace water that was lost due to evaporation.

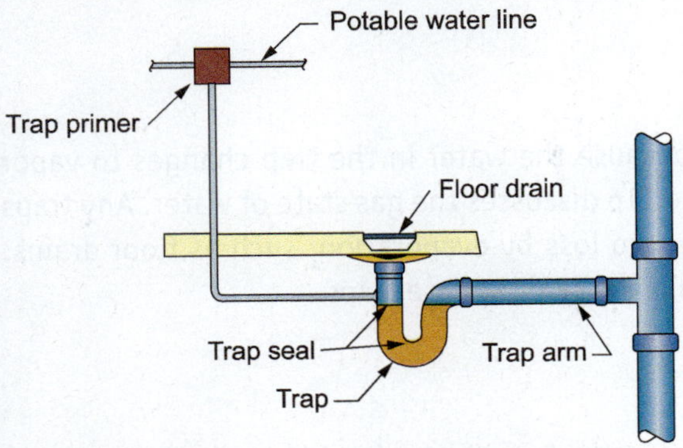

TRAP INSTALLATION

General instructions for trap installations include the following points:

1. Install the trap without stresses or strains from the fixture or the rough plumbing.
2. Support all fixtures with proper brackets and hangers.
3. Do not install traps where freezing could occur.
4. Do not reduce the size of a standpipe at the connection to a trap. The standpipe should be 18 inches to 48 inches in length. Remember that a laundry machine discharge is pumped—do not undersize the standpipe.
5. Do not connect a trap to a smaller pipe than the trap outlet.
6. Do not connect a fixture outlet to a trap smaller than the outlet.
7. The maximum length of a fixture tailpiece to the trap should be 24 inches. The longer the tailpiece, the more fouling area there is.
8. The minimum horizontal distance from the trap outlet to the face of the tee or tee-wye to which it connects should not be less than twice the diameter of the trap. This minimum horizontal length helps break self-siphonage and maintain the seal.
9. Be sure the trap is free from vibrations.

Connections Permitted to Traps

The following connections are generally permitted between trap parts or at the trap outlet:

- Slip joints (provided they are readily accessible)
- Ground joints (e.g., metal to metal)
- Solid connections (solvent cement, soldered, threaded, caulked)

REVIEW QUESTIONS

True or False

1. __________ A plumbing trap is a device that prevents the passage of sewer gas into the building.
2. __________ The typical trap seal is 2 inches deep.
3. __________ Most of the causes of trap seal loss can be minimized or prevented by proper drainage and venting installations.
4. __________ A trap primer is a device that maintains the water seal in a trap.
5. __________ The maximum vertical distance from a fixture outlet to a trap is limited to prevent the water momentum from washing out the trap.

Chapter 20

LEARNING OBJECTIVES

The student will:

- Describe different loads placed on fasteners.
- Describe installation options for fixture supports for common fixtures and accessories.

Fixture Supports

Long before the first hole is drilled and the first piece of pipe is laid, the decision as to how fixtures and appliances will be supported and installed must be made. These decisions will be influenced by the type of fixture selected, the intended purpose of the building, and the building components to which the fixture is to be mounted. Since there are countless variations of fixtures and each one of them has a multitude of installation options, this chapter will focus on some of the more common methods of support. Always read, understand, and follow the directions before you begin the installation; it is all too easy to accidentally choose the wrong support or misadjust the spacing or heights of the mounts. This can be a costly error since most fixture supports are hidden from the consumer by wall coverings, so it is crucial that everything be installed correctly during the rough-in phase of construction. Fixtures must be firmly attached to the building structure so that they can support the weight of both the fixtures and any additional loads placed on them.

FASTENERS

Before discussing the wide variety of individual fixture supports, it is important to discuss something that they all have in common—the fasteners. In many instances, the manufacturer of the support or fixture will specify what type of fastener to use, but this recommendation is usually based on an assumption that the structure is wood, concrete, or drywall; your application may be different. In order to select the appropriate fastener, the following factors must be taken into account.

Strength Requirements

Depending on how the fixture is configured, the fastener may have to support the weight of the fixture, the carrier, and possibly its user. In simpler applications, the fastener may simply have to keep the fixture from shifting. In Chapter 6 of *Plumbing 101*, the different ways forces could be applied (compression, tension, and shear) were discussed. See Figure 20-1. In practical applications, a combination of any of the following would be applied, but for clarity each one is considered separately.

FIGURE 20-1 Example of the three categories of loads.

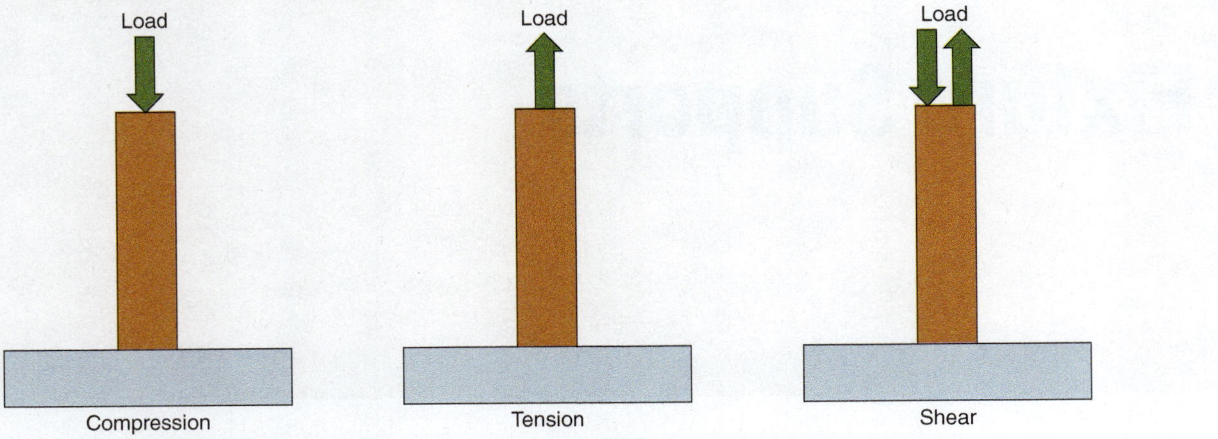

Compression

A compression load is one that tends to push the material together, crush it, or make it smaller. A concrete wedge anchor or plastic insert anchor are good examples of how compressive forces hold them into the concrete. This suits concrete well because concrete is strong in compression but weak in tension.

Tension

A tensile load is one that tends to pull a material apart. Almost all threaded fasteners are subjected to tensile loads. The head of the bolt and the nut are pressing two or more pieces of material together, but the bolt itself is under a tension load.

Shear

Shear loads act to make the parts of the material slide past each other. This usually involves the load being applied perpendicular to the fastener.

Ease of Assembly and Disassembly

In the competitive, fast-paced world of construction, time is money, so many fasteners are chosen by how fast or easy it is to install the fastener. In the past, tradesmen used nails because they were fast to install and only a hammer was required. However with the compact powerful battery-operated drills available today, most people use screws instead. A screw has more holding power and is less likely to pull free over time than a nail. Screws are also easier to remove.

FIXTURES

Water Closet

One of the most common fixtures found in all buildings is the water closet, or toilet. These fixtures can be for private or public use and in residential, commercial, industrial, or penal settings. Regardless of the application, water closets will be either wall-hung or floor mounted.

Floor Mounted

Floor mounted is the most commonly used residential water closet. It typically requires only a closet flange to be connected. The closet flange serves two primary functions. The first is that it secures the piping to the building; the second function is to connect the water closet to the pipe and make a watertight seal. Closet flanges are available in a wide variety of materials and sizes. Cast iron was traditionally the most commonly used material for residential and commercial applications, but plastics such as PVC or ABS have become the most commonly used today. While the plastic flanges are not as strong as cast iron flanges, plastic is less likely to corrode over time. Closet flanges are normally sized to fit either into or on the outside of a 3-inch or 4-inch pipe. Many plumbers feel that the 3 inches inside the pipe is too constrictive, so they prefer the type that mounts to the outside of the pipe or a 4-inch flange. Figure 20-2 shows a side-by-side comparison of two.

Closet flanges are installed during the rough-in phase of the project and normally have the pipe opening blanked off to all the DWV system to be pressure tested. The floor flange must be set on and securely anchored to the building structure. Figure 20-3 shows a closet flange attached to a wood subfloor with woodscrews. Do not use drywall screws—the heads are too small and will pull through the hole. On a slab job, the concrete usually is sufficient to hold the pipe in place.

Wall-Hung

As their name implies, wall-hung water closets do not sit on the floor; they are securely mounted to the wall. This has the advantage of being easier to mop underneath. However, this installation requires a very strong rigid mounting device, known as a carrier. Like the closet flange, the purpose of the carrier is to secure the piping to the building and secure the water closet to the building. Since carriers are mounted into the wall system, they must be installed during the rough-in phase of

FIGURE 20-2 Section view of inside the pipe (right) and outside the pipe (left) closet flanges.

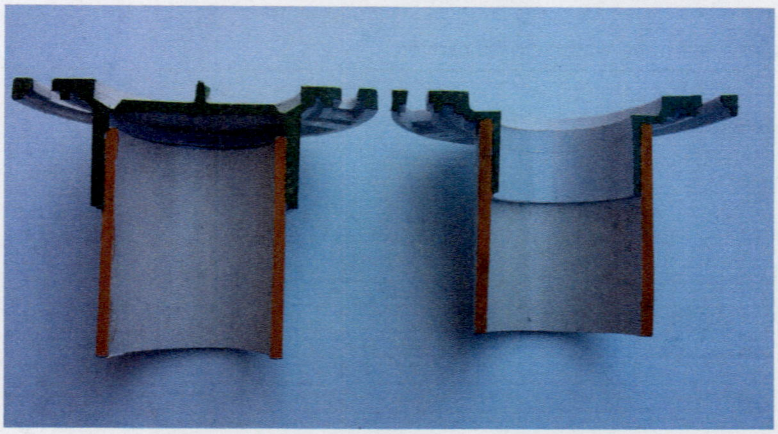

FIGURE 20-3 This closet flange is secured to the subfloor with corrosion-resistant screws.

Photo by Ed Moore

construction. Because a great deal of strength is required to support the weight of the fixture and the user, carriers are typically made of cast iron.

When ordering a wall-hung water closet carrier it is important to match it with the water closet itself. Some water closets have four mounting bolts while others have three. It is also important to identify the inlet opening size (usually either 3 in. or 4 in.) and how it will be connected to any additional fixtures. It is very common in commercial settings to have a battery of fixtures, and selecting the correct flow pattern is essential. Figure 20-4 shows a wall-hung water closet carrier.

Urinal

Like water closets, urinals can be wall-mounted or floor-mounted. The installation of floor-mounted urinals will need to be carefully coordinated with the flooring contractor to determine the final height of the fixture and how the flooring will join the edge of the fixture. Depending on whether the job requires it, urinals can be supported by carriers similar to those mentioned earlier for the water closets. Since urinals weigh less and are not intended to support the weight of the user, the carriers are not as heavy as those for

FIGURE 20-4 Typical wall-hung water-closet carrier.

Courtesy of Zurn Industries, LLC

FIGURE 20-5 Wall-hung urinal carrier.

Photo by Ed Moore

water closets. Figure 20-5 illustrates a urinal carrier that would be concealed behind the finished wall.

Lavatory and Service Sinks

Lavatories are typically either wall-hung or surface-mounted on a countertop. The installation of countertop-mounted units should be coordinated with the countertop installer to determine who is responsible for cutting the hole in the counter. Since cutting tops made of marble or granite requires specialized tools and training, it is best left to the contractor.

Wall-hung lavatories can use carriers similar to the one shown for urinals. Some have additional arm attachments that extend out from the wall. These arms can be either concealed or bracket type. The concealed arms are designed to be inserted into the lavatory or service sink, while the bracket style is exposed under the sink. The additional arms help support the front edge of the sink should anyone push down on it. Figure 20-6 shows a wall-mounted carrier with the concealed style

FIGURE 20-6 Wall-hung carrier for lavatory.

Courtesy of Zurn Industries, LLC

arms. Do not assume that pedestal sinks can be substituted for a wall-hung sink. Many pedestal sinks have mounting brackets that secure them to the wall but require the additional support of the pedestal beneath them. Whether the lavatory is wall-mounted or a pedestal, the mounting brackets or supports should rigidly attach to the building studs or cross supports. Drywall or paneling should not be used as an anchor point.

Floor-mounted service or utility sinks are fairly common in residential garages and some light commercial applications. These sinks typically are supported by four legs and are usually secured to the wall. The legs can be either stamped metal or molded plastic and are secured with screws. Once attached to the sink base, the legs can then be secured to the floor for additional stability. Figure 20-7 illustrates a floor-mounted utility sink.

Shower Pan

Shower pans and shower floors are required to be adequately supported, waterproof, and sloped toward the shower outlet. Manufactured shower pans are typically fiberglass constructed and have a threshold and sloped floor incorporated in them. Like bathtubs, they are installed on a level subfloor and attached to the surrounding wall studs with corrosion-resistant fasteners. Figure 20-8 is an example of a premolded fiberglass shower pan.

In many cases, standard shower pans are not suitable. Either the desired shape is not available or the customer wants a tiled floor. If this is the case, the

FIGURE 20-7 Utility sink.

Photo by Ed Moore

FIGURE 20-8 Premolded shower pans are attached to the building structure similar to bathtubs.

proposed shower area is covered with a waterproof membrane. This membrane is required to extend a minimum of 2 inches above the threshold and be fastened securely to the drain. Figure 20-9 shows a rubber shower pan liner that is installed beneath a tiled floor. A common mistake many plumbers make is not incorporating the code-required slope toward the drain.

FIGURE 20-9 The plumber must install the safety pan liner and drain and make certain it is watertight. Work closely with the tile installer for a smooth installation.

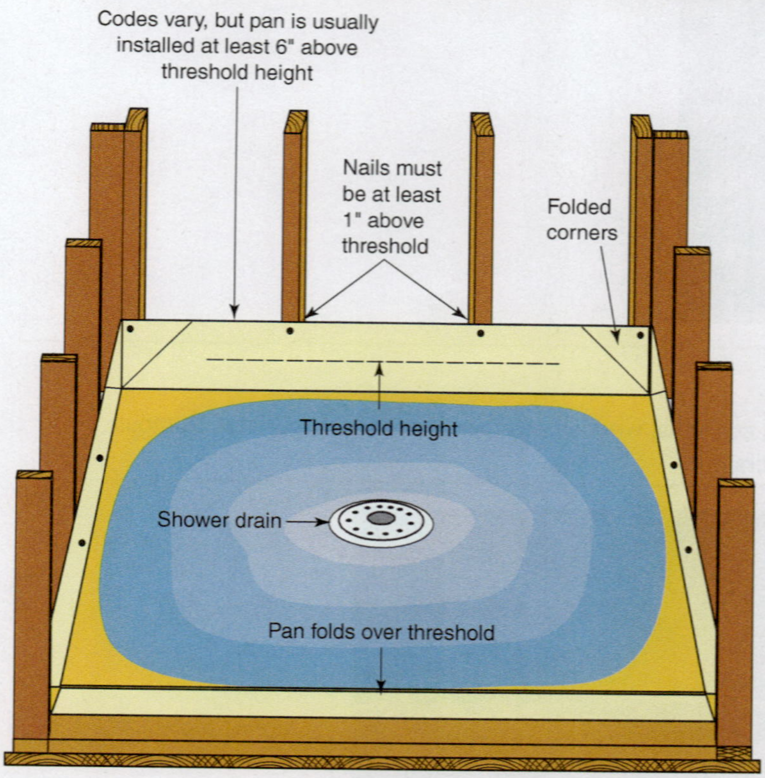

Handrails and Grab Bars

Handrails and grab bars are required by the Americans with Disabilities Act (ADA) for handicap toilet stalls and showers. These handrails must be securely fastened to solid structural members in the building so that they can support the weight of the individual using them. In masonry construction, this is usually accomplished by using concrete anchors. Figure 20-10 illustrates a set of handrails that are attached to a cinderblock wall. When wood framing construction is used, it is essential that the wood studs be used to anchor the rails; drywall is not strong enough. In these cases, it is customary to incorporate additional supports into the wall construction during the rough-in stage. Figure 20-11 illustrates a 2″ × 6″ inlaid in the wall studs to provide a suitable base for a handrail.

FIGURE 20-10 This handrail is mounted to a masonry wall with stainless steel screws and plastic inserts. The cover plate (left side of image) will cover screw heads once in place.

Photo by Ed Moore

FIGURE 20-11 This additional wooden member provides a larger area to mount a handrail.

Photo by Ed Moore

REVIEW QUESTIONS

True or False

1. __________ Shower pan liners must extend a minimum of 2 inches above the threshold.
2. __________ Urinal carriers are heavier duty than carriers used for water closets.
3. __________ When installing handrails, drywall and plaster are good anchor points.
4. __________ The fasteners that come with the fixture should always be used.
5. __________ All carriers for lavatories have concealed arms for support.

Chapter 21

LEARNING OBJECTIVES

The student will:

- Describe the equipment used for hoisting and rigging.
- Describe safe rigging practices and the use of knots and hitches.

Hoisting and Rigging

A variety of devices are available for hoisting applications. You must consider cost, distance of the move, weight of the load, number of loads, and whether the hoist is indoors or outdoors before you can select the best method for the job to be done. This chapter covers some of the choices of hoisting devices that are available and are often used in plumbing applications.

CRANES

Because of a crane's ability to reach, rotate, and lift heavy loads, it is a common fixture on many construction sites. Construction crews use cranes to lift and move various building materials such as structural steel and concrete. Cranes are also used to position large appliances and fixtures.

Construction cranes are large pieces of equipment that can be designed to handle loads from clamshell buckets for excavation to lifting loads up to 300 tons or more. The distance to which they can reach is often referred to as the boom length or swing. The boom length can vary from 25 feet to 200 feet or more. Generally speaking, the lifting portion of the crane is mounted on a truck or tractor, which enables it to be moved and maneuvered into position. Some cranes are so large that they are brought to the jobsite in sections and then reassembled.

HOISTS

Medium to heavy loads can be lifted with a hoist. A hoist is a device that is supported above the load to be lifted and is usually considered to be human-portable. Chain-fall hoists are geared-drum devices that raise a lifting chain by pulling on a lighter hand chain that operates the gear train. There is also a lever-operated hoist that uses a lever handle instead of a hand chain. Other hoists are operated by electric motors operated by a handheld pendant. See Figure 21-1.

FIGURE 21-1 This stationary hoist is equipped with a manually operated chain-fall hoist and an electric motor driven hoist.

Photo by Ed Moore

These hoists are plainly marked with their rated capacity so that you can apply them properly. They will hold the load in any position without requiring holding effort by the operator.

A "come-along," a mechanical device usually consisting of a chain or cable attached at each end, is often used in the plumbing industry to facilitate movement of relatively lightweight loads through leverage. Since it lifts through a ratcheting motion, the load may tend to drop down a fraction of an inch after each stroke of the handle. See Figure 21-2.

BLOCK AND TACKLE

Lighter loads can be moved with pulleys and ropes, also called block and tackle. A series of pulleys and ropes can support a load in such a way that the pull required on a rope will be much less than the weight of the load.

Pulleys can be set up in single or multiple arrangements and in snatch block configuration. A snatch block is a pulley with a hook connected in such a way that the hook can be opened on one side so that the run of the rope can be put on the pulley.

Center of Gravity

Whenever an object is lifted, its weight will be distributed over the lifting points. If the object is uniform in construction and its weight is distributed evenly, then each lifting point has the same load. If more of the object's weight is toward one side, then that side's lifting point would have more weight. Figure 21-3 illustrates this concept.

FIGURE 21-2 A single "come-along" moves lighter loads through leverage.

FIGURE 21-3 When more weight is located to one side, the tension will be greater in that side's sling.

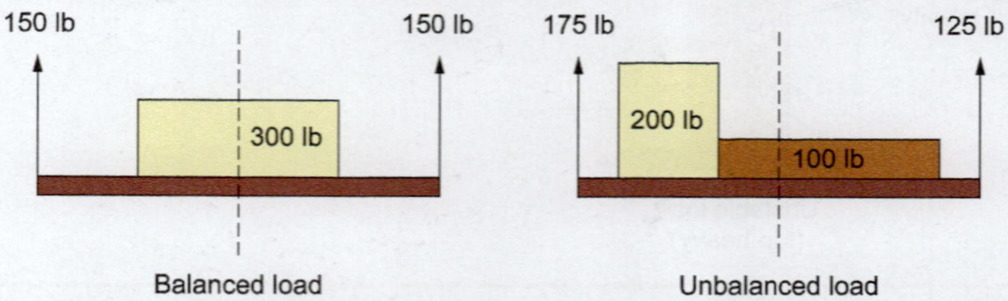

When lifting an object with a crane or hoist, the point of attachment must be located over the center of gravity of the object, otherwise one side will lift up first and the load could shift. Center of gravity is the balancing point of the load. While there are complex math formulas to determine the center of gravity, the easiest method is trial and error. First, look at the load and approximate the balancing point. Next, configure all of the rigging equipment so that the lifting hook is directly over the expected center of gravity. Slowly lift the object, to see if one side lifts off the ground before the other, and then set the load down. Next, reconfigure the rigging equipment so that the lifting hook is closer to the side that remained on the ground. Repeat until the load remains level. Figure 21-4 illustrates this concept.

Irregular loads should be lifted on sturdy pallets or platforms. The following procedure is typically used:

1. Tie each item solidly to the platform, and arrange the hoisting lines to maintain the platform level throughout the lift.

FIGURE 21-4 When the lifting hook is not over the center of gravity, the load will not lift evenly.

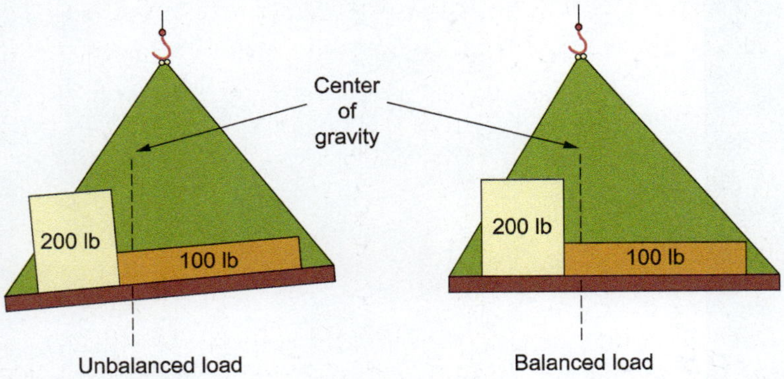

FIGURE 21-5 When the center of gravity is above the lifting points, the object can become top-heavy and tip over.

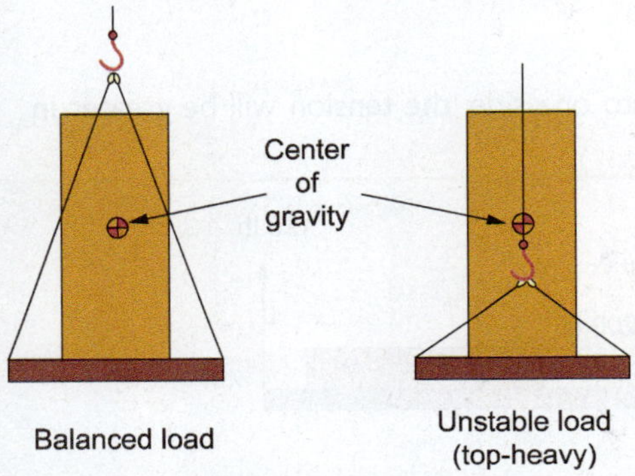

2. The rigging should attach to the base structure of the pallet or platform.
3. Retain any shipping crates or protective covering on the load until the product is ready to be installed and piped.
4. Do not use such protective covering or crating as the anchors to connect hoisting lines because these coverings usually will not withstand the forces developed in hoisting.
5. If the object is being lifted by a lifting hook and slings, make certain that the center of gravity is within the triangle formed by the slings. Otherwise, the object may be top-heavy and flip. See Figure 21-5.

RIGGING CONCEPTS

According to the dictionary, rigging is the use of tackle, equipment, or machinery fitted for a specific purpose. Rigging in the plumbing industry involves the proper use of knots, hitches, pulleys, hoists, and similar equipment for the safe and proper transport of heavy equipment and materials. In general, the lightest, quickest way to handle a load will be the most economical method.

The initial required knowledge for rigging is safety awareness and planning ahead for safety. The specific questions that must be considered before moving a load are these:

- What type of load must be moved?
- How heavy is the object?
- What is the shape of the object?
- What path must the load follow?

A heavy, irregular, or round object will be more difficult to lift than a rectangular-shaped object. The center of gravity on loads that contain liquids will shift as the object is being moved; this could cause a dangerous imbalance. Occasionally an object has to be relocated, so a special crate or hoisting supports may have to be built to make the task safer.

RIGGING METHOD

Ask and obtain answers to these questions before beginning a hoisting job:

- What is the best method to lift the load? A bathtub can be transported by manual methods or by using simple or complex machinery. A lavatory would simply be carried to the site. Pipe or heavier material would have to be hoisted.
- What tools, machines, and rigging are required to transport the load? Consider cranes, forklifts, chain falls, and other hoisting devices for heavier loads.
- Will the equipment require special training for the operator? For example, forklifts and cranes require special permits for operation. Check with your employer or your local OSHA office to learn where this training/certification is available.
- Will special equipment and operators be more economical than regular equipment?
- Will special equipment require special treatment for concerns such as inclement weather?
- Who is responsible for moving the load?

Only one person must be in charge of the lift and it is his or her job to ensure the safety of all involved. Otherwise quick decisions cannot be made and job safety will be reduced.

Slings

Rigging also may involve the use of wire rope slings, webbed slings, and other devices in preparation for lifting objects. Most wire rope slings used in plumbing have an eye splice on each end. The most common wire rope diameters are $\frac{3}{8}$ inch, $\frac{1}{2}$ inch, and $\frac{5}{8}$ inch. Sling lengths most commonly used are 3 feet, 4 feet, 6 feet, 8 feet, 12 feet, and 20 feet.

FIGURE 21-6 Both wire rope slings and nylon web slings are popular for lifting. Nylon web slings are less likely to damage surface finishes.

Webbed nylon slings are also frequently used as rigging. Normally, you will see 2-inch- and 3-inch-wide double-thickness choker-type slings used for rigging. The length of these slings has no bearing on the load capacity. Figure 21-6 shows both wire rope and webbed nylon slings.

Before using webbed slings, be sure to check for fraying, torn fibers, or knots. A sling with a knot or two slings knotted together have only 25 percent of the rated capacity, so never tie them into a knot. In addition to obvious holes and ripping in the webbing, you should look for the types of damage listed in Table 21-1.

- All wire ropes, chains, hooks, sheaves, and drums must be inspected before each lift.
- In chains, look for bent links, cracked welds, nicks, and gouges. Only one link has to fail for an accident to occur.
- When in doubt about the condition of any rigging, check with your supervisor.
- Wear gloves to prevent injury. Keep your hands out of pinch points and stay clear of the rigging as it tightens under the weight of the load.
- Hands or feet cannot be used for spooling cables onto drums.
- If a slack line occurs, check the seating of the rope in the sheaf or drum before proceeding.
- Note that increasing the angle from the vertical between legs of a two-leg sling increases the load stress on the legs.
- The load must be properly set in the throat of the hook. Loading toward the point (except in grab hooks) leads to spreading of the hook.

TABLE 21-1 Visual Indications of Damage to Webbing

Type of Webbing	Heat	Chemical	Molten Metal or Flame	Paint and Solvents
Nylon and Cordura	In excessive heat, nylon becomes brittle and has a shriveled, brownish appearance. Fibers will break when flexed. Should not be used above 200°F.	Change in color usually appearing as a brownish smear or smudge. Transverse cracks when belt is bent over a mandrel. Loss of elasticity in belt.	Webbing strands fuse together. Hard shiny spots. Hard and brittle feel. Will not support combustion.	Paint that penetrates and dries restricts movement of fibers. Drying agents and solvents in some paints will appear as chemical damage.

KNOTS AND HITCHES

Fiber rope used to lift significant loads should be made of hemp, nylon, or polypropylene. Plastic materials are stronger and more resistant to abrasion, but they stretch more than hemp. Plastic materials are also more resistant to the aging effects of moisture and sunlight.

There are three parts of a rope referred to in connection with knots or hitches. They are listed below:

- End: the end of the rope beyond the knot or hitch
- Bight: the portion of the rope that is actually in the knot or hitch
- Standing rope: the load-carrying portion of the rope

The extreme end of a rope should be bound with string or wire so that it cannot unravel. This is called *whipping*. See Figure 21-7. Whipping not only keeps laid rope from unraveling but also secures the outer jacket to the inner core on braided rope.

Knots

1. One of the most useful knots to know is the bowline. The bowline knot forms a loop at the end of the rope. This knot has the advantage of being easy to untie and the loop will not slip. The following are the steps to tie a bowline. Use the end of the rope to make an overhand loop onto the standing part. Allow enough rope to make the size of loop required. A loop formed in a rope is said to be underhand if the end goes below the standing part. It is said to be overhand if the end goes above the standing part.
2. Feed the end of the rope up from beneath the loop.
3. Pass the end of the rope around the backside of the standing part of the rope.

FIGURE 21-7 Steps for whipping a rope.

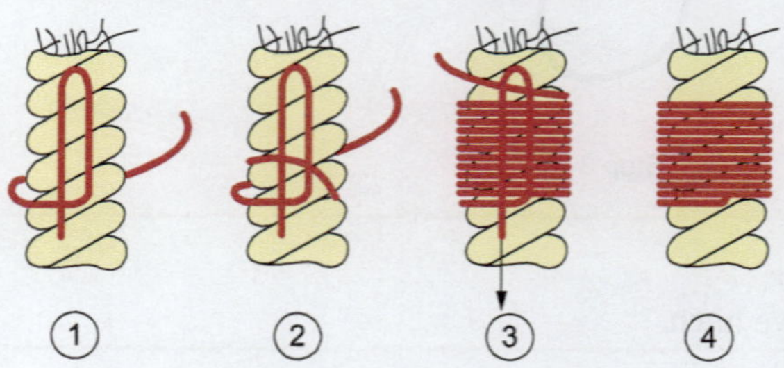

1. Form a loop with the whipping rope and lay it along the rope.

2. Wrap the whipping rope tightly around the rope, making the wraps close together.

3. Feed the end of the whipping rope through the end of the loop and pull the lower end. The loop will pull the loose end under the wraps to secure them.

4. Trim the loose ends.

4. Pass the end of the rope back down through the original loop. While holding the end of the rope and the second loop, pull on the standing end. The knot should tighten to form a secure loop.

Caution: Most people make the mistake of losing track of the first loop for the knot and the second loop. See Figure 21-8 for an illustration of these steps.

Hitches

Knots support, whereas hitches grip. This means that knots remain tied even if the load is slack, but hitches require the pull of the load to remain tight and secure. Hitches are used to connect to a load where it is important to release the hitch quickly, and so the rope can be immediately reused at or near the same place for the next load. Thus, hitches are temporary (for this load only) and knots usually are meant to be in place for some time.

Two of the more popular hitches are the clove hitch and the timber hitch. The clove hitch (see Figure 21-9) is the best hitch to use when the rope is being pulled at an angle to the object. The timber hitch works extremely well when the rope is pulled perpendicular to the object, such as in Figure 21-10.

Steps for Tying a Clove Hitch

1. Pass the end of the rope around the load and over the standing part.
2. Pass the end around the load again.

FIGURE 21-8 Steps for tying a bowline.

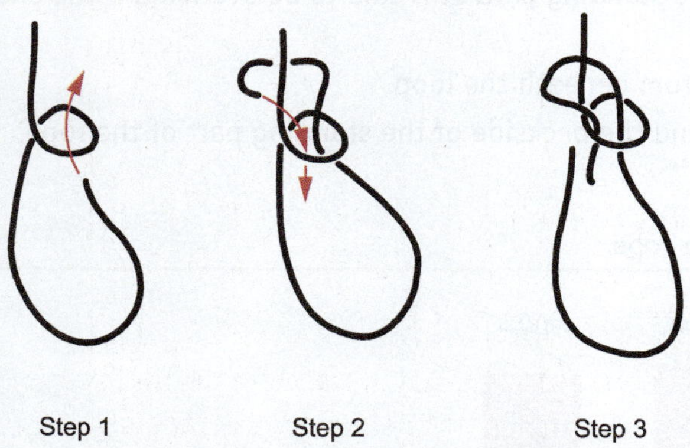

FIGURE 21-9 Steps for tying a clove hitch.

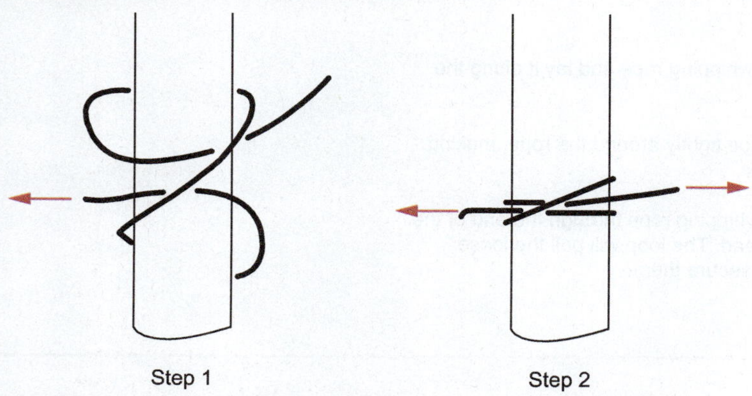

FIGURE 21-10 Steps for tying a timber hitch.

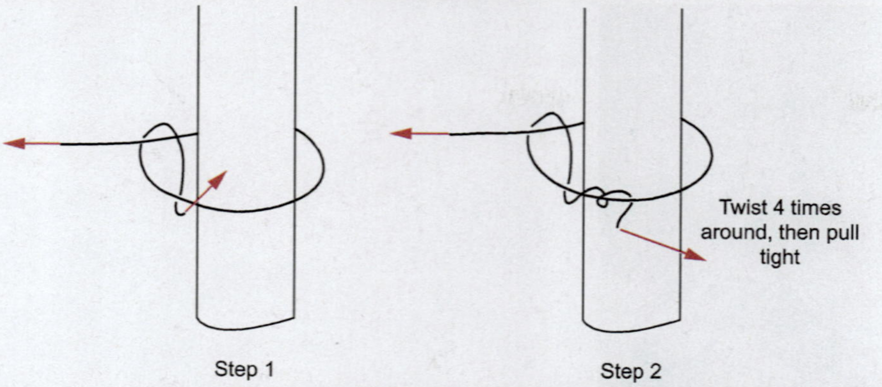

3. Feed the end under the bight.
4. Pull tight. The standing rope and the end should be parallel to each other. (See Figure 21-9.)

Steps for Tying a Timber Hitch

1. Pass the rope around the load.
2. Pass the end around the standing rope and then wrap it three to four times back around itself.
3. Pull tight so that the end of the rope is now trapped between the load and itself. (See Figure 21-10.)

Grasping Loads with Hitches

In addition to the clove hitch and the timber hitch, a number of hitches can be made with wire rope, nylon slings, or fiber rope with loops at the ends. Each has its own advantages and disadvantages. Some are better at grasping the load; others give larger lifting capacity, better load stability, and ease of use.

Single Leg Vertical Hitch

The single leg vertical hitch is the simplest to connect and calculate the lift capacity. As shown in Figure 21-11, it involves a single point of contact with the object. In this figure, a shackle is used to join the lifting eye of the sling to the eye bolt on the motor. Since there is only one sling supporting the object, the weight of the object is equal to the lifting tension in the sling used. The disadvantage to this lifting arrangement is that the load must be perfectly balanced.

Choker Hitch

The choker hitch will tighten around the object as it is being lifted. This is a great advantage when lifting a bundle of pipe or other items. It does, however, reduce the lift capacity of the sling to about $\frac{3}{4}$ of its rating. This reduction is due to the sharp bend in the sling. See Figure 21-12.

Basket Hitch

The basket hitch forms a cradle that the object lies in. Although this hitch is simple and doubles the lifting capacity of the sling, it should not be used on objects that are difficult to balance. This hitch does not grasp the load. See Figure 21-13.

FIGURE 21-11 Single leg hitch used to lift electric motor.

Photo by Ed Moore

FIGURE 21-12 Choker hitch grips the load better but has reduced capacity.

Photo by Ed Moore

FIGURE 21-13 Basket hitch does not grip the load.

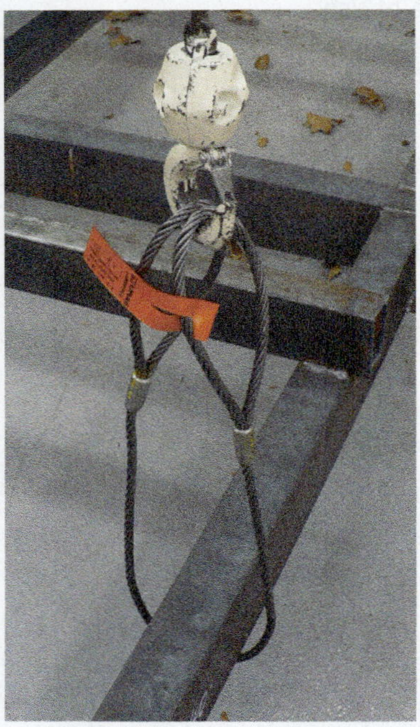

Photo by Ed Moore

Bridle Hitch

A bridle hitch can be formed by both a steel lifting ring and two or more lifting slings that are permanently attached, or by a shackle and two independent slings. The bridle hitch allows for two or more lifting points to be connected to one hook for lifting. Problems arise when the two connecting points become separated. The wider the lifting points, the more the two slings are trying to crush the part than lift it. As a general rule, you never want the lifting points any farther apart than the length of the slings. The following formula allows for the calculation of the tension in each of the lifting legs.

$$T = \frac{W}{N} \times \frac{L}{V}$$

T = tension in each sling (in pounds)
W = weight of the load (in pounds)
N = number of slings used
L = length of the slings used (in feet or inches)
V = vertical height above the lifting points where the slings join (in feet or inches)

EXAMPLE 1

What would the tension in the slings be if a 1,000-pound load was being lifted by two 10-foot slings and the lifting hook connection was 9 feet above the lifting points? See Figure 21-14.

$$T = \frac{W}{N} \times \frac{L}{V}$$

$$T = \frac{1{,}000 \text{ lbs.}}{2} \times \frac{10'}{9'}$$

$$T = 500 \times 1.11 = 555 \text{ pounds}$$

If the lifting points were spread farther apart, the vertical component of the equation would decrease. The following example illustrates how a wide sling angle greatly increases the tension in the slings and thereby reduces the lifting capacity.

FIGURE 21-14 Load distribution for Example 1.

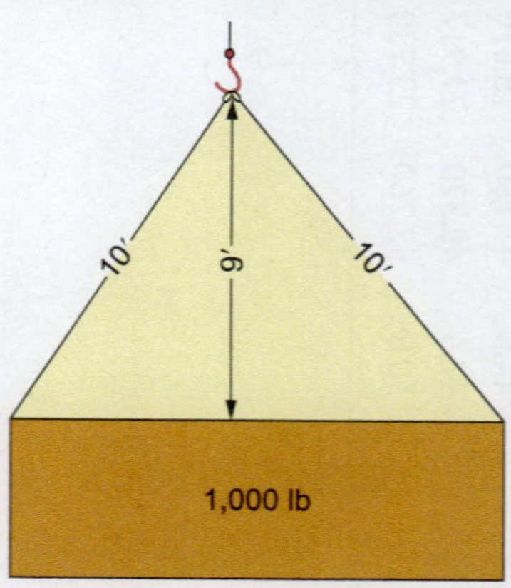

TABLE 21-2 Load Capacities for Various Hitch Types

Flemish Eye Loop		Vertical Lift		Choker Hitch		Basket Hitch		Two-Part Bridle Sling (30°)		Two-Part Bridle Sling (60°)		Two-Part Bridle Sling (90°)		Two-Part Bridle Sling (120°)	
Inches	mm	Pounds	kg	Pounds	kg	Pounds	kg	Pounds	kg	Pounds	kg	Pounds	kg	Pounds	kg
$\frac{1}{4}$	6.4	920	417	700	318	1,840	835	1,780	807	1,600	726	1,300	590	920	417
$\frac{3}{8}$	9.0	2,020	916	1,520	690	4,040	1,833	3,900	1,769	3,500	1,587	2,860	1,297	2,020	916
$\frac{1}{2}$	12.7	3,740	1,696	2,800	1,270	7,480	3,393	7,220	3,275	6,480	2,939	5,280	2,395	3,740	1,696
$\frac{5}{8}$	16.0	5,600	2,540	4,200	1,905	11,200	5,080	10,840	4,917	9,700	4,400	7,920	3,592	5,600	2,540
$\frac{3}{4}$	19.0	8,080	3,665	6,060	2,748	16,160	7,330	15,640	7,094	14,020	6,359	11,460	5,198	8,080	3,665
$\frac{7}{8}$	22.2	10,920	4,953	8,180	3,710	21,840	9,906	21,100	9,571	18,920	8,582	15,480	7,022	10,920	4,953
1	25.4	14,180	6,432	10,680	4,844	28,360	12,864	27,400	12,430	24,600	11,158	20,060	9,099	14,180	6,432
$1\frac{1}{8}$	28.6	16,660	7,557	12,500	5,670	33,320	15,114	32,300	14,651	28,900	13,109	23,500	10,660	16,660	7,557
$1\frac{1}{4}$	31.8	20,740	9,407	15,540	7,049	41,480	18,815	40,120	18,198	36,000	16,330	29,300	13,290	20,740	9,407
$1\frac{3}{8}$	35.0	25,340	11,500	19,000	8,618	50,680	22,988	49,000	22,226	43,880	19,904	35,840	16,257	25,340	11,494
$1\frac{1}{2}$	38.0	30,620	13,889	22,960	10,415	61,240	27,778	59,500	29,989	53,040	24,058	43,300	19,640	30,620	13,889
$1\frac{5}{8}$	41.3	35,900	16,284	26,920	12,211	71,800	32,568	70,000	31,751	62,180	28,204	50,760	23,025	35,900	16,284
$1\frac{3}{4}$	44.5	41,160	18,670	30,860	13,998	82,320	37,340	80,000	36,287	71,280	32,332	58,200	26,400	41,160	18,670
$1\frac{7}{8}$	47.6	48,320	21,917	36,240	16,438	96,640	43,833	93,600	42,456	83,860	37,956	68,320	30,990	48,320	21,918
2	50.8	52,760	23,930	39,560	17,944	105,520	47,863	102,400	46,448	91,380	41,450	74,600	33,840	52,760	23,932

FIGURE 21-15 Load distribution for Example 2.

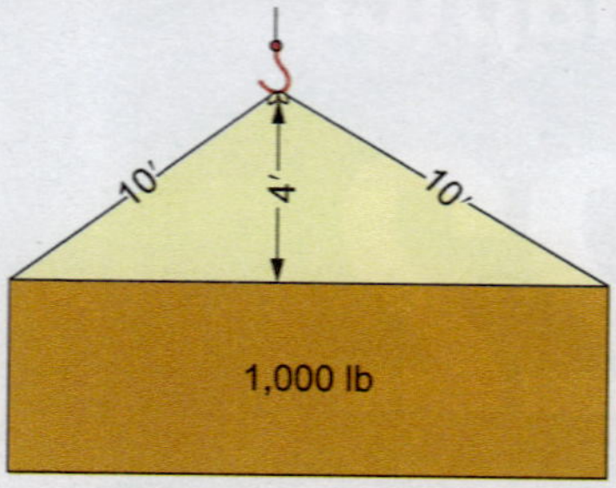

EXAMPLE 2

What would the tension in the slings be if a 1,000-pound load was being lifted by two 10-foot slings and the lifting hook connection was 4 feet above the lifting points? See Figure 21-15.

$$T = \frac{W}{N} \times \frac{L}{V}$$

$$T = \frac{1{,}000 \text{ lbs.}}{2} \times \frac{10'}{4'}$$

$$T = 500 \times 2.5 = 1{,}250 \text{ pounds}$$

This demonstrates that you cannot always expect the load on the slings to be one-half of the weight when calculating the lifting capacity. Table 21-2 shows some load capacities for wire rope slings used in various hitches, but always check with the manufacturer of the equipment you are using before calculating a lift.

REVIEW QUESTIONS

True or False

1. ________ A choker hitch has less lifting capacity than a basket hitch.
2. ________ The wider the angle between the legs of the bridle hitch, the lower the lifting capacity.
3. ________ A web sling gets its strength based on its width, not its length.
4. ________ A basket hitch has twice the lifting capacity of a vertical leg hitch.
5. ________ Knowing the weight of the object is all that is needed to make a safe lift.

Chapter 22

LEARNING OBJECTIVES

The student will:

- Summarize an organized approach and methods for safe material moving on the job.
- Describe types of ladders, scaffolds and scaffolding methods, and related safety practices.
- Define what constitutes a confined space and discuss confined space safety practices.
- Define what constitutes a trench and discuss trench safety practices.

Safety Concepts

Safety practices for the construction industry are regulated by the federal Occupational Safety and Health Administration (OSHA). States and territories may elect to develop their own unique occupational safety and health programs. If so, these state plans are approved and monitored by OSHA, which provides up to 50 percent of an approved plan's operating costs. A state plan, including the job safety and health standards that employers are required to meet, must be at least as effective as OSHA standards. Benefits of a state plan include coverage for public sector employees and the opportunity to create unique standards or to develop innovative programs that address the types of hazards specific to each state's workplaces.

It is highly recommended that you receive safety training from OSHA-certified instructors on the topics covered in this chapter. A full listing of OSHA training courses can be found on OSHA's website at http://www.osha.gov/dte.

PREVENTING INJURY AND LOSS

Nobody wants accidents to happen. A permanent injury or the death of an employee can cause the loss of an entire business. To prevent such losses, the company may not have to spend a lot of money on training and good equipment. All employees need to use good common sense and apply recognized prevention principles.

There are reasons why accidents happen, but they boil down to this: something goes wrong somewhere. It may take some thought and perhaps the help of trained safety professionals to figure out what went wrong, but an accident always has a cause. Once you know why an accident happened, it is possible to prevent future incidents. You need some basic facts and a plan with guidelines to prevent repeated accidents.

SAFETY

Safety is the first requirement of any of our job activities, and this is an especially demanding requirement when we are working to move heavy loads at the jobsite. Safety and using safe practices may seem to be common sense to some workers, but remember that it is too late after someone is injured or material is damaged to make up for a careless or reckless action!

Safe practices minimize risk, danger, and chance of injury. The safe worker, therefore, will have these two important traits to enhance safety:

- The ability to recognize, avoid, and minimize hazards at all times
- A sense of priority to plan for safety

Safety consciousness is a term that really means awareness. The worker who is aware of himself and others will concentrate on these items when rigging and hoisting:

- Safety awareness means that you keep your mind on the job at hand. Do not permit yourself to daydream about other things while the lifting is going on!
- Human nature is such that we all respond to positive ideas better than negative ones. When discussing safety with coworkers, stress the positive aspects of safety. Rather than concentrating on death and injury, talk about completing the task on or ahead of schedule; discuss concerns for materials properly placed ready for the next job phase; or address the need for all the workers to be available for the next job phase.

Safety is economical for all concerned. It minimizes costs for everyone: the worker, the employer, the building owner, and the community at large.

SAFETY GUIDELINES FOR HOISTS AND CRANES

Listed below are general guidelines for safe operation of hoists and cranes. Your employer may require additional safety guidelines to meet its specific safety needs.

The proper installation, operation, testing, and maintenance of cranes and hoisting devices are a continuing responsibility of the owner and user. All hoists and cranes should be inspected as required by OSHA guidelines. This includes annual as well as daily pre-use inspections. These inspections should be documented, signed, and dated. Special attention should be paid to load hooks, ropes, brakes, and limit switches.

- The safe load capacity of each hoist should be clearly posted on the hoist body.
- All employees working with hoisting apparatus should be trained on safe lifting/rigging practices and operating rules. The operator is responsible for compliance with safety procedures and for maintaining safe operating conditions of the lifting equipment.
- A load should be lifted only when it is directly under the hoist.
- All hoists should be attached to their supports and have adequate capacity for the maximum loads to be hoisted.
- All lifting hooks must have operating safety latches.
- All slings must be inspected prior to use.
- Each control cord should be clearly marked "hoist" or "lower."
- Equipment should be kept away from energized power lines.

- When a crane is being used, standard hand signals should be posted at the site. Employees operating the crane should be trained in the hand signals as per construction industry guidelines.
- Only trained and certified employees should be allowed to operate any hoisting or crane device.

FORKLIFT SAFETY

Forklift operation requires a special certification. OSHA Standard 1910.178 requires that employers certify that operators of forklifts have been trained and their skills evaluated on the equipment that they will be using on the jobsite with equipment specific to that job. Proof of certification must be provided and must include the name of the operator, the date of training, the date of evaluation, and the trainer/evaluator's name. Certification from one company cannot be transferred to another company.

Forklifts can haul and dump tubs of material, carry containers of molten metal, and transport pallets of heavy products. Forklifts can be dangerous to people and property when operated incorrectly. Suggested requirements for drivers are: above-average vision, hearing, and health; a mature attitude; a good vehicle driving record; a positive safety attitude; and completion of a forklift operator training course every 2 years.

Following are some general safety guidelines for forklift operation:

- When carrying a load, drive up a ramp or grade. Never drive forward down a ramp or grade when you are carrying a load. Never make a turn while your forklift is on a ramp. Lower the forks to keep the center of gravity low.
- Always use a proper dock board when loading a vehicle from a dock. Keep the forklift away from the edge of the loading dock.
- Make sure the parking brake is set and wheels are chocked on the vehicle being loaded.
- Place the forks all the way under the load. Space forks apart so they fit the load being lifted. This will help to maintain proper balance and prevent the load from falling. Never lift a load that appears to be unstable. Use belts to secure the load onto the forks.
- Center the forks beneath the load being lifted. Lifting an off-center load can cause the load to fall. Tilt the uprights slightly back when raising and carrying a load.
- Do not carry any riders unless the forklift is specifically designed for them. Always keep hands and feet inside. Never speed or allow unauthorized persons to drive a forklift.
- Never smoke when refueling or when checking the battery of a forklift. Always turn off the engine when refueling.
- Use a properly secured safety platform when the truck is to be used as a lifting device.
- Never carry loads that obstruct your view.

- When the forklift is parked, fully lower the forks, put the controls in neutral, turn off the engine, set the parking brake, and remove the key.
- When turning, reduce your speed and maneuver carefully.
- Stay a safe distance away from other forklifts. Never drive side by side.
- At blind corners, stop the forklift and sound the horn.
- Know where low clearances, pipes, sprinklers, or low doorways are located.
- Completely inspect the forklift prior to any operation of the unit. If you find anything wrong, report it to your supervisor or maintenance department.

Modifications and additions to forklifts that affect the capacity or safe operation of the vehicle should not be performed by the owner or user without the manufacturer's prior written approval. Any modifications or additions alter the geometry and physical forces applied to lifting. Serious consequences can and have occurred. Examples of prohibited modifications are attachments that extend the length of the forks or adding counterbalances to the rear. Additional precautions are also needed when using a forklift as a personnel hoist.

SAFE RIGGING PRACTICES

The following are important safety practices for rigging:

- Cable clips should be installed in accordance with standards listed on their packaging or in the manufacturer's instructions.
- The weight of the load should be determined to select the proper size of choker.
- Sharp edges of the material to be rigged should be protected to prevent damage to the choker.
- Softeners (padding) should be used to prevent material damage.
- Tag lines should be used when hoisting and rigging loads. A tag line is used only for steering the load, not lifting or balancing.
- Material or equipment rigging should not be rigged from structural points that are unstable (such as unfinished work, handrails, or conduit).

LADDERS

Ladders usually consist of two side rails joined at regular intervals by crosspieces called steps, rungs, or cleats, on which a person may step when ascending or descending. There are variations called straight ladders, extension ladders, and step ladders, as well as a variety of specialty ladders designed for specific uses. The three basic ladders used in plumbing work are described in detail below.

Straight Ladders

Straight ladders range in length from 4 feet to 15 feet. They are made with straight sides (of wood, fiberglass, or metal) with rungs, usually 12 inches on center, between them. The rungs may vary from 12-inch wide for short ladders to 20-inch to 30-inch wide for long ladders. In larger sizes, wider runged ladders become very heavy, so they are difficult to move about or transport. Straight ladders must be leaned against a wall for support. Whenever the ladder is going to be used as a means to access an elevated work area, it is good practice to secure the ladder to the structure.

Extension Ladders

Extension ladders are made up of two or more straight ladders that are nested together and arranged so that one member can slide over the other to make a ladder of adjustable length. They are available in lengths that will reach up to 40 feet. In larger sizes, they also are very difficult to transport and set up. Extension ladders also have to be leaned against a support. Figure 22-1 shows an extension ladder on the left side and a straight ladder on the right side of the drawing. Notice how they must be set up at the correct angle to be safe.

Step Ladders

Step ladders are made of two pieces hinged together at the top. When the two halves are folded together, they are convenient to carry or transport. When opened, the triangular form is fairly stable (see Figure 22-2). These ladders are self-supporting, so it is possible to use them in the middle of a room. They are available in lengths from 2 feet to at least 14 feet in height.

FIGURE 22-1 Straight and extension ladders must be properly angled to avoid slippage or falling.

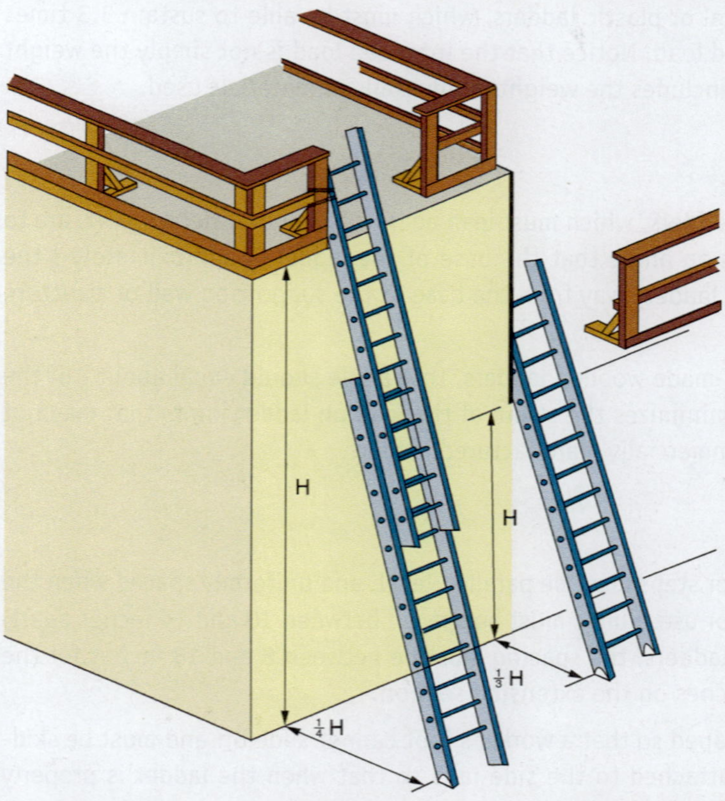

FIGURE 22-2 Step ladders must be opened and locked to perform correctly.

Photo by Ed Moore

OSHA GUIDELINES FOR LADDERS

Loads

Self-supporting (foldout) and non–self-supporting (leaning) portable ladders must be able to support at least four times the maximum intended load, except extra–heavy-duty metal or plastic ladders, which must be able to sustain 3.3 times the maximum intended load. Notice that the intended load is not simply the weight of the worker. It also includes the weight of any tools or materials used.

Angle

Non–self-supporting ladders, which must lean against a wall or other support, are to be positioned at such an angle that the base of the ladder is approximately $\frac{1}{4}$ the working length of the ladder away from the base of the supporting wall or structure (see Figure 22-1).

In the case of job-made wooden ladders, that angle should equal about $\frac{1}{8}$ of the working length. This minimizes the strain of the load on ladder joints that may not be as strong as on commercially manufactured ladders.

Rungs

Ladder rungs, cleats, or steps must be parallel, level, and uniformly spaced when the ladder is in position for use. Rungs must be spaced between 10 and 14 inches apart. For extension trestle ladders, the spacing must be between 8 and 18 inches for the base, and 6 and 12 inches on the extension section.

Rungs must be shaped so that a worker's foot cannot slide off and must be skid-resistant. Rungs are attached to the side rails so that when the ladder is properly placed, the tread will be parallel to the ground.

IN THE FIELD

Most step ladders are intended to support only one person. The horizontal cross members on the opposite side are *not* steps.

Slipping

Ladders are to be kept free of oil, grease, wet paint, and other slipping hazards. Wood ladders must not be coated with any opaque covering, except identification or warning labels on one face only of a side rail.

Other Requirements

Fold-out or step ladders must have a metal spreader or locking device to hold the front and back sections in an open position when in use.

When two or more ladders are used to reach a work area, they must be offset with a landing or platform between the ladders. The area around the top and bottom of the ladder must be kept clear. Ladders must not be tied or fastened together to provide longer sections unless they are specifically designed for such use. Never use a ladder for any purpose other than the one for which it was designed. Figure 22-3 shows a very common practice that is dangerous; ladders should never be used as a walk board.

FIGURE 22-3 Never use a ladder as a walk board. This can damage the ladder and cause personal injury and property damage.

Proper Selection of Ladders

The following are three basic guidelines for the proper selection of ladders:

- Select a ladder of proper duty rating to support the combined weight of user and materials.
- Select a ladder of proper length to safely reach the desired height.
- Select nonmetal ladders for use on or around exposed electrical elements.

Inspection Before Each Use

Prior to each use of a ladder, the user should follow an inspection procedure, which includes the following:

1. Inspect thoroughly for missing or damaged components. Never use a damaged ladder and never make temporary repairs.
2. Inspect thoroughly for loose fasteners. Make sure all working parts are in good working order. Lubricate if necessary.
3. Clean ladder of all foreign material (wet paint, mud, snow, grease, oil, etc.).
4. Destroy ladder if damaged, worn, or exposed to fire or chemicals. Bring the ladder to the shop and tag for inspection. Put a note on your daily report, and management will make the decision regarding destruction.

Ladder Accidents

Accidents on ladders may arise from the following causes:

- Overreaching
- Standing on the top rung or step
- Failure to secure the ladder
- Defective ladder
- Carrying large loads while climbing the ladder
- Using a metal ladder near electrical circuits or wires

All of the above causes could be called judgment errors attributed to lack of safety awareness or foresight. Planning before the job starts can help avoid these situations.

Before each use of a ladder, consider the following:

- Metal ladders conduct electricity. Keep away from electrical circuits or wires.
- Consult manufacturer for use in chemical or other corrosive environments.
- Use the ladder only as outlined in instructions. Ladders are designed for one person only.
- Do not use in high winds or during a storm.
- Keep shoes clean. Leather shoes should not be used.
- Never leave a ladder set up and unattended.

Proper Ladder Setup and Use

- Get help in setting up a ladder if possible.
- Do not place on unstable, loose, or slippery surfaces. Do not place in front of unlocked doors. Ladders are not intended to be used on scaffolds.
- Secure the base section before raising the ladder to the upright position. Do not raise or lower the ladder with the fly section extended.
- Extend and retract fly section only from the ground when no one is on the ladder.
- Do not overextend the ladder. A minimum overlap of sections is required as follows:
    - Ladder size up to and including 32 feet—3-foot overlap
    - Over 32 feet up to and including 36 feet—4-foot overlap
    - Over 36 feet up to and including 48 feet—5-foot overlap
    - Sizes over 48 feet—6-foot overlap
- Position the ladder against an upper support surface. Make sure the ladder does not lean to the side. The ladder must make a 75° angle with the ground.
- Erect the ladder approximately 3 feet beyond the upper support point.
- Check that the top and bottom of the ladder are properly supported. Make sure rung locks are engaged before climbing.
- Face the ladder when climbing up or down. Maintain a firm grip. Use both hands in climbing.
- Keep your body centered between the side rails. Do not overreach. Get down and move the ladder as needed.

Proper Ladder Care and Storage

When not in use, ladders should hang on racks with supports every 6 feet. Wooden ladders should be treated with wood preservative; however, they should never be painted. Protect wooden ladders from exposure to the elements but allow good ventilation. Keep wooden ladders away from heat and moisture.

SCAFFOLDS

On most jobs, piping and/or equipment must be installed above floor level. Generally, you will find that ladders are inadequate to support material or to provide a safe surface for workers performing maximum-effort tasks. For such tasks, it is necessary to use a work platform called a scaffold at the place where the work is to be done.

Pre-Manufactured (Bridgework)

Pre-manufactured scaffolding can be made from standard manufactured frames assembled into the required structure or made from lumber components to provide the required work surface. The pre-manufactured scaffold system is usually more economical and safer, as industry standards clearly define load limits and safe methods of assembly and support. Pre-manufactured scaffolds can be purchased or rented in a variety of styles and sizes. The basic assembly is an end frame and diagonal braces connected to the next end frame. This system can be repeated vertically and laterally as required to provide a scaffold structure of any extent.

Accessories

Optional equipment available for scaffolds includes wheels, base plates, extension jacks, corner braces, side arm supports, floor planks, and guard rails. The end frames (and diagonals) are made of tubular steel or aluminum and can be connected together with pins for rapid assembly. The frames are made in units 30 inches wide for scaffolds that can be moved through standard doorways and 60 inches wide for general-purpose, heavy-duty applications. Standard frame height is 5 feet, but many other heights are available for a variety of purposes.

The work surface is usually assembled with 2″ × 10″ scaffold-grade planks. The complete top of the scaffold should be planked to minimize risk. The underside of the planks should be cleated to be sure that the planks cannot slide off the frame. Do not use planks that overhang the frames by more than 6 inches.

Guidelines for Safe Use of Scaffolds

Scaffolds, by their very nature, present a danger of falling or being struck by something falling. Because this possibility exists, certain safety precautions must be kept in mind when working on or around scaffolds:

- When erecting a scaffold, be sure it is capable of supporting at least four times the maximum load, including the weight of materials, workers, and the scaffold itself. The height must not exceed four times the minimum base dimensions as well. Footings should be sound and rigid.
- Check the scaffolding for damage prior to use. Damaged scaffolding should not be used.
- Planking should be at least 2″ × 10″ boards, of scaffold grade, placed together to help keep materials and tools from falling. Choose planks that are straight-grained and free of shakes, large or loose knots, and other defects. Extend the planks beyond the center line of supports from 6 to 12 inches, and cleat or otherwise fasten so the planking stays in place.
- Always use a safe means of access when climbing a scaffold, such as a fixed or portable ladder, ramp, runway, or stairway. Climbing on cross braces is never acceptable.
- While using a mobile scaffold, be certain to lock the wheels before beginning use. Do not ride or allow anyone to ride on scaffolding while it is being moved unless the scaffolding is constructed of a specific alloy designed for occupied horizontal travel. All material and equipment should be removed or secured before moving the scaffold. Do not try to move a rolling scaffold without sufficient help. Be aware of holes in floors and overhead obstructions.
- While working on a scaffold, do not allow tools and materials to accumulate in a manner that creates a hazard.

- While working on a scaffold 10 feet or more above the ground, it must be equipped with guardrails, including a toe board. Wear a safety belt and life line anytime you are 6 feet or more above the floor. When working near overhead electrical power lines, a minimum of 10 feet of clearance must be maintained. Clearance will increase depending on voltage.
- Always wear hard hats and other appropriate personal protective equipment.

Scaffold Hazards

Accident records show that the following are the most common causes of accidents involving ladders and scaffolds:

- Overextending or overreaching
- Horseplay
- Jumping off the ladder or scaffold
- Failure to build level and plumb
- Overloading with tools or materials
- Standing on an unsupported extension of the platform
- Using the ladder or scaffold for purposes other than intended
- Failure to prevent swaying
- Failure to secure platform or side rails

Scissor Lifts and Power Scaffolds

Scissor lifts and power scaffolds are designed to be of a minimum size when retracted, for greatest range of applications in crowded aisles, passing through small doors, and/or clearing minimum headroom obstructions, combined with maximum possible extended height and lifting capacity. These devices may be gas- or electric-powered. Scissor lifts and power scaffolds range from small battery-powered models with dimensions that allow them to pass through 3-foot doors, with 300 pounds of lifting capacity to 15-foot or 20-foot height, to models that can reach to 100 feet and others that can lift a ton.

Power scaffolds or cherry pickers are based on a rotating single boom rising from a massive base. These units permit the worker on the platform to reach objects that are well outside the space immediately above the base. One very large machine of this type can reach 80 feet from the center of the base, with a capacity of 600 pounds. It is important to know that its load capacity decreases the farther it extends and when at a low angle to the horizon.

Scissor lifts are designed like a pair of scissor arms linked together (see Figure 22-4). These hoists can lift a greater load, but the working area has to be directly above the machine.

As in many other things, the problems that must be solved on the job determine the model that should be used.

On sites where the work requiring a scissor lift or power scaffold is not on ground level, be sure the building is rated for the additional weight of this equipment. In addition, be sure there is adequate ventilation if a gas-powered device is used.

FIGURE 22-4 Scissor lifts are a great time-saver.

© Asier Romero/ShutterStock.com

These products are expensive, but they are available for rent (hourly, daily, weekly, etc.), so they can produce very efficient performance in the appropriate job situation, without incurring exorbitant costs.

Summary

The important concept to be aware of in scaffold work is that you and your coworkers are above the ground or floor level. Thus, you, the equipment, and the tools you are working with are defying the law of gravity when you are on the work platform. Care in performing the work must be observed at all times to avoid personal injury or equipment damage.

Remember that a scaffold is erected to provide a safer place to work than ladders, but scaffolds still involve dangers that must be carefully considered and planned for. With such preparation and with careful attention to details, scaffold work can be safe, efficient, and productive.

TRENCHING AND SHORING

Consult and understand OSHA regulations and safety procedures prior to any trenching and excavation work. There is considerable danger at all times when working in a trench. Three simple facts about safety *must* be realized when doing trench work.

- A person does not have to be totally buried to suffocate. One must only be buried to chest level to suffocate; if the chest and lungs cannot expand, breathing will stop.
- A person can live only minutes without breathing. If breathing is hindered for even a few minutes, the person may suffer brain damage.
- No job is so important that safety can be ignored.

The following section will address some of the hazards most associated with trench work, which are as follows:

- Cave-ins
- Materials (tools, pipe, or debris) falling into the trench
- Machinery falling into the trench
- Debris falling from digging machinery
- Power line hazards
- Unprotected jobsites that jeopardize the safety of passersby

IN THE FIELD

OSHA defines a "competent person" as "one who is capable of identifying existing and predictable hazards in the surroundings or working conditions which are unsanitary, hazardous, or dangerous to employees, and who has authorization to take prompt corrective measures to eliminate them" [29 CFR 1926.32(f)]. Through training and/or experience, a competent person, knowledgeable of applicable standards, is capable of identifying workplace hazards relating to the specific operation and has the authority to correct them. Some OSHA standards add additional specific requirements that must be met by the competent person.

Cave-Ins

The degree of cave-in potential relates to the excavation depth, soil condition, the ditch and/or shoring design, and the weather conditions. The following are the two most common methods of preventing cave-ins.

Angle of Repose

One method to avoid cave-ins is to slope trench walls equal to or less than the angle of repose. By definition, the angle of repose is *the greatest angle above the horizontal plane at which a material will lie without sliding*. These angles were calculated based upon studies relating to the following factors:

- Depth of excavation
- Possible water content variation of the soil during ditch exposure
- Possible change of the stability of the ditch wall or excavated material due to exposure to the elements during excavation (heat or freezing)
- Loading of additional material on the excavation walls (shoring materials, pipe, excavated soil, etc.)
- Vibration from machinery, blasting, or traffic

Figure 22-5 shows the concept of angled sides. Unfortunately, there are some problems with using sloped sides with trench work:

- The amount of earth to be moved can be much greater than with the shoring method.
- The amount of space required between obstructions is greater than with the shoring method.

As the soil becomes softer or less able to stick together, the angle of repose becomes larger. This excavation method requires very large trenches for many common earth materials. Therefore, if conditions are crowded, this method cannot be used economically. *Plumbing 301* goes into more detail on how to determine the soil type and the appropriate slope.

Shoring

An alternative to angle of repose excavation is shoring. Any shoring system must be engineered by a competent person. Shoring involves the temporary placement of supporting members to hold the trench walls in place. Shoring retains the walls of earth, which would otherwise collapse during trenching operations. Consult OSHA requirements.

Shoring may be performed using one of the following methods:

- **Timber bracing:** Wooden stanchions placed intermittently are supported by wooden braces.
- **Light timber sheathing:** Continuous, close-set wooden stanchions are held by wooden cross-members.
- **Trench box or shield:** A box of wood or metal is moved along the trench as the pipe is placed. See Figure 22-6.

FIGURE 22-5 The angle of repose for the soil dictates what the slope of the sides must be.

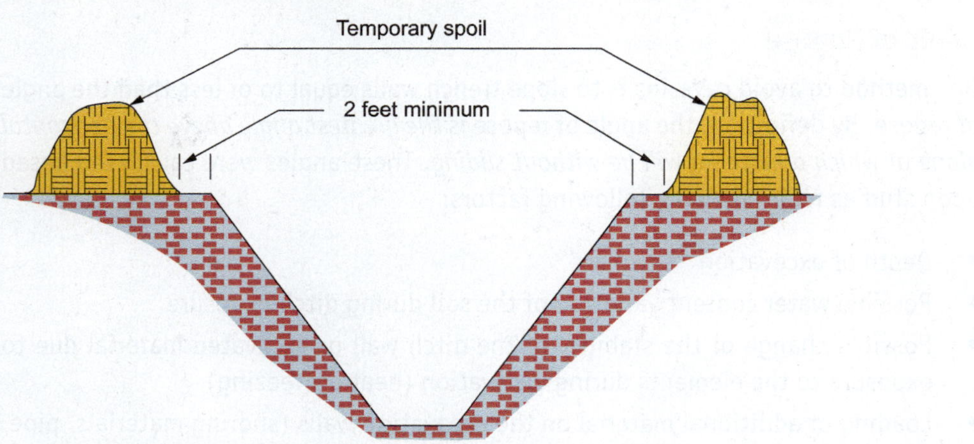

FIGURE 22-6 Trench shields support trench walls.

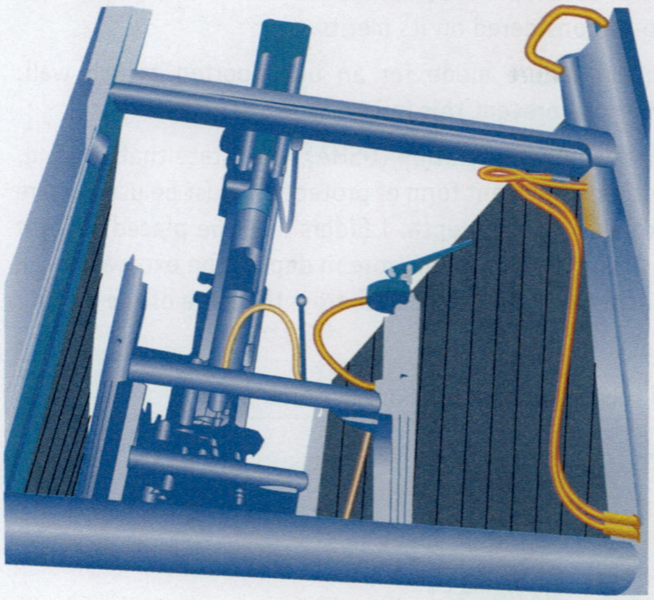

- **Hydraulic shoring:** Hydraulic jacks, usually made of aluminum, are activated by a hydraulic pump. The wall supports and jacks may be one unit.
- **Screw jack:** Metallic screw device with a foot support on each end holds timber boards against the ditch face. See Figure 22-7.
- **Sheet piling:** Sheathing that forms a continuous support wall, or a row of timber, concrete, or steel pilings are driven below the bottom of the excavation to assure a tightly assembled wall to resist earth pressure.

FIGURE 22-7 Adjustable jacks support the sides of trench walls and prevent cave-ins.

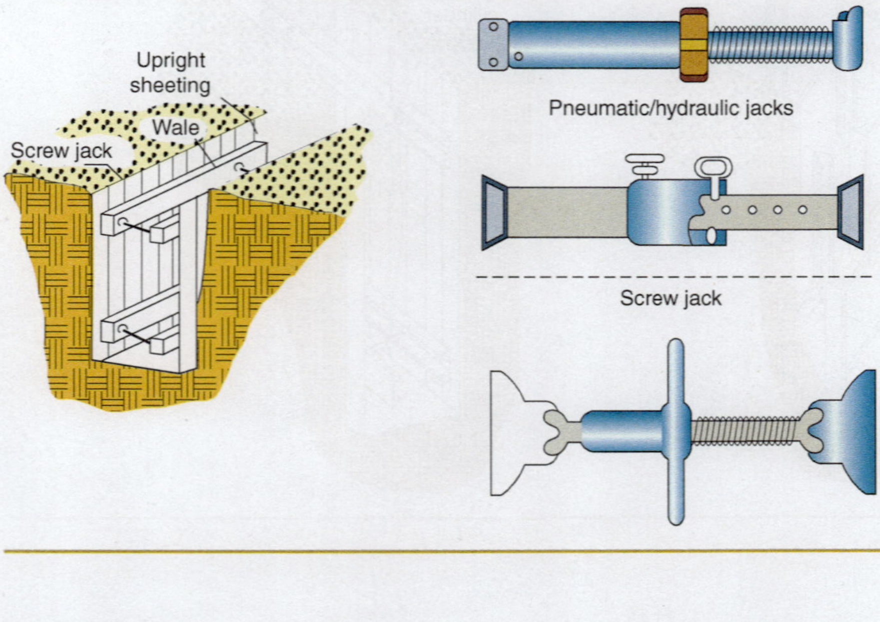

A fairly shallow trench in open country might be dug using angle of repose walls, whereas a deeper trench or one in built-up areas where space is restricted would have to be shored. Each case has to be considered on its merits.

Figure 22-8 shows a typical failure mode for an unsupported trench wall. Figure 22-9 shows supports that can prevent this failure type.

Occupational Safety and Health Administration (OSHA) rules state that shoring, sloping of the ditch, or some other equivalent form of protection must be used where the walls of an excavation exceed 4 feet in depth. Ladders must be placed at least every 25 feet of lateral travel for ditches 4 feet or more in depth. The excavated soil or spoil material must be deposited at least 2 feet back from the edge of the ditch.

FIGURE 22-8 Typical sidewall failure occurs at the bottom of the trench.

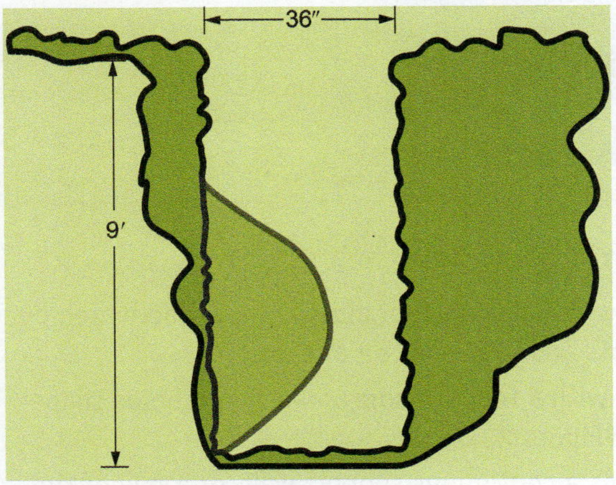

FIGURE 22-9 Proper supports for trench walls.

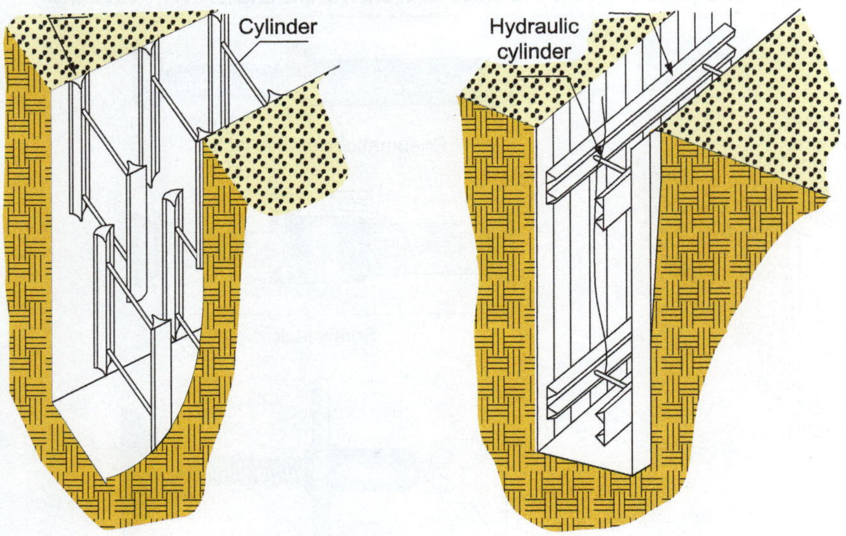

Materials and Tools Falling into the Ditch

Any pipe or tool used on the job should be placed at least 2 feet away from the edge of the ditch. Anyone who has ever dug a hole realizes that when objects are placed too close, they eventually fall in. For this reason, all workers should be wearing hard hats to protect against debris or tools that may be dropped into the trench. Also, when pipe is transported into the trench, it should not be dropped or rolled into the trench, but lifted and properly supported so that the spoil material will not topple into the excavation. When necessary, arrange for additional assistance if a tool or a pipe is too heavy for the worker's capabilities. This eliminates the possibility of dropping the item and endangering the trench worker's life. Remember that you should not stack material by the edge of the ditch. Transport pipe only when it is required.

Excavated Debris Falling into the Trench

All materials and debris should be located at least 2 feet away from the trench edge. This will prevent debris from falling into the trench and will also prevent overloading the trench wall with extra weight.

Machinery Falling into the Excavation

Machinery falling into an excavation is an all-too-frequent occurrence. The primary concern for the plumber is to avoid injury by staying out of the reach of the equipment at all times. An unintentional muscle contraction by an operator could severely injure nearby workers, both in and out of the trench.

Another good reason for maintaining a safe distance is that machinery may vibrate significantly. This vibration is transmitted to the soil and may induce a cave-in.

Debris Falling from Working Machinery

Keeping a good distance from machinery, especially if in the trench, reduces the possibility of debris falling into the trench from working machinery. Hard hats and goggles provide protection against some falling debris, but they are no match for a large section of pipe or heavy load being lifted.

Machinery Hitting Power Lines and Similar Utilities

Overhead and underground lines must be located prior to any excavation. The law requires contacting the utility locator in your area at least 72 hours in advance. Emergency locator services are available. Field personnel must record all reference numbers as required.

IN THE FIELD

Pay extra attention to moving large, wide loads during periods of high winds. A gust of wind can easily cause a crane to swing a load into a power line. It has also been known to knock a person carrying a sheet of plywood or other flat object off balance.

Underground utilities include all water, gas, telephone, electric, steam, storm, sprinkler, and other equipment piping. Most of these lines can be located by contacting local utility companies. Severe fines are often imposed on those who fail to contact utilities prior to digging and accidentally cut utility lines. Markings are generally within 2 feet of obstructions. Document markings prior to digging in the event of missed obstructions.

Overhead utilities include power, lighting, and telephone cables, as well as trees and similar hazards. Smart technicians maintain a constant vigil on these lines during equipment operation.

In the event that a power line is struck, maintain personal clearance from the machine. Do not touch the machine, the operator, or any wiring until it has been ascertained that the line is no longer live. The technician must have all utility numbers and a phone or communication radio immediately accessible in case of emergency. He must also have proper first aid equipment and knowledge of first aid treatment.

Passersby Falling into an Open Trench

Any experienced technician can verify that an open-ditch jobsite is an invitation to curious onlookers. An open ditch entices the passerby to examine its depth. Most often, children fall prey to their curiosity and accidents occur.

It is the responsibility of the workers to properly secure the jobsite at all times, especially at night. Cover the ditch with timber, plywood, or similar material capable of holding weight. If there is a possibility of rain, position this protective sheathing to shield the ditch from rain water as well as to protect the public against potential falls. Use reflective barricades and lights around the trench for additional protection.

In general, it is the responsibility of workers to make sure that the entire trenching and shoring operation is properly performed. Consult OSHA regulations for additional information.

Keep a well-stocked first-aid kit and emergency numbers handy in the event of an accident. Get first-aid and CPR training—it may save a life.

Trenches should only be opened as necessary. Jobsite coordination is critical. When done correctly, good pre-planning will have material, equipment, shoring materials, workers, and inspection properly organized to minimize the time that the trench is open.

If an inspector allows, an excavation may be closed daily and re-excavated the next day.

PREPARING FOR EXCAVATION WORK

With so many potential hazards related to excavation work, many companies use master checklists to help their employees minimize risk on the construction site. Table 22-1 is a sample checklist of many of the tasks that should be completed prior to starting excavation work.

TABLE 22-1 Trench Safety Checksheet

Project: Date: Weather:

Trench Depth: Length: Width: Type of Protective System: Soil Type:

Yes	No	N/A	Excavation
			Excavations and protective systems inspected by Competent Person daily, before start of work.
			Competent Person has authority to remove workers from excavation immediately.
			Surface encumbrances supported or removed.
			Employees protected from loose rock or soil.
			Hard hats worn by all employees.
			Spoils, materials, and equipment set back a minimum of 2′ from edge of excavation.
			Barriers provided at all remote excavations, wells, pits, shafts, etc.
			Walkways and bridges over excavations 6′ or more in depth equipped with guardrails.
			Warning vests or other highly visible PPE provided and worn by all employees exposed to vehicular traffic.
			Employees prohibited from working or walking under suspended loads.
			Employees prohibited from working on faces of sloped or benched excavations above other employees.
			Warning system established and used when mobile equipment is operating near edge of excavation.
			Utilities
			Utility companies contacted and/or utilities located.
			Exact location of utilities marked when near excavation.
			Underground installations protected, supported, or removed when excavation is open.
			Wet Conditions
			Precautions taken to protect employees from accumulation of water.
			Water removal equipment monitored by Competent Person.
			Surface water controlled or diverted.
			Inspection made after each rainstorm.
			Hazardous Atmosphere
			Atmosphere tested when there is a possibility of oxygen deficiency or build-up of hazardous gases.
			Oxygen content is between 19.5% and 23.5%.
			Ventilation provided to limit flammable gas build-up to 20% of lower explosive limit of the gas.
			Testing conducted to ensure that atmosphere remains safe.
			Emergency Response Equipment readily available where a hazardous atmosphere could or does exist.
			Employees trained in the use of Personal Protective and Emergency Response Equipment.
			Safety harness and life line individually attended when employees enter deep confined excavation.

Signature of Competent Person,

Date

CONFINED SPACES

OSHA defines a confined space as a space that is large enough for a person to enter and perform work, but has restricted means of entering or exiting. Examples of confined spaces would be manholes, pipelines, tanks or vaults, pumping station pits, and some trenches or tunnels. In all cases, these are areas that are not designed for continuous human occupancy. Many of these areas are classified as "permit-required" confined spaces. These areas contain one or more atmospheric or physical hazards.

Atmospheric hazards usually include at least one or more of the following situations. Flammable or explosive materials (including solvents) are present to such a level that they can ignite. OSHA typically defines this level of concentration as being 10 percent or more of the substance's lower flammable limit. The lower flammable limit is the lowest mixture of the substance and air at which the material will ignite. Inert gases or gases that can cause asphyxiation are another form of atmospheric hazard. In this case the gases may not be flammable or explosive but they lack the necessary level of oxygen that humans need to survive. Examples of these gases are refrigerants used for cooling and shielding gases used in welding, such as argon. Another hazardous situation could simply be that the air within the space does not have sufficient oxygen to support life. This level is usually considered to be less than 19 percent by volume. This particular situation is a prime example of why confined spaces must be continuously monitored while work is ongoing. As the air is breathed or when processes such as welding are occurring, the oxygen levels will decrease. Without constant monitoring, workers will slowly suffocate.

Physical hazards can include a variety of conditions, but they usually can be considered involving a fall, electrocution or becoming engulfed or trapped. In the case of falls, this may be the result of a worker slipping on a wet surface as he enters the space or while he is working. In many cases, this can be avoided by providing adequate lighting so that workers can see wet or slippery surfaces. In other situations, workers maybe working from either a suspended or elevated platform or by being suspended by ropes. Because the confined space is small or below ground level, it may not be obvious that it is really an elevated workspace. The possibility of electrocution can greatly be reduced by having a competent electrician who is familiar with jobsite lockout/tagout procedures cut off power to any equipment or energy sources that are present. However, workers still need to make sure that the tools and equipment that they are using are properly grounded and maintained so that they are not a possible source of electrocution. Engulfment is a situation when workers are "swallowed" or entrapped. This may occur when pumps are not properly locked out and turned on or when piping under repair spills its content into the space. All too often, workers are injured when they open the system and steam or water enters the space at such a rate that escape is impossible. Soil from excavation or cave-ins can also be a source or cause of engulfment.

Whenever a confined space is accessible, there should be an attendant present. This person will be responsible for monitoring the workers and jobsite conditions and making sure no unauthorized persons enter the work space. If the attendant recognizes a potential danger, he should immediately notify those working in the space and either assist them in exiting or call rescue workers to help.

REVIEW QUESTIONS

True or False

1. __________ A ladder should never be used as a walk board.
2. __________ Painting wooden ladders may cover up defects and lead to accidents.
3. __________ Never climb up the cross bracing on scaffolds.
4. __________ Step ladders should only be used when opened completely.
5. __________ Three-foot overlap is the standard for extension ladders, regardless of size.
6. __________ Angle of repose is the greatest angle above the horizontal at which a given material will lie without sliding.

Fill in the Blank and Short Answer

7. __________ is the first requirement of any of our job activities.
8. What two traits will a safe worker have?
9. Give two general safety guidelines for forklift operation.
10. How often should rigging equipment be checked?
11. What must a person have to operate a forklift?

Chapter 23

LEARNING OBJECTIVES

The student will:

- Summarize the purpose and demonstrate the use of rough-in and specification sheets.
- Demonstrate how to draw and read single line drawings of fixtures and simple plumbing pipe diagrams.
- Demonstrate the use of single line drawing as it applies to larger equipment, piping, and projects.
- Discuss how isometric views differ from orthographic views.

Rough-In Sheets, Single Line Drawings, and Isometric Drawings

SCALE RULER REVIEW

For obvious reasons, a building or other large objects cannot be drawn on paper to their full size; this means that they must be shrunk to a suitable size. This reduction ratio is known as the scale and for the object to remain in proportion, the scale must be applied the same in both the *x* and *y* axis. The person who creates the drawing should pick a scale that allows the user to see the entire object with enough detail to understand its purpose, but also be on a reasonable size piece of paper. Once the reduction ratio, or scale, has been selected, the drawer performs a series of calculations to determine what size line represents the actual length of the object. To make this cumbersome task easier, a scale ruler was devised that has the 1-inch length reduced to a smaller proportion. For example, the $\frac{1}{4}$ scale has the 1-inch space divided into four equal divisions with each one numbered 1 through 4. This allows the user to quickly count the $\frac{1}{4}$ of an inch as if they were a full size inch. When this concept is used with the idea that a portion of an inch is equal to 1 foot, the user can now quickly draw a line that would represent multiple feet within a single inch. Figure 23-1 shows a scale ruler marked as $\frac{1}{4}'' = 1'$. Since there could be any number of scales used in the construction trades, a large number of scale rulers would be necessary. To reduce the number of rules needed, a special triangle shaped rule was created. This rule consists of three straight edges that are marked with eleven different scales. The two most commonly used scales are fractional forms, $\frac{1}{4}''$ equals $1'$ or $\frac{1}{8}''$ equals $1'$.

ROUGH-IN SHEETS

Wholesalers distribute a large variety of products, many with similar functions but with slight differences in dimensions, applications, auxiliary equipment, and so on. In addition, similar products are made by different manufacturers, and the major industry manufacturers make hundreds of thousands of plumbing products.

In view of the many choices possible, the technician responsible for installing a given fixture or appliance needs detailed guidance to perform a proper installation for the specific product.

FIGURE 23-1 Drafting scales allow measurements to be taken easily from scale drawings.

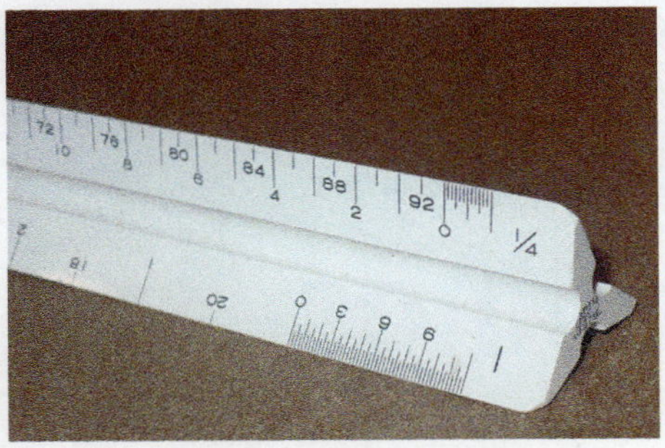

Photo by Ed Moore

This detailed guidance is usually provided in the form of rough-in sheets. Rough-in sheets are scaled drawing sheets of the specific product showing pertinent dimensions and the proper orientation of the product when installed. Manufacturers provide a separate sheet for each of their products. These individual sheets are sometimes arranged into entire catalogs. Rough-in sheets are also available to be downloaded from most manufacturers' websites. Whether you download a file or copy it from a catalog, you should make sure that the rough-in data you have is up to date.

Figure 23-2 shows a rough-in sheet for a two-piece round-front siphon-action jetted-bowl water closet. The sheet includes the manufacturer's name, product name (including descriptive wording), product drawing, model number, important dimensions governing installation, and dimensions of the product.

In the case of water closets, rough-in sheets often show waste outlets and water inlets, fixture dimensions, hanger and support locations, and clearances to walls and partitions.

Figure 23-3 shows a rough-in sheet for a bath/shower diverter assembly. This drawing shows product dimensions, locations of connections, the size and locations of elements that penetrate the wall, locations of associated items (spout, diverter, shower arm), and a drawing of the product itself. Thus, all the details of a shower diverter replacement can be considered, prior to actually removing the original product, to make sure that minimum damage is done to the wall.

Rough-in sheets (sometimes called specification sheets) are also available for fittings, valves, pipe, and other products. Rough-in sheets fulfill these needs:

- They assist the project designer in selecting and specifying the best product for an application.
- After approval by the architect or engineer, they provide assurance to everyone that the product is acceptable for the project.
- They assist the installing technician in making a proper rough-in for satisfactory connections to the device.
- They assist everyone involved in determining required clearances for the product.

Rough-in sheets provide the information needed to properly install the connecting piping so that the fixture or appliance will fit as intended when it is installed.

FIGURE 23-2 Water closet rough-in sheet.

American Standard

CADET® 3 ROUND FRONT TOILET with the FLUSH RIGHT™ SYSTEM

VITREOUS CHINA

CADET® 3 ROUND FRONT TOILET with the FLUSH RIGHT™ SYSTEM

❑ **2384.012**
- Vitreous china
- Low-consumption (6.0 Lpf/1.6 gpf)
- EverClean™ silver-based ceramic glazed
- Featuring the ***Flush Right™ System***
- Space saving round front siphon action jetted bowl
- Fully-glazed 2-1/8" trapway with 2" ball pass
- Generous 9" x 8" water surface area
- Close-coupled tank with flat tank cover for superior storage
- Oversized 3" flush valve with chemical resistant flapper
- Chrome trip lever
- New Speed Connect™ tank-to-bowl coupling system
- New sanitary dam on bowl with four point tank stabilization
- 2 color-matched bolt caps
- 100% factory flush tested

❑ **3011.016** Round Front Bowl
❑ **4021.016** Tank

Nominal Dimensions:
28-1/4" x 17-1/4" x 28-7/8"
(718 x 438 x 733mm)

Fixture only, seat and supply by others

Alternative Tank Configurations Available:

❑ **4021.500** Tank complete with Aquaguard Liner
❑ **4021.600** Tank complete with tank cover locking device
❑ **4021.700** Tank complete with Aquaguard Liner and tank cover locking device
❑ **4021.800** Tank complete with trip lever located on right side
❑ **4021.900** Tank complete with tank cover locking device and trip lever located on right side

Compliance Certifications - Meets or Exceeds the Following Specifications:
- ASME A112.19.2M for Vitreous China Fixtures - includes Flush Performance, Ball Pass Diameter, Trap Seal Depth and all Dimensions
- Los Angeles Department of Water and Power Supplementary Purchase Specification for Non-Adjustability

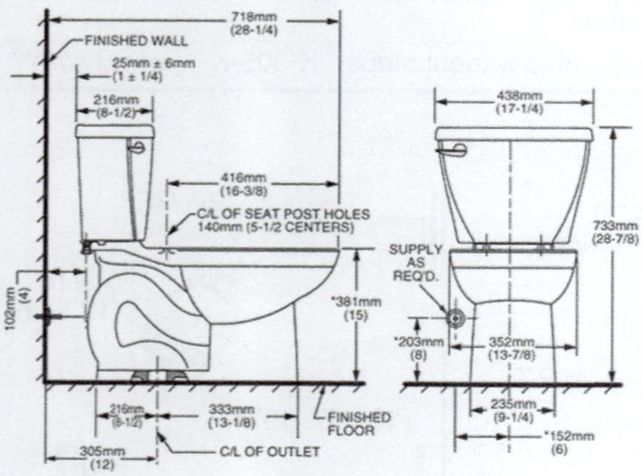

NOTES:
THIS TOILET IS DESIGNED TO ROUGH-IN AT A MINIMUM DIMENSION OF 305MM (12") FROM FINISHED WALL TO C/L OF OUTLET.
SUPPLY NOT INCLUDED WITH FIXTURE AND MUST BE ORDERED SEPARATELY.
* DIMENSION SHOWN FOR LOCATION OF SUPPLY IS SUGGESTED.
IMPORTANT: Dimensions of fixtures are nominal and may vary within the range of tolerances established by ANSI Standard A112.19.2.
These measurements are subject to change or cancellation. No responsibility is assumed for use of superseded or voided pages.

To Be Specified	
❑ Color: ❑ White ❑ Bone ❑ Silver ❑ Linen ❑ Black	
❑ Seat #5330.010	Round front Champion® Slow Close solid plastic seat and cover with easy lift off feature
❑ Seat #5322.011	Round front "Rise and Shine" solid plastic seat and cover with easy clean lift-off hinge system
❑ Seat #5308.014	Round front Laurel molded wood seat and cover
❑ Supply with stop:	

Cadet® 3 Toilet
featuring the Flush Right™ system

TB-22

11/05

FIGURE 23-3 Tub and shower diverter rough-in sheet.

KOHLER®
Roughing-In

K-T14420, K-304, K-305, K-306
PURIST™ RITE-TEMP™
PRESSURE-BALANCING FAUCETS

Product Information

Applicable faucet trim only:	
Bath and shower cross handle* (shown)	K-T14420-3
Bath and shower lever handle*	K-T14420-4
Shower cross handle*	K-T14421-3
Shower lever handle*	K-T14421-4
Non-diverter valve cross handle*	K-T14423-3
Non-diverter valve lever handle*	K-T14423-4
Shower head, arm and flange	K-14424
35° Bath Spout	K-14426
90° Bath Spout	K-14427
*Rite-Temp valve must be ordered to complete faucet.	

Applicable Rite-Temp valve only:		
Rite-Temp valve without stops	K-304-K	not shown**
Rite-Temp valve with stops	K-304-KS	not shown**
Rite-Temp valve with CPVC female stops	K-304-KP	not shown**
Rite-Temp valve without stops	K-305-K	not shown**
Rite-Temp valve with stops	K-305-KS	not shown**
HiFlow Rite-Temp valve with integral stops	K-306-KS	not shown**
HiFlow Rite-Temp thin wall installation kit	88526	not shown**
**Required for configuration illustrated.		
ADA Cross and lever handles are ADA compliant		

⚠ **WARNING: Risk of damage to the K-306-KS valve assembly.** When using the K-306-KS valve in a fiberglass or acrylic installation, use the thin wall installation kit (88526).

Installation Notes

Avoid cross-flow conditions. Do not install shut-off device on either valve outlet.
Cap shower outlet if deck-mount spout, diverter, or handshower is connected to spout outlet.
Install straight pipe or tube drop of 7" (17.8cm) to 18" (45.7cm) with single elbow between valve and wall-mount spout.

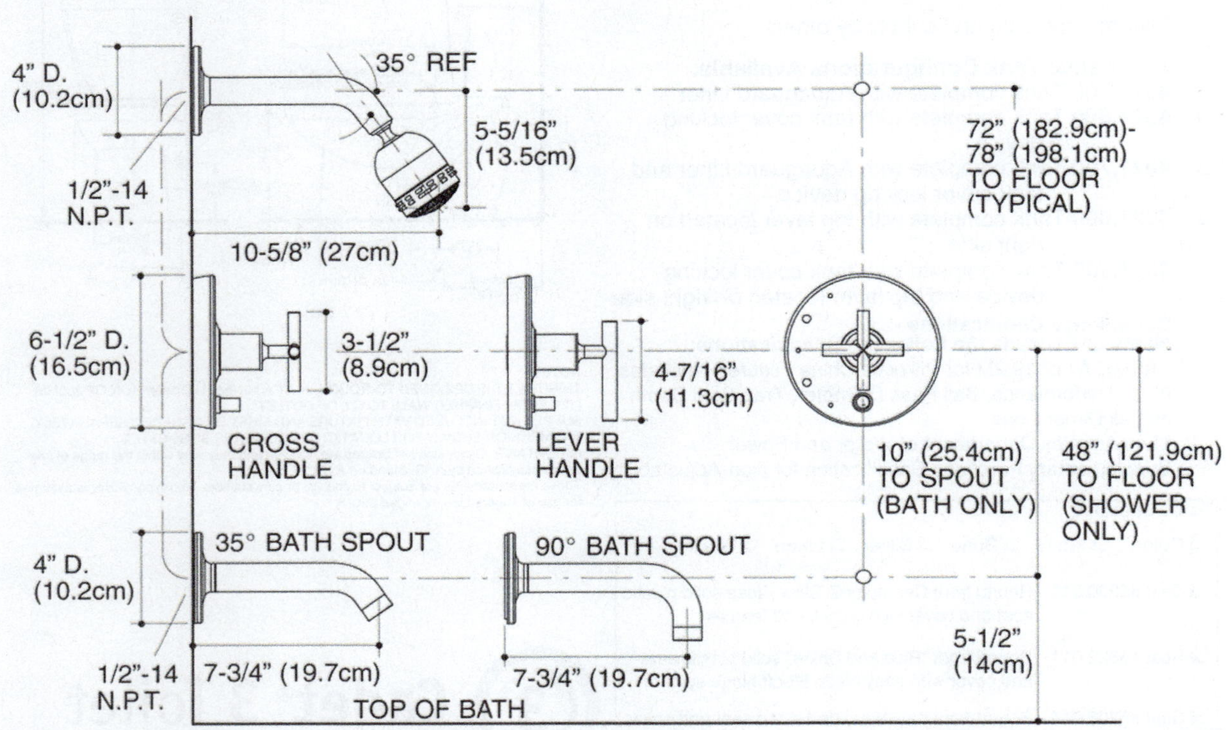

Therefore, it is necessary to study and properly interpret the rough-in sheets so the job will progress in a smooth manner. A proper rough-in means that almost anyone with the necessary skills can finish the job.

It is possible that the work of other trades can create problems, even if all rough-in dimensions are accounted for properly. For example, if a tile wall covering is thicker than usual, the plumbing rough-in may not be satisfactory to mount tub faucet handles, related trim, or the spout. Discuss these critical items with the job superintendent and other trade technicians on the job to minimize such difficulties. Occasionally, plumbing fixtures may have requirements for other trades. In the case of Figure 23-5, the rough-in sheet shows electrical requirements for this water closet. Such requirements must be shared with the other trades on the jobsite to help prevent conflicts. It is fairly common to find electrical requirements for water heaters, pumps, flush valves, and electronic faucets on the plumbing rough-in sheets.

EXAMPLE 1

If a restroom is to be equipped with three water closets as shown in Figure 23-4, we note that the fixtures are floor-mounted. The plans and specifications must be examined for fixture type and the rough-in sheet must be checked for the dimensions of the waste and water connections.

If the fixture specified is the Kohler Hatbox™ K-3492, Figure 23-5 shows the necessary dimensions.

Figure 23-5 shows that the rim of the bowl is $16\frac{5}{8}$-inches high (instead of 14 in.), for easier use by elderly or handicapped persons and that the waste opening is 12 inches from the finished wall (check with the tile installer).

INSTALLATION TOLERANCES

Rough-in sheets show the variations permissible in certain installation dimensions and details for the product. Such permissible variations are called *tolerances*. The tolerances allow for minor errors in the installation. This means that even though the technician should work as accurately as possible, he or she would not have to correct small deviations, thereby saving in the cost of the rough-in installation. Install all products as the rough-in sheets indicate and be sure that building materials installed by other trades do not exceed the tolerances in the product.

FIGURE 23-4 Battery of water closets.

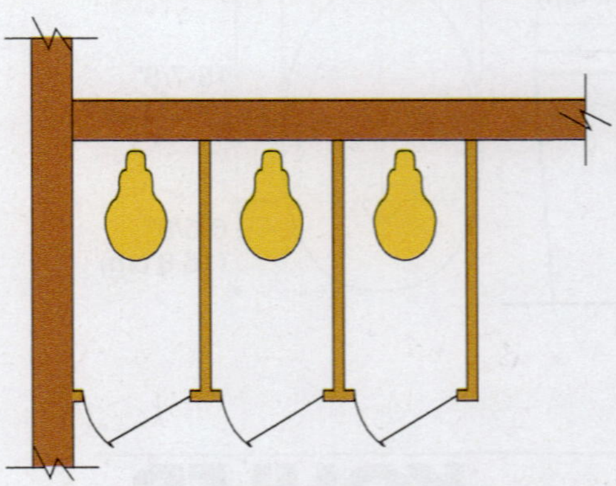

FIGURE 23-5 Water closet rough-in sheet.

KOHLER® Roughing-In

K-3492
PURIST®
HATBOX ™ TOILET

Product Information

Fixture:	
Configuration	1-piece, elongated
Water per flush	1.6 gallons (6 L)
Passageway	2-1/8″ (5.4 cm)
Water area	12″ (30.5 cm) x 9-1/2″ (24.1 cm)
Water depth from rim	4-7/8″ (12.4 cm)
Included Components:	
Seat	1022679

Pump	HP	VAC	Hz	A
	1/5	120	60	8.5

Required Electrical Service

Dedicated circuit required, protected with Class A Ground-Fault Circuit-Interrupter (GFCI):	
Receptacle	120 VAC, 15 A, 60 Hz

Installation Notes

Install this product according to the installation guide.

Fixture dimensions are nominal and conform to tolerances in ASME Standard A112.19.2.

Both the electrical and water supplies can be installed on either the wall or the floor. Follow the information for your particular installation.

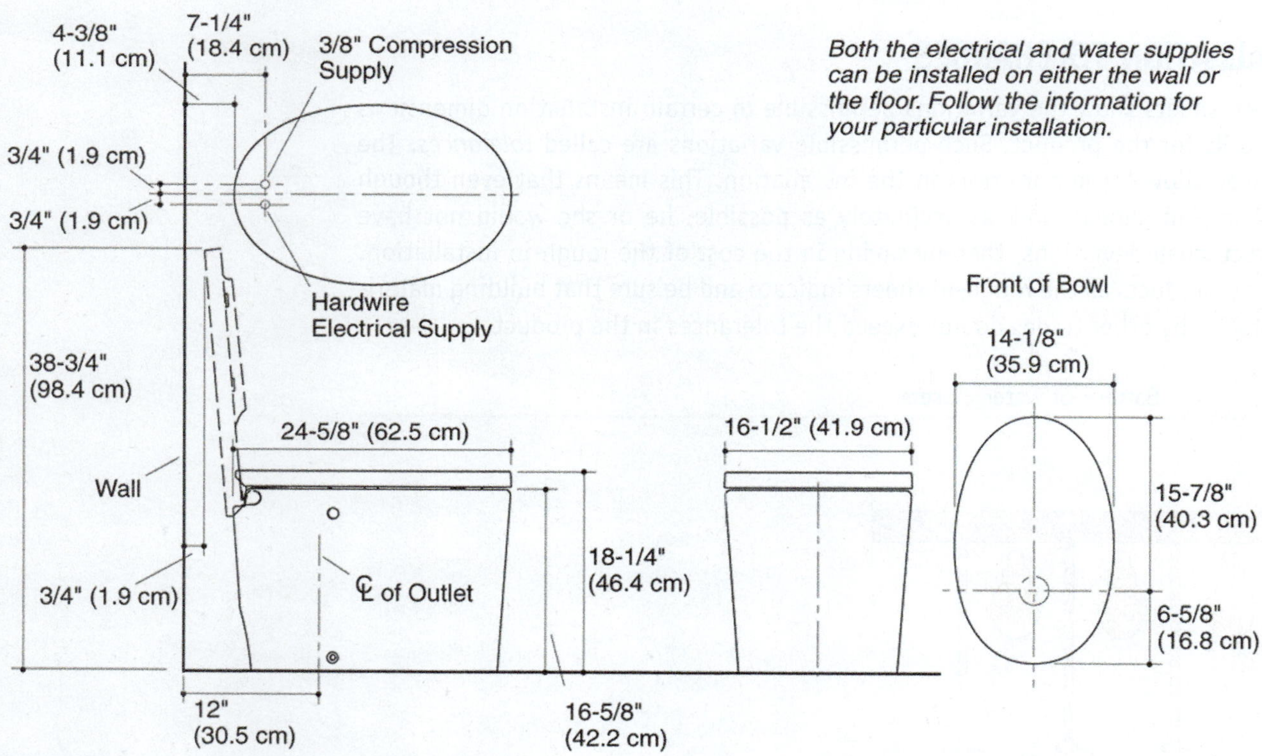

PURIST® HATBOX ™ TOILET
1035530-1-C

KOHLER®

EXAMPLE 2

Figures 23-6 through 23-9 show specifications for various fixtures. Use these diagrams to design a bathroom (Figure 23-6) using the following fixtures:

- Water closet: 1.6 gpf, American Standard Cadet™, elongated (Figure 23-7)
- Bathtub: American Standard Evolution™ Whirlpool (Figure 23-8)
- Lavatory: American Standard Cadet™ Oval Countertop Sink (Figure 23-9)

Note that in a typical installation, you would also need to include a diverter, waste and overflow, and wall surround.

The drawings show that the water closet (Figure 23-7) has a 12-inch waste opening dimension from the rear wall. Also, the water supply should be 8 inches above the finished floor and 6 inches left of the bowl centerline.

In Figure 23-8, the whirlpool bathtub unit is dimensioned for a right-hand unit. Remember that right-hand and left-hand units are mirror images, so you measure and install the tub the same way for either right-hand or left-hand alignment. Thus, the waste opening is 11 inches from the rough end wall, 16 inches from the rough rear wall. These dimensions are from the rough wall because the tub is set against the rough wall. The waste and overflow must be for a $21\frac{1}{2}$-inch high tub.

The lavatory (Figure 23-9) can be positioned after the vanity cabinet details have been determined. The rim height of this lavatory is $33\frac{1}{4}$ inches. The drawing matches the drain opening and water supply openings to the countertop height. Thus, the center of the waste would be $12\frac{7}{8}$ inches below the countertop and the water supplies should be $10\frac{3}{4}$ inches below the rim. These openings would be centered on the lavatory. If the lavatory is off-center in the cabinet (a common design decision), the waste and water opening would be similarly off-center. If these openings are not positioned properly, expensive fittings and additional time may be required to install the fixture.

IN THE FIELD

Make sure to add insulation before installing plumbing fixtures on outside walls. In the case of a one-piece shower, it may be impossible to add insulation after the install. The added insulation will increase user comfort and save a considerable amount of energy during the life of the product.

FIGURE 23-6 Plan view of full bath.

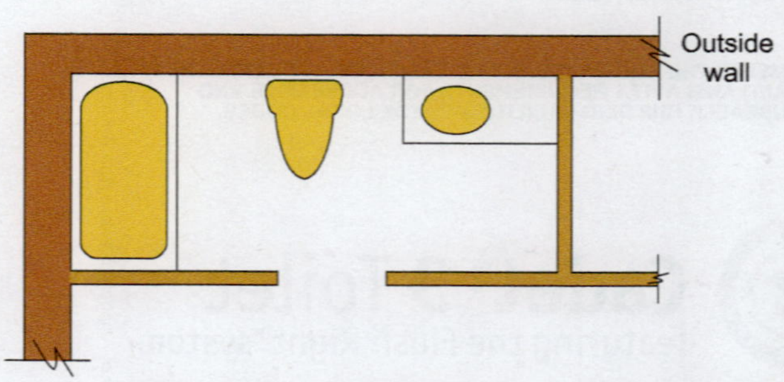

FIGURE 23-7 Water closet rough-in sheet.

American Standard

CADET® 3 RIGHT HEIGHT™ ELONGATED TOILET with the FLUSH RIGHT™ SYSTEM

VITREOUS CHINA

CADET® 3 RIGHT HEIGHT™ ELONGATED TOILET with the FLUSH RIGHT™ SYSTEM

❑ **2386.012**
- Vitreous china
- Low-consumption (6.0 Lpf/1.6 gpf)
- EverClean™ silver-based ceramic glazed
- Featuring the ***Flush Right™ System***
- 16-1/2" rim height for accessible application
- Elongated siphon action jetted bowl
- Fully-glazed 2-1/8" trapway with 2" ball pass
- Generous 9" x 8" water surface area
- Close-coupled tank with flat tank cover for superior storage
- Oversized 3" flush valve with chemical resistant flapper
- Chrome trip lever
- New Speed Connect™ tank-to-bowl coupling system
- New sanitary dam on bowl with four point tank stabilization
- 2 color-matched bolt caps
- 100% factory flush tested

❑ **3016.016** Right Height™ Elongated Bowl
❑ **4021.016** Tank

Nominal Dimensions:
30-1/4" x 17-1/4" x 30-1/4"
(768 x 438 x 768mm)

Fixture only, seat and supply by others

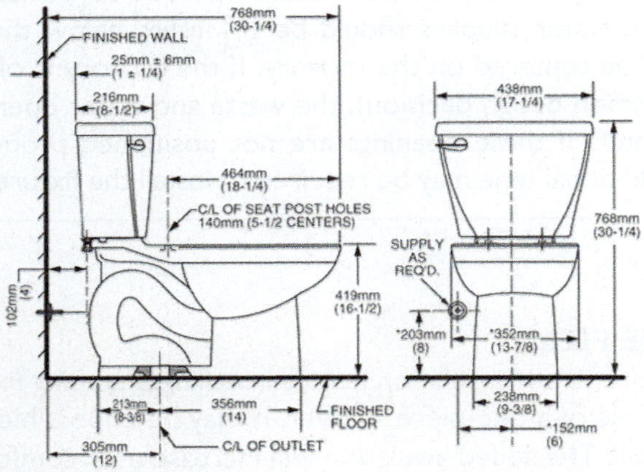

Alternative Tank Configurations Available:

❑ **4021.500** Tank complete with Aquaguard Liner
❑ **4021.600** Tank complete with tank cover locking device
❑ **4021.700** Tank complete with Aquaguard Liner and tank cover locking device
❑ **4021.800** Tank complete with trip lever located on right side
❑ **4021.900** Tank complete with tank cover locking device and trip lever located on right side

Compliance Certifications - Meets or Exceeds the Following Specifications:
- ASME A112.19.2M for Vitreous China Fixtures - includes Flush Performance, Ball Pass Diameter, Trap Seal Depth and all Dimensions
- Los Angeles Department of Water and Power Supplementary Purchase Specification for Non-Adjustability

NOTES:
THIS TOILET IS DESIGNED TO ROUGH-IN AT A MINIMUM DIMENSION OF 305MM (12") FROM FINISHED WALL TO C/L OF OUTLET.
SUPPLY NOT INCLUDED WITH FIXTURE AND MUST BE ORDERED SEPARATELY.
* DIMENSION SHOWN FOR LOCATION OF SUPPLY IS SUGGESTED.
IMPORTANT: Dimensions of fixtures are nominal and may vary within the range of tolerances established by ANSI Standard A112.19.2.
These measurements are subject to change or cancellation. No responsibility is assumed for use of superseded or voided pages.

To Be Specified

❑ Color: ❑ White ❑ Bone ❑ Silver ❑ Linen ❑ Black

❑ Seat #5325.010 Elongated Champion® Slow Close solid plastic seat and cover with easy lift off feature

❑ Seat #5324.019 Elongated "Rise and Shine" solid plastic seat and cover with easy clean lift-off hinge system

❑ Seat #5311.012 Elongated Laurelmolded wood seat and cover

❑ Supply with stop:

MEETS THE AMERICANS WITH DISABILITIES ACT GUIDELINES AND ANSI A117.1 REQUIREMENTS FOR ACCESSIBLE AND USEABLE BUILDING FACILITIES-CHECK LOCAL CODES.

Cadet® 3 Toilet
featuring the Flush Right™ system

TB-23

11/05

FIGURE 23-8 Whirlpool tub rough-in sheet.

American Standard

EVOLUTION™ 5' X 36"
WHIRLPOOLS BY AMERICAN STANDARD
HIGH GLOSS ACRYLIC

EVOLUTION™ 5' X 36"
Bathing design with form-fitted backrest and molded-in armrests

❑ **Model# 2771V (specify color)**
- Acrylic with fiberglass reinforcement
- Deep Soak Overflow Drain for maximum water depth
- 1.4 HP self-draining pump/motor
- Eight flow adjustable jets including:
  Two adjustable Ideal Flow Comfort Jets™
  Six adjustable Comfort Jets™
- Two silent air volume controls
- Side-mounted On/Off control
- Complete factory installed and tested whirlpool system
- Self leveling feet
- Designed for undermount installation option

❑ **Model# 2771VC (specify color)**
- Same as above with the EverClean™ whirlpool system featuring Antimicrobial AlphaSan®

❑ **Model# 2771VAC EverClean™ Air Spa**
- Features Antimicrobial AlphaSan®
- 1.25 HP single speed pump/motor
- side mounted air switch On/Off
- Eight Comfort Air jet system
- Quick connect EZ Install heater system (must order EZ Install heater separately)

❑ **Model# 2771VA Air Spa**
- Same as 7236 VAC without Antimicrobial AlphaSan®

Nominal Dimensions:
60" x 36" x 19-3/4"
(1524 x 914 x 502mm)

Components Required:

❑ Bath Filler*: (specify finish)

❑ Bath Drain: 1599.205.XXX (specify color)

❑ Five Foot Apron: 9261.019.XXX (specify color)

❑ E-Z-Install™ Whirlpool Heater - EZHEAT-100

❑ Tile Bead Kit: TILE-BEAD

* See American Standard Faucet Selector

For color/finish availability and other faucet choices, please see the Product Selector and Pricing Guide.

Compliance Certifications -
Meets or Exceeds the Following Specifications:
- Underwriters Laboratories - Hydromassage Bathtubs UL 1795
- ASME A112.19.7M for Whirlpool Appliances
- ANSI Z124.1 for Plastic Bathtubs
- ASTM E162 for Flammability
- NFPA 258 for Smoke Density

NOTE: Roughing-in dimensions on reverse side of page.

NOTE: E-Z-Install™ Whirlpool Heater is designed to maintain water temperature during operation. It will not heat cold water to desired bathing temperature.

FIGURE 23-8 Whirlpool tub rough-in sheet. (Continued)

American Standard

EVOLUTION™ 5' X 36"
WHIRLPOOLS BY AMERICAN STANDARD
HIGH GLOSS ACRYLIC

Whirlpool shown

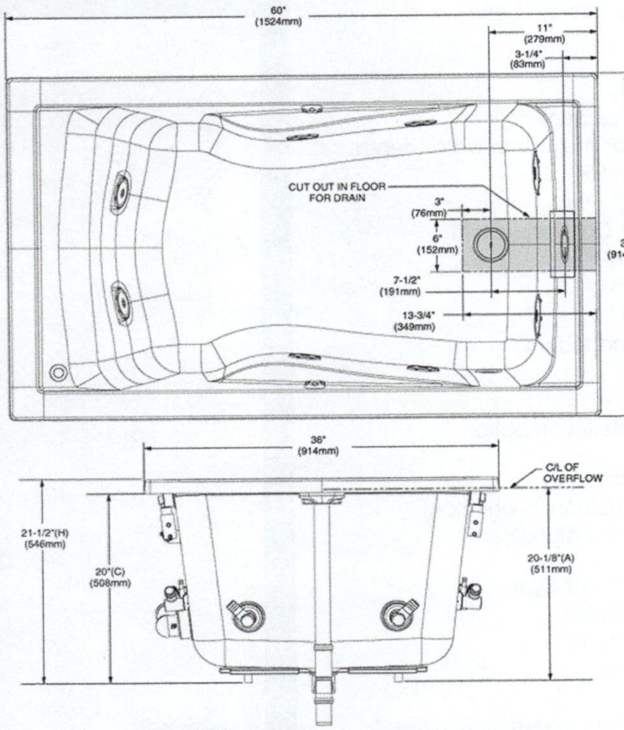

DIMENSIONS OF FIXTURES ARE NOMINAL AND MAY VARY WITHIN THE RANGE OF TOLERANCES ESTABLISHED BY ANSI STANDARDS Z124.1 AND A112.19.7.

GENERAL SPECIFICATIONS FOR 2771V WHIRLPOOL

INSTALLED SIZE	60 x 36 x 21-1/2 In. (1524mm x 914 mm x 546mm)
WEIGHT	75 Lbs. (34 Kg.)
WEIGHT w/WATER	800 Lbs. (363 Kg.)
GAL. TO OVERFLOW	87 Gal. (329 L.)
MIN. OPERATING GAL.	50 Gal. (189 L.)
BATHING WELL AT SUMP	41-1/4 x 26-1/4 In. (1048mm x 667mm)
BATHING WELL AT RIM	52-1/2 x 30 In. (1334mm x 762mm)
WATER DEPTH TO OVERFLOW	18-1/2 In. (470mm)
FLOOR LOADING (PROJECTED AREA)	53 Lbs./Sq. Ft. (260 Kgs/Sq. m.)

WHIRLPOOL ELECTRICAL SPECIFICATIONS

PUMP	1.4 HP, 9.7 AMPS, 120V.

NOTES:

BATH CAN BE INSTALLED EITHER ABOVE OR BELOW FLOOR LINE AS PIER, ISLAND, OR PENINSULA TYPE INSTALLATION. SUPPORT OF BATH BY SUMP BEDDING. SUPPORT AND ENCLOSING MATERIAL BY OTHERS.

REFER TO INSTALLATION INSTRUCTIONS SUPPLIED WITH WHIRLPOOL FOR ADDITIONAL INFORMATION.

■ OPTIONAL 3 WALL TILING BEAD. KIT No. TILE-BEAD AND APRON No. 5FT-APRON AVAILABLE AND MUST BE ORDERED SEPARATELY.

FITTINGS NOT INCLUDED AND MUST BE ORDERED SEPARATELY.

PROVIDE SUITABLE REINFORCEMENT FOR ALL SUPPORTS.

IMPORTANT:
Dimensions of fixtures are nominal and may vary within the range of tolerances established by ANSI Standard Z124.1 These measurements are subject to change or cancellation. No responsibility is assumed for use of superseded or voided leaflet.

3/06 EV-6

FIGURE 23-9 Lavatory rough-in sheet.

American Standard

BARRIER FREE

CADET™ OVAL COUNTERTOP SINK

VITREOUS CHINA

CADET OVAL COUNTERTOP SINK

- European-styled interior bowl
- Vitreous china construction
- Self-rimming oval countertop
- Front overflow
- Supplied with cutout template
- Front overflow

- ☐ **0419.444** Faucet holes on 4" (102mm) centers
- ☐ **0419.888** Faucet holes on 8" (203mm) centers (illustrated)
- ☐ **0419.111** Center hole only

Nominal Dimensions:
21" x 17-5/8" (533 x 448mm)

Bowl sizes:
16-1/2" (419mm) wide,
11" (279mm) front to back,
5-5/8" (143mm) deep

Compliance Certifications - Meets or Exceeds the Following Specifications:
- ASME A112.19.2M for Vitreous China Fixtures
- CAN/CSA B45 series

To Be Specified

- ☐ Color:
- ☐ Faucet*:
- ☐ Faucet Finish
- ☐ Supplies:
- ☐ 1-1/4" Trap:

* See faucet section for additional models available

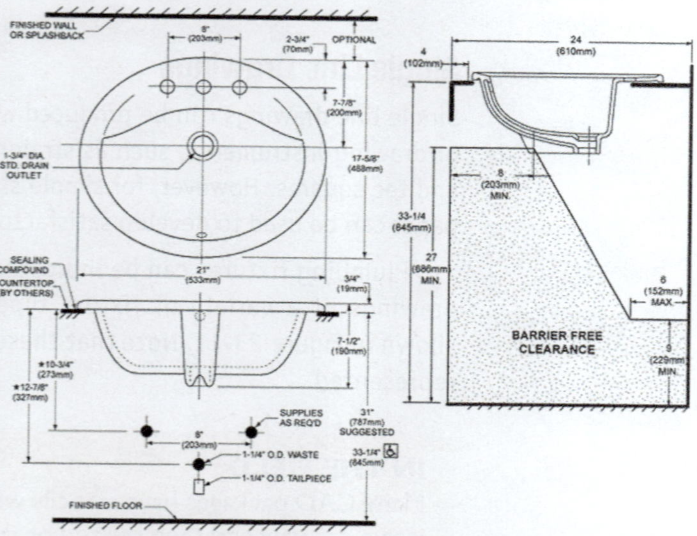

NOTES:
★ DIMENSIONS SHOWN FOR LOCATION OF SUPPLIES AND "P" TRAP ARE SUGGESTED.
▲ FOR COUNTERTOP CUTOUT USE TEMPLATE SUPPLIED WITH SINK.
FITTINGS NOT INCLUDED AND MUST BE ORDERED SEPARATELY.
IMPORTANT: Dimensions of fixtures are nominal and may vary within the range of tolerances established by ANSI Standard A112.19.2. These measurements are subject to change or cancellation. No responsibility is assumed for use of superseded or voided pages.

MEETS THE AMERICAN DISABILITIES ACT GUIDELINES AND ANSI A117.1 ACCESSIBLE AND USEABLE BUILDINGS AND FACILITIES - CHECK LOCAL CODES.
Install lavatory 864mm (34") from finished floor.
Lavatory installed 51mm (2") minimum from front edge of countertop provides 686mm (27") knee clearance area.

BS-48

Revised 3/03

Many rough-in sheets provide information necessary for selection of fixtures (when properly roughed-in) that comply with requirements of the Americans with Disabilities Act (ADA). Both traditional and special-design lavatories can be installed to comply with ADA knee and toe clearance requirements. Similarly, barrier-free water closets are available for both residential and commercial settings. See Figure 23-10.

Special pipe fittings are also described by rough-in sheets. For minimum-tolerance work, consult the manufacturer's drawings for precise dimensions. Usually, the critical dimensions are the horizontal laying length, vertical rise, and radius of the bend. The rough-in sheet also indicates standard sizes available. Thus, for example, you can plan the sequence of fittings required for a 3-inch stack connecting to a 5-inch building drain.

CREATING DRAWINGS AND DIAGRAMS

Rough-in sheets contain detailed drawings of particular items that are to be installed in a plumbing system. We also need a drawing showing the plumbing system itself. Such drawings range from very simple sketches to elaborate drawings prepared by a professional engineer. In this chapter, we describe methods to develop quick, accurate sketches that will be useful in completing the plumbing installation efficiently and economically. Such sketches are called *single line drawings*.

Single Line Drawings

Single line drawings can be produced with computer aided drawing (CAD) programs or drawing instruments, such as straight edges, triangles, pencils, drawing boards, and tee squares. However, for simple sketches, a straight edge, pencil, and piece of paper can be used to develop satisfactory sketches for simple piping layouts.

Plumbing fixtures can be indicated by using templates that permit very quick drawings of a variety of fixture types. Examples of typical fixture figures are shown in Figure 23-11. Note that these drawings are symbolic pictures of the item represented.

IN THE FIELD

Many CAD packages have stencils with pre-drawn figures that can be imported into a drawing and then resized or rotated. See Figure 23-12.

Piping is usually depicted by the following methods. However, each architect is different, so always look at the drawing legend to confirm.

- Cold water: long segment, space, short segment, space, long segment, and so on
- Hot water: long segment, space, short segment, short segment, space, long segment, and so on
- Drain: solid line
- Vent: dashed line

FIGURE 23-10 ADA water closet rough-in sheet.

American Standard

♿ **BARRIER FREE**

AFWALL™ ADA RETROFIT ELONGATED FLUSH VALVE TOILET

VITREOUS CHINA

AFWALL™ ADA RETROFIT ELONGATED*

- Vitreous china
- Low-consumption (6.0 Lpf/1.6 gpf)
- Off-floor mounting
- Elongated bowl
- Direct-fed siphon jet action
- Fully-glazed 2" ballpass trapway
- 10" x 12" water surface area
- Condensation channel
- 1-1/2" inlet spud
- 100% factory flush tested

❑ **2294.011** Top spud
❑ **2296.019** Top spud with slotted rim for bedpan holding (White only)

Afwall ADA 1.6 replaces standard wall-hung 3.5 or 1.6 models and roughs-in at 410mm (16-1/8") rim height **to meet ADA with no need to modify the existing carrier or through wall supply** (providing centerline of current outlet is 133mm [5-1/4"] above finished floor).

American Standard kit #736046-100 Flush Valve Conversion Kit containing a 25mm (1") chrome-plated street ell, a vacuum breaker, and a 330mm (13") tail piece included to meet minimum 152mm (6") code between bowl and vacuum breaker.

Recommended working pressure--between 30 psi at valve when flushing and 80 psi static

Compliance Certifications -
Meets or Exceeds the Following Specifications:
- ASME A112.19.2M (and 19.6M) for Vitreous China Fixtures - includes Flush Performance, Ball Pass Diameter, Trap Seal Depth and all Dimensions

Nominal Dimensions: 635 x 375 x 410mm (25" x 14-3/4" x 16-1/8")

Seat, carrier, bolt caps, and flush valve by others.

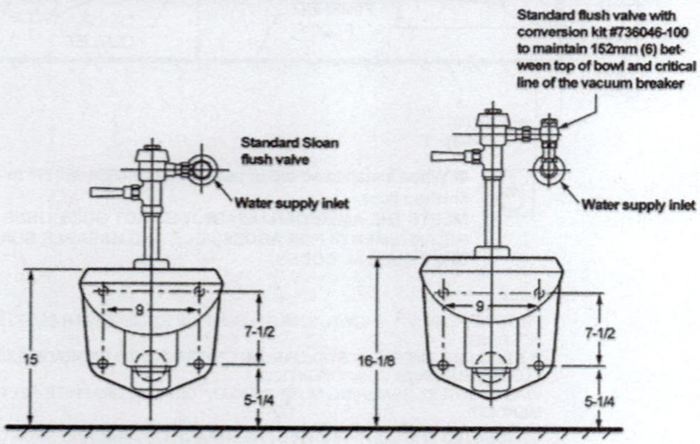

Standard Afwall Afwall ADA1.6

To Be Specified
- ❑ Color: ❑ White ❑ Bone ❑ Silver ❑ Shell ❑ Black
- ❑ Seat: Olsonite #95 open front seat less cover
- ❑ Seat: Church #9500C open front seat less cover
- ❑ Alternate Seat:
- ❑ Kit to convert existing flush valve from 3.5 gpf to 1.6 gpf:
  - ❑ Coyne Delaney: Kit No. F144-1.6ACQ Cap No. F159-1.6AQ
  - ❑ Sloan Royal: Kit No. A-41-A 1.6 Gal Closet Kit LC Cap No. A-72 C.P.
  - ❑ Zurn: Kit No. Z6000EC-WSI Cap No. Z6000LL for WSI flush valve
- ❑ Flush Valve: Sloan Royal #111 or equal
- ❑ Carrier Fitting J.R. Smith 210L or equal

NOTE: Roughing-in information shown on reverse side of page.

* Patent Pending

SPS 2294/2296

COM/INS-002

Revised 6/95

FIGURE 23-10 ADA water closet rough-in sheet. (Continued)

American Standard

AFWALL™ ADA RETROFIT ELONGATED FLUSH VALVE TOILET

VITREOUS CHINA

2294.011/2296.019

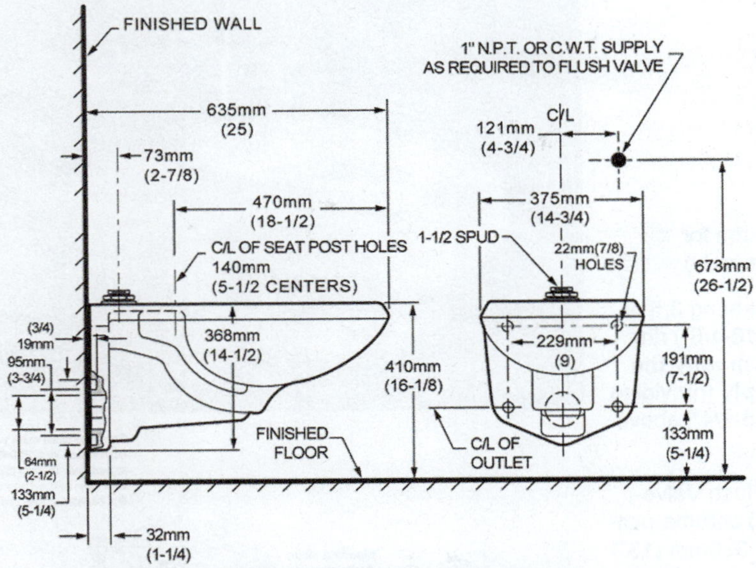

● When installed so top of seat is 432 to 483mm (17" to 19") from the finished floor.
MEETS THE AMERICAN DISABILITIES ACT GUIDELINES AND ANSI A117.1 REQUIREMENTS FOR ACCESSIBLE AND USEABLE BUILDING FACILITIES-CHECK LOCAL CODES.

NOTES:
PRODUCT 2294.011 SHOWN, 2296.019 SAME AS EXCEPT WITH SLOTTED RIM FOR BED PAN HOLDING.
* REQUIRES AMERICAN STANDARD KIT 736046-100 TO UPGRADE EXISTING INSTALLATION OF FLUSH VALVE AND COMPONENTS.
WASTE OUTLET SEAL RING MUST BE NEOPRENE OR GRAPHITE-FELT (WAX RING NOT RECOMMENDED).

SUGGESTED 1/16 CLEARANCE BETWEEN FACE OF WALL AND BACK OF BOWL.

TO COMPLY WITH AREA CODE GOVERNING THE HEIGHT OF VACUUM BREAKER ON THE FLUSH VALVE, THE PLUMBER MUST VERIFY DIMENSIONS SHOWN FOR SUPPLY ROUGHING.

FLUSH VALVE NOT INCLUDED WITH FIXTURE AND MUST BE ORDERED SEPARATELY.

CARRIER FITTING AS REQUIRED TO BE FURNISHED BY OTHERS.

PROVIDE SUITABLE REINFORCEMENT FOR ALL WALL SUPPORTS.

IMPORTANT: Dimensions of fixtures are nominal and may vary within the range of tolerance established by ANSI Standard A112.19.2
These measurements are subject to change or cancellation. No responsibility is assumed for use of superseded or voided pages.

SPS 2294/2296

FIGURE 23-11 Plumbing fixture symbols—residential.

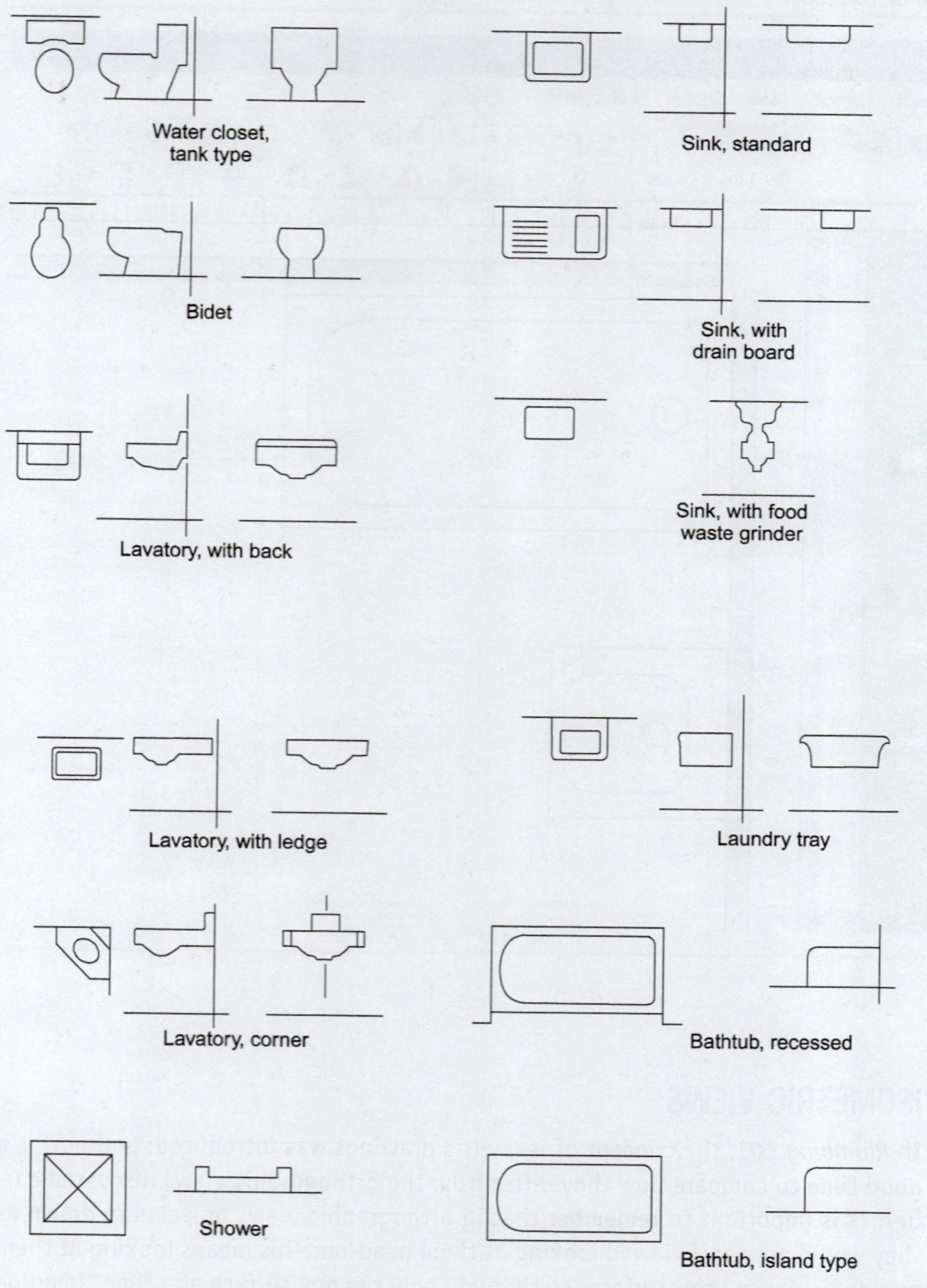

In a larger sense, the technical literature also provides drafting aids, in that codes, rough-in sheets, specifications, standards, and data tables have an effect on piping and fixture layouts.

FIGURE 23-12 CAD programs have symbols that can be pulled into scale drawings and resized or rotated.

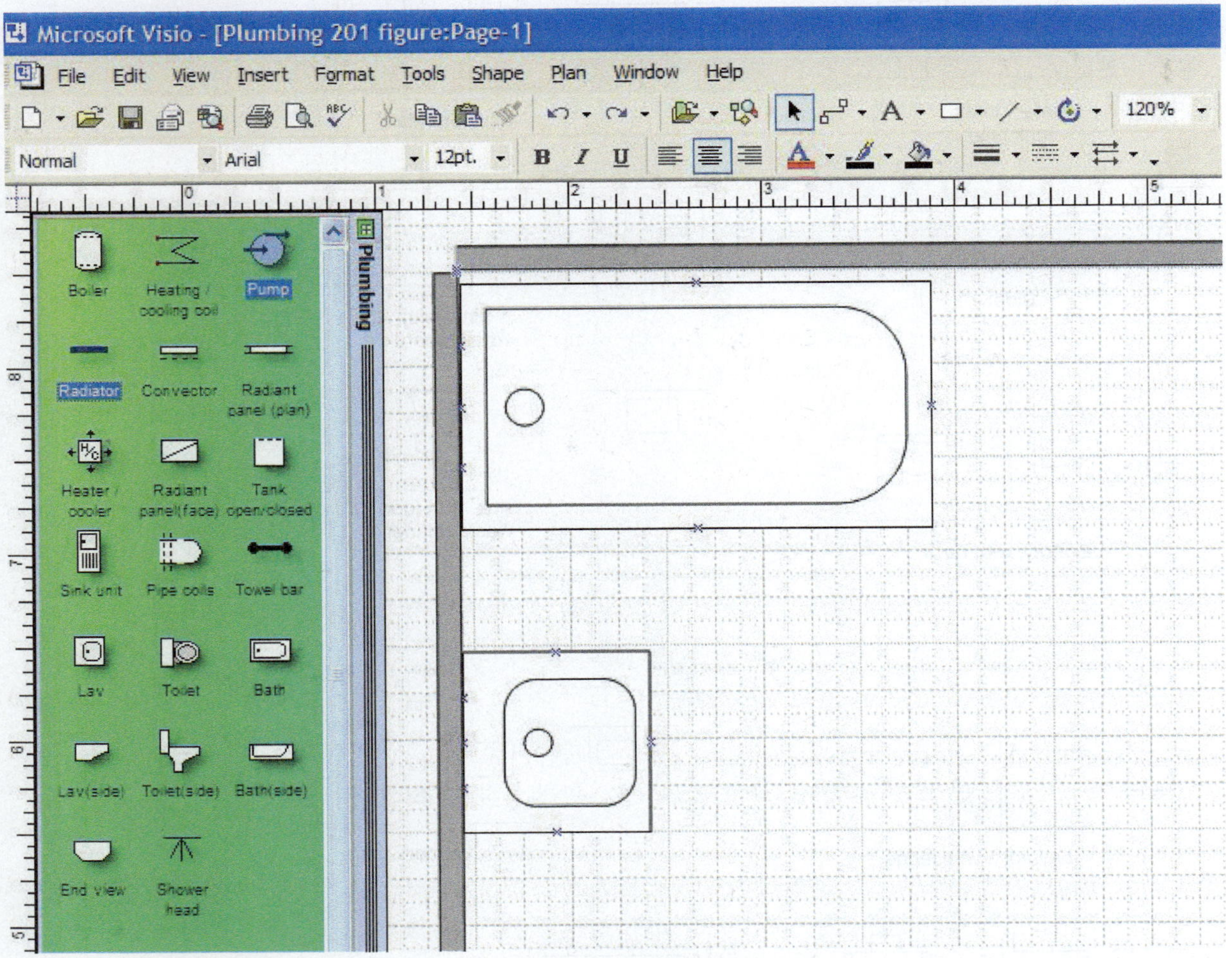

Source: Microsoft® Visio®.

ISOMETRIC VIEWS

In *Plumbing 101*, the concept of isometric drawings was introduced; and now is a good time to compare how they differ from the orthographic views discussed earlier. It is important to remember that in orthographic views objects are drawn as they would appear if viewed looking at them head-on. This means looking at them perpendicular to their surface, so that you only see one surface at a time. Imagine looking at this book while it is closed but all you can see is the cover. You could easily describe how tall the book is and how wide, but you could not determine how thick the book is. To see all three dimensions—height, width, and thickness—you would have to view the book at an angle. What you are viewing now is an isometric view of the book. All three dimensions of the book are visible, but none of them represent what the true shape is. It is for this reason; the shapes that were used in the orthographic views (see Figure 23-11) cannot be used in isometric views of a piping system.

In the following example, isometric views illustrate how the piping is routed to each fixture; they help the technician understand how all the fixtures are connected and how water will flow throughout the system. Since the angles and lengths are distorted in isometric drawings, it would not look correct to use the fixture symbols just discussed; only the piping and traps are shown. Also remember from previous discussions, that lines drawn vertically represent a change in elevation. Any lines drawn at an angle represent horizontal pipes with a slope.

EXAMPLE 3

Figure 23-13 shows a simple sketch of a residential arrangement, equipped with a kitchen sink, water closet, lavatory, and bathtub. The drawing also indicates the drainage and water connections for the building. Exact locations of the fixtures must be determined and shown on a carefully made scale drawing. We must determine the local code provisions and the particular fixtures to be used. We also must know if the owner has any requirements beyond basic code provisions, such as the type and thickness of wall and floor coverings, more space around fixtures, special mounting heights, accommodations for the physically challenged, and so on.

Figures 23-24 through 23-27 show rough-in sheets for the four fixtures for this job. We will use information from these sheets to make the single line detail drawings for the job.

First, make a plan view of the wall layout of the bathroom. Use $\frac{1}{4}$-inch or $\frac{1}{2}$-inch scale. Then place the fixtures in the desired arrangement. Finally, show the dimensions to drain and water openings so that the installing technician will have enough information to rough-in the installation. See Figure 23-14.

Figures 23-15, 23-16, and 23-17 show elevation views of this bathroom, but they are seldom needed to convey essential information. Elevation views are generally shown in the rough-in sheets.

An isometric drawing of the piping for this job is shown in Figure 23-18. The isometric sketch is always helpful to show the three-dimensional relationships, but it might not be made for a job of this type, depending upon the skill and experience of the installing technician.

The kitchen layout is developed next. Draw the plan view of the room and locate the sink. Figure 23-19 shows the side elevation of the kitchen sink, and Figure 23-20 shows the dimensions of waste and water openings. Sinks are frequently centered on a window. If so, note on the drawing how the drain or vent is to be provided. See Figure 23-21.

The dishwasher connection into the kitchen sink is shown in Figure 23-22. Notice how the dishwasher drain pipe is routed to form a high loop before it connects to the sink tailpiece. Figure 23-23 shows the entire residential drainage and venting system in an isometric sketch.

These single line drawings are prepared for clarification for the installing technician, coordination with other crafts on the job, and approval of the owner. It commonly happens that a fixture or pipe line cannot be located as shown on the original house plans. Developing the sketches as described may disclose the conflict. The sooner the conflict is recognized, the easier it is to solve.

EXAMPLE 3
continued

Make these sketches for your jobs. Go over the details with the superintendent or other technicians on the job. Be sure the owner knows about and signs off on any changes that have to be made. Install the job according to the sketch and keep the drawings with the job records for future reference.

FIGURE 23-13 Residential layout.

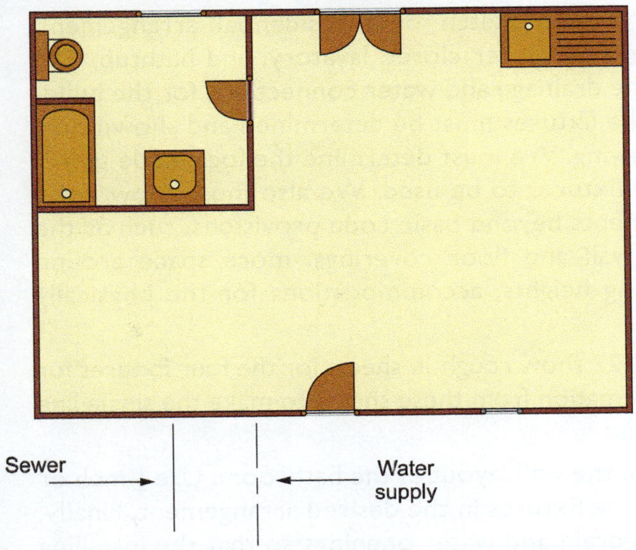

FIGURE 23-14 Plan view of residential bathroom (not to scale).

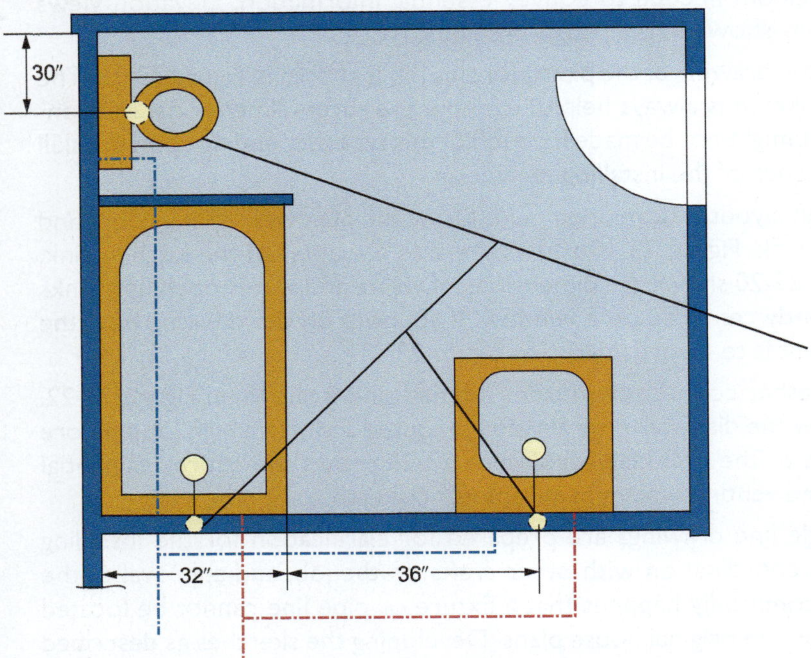

FIGURE 23-15 Elevation view of tub and shower (not to scale).

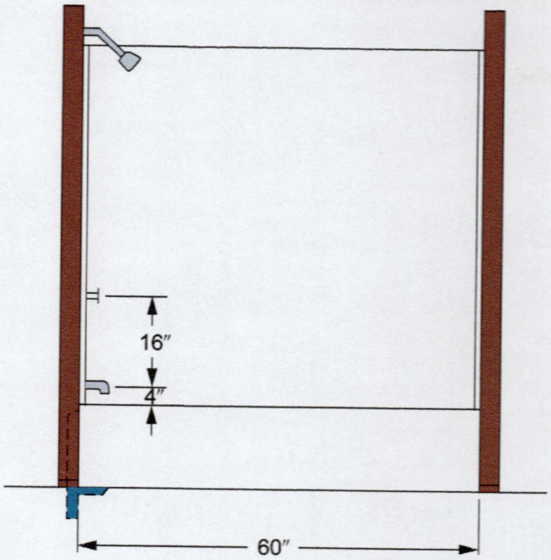

FIGURE 23-16 Elevation view of lavatory (not to scale).

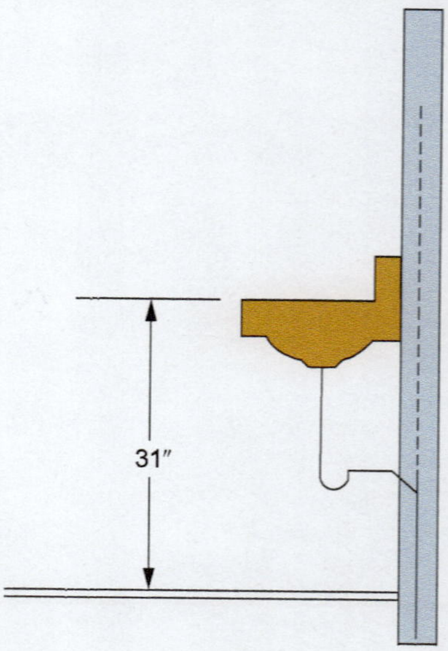

COMMERCIAL PROJECTS

Single line drawing of commercial projects is an extension of the techniques used for residential work. Such projects usually involve more fixtures of similar types than we find in residences, except that bathing fixtures are seldom used. Figure 23-28 shows typical commercial fixture simplified symbols.

FIGURE 23-17 Elevation view of water closet (not to scale).

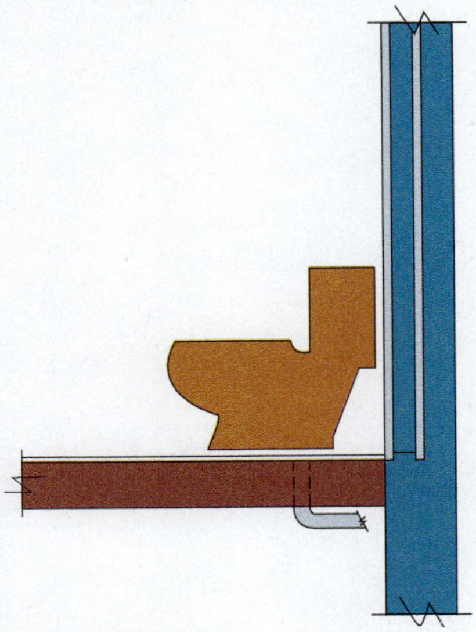

FIGURE 23-18 Isometric drawing of bath drainage piping (not to scale).

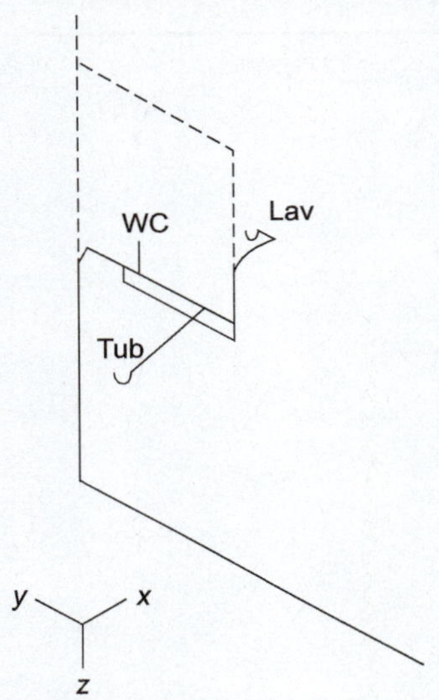

Figure 23-29 shows a plan view of a department store restroom containing five water closets and five lavatories. Figure 23-30 shows the plan view of the water closets (top portion) and the elevation view of the closet carriers (bottom portion).

FIGURE 23-19 Side elevation view of sink (not to scale).

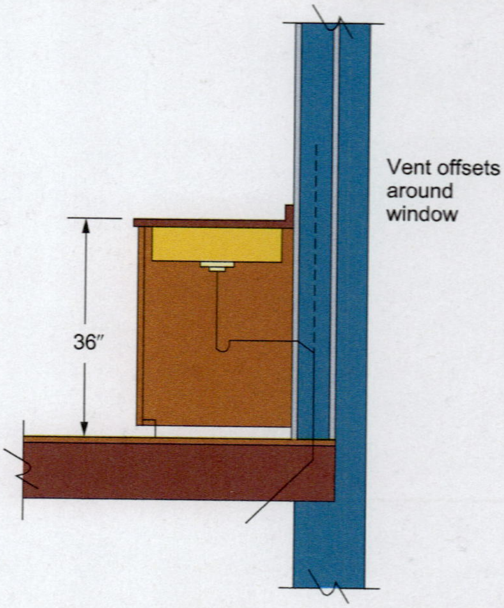

FIGURE 23-20 Elevation view of kitchen sink—styling and doors removed (not to scale).

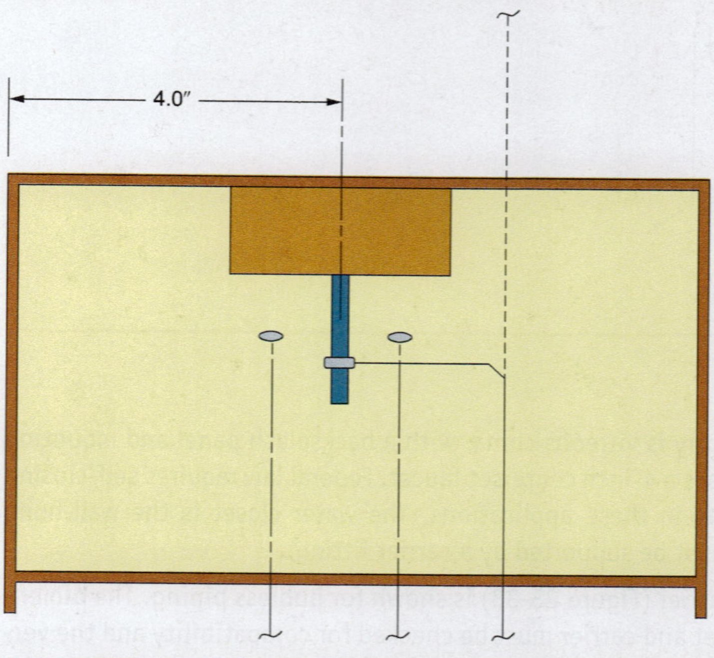

Figures 23-31 and 23-32 show the rough-in sheets for the fixtures. Figure 23-32 has a note that the elevation can be changed to comply with ADA guidelines.

FIGURE 23-21 Isometric drawing of sink drainage piping.

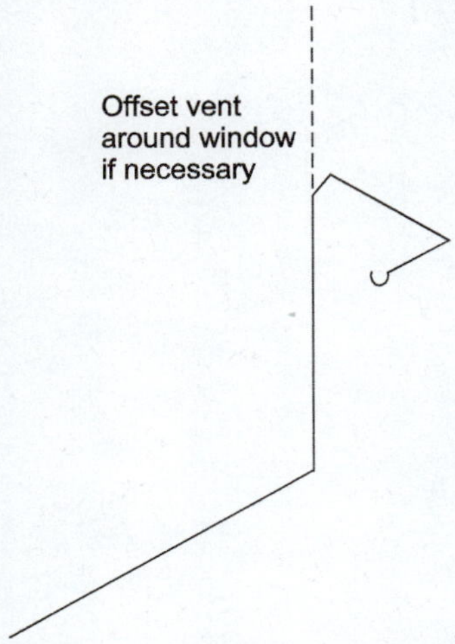

FIGURE 23-22 Isometric drawing of dishwasher and kitchen sink drainage piping.

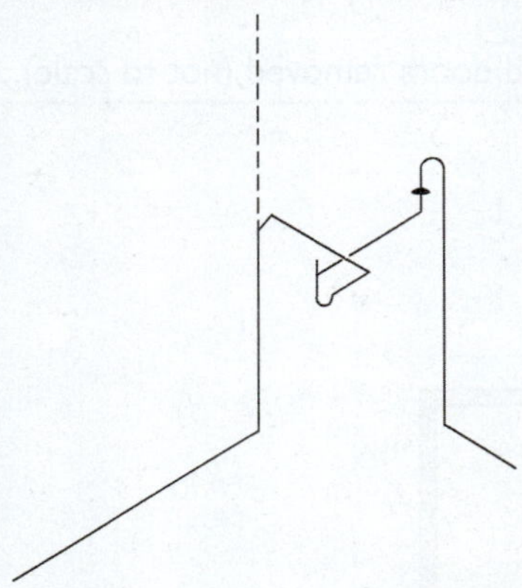

Note that the lavatory is vitreous china, with a backsplash panel and mounting bracket. The lavatory uses a 4-inch centerset faucet. Federal law requires self-closing or self-metering faucets in these applications. The water closet is the wall-hung type, which means it must be supported by a carrier fitting.

The water closet carrier (Figure 23-33) is shown for hubless piping. The dimensions of the water closet and carrier must be checked for compatibility and the vertical dimension of the bowl outlet must be taken from Figure 23-33 to set the carrier mounting plate properly.

Study the rough-in sheets to be sure that carrier and bowl match, faucet and lavatory match, and pipe sizing is correct for the fixtures and devices being used.

FIGURE 23-23 Isometric drawing of complete DWV system.

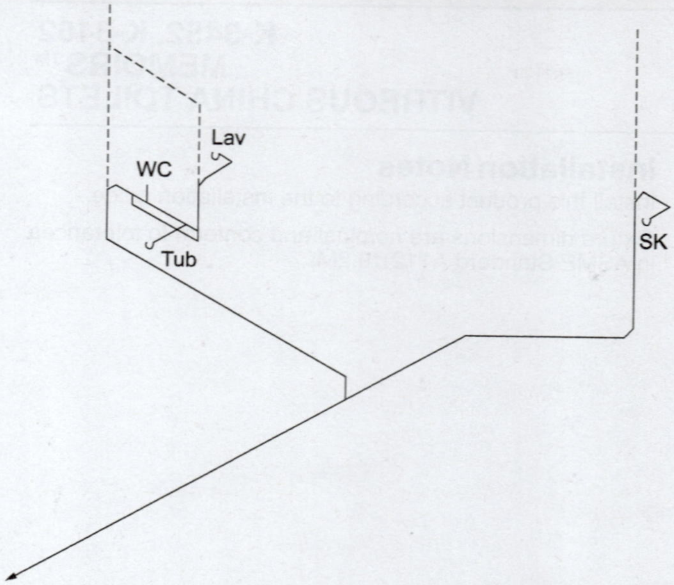

Figure 23-30 also shows the plan of the water closets and an elevation sketch of the piping for same water closets. The fixtures were drawn using a template, assuring uniformity and speed. The fixture locations are dimensioned so the rough piping locations can be determined. In the arrangement shown, each water closet is individually vented, but other venting methods may be used.

An isometric drawing is shown in Figure 23-34. This drawing shows drainage and venting only. To add water piping to this view would make it more confusing and reduce, rather than enhance, its usefulness. Water supply piping is seldom a problem in these applications, so the plan view representation is usually sufficient. Some building departments ask for isometric views of the water piping (properly sized) as a requirement to issue a permit. If this is the case in your area, a separate water piping isometric is usually preferred to combining water, waste, and vent in a single sketch.

The battery of five lavatories is shown in plan view in Figure 23-35 and elevation view in Figure 23-36. Each lavatory is individually vented (to make the installation similar to the water closets), but other arrangements are possible. The isometric view is shown in Figure 23-37.

You should practice making plan views, elevation views, and isometric drawings of fixture layouts in buildings with which you are familiar. A tabletop or board, straight edge, triangle (30°, –60°, –90°), fixture template, pencil, and paper are all you need to sharpen these skills. Use the symbols shown in Figure 23-28 to standardize commercial fixtures.

FIGURE 23-24 Water closet rough-in sheet.

KOHLER®
Roughing-In

K-3452, K-3462 MEMOIRS™ VITREOUS CHINA TOILETS

Product Information

Fixture:	
Configuration	2-piece, round front
Water per flush	1.6 gallons (6L)
Passageway	2" (5cm)
Water area	9-3/4" (24.8cm) x 9" (22.9cm)
Water depth from rim	5-3/8" (13.7cm)
Seat post hole centers	5-1/2" (14cm)
Included components:	
Bowl	K-4257
Tank with classic design	K-4454
Tank with stately design	K-4464
Tank cover, classic design	84406
Tank cover, stately design	84407
Trip lever	K-9439

Installation Notes

Install this product according to the installation guide.

Fixture dimensions are nominal and conform to tolerances in ASME Standard A112.19.2M.

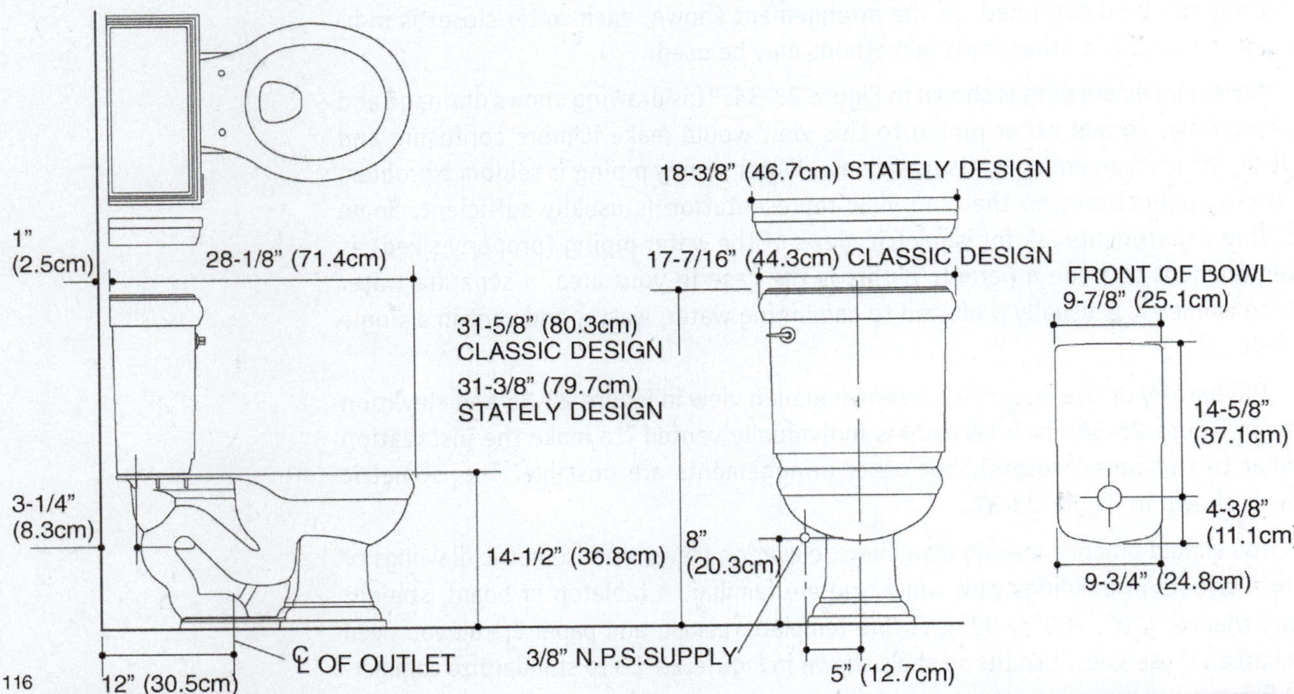

116

K-3452, K-3462 Memoirs™ Toilets
115754-1-**CC**

KOHLER®

FIGURE 23-25 Bathtub rough-in sheet.

KOHLER
Roughing-In®

K-721, K-722
MEMOIRS™
ENAMELED CAST IRON BATH

Ordering Information

Applicable product:		
Left outlet	K-721	shown
Right outlet	K-722	not shown
Accessories/hardware:		
Drain	K-7161-AF	recommended

Product Information

ADA compliant.			
Fixture*:	basin area	top area	weight
Bathing well	45" x 23"	54" x 25"	350 lbs.
	water depth	capacity	
To overflow	11-5/8"	52 gal.	
* Approximate measurements for comparison only.			

Installation Notes

Refer to installation instructions included with fixture before beginning installation.

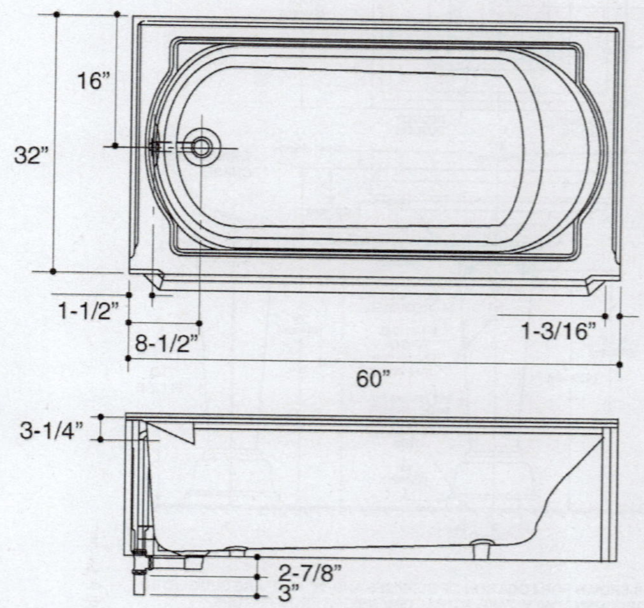

Roughing-In Notes
Fixture dimensions are nominal and conform to tolerances in ANSI/ASME Standard A112.19.1.
No change in measurements if connected with drain illustrated.

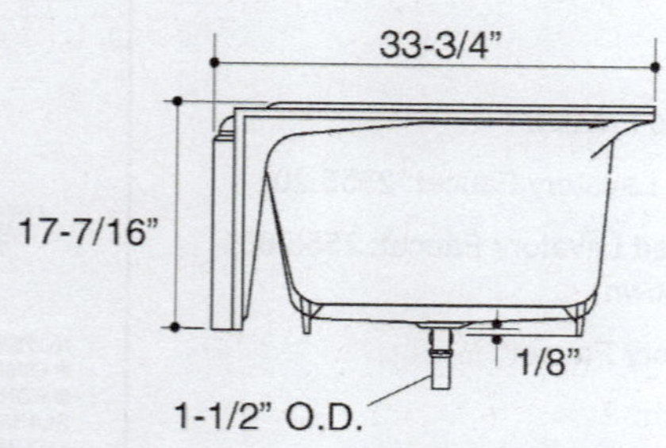

316

K-721, K-722 Memoirs™ Bath
115728-1-**AA** (A)

KOHLER®

Courtesy of Kohler Company

FIGURE 23-26 Pedestal sink rough-in sheet.

American Standard

TOWN SQUARE™ 27" PEDESTAL SINK

FIRECLAY

TOWN SQUARE 27" PEDESTAL SINK

- Historic American classic design
- Clean straight lines with classic ogee curves
- Rear overflow
- Supplied with mounting kit
- Design-matched to Town Square fixture and faucets
- Right Height™ for greater comfort during use
- Made from fireclay

☐ **0780.400** Faucet holes on 4" (102mm) centers
☐ **0780.800** Faucet holes on 8" (203mm) centers
☐ **0780.100** Center hole only

Nominal Dimensions:
27" x 21" x 36" (686 x 533 x 914mm)

Separate Components:
☐ **0780.004** Pedestal top 4" (102mm) centers
☐ **0780.008** Pedestal top 8" (203mm) centers
☐ **0780.001** Center hole only
☐ **0031.000** Pedestal only

Bowl sizes:
15" (381mm) wide, 12-1/2 (318mm) front to back, 6-1/2" (165mm) deep

Compliance Certifications - Meets or Exceeds the Following Specifications:
- ASME A112.19.9M for Non-Vitreous Ceramic
- CAN/CSA B45 Series

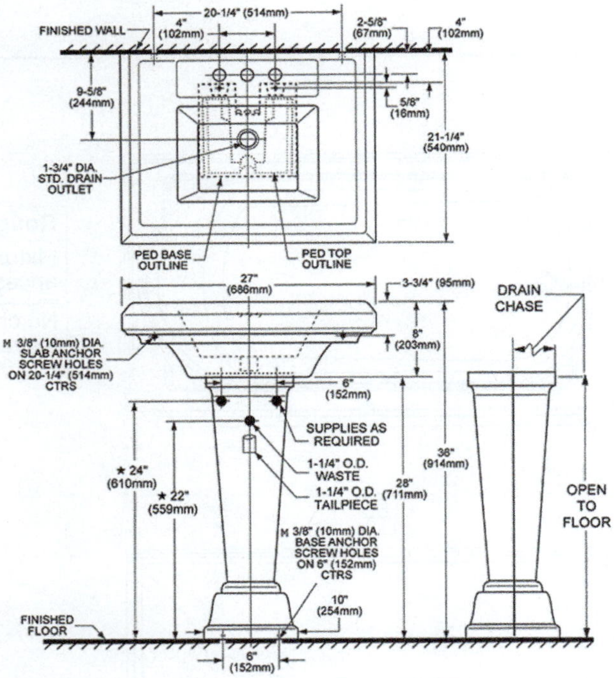

To Be Specified

☐ Color:

☐ Town Square Faucet:

☐ Centerset Lavatory Faucet: 2555.201

☐ Widespread Lavatory Faucet: 2555.801 (shown)

☐ Other Lavatory Faucet*:

☐ Faucet Finish:

☐ Supplies:

☐ 1-1/4" Trap:

* See faucet section for additional models available

NOTES:
★ DIMENSIONS SHOWN FOR LOCATION OF SUPPLIES AND "P" TRAP ARE SUGGESTED.
■ FOR REINFORCEMENT ONLY. TAKE ACTUAL DIMENSIONS FROM FIXTURE.
SLAB/PEDESTAL ANCHORS AND MOUNTING FEATURES TO BE FURNISHED BY OTHERS.
FITTINGS NOT INCLUDED AND MUST BE ORDERED SEPARATELY.
PROVIDE SUITABLE REINFORCEMENT FOR ALL WALL SUPPORTS.
INSTALLATION INSTRUCTIONS SUPPLIED WITH LAVATORY.
IMPORTANT: Dimensions of fixtures are nominal and may vary within the range of tolerances established by ANSI Standard A112.19.9M.
These measurements are subject to change or cancellation. No responsibility is assumed for use of superseded or voided pages.

BS-15

Revised 10/02

Courtesy of American Standard Bath & Kitchen

FIGURE 23-27 Kitchen sink rough-in sheet.

ELKAY®
SPECIFICATIONS

Celebrity® Single Bowl Sink
CR Series

GENERAL
Good quality Celebrity sink seamlessly drawn of #20 gauge, type 304 (18-8) nickel bearing stainless steel. Self-rimming.

DESIGN FEATURES
Bowl Depth: 7".

Coved Corners: 3" radius.

Bowl Recess: 3/16" below outside edge of sink. CR2522 and CRS3322 have 1/4" drop ledge.

Finish: Exposed surfaces are machine polished to a bright finish.

Underside: Fully undercoated.

OTHER
Drain Opening: 3-1/2".

Faucet Holes: Available in 3 or 4, 1-1/2" diameter faucet holes, 4" center to center.

NOTE: Unless otherwise specified, sink is furnished with 4 faucet holes as shown.

These sinks comply with ANSI Standard A112.19.3M.

These sinks are listed by the International Association of Plumbing and Mechanical Officials as meeting the requirements of the Uniform Plumbing Code.

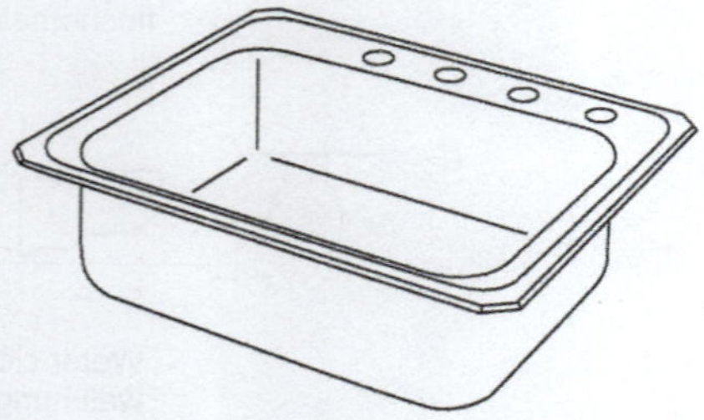

Model CR25214

SINK DIMENSIONS (INCHES)*

Model Number	Overall		Inside Bowl			Cutout in Countertop (1 1/2" Radius Corners)		No. of 1 1/2" Dia. Faucet Holes 4" Centers	Minimum Cabinet Size	Ship. Wt. Lbs.
	L	W	L	W	D	L	W			
CR1721	17	21 1/4	14	15 3/4	7	16 3/8	20 5/8	3	21	9 1/2
CR2521	25	21 1/4	21	15 3/4	7	24 3/8	20 5/8	3 or 4	30	11 3/4
CR2522	25	22	21	15 3/4	7	24 3/8	21 3/8	3 or 4	30	12
CR3122	31	22	28	15 3/4	7	30 3/8	21 3/8	3 or 4	36	15 1/4
CRS3322	33	22	28	15 3/4	7	32 3/8	21 3/8	3 or 4	36	15 1/2

*Length is left to right. Width is front to back.

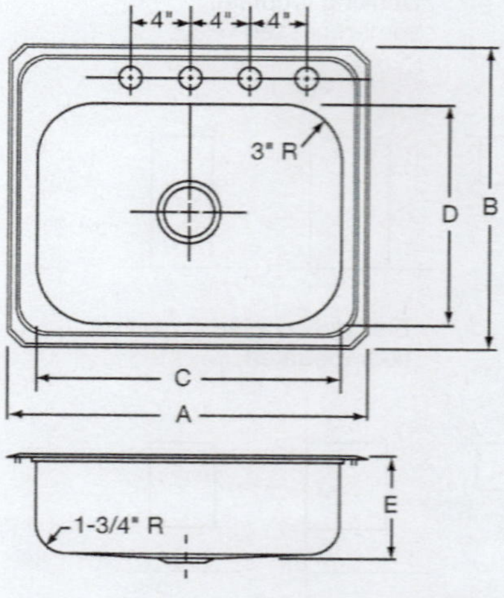

ALL DIMENSIONS IN INCHES, TO CONVERT TO MILLIMETERS MULTIPLY BY 25.4.

In keeping with our policy of continuing product improvement, Elkay reserves the right to change product specifications without notice.

This specification describes an Elkay product with design, quality and functional benefits to the user. When making a comparison of other producers' offerings, be certain these features are not overlooked.

Elkay
www.elkayusa.com

2222 Camden Court
Oak Brook, IL 60523

Printed in U.S.A.

(Rev. 1/04) 1-12D

FIGURE 23-28 Plumbing fixture symbols—commercial.

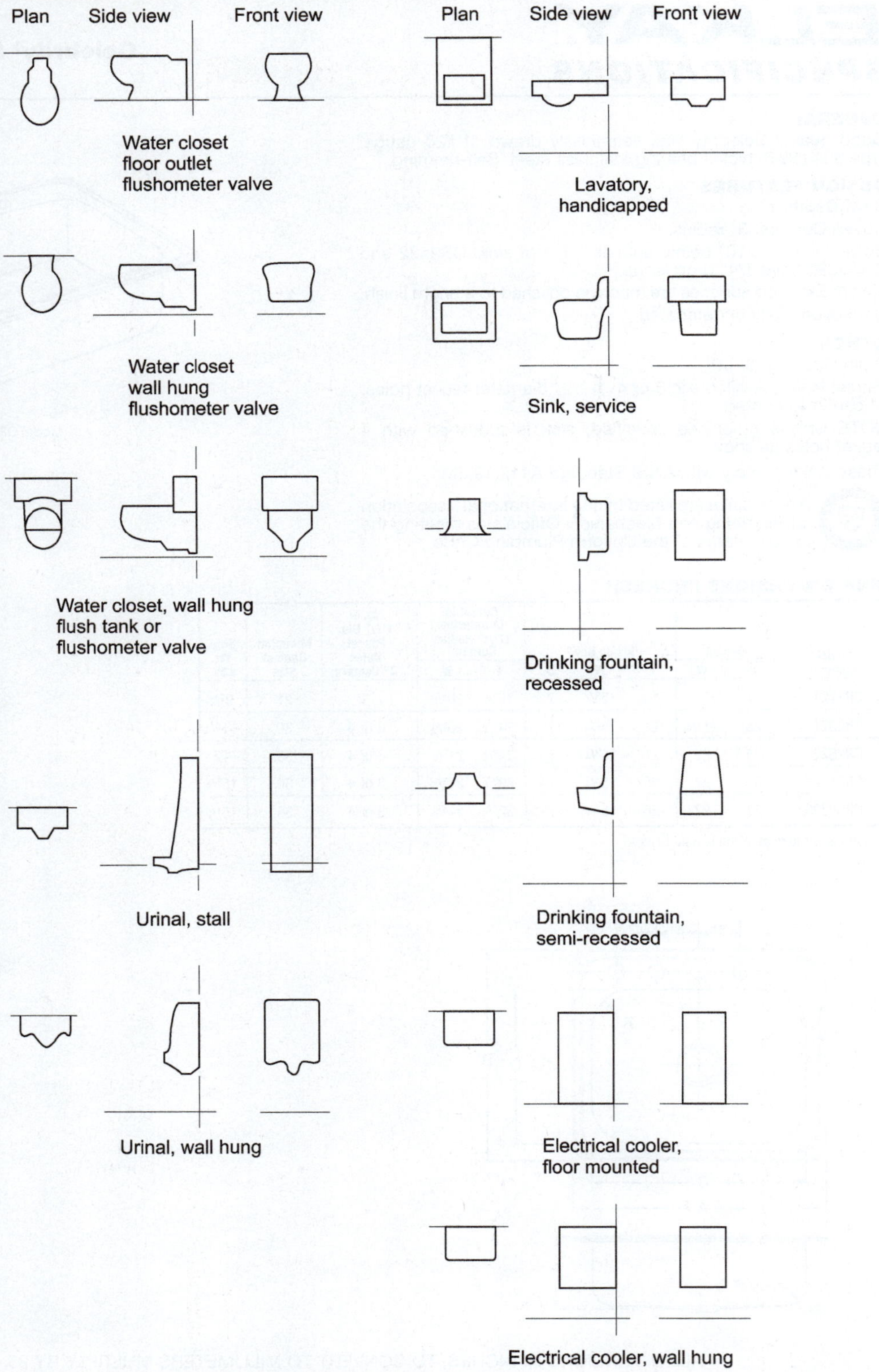

FIGURE 23-29 Plan view of a commercial bathroom—battery of fixtures.

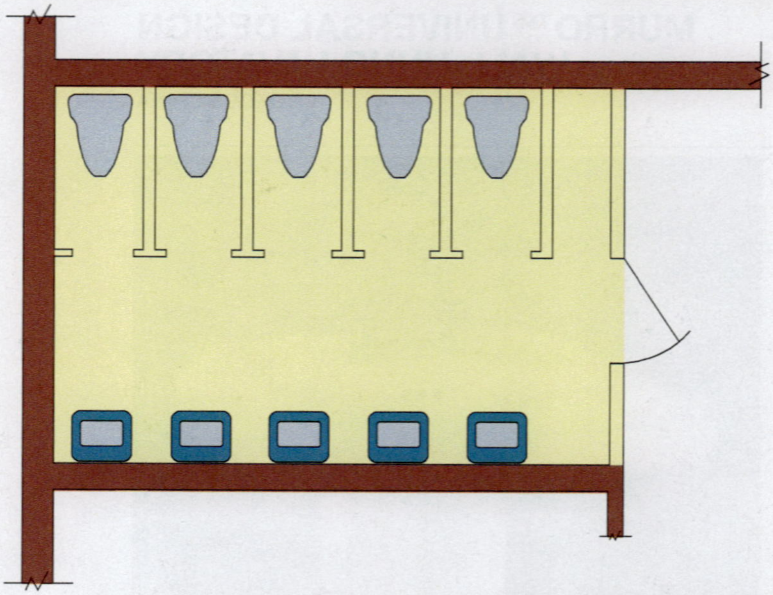

FIGURE 23-30 Plan view of water closets and an elevation view of carriers and vent pipe.

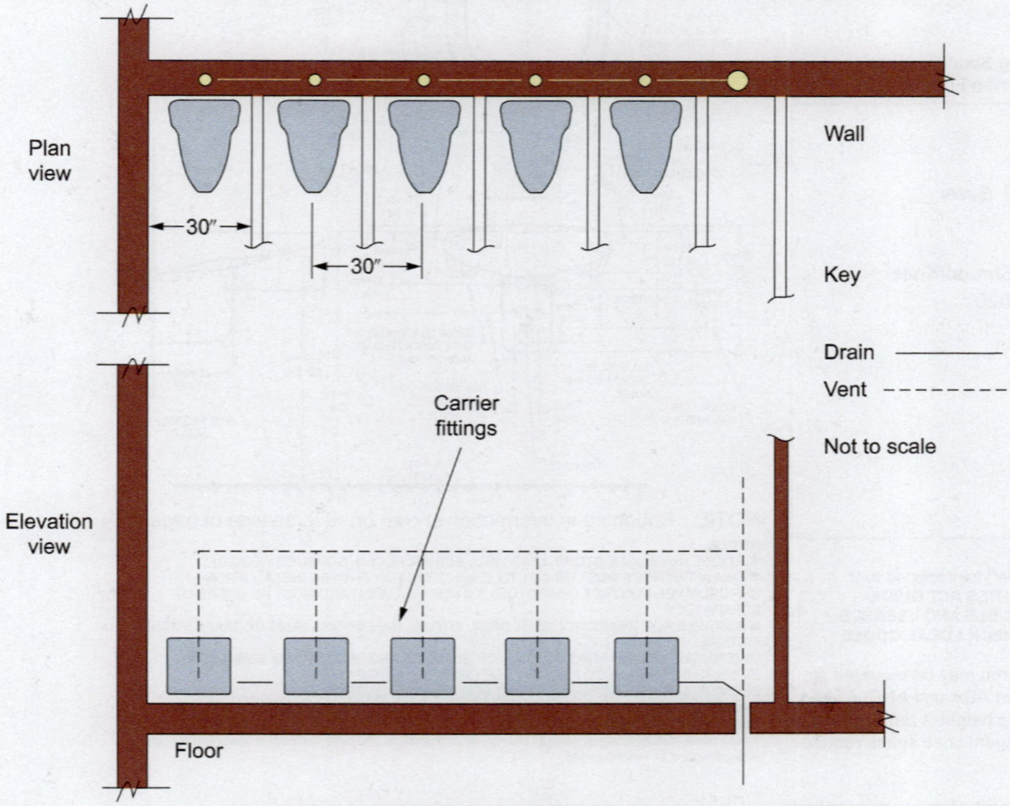

FIGURE 23-31 Lavatory rough-in sheet.

American Standard

BARRIER FREE

MURRO™ UNIVERSAL DESIGN WALL-HUNG LAVATORY

VITREOUS CHINA

MURRO™ UNIVERSAL DESIGN WALL-HUNG LAVATORY

- Vitreous china
- Rear overflow
- Recessed self-draining deck
- For concealed arm or wall support
- Shown with optional vitreous china shroud/ knee contact guard 0059.020 available

❑ **0954.000 Faucet holes on 102mm (4") Ctrs** (Illustrated)
❑ **0954.023 Faucet holes on 102mm (4") Ctrs**
• Extra right-hand hole
❑ **0954.021 Faucet holes on 102mm (4") Ctrs**
• Extra left-hand hole
❑ **0958.000 Faucet holes on 203mm (8") Ctrs**
❑ **0955.000 Center hole only**
❑ **0955.023 Center hole faucet**
* Extra right-hand hole
❑ **0955.021 Center hole faucet**
* Extra left-hand hole

Nominal Dimensions: 559mm (22") deep, 540mm (21-1/4") wide

Bowl sizes:
394mm (15-1/2") wide, 343mm (13-1/2") front to back, 127mm (5") deep

❑ **0059.020 Shroud/Knee Contact Guard** (Vitreous China)

Compliance Certifications - Meets or Exceeds the Following Specifications:
- ASME A112.19.2 for Vitreous China Fixtures

To Be Specified

Color: ❑ White ❑ Bone ❑ Silver ❑ Shell ❑ Black

❑ Optional Vitreous China Shroud/Knee Contact guard: 0059.020

❑ Faucet*:

❑ Faucet Finish:

❑ Supplies:

❑ 1-1/4" Trap:

❑ Nipple:

* See faucet section for additional models available

Top of front rim mounted 864mm (34") from finished floor. **MEETS THE AMERICAN DISABILITIES ACT GUIDE-LINES AND ANSI A117.1 ACCESSIBLE AND USEABLE BUILDINGS AND FACILITIES - CHECK LOCAL CODES.**

For Universal Design Options, top of rim may be mounted at 813mm (32") from finished floor to meet ADA and ANSI A117.1 requirements. A 864mm (34") mounting height is required for California Title 24 and other more stringent knee space requirements. Check local codes.

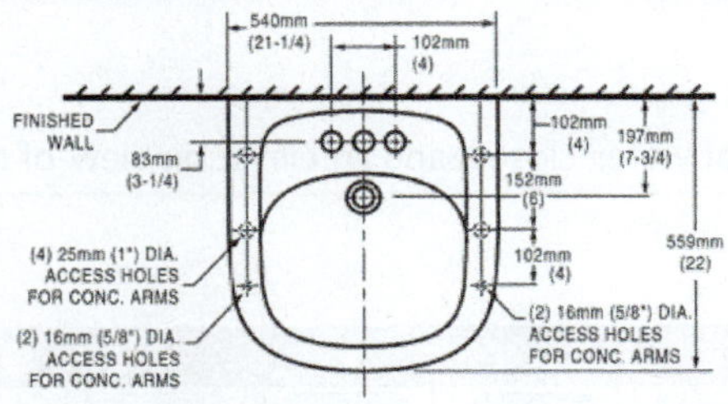

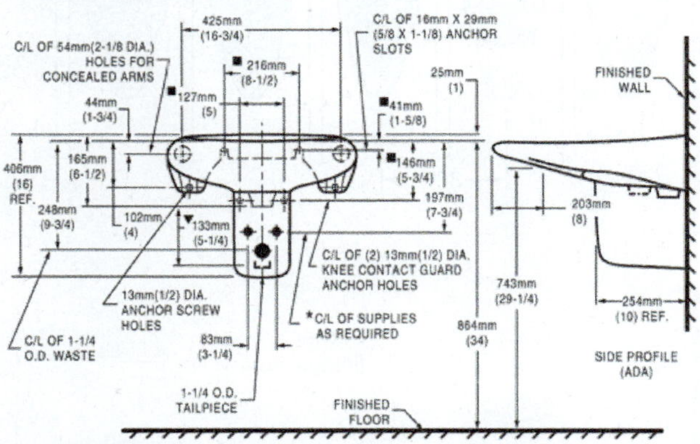

NOTE: Roughing-in information shown on reverse side of page

NOTES:
* LOOSE KEY ANGLE STOPS, LESS WALL ESCUTCHEONS. SUPPLIES REQUIRED.
▼ DRAIN TAILPIECE MUST BE CUT TO 67mm (2-5/8) FOR PROPER INSTALLATION.
SHROUD/KNEE CONTACT GUARD 0059.020 NOT INCLUDED AND MUST BE ORDERED SEPARATELY.
■ SUITABLE FOR REINFORCEMENT ONLY, ACTUAL DIMENSIONS MUST BE TAKEN FROM FIXTURE.
DIMENSIONS SHOWN FOR LOCATION OF SUPPLIES AND "P" TRAP ARE SUGGESTED.
FITTINGS NOT INCLUDED AND MUST BE ORDERED SEPARATELY.
PROVIDE SUITABLE REINFORCEMENT FOR ALL WALL SUPPORTS.
IMPORTANT: Dimensions of fixtures are nominal and may vary within the range of tolerances established by ANSI Standard A112.19.2.
These measurements are subject to change or cancellation. No responsibility is assumed for use of superseded or voided pages.

CI-51

Revised 9/04

FIGURE 23-31 Lavatory rough-in sheet. (Continued)

American Standard

BARRIER FREE

MURRO™ UNIVERSAL DESIGN WALL-HUNG LAVATORY

VITREOUS CHINA

0954.023 4" CTRS EXTRA RIGHT HAND HOLE

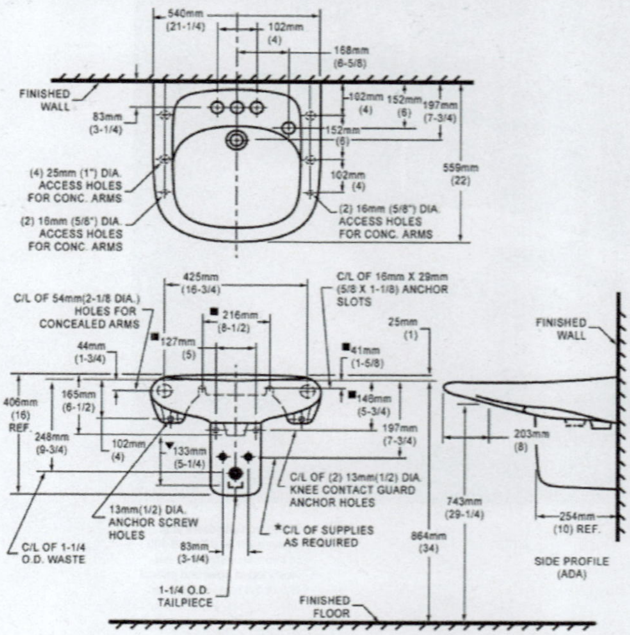

0954.021 4" CTRS EXTRA LEFT HAND HOLE

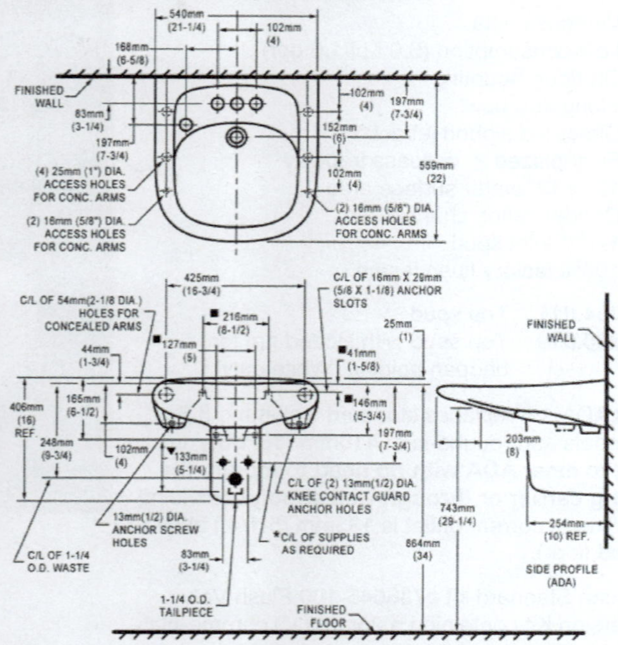

0955.023 CENTER HOLE WITH EXTRA RIGHT HAND HOLE

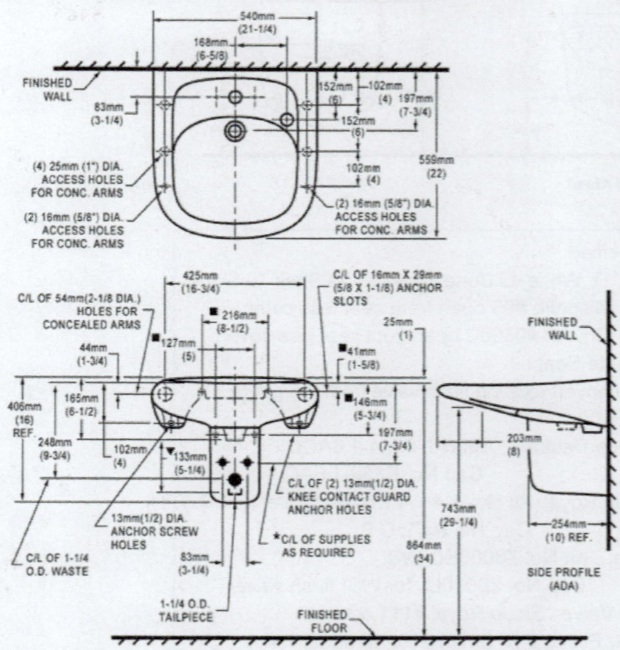

0955.021 CENTER HOLE WITH EXTRA LEFT HAND HOLE

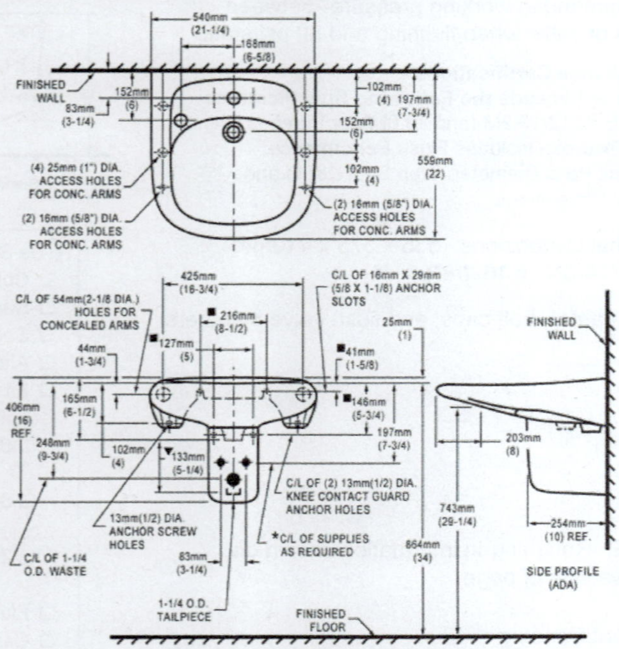

CI-52

FIGURE 23-32 ADA water closet rough-in sheet.

American Standard

BARRIER FREE

AFWALL™ ADA RETROFIT ELONGATED FLUSH VALVE TOILET

VITREOUS CHINA

AFWALL™ ADA RETROFIT ELONGATED*

- Vitreous china
- Low-consumption (6.0 Lpf/1.6 gpf)
- Off-floor mounting
- Elongated bowl
- Direct-fed siphon jet action
- Fully-glazed 2" ballpass trapway
- 10" x 12" water surface area
- Condensation channel
- 1-1/2" inlet spud
- 100% factory flush tested

❑ **2294.011** Top spud
❑ **2296.019** Top spud with slotted rim for bedpan holding (White only)

Afwall ADA 1.6 replaces standard wall-hung 3.5 or 1.6 models and roughs-in at 410mm (16-1/8") rim height **to meet ADA with no need to modify the existing carrier or through wall supply** (providing centerline of current outlet is 133mm [5-1/4"] above finished floor).

American Standard kit #736046-100 Flush Valve Conversion Kit containing a 25mm (1") chrome-plated street ell, a vacuum breaker, and a 330mm (13") tail piece included to meet minimum 152mm (6") code between bowl and vacuum breaker.

Recommended working pressure--between 30 psi at valve when flushing and 80 psi static

Compliance Certifications -
Meets or Exceeds the Following Specifications:
- ASME A112.19.2M (and 19.6M) for Vitreous China Fixtures - includes Flush Performance, Ball Pass Diameter, Trap Seal Depth and all Dimensions

Nominal Dimensions: 635 x 375 x 410mm (25" x 14-3/4" x 16-1/8")

Seat, carrier, bolt caps, and flush valve by others.

NOTE: Roughing-in information shown on reverse side of page.

* Patent Pending

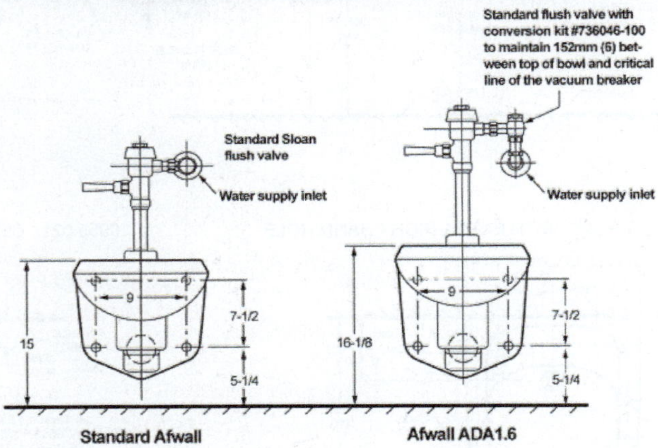

To Be Specified

❑ Color: ❑ White ❑ Bone ❑ Silver ❑ Shell ❑ Black
❑ Seat: Olsonite #95 open front seat less cover
❑ Seat: Church #9500C open front seat less cover
❑ Alternate Seat:
❑ Kit to convert existing flush valve from 3.5 gpf to 1.6 gpf:
  ❑ Coyne Delaney: Kit No. F144-1.6ACQ
  Cap No. F159-1.6AQ
  ❑ Sloan Royal: Kit No. A-41-A 1.6 Gal Closet Kit LC
  Cap No. A-72 C.P.
  ❑ Zurn: Kit No. Z6000EC-WSI
  Cap No. Z6000LL for WSI flush valve
❑ Flush Valve: Sloan Royal #111 or equal
❑ Carrier Fitting J.R. Smith 210L or equal

SPS 2294/2296

COM/INS-002

Revised 6/95

FIGURE 23-32 ADA water closet rough-in sheet. (Continued)

American Standard

AFWALL™ ADA RETROFIT ELONGATED FLUSH VALVE TOILET

VITREOUS CHINA

2294.011/2296.019

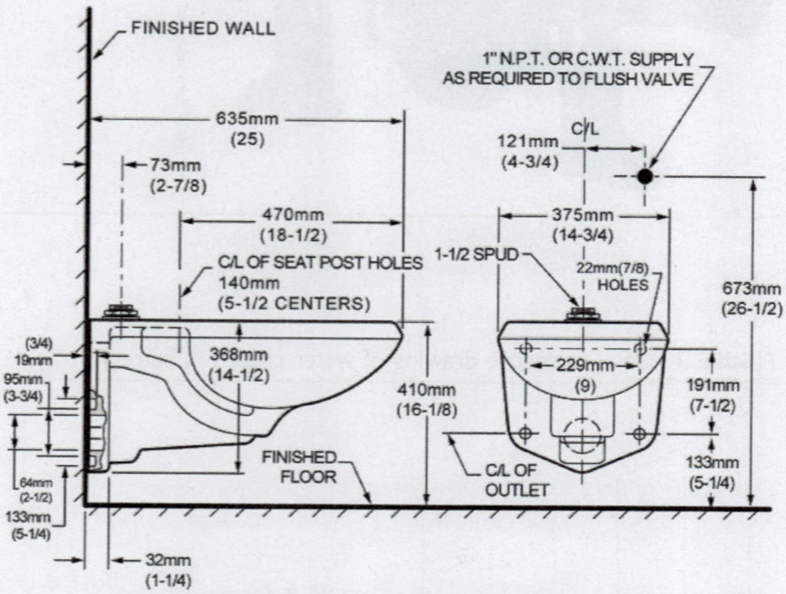

● When installed so top of seat is 432 to 483mm (17" to 19") from the finished floor.
MEETS THE AMERICAN DISABILITIES ACT GUIDELINES AND ANSI A117.1 REQUIREMENTS FOR ACCESSIBLE AND USEABLE BUILDING FACILITIES-CHECK LOCAL CODES.

NOTES:

PRODUCT 2294.011 SHOWN, 2296.019 SAME AS EXCEPT WITH SLOTTED RIM FOR BED PAN HOLDING.

* REQUIRES AMERICAN STANDARD KIT 736046-100 TO UPGRADE EXISTING INSTALLATION OF FLUSH VALVE AND COMPONENTS.

WASTE OUTLET SEAL RING MUST BE NEOPRENE OR GRAPHITE-FELT (WAX RING NOT RECOMMENDED).

SUGGESTED 1/16 CLEARANCE BETWEEN FACE OF WALL AND BACK OF BOWL.

TO COMPLY WITH AREA CODE GOVERNING THE HEIGHT OF VACUUM BREAKER ON THE FLUSH VALVE, THE PLUMBER MUST VERIFY DIMENSIONS SHOWN FOR SUPPLY ROUGHING.

FLUSH VALVE NOT INCLUDED WITH FIXTURE AND MUST BE ORDERED SEPARATELY.

CARRIER FITTING AS REQUIRED TO BE FURNISHED BY OTHERS.

PROVIDE SUITABLE REINFORCEMENT FOR ALL WALL SUPPORTS.

IMPORTANT: Dimensions of fixtures are nominal and may vary within the range of tolerance established by ANSI Standard A112.19.2
These measurements are subject to change or cancellation. No responsibility is assumed for use of superseded or voided pages.

SPS 2294/2296

FIGURE 23-33 Water closet carrier with a side horizontal discharge.

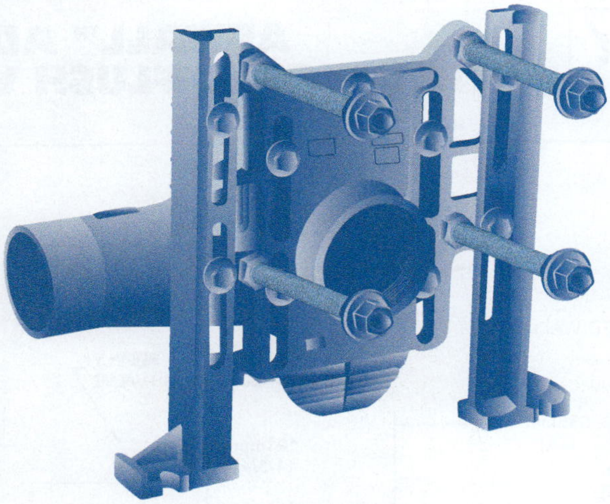

Courtesy of Zurn Industries, LLC

FIGURE 23-34 Isometric drawing of water closets in battery.

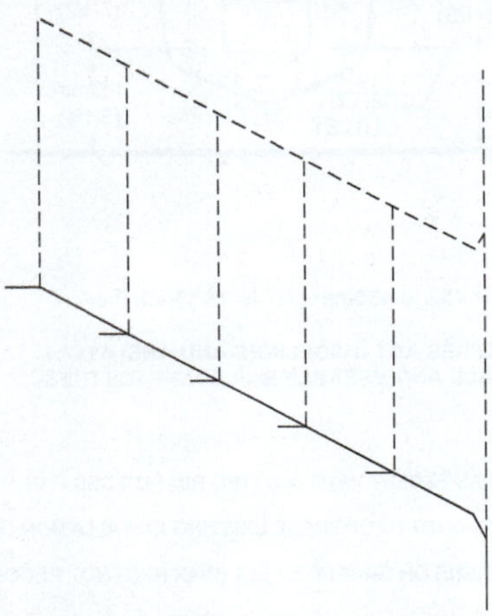

FIGURE 23-35 Plan view of lavatories.

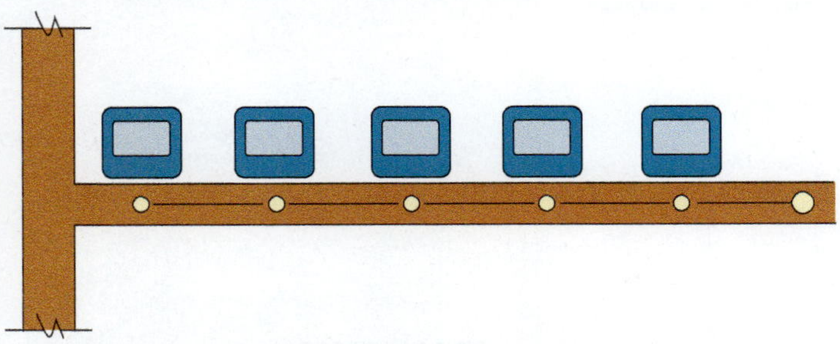

FIGURE 23-36 Elevation view of lavatories.

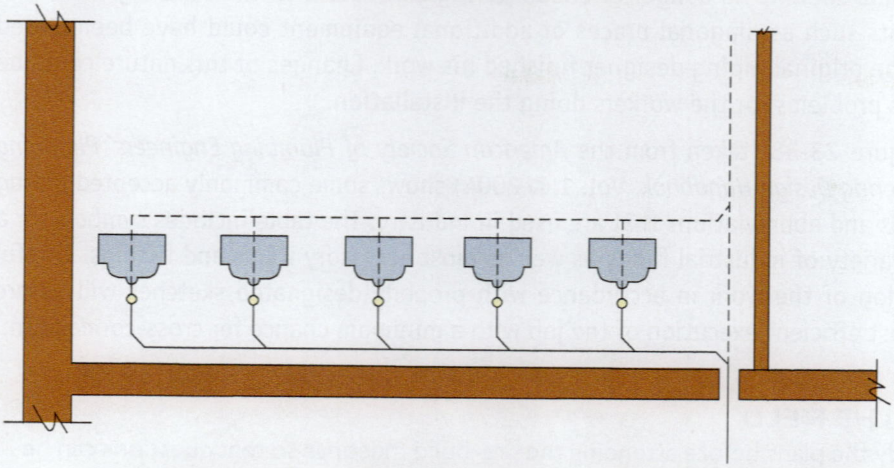

FIGURE 23-37 Isometric drawing of lavatory layout.

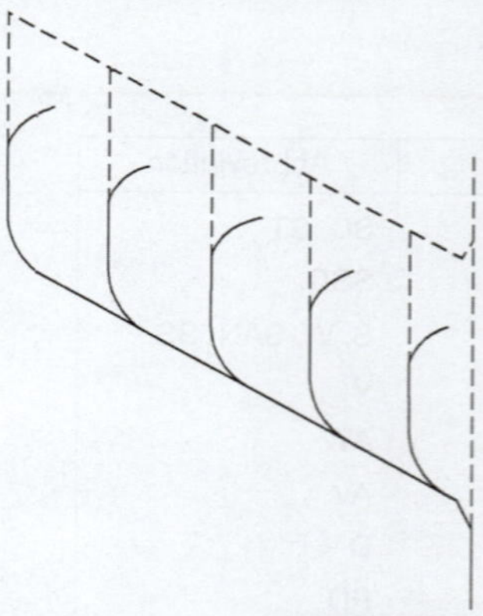

INDUSTRIAL PIPING LAYOUTS

Typical industrial piping layouts include systems for gases or liquids that are not encountered in residential or commercial buildings. Because these systems are larger and contain more individual components and elements, carefully drawn plans are often necessary. When close tolerances are required and when conflicts at turns or branch takeoffs are possible, such plans are mandatory. Since there could be a wide variety of fluids or gases in the system, the piping must also identify the product contained to help guard against accidental cross-connections.

IN THE FIELD

You should always make certain that existing piping has been depressurized and cleared for work before starting new work. Pipes that contained chemicals may need to be flushed or purged before new work can begin. There have been cases where workers have used torches to cut into existing lines that still contained fumes of benzene (or other flammable materials) and caused an explosion.

One of the first things the technician should do is to check with the foreman to make sure there is no danger of encountering conflicts with the building structure. Elements such as diagonal braces or additional equipment could have been added after the original piping designer finished his work. Changes of this nature could be serious problems for the workers doing the installation.

Figure 23-38 (taken from the *American Society of Plumbing Engineers' Plumbing Engineering Design Handbook*, Vol. 1 © 2004) shows some commonly accepted piping symbols and abbreviations that are used in industry. The table includes symbols for a large variety of industrial fluids as well as most accessory items and fittings. Careful execution of the work in accordance with properly designated sketches will assure the most efficient execution of the job with a minimum chance for cross-connection.

IN THE FIELD

Study the plans before attending the pre-build meetings so that questions can be addressed early on in the project. Make sure a contact person is named for each trade performing work.

FIGURE 23-38 Piping symbols and abbreviations.

Symbol	Description	Abbreviation
——— SD ———	Storm drain, rainwater drain	SD, ST
— ——— SSD ———	Subsoil drain, footing drain	SSD
——— SS ———	Soil, waste, or sanitary sewer	S, W, SAN, SS
— — — — — —	Vent	V
——— AW ———	Acid waste	AW
— — – AV – — —	Acid vent	AV
——— D ———	Indirect drain	D
——— PD ———	Pump discharge line	PD
——— – ———	Cold water	CW
——— – – ———	Hot water supply (140°F)	HW
——— – – – ———	Hot water recirculating (140°F)	HWR
——— TW ———	Tempered water (temp. °F)	TEMP. HW, TW
——— TWR ———	Tempered water recirculating (temp. °F)	TEMP. HWR, TWR
——— DWS ———	(Chilled) drinking water supply	DWS
——— DWR ———	(Chilled) drinking water recirculating	DWR
——— SCW ———	Soft cold water	SCW
——— CD ———	Condensate drain	CD
——— DI ———	Distilled water	DI
——— DE ———	Deionized water	DE
——— RO ———	Reverse osmosis water	RO
——— CWS ———	Chilled water supply	CWS
——— CWR ———	Chilled water return	CWR
——— LS ———	Lawn sprinkler supply	LS
——— F ———	Fire protection water supply	F
——— G ———	Gas—low-pressure	G
——— MG ———	Gas—medium-pressure	MG
——— HG ———	Gas—high-pressure	HG
— — – GV – — —	Gas vent	GV

FIGURE 23-38 Piping symbols and abbreviations. (Continued)

Symbol	Description	Abbreviation
——— FOS ———	Fuel oil supply	FOS
——— FOR ———	Fuel oil return	FOR
— — — FOV — — —	Fuel oil vent	FOV
——— LO ———	Lubricating oil	LO
— — — LOV — — —	Lubricating oil vent	LOV
——— WO ———	Waste oil	WO
— — — WOV — — —	Waste oil vent	WOV
——— O_2 ———	Oxygen	O_2
——— LO_2 ———	Liquid oxygen	LO_2
——— A ———	Compressed air	A
——— X#A ———	Compressed air—X#c	X#A
——— MA ———	Medical compressed air	MA
——— LA ———	Laboratory compressed air	LA
——— HPCA ———	High pressure compressed air	HPCA
——— HHWS ———	(Heating) hot water supply	HHWS
——— HHWR ———	(Heating) hot water return	HHWR
——— V ———	Vacuum	VAC
——— NPCW ———	Non-potable cold water	NPCW
——— NPHW ———	Non-potable hot water	NPHW
——— NPHWR ———	Non-potable hot water return	NPHWR
——— MV ———	Medical vacuum	MV
——— SV ———	Surgical vacuum	SV
——— LV ———	Laboratory vacuum	LV
——— N_2 ———	Nitrogen	N_2
——— N_2O ———	Nitrous oxide	N_2O
——— CO_2 ———	Carbon dioxide	CO_2
——— WVC ———	Wet vacuum cleaning	WVC
——— DVC ———	Dry vacuum cleaning	DVC
——— LPS ———	Low-pressure steam supply	LPS
— — — LPC — — —	Low-pressure condensate	LPC
——— MPS ———	Medium-pressure steam supply	MPS
— — — MPC — — —	Medium-pressure condensate	MPC
——— HPS ———	High-pressure steam supply	HPS
— — — HPC — — —	High-pressure condensate	HPC
— — — ATV — — —	Atmospheric vent (steam or hot vapor)	ATV
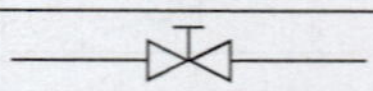	Gate valve	GV
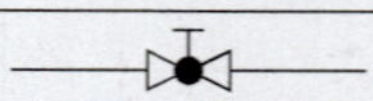	Globe valve	GLV
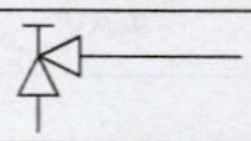	Angle valve	AV
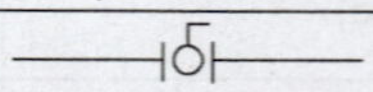	Ball valve	BV
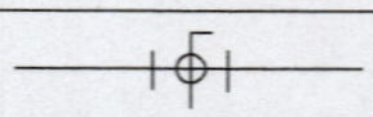	Butterfly valve	BFV
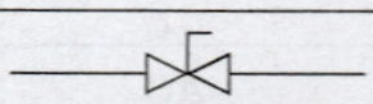	Gas cock, gas stop	
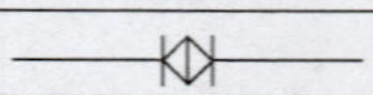	Balancing valve (specify type)	BLV
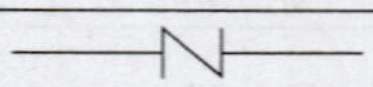	Check valve	CV
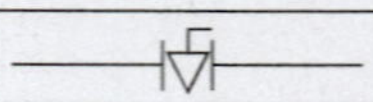	Plug valve	PV

FIGURE 23-38 Piping symbols and abbreviations. (Continued)

Symbol	Description	Abbreviation
S	Solenoid valve	
M	Motor-operated valve (specify type)	
	Pressure-reducing valve	PRV
	Pressure-relief valve	RV
	Temperature-pressure-relief valve	TPV
	Backflow preventer	RZBP
	Hose bibb	HB
	Recessed-box hose bibb or wall hydrant	WH
	Valve in yard box (valve type symbol as required for valve use)	YB
	Union (screwed)	
	Union (flanged)	
	Strainer (specify type)	
	Pipe anchor	PA
	Pipe guide	
	Expansion joint	EJ
	Flexible connector	FC
	Tee	
	Concentric reducer	
	Eccentric reducer	
	Aquastat	
FS	Flow switch	FS
PS	Pressure switch	PS
	Water hammer arrester	WHA
	Pressure gauge with gauge cock	PG
	Thermometer (specify type)	
AV	Automatic air vent	AAV
	Valve in riser (type as specified or noted)	
	Riser down (elbow)	
	Riser up (elbow)	
	Air chamber	AC
	Rise or drop	
	Branch—top connection	

FIGURE 23-38 Piping symbols and abbreviations. (Continued)

Symbol	Description	Abbreviation
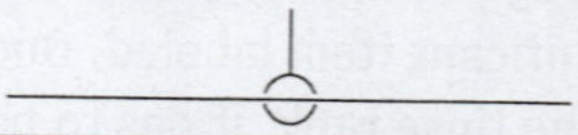	Branch—bottom connection	
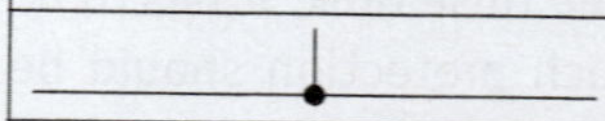	Branch—side connection	
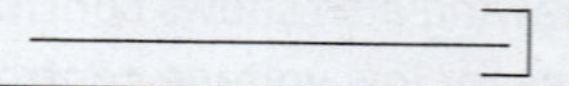	Cap on end of pipe	
	Cleanout plug	CO
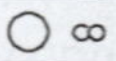	Floor cleanout	FCO
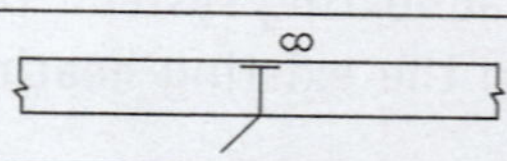	Wall cleanout	WCO
	Yard cleanout or cleanout to grade	CO
	Drain (all types) (specify)	D
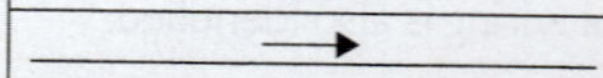	Pitch down or up—in direction of arrow	
	Flow—in direction of arrow	
	Point of connection	POC
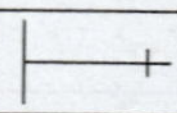	Outlet (specify type)	
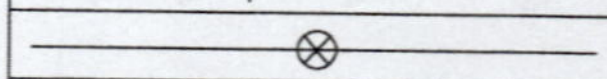	Steam trap (all types)	
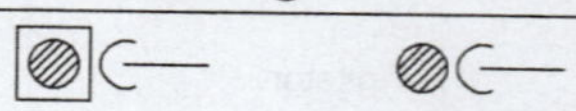	Floor drain with p-trap	FD

Figure 23-39 shows some examples of symbols for special fixtures. These symbols, plus the others discussed in this chapter, are typical of those frequently used in industrial plans. Careful study of such plans will enable you to understand most of the job requirements before beginning the work. Consultation with the owner,

FIGURE 23-39 Examples of plumbing symbols used in industrial settings.

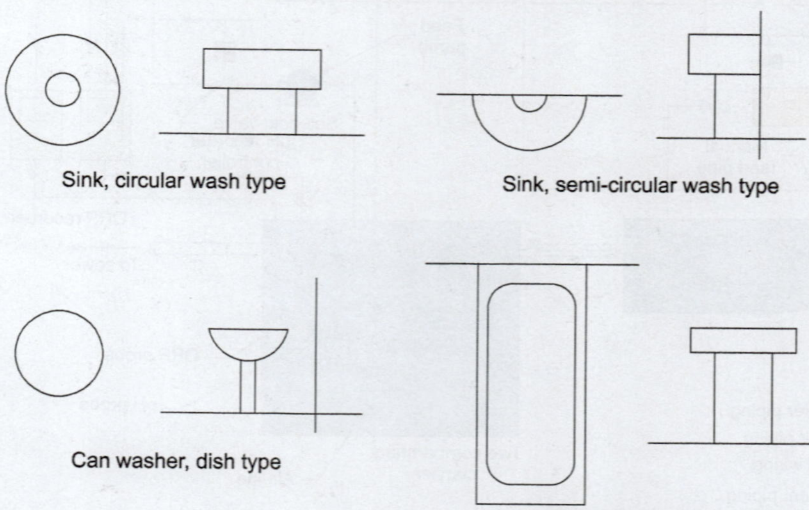

general contractor, designer, or architect is also usually required to develop a complete understanding of what must be done.

Figure 23-40 shows the neutralizing operation of a sewage treatment plant. The drawing includes a key to the symbols used, with each significant item labeled. One omission should be noted: if potable water is supplied to the rinse tank, it has to be protected with an appropriate backflow preventer and such protection should be shown on the drawing. It is important to notice that this drawing also shows control wiring and air lines. Some control engineers use dashed lines for low voltage control or air lines, so don't automatically assume it is a vent.

Figure 23-41 shows a portion of a plan for an industrial heating system. The work to be performed is to add piping and equipment to the existing heating

FIGURE 23-40 Detail of sewage treatment plan layout. Notice control wiring is also identified.

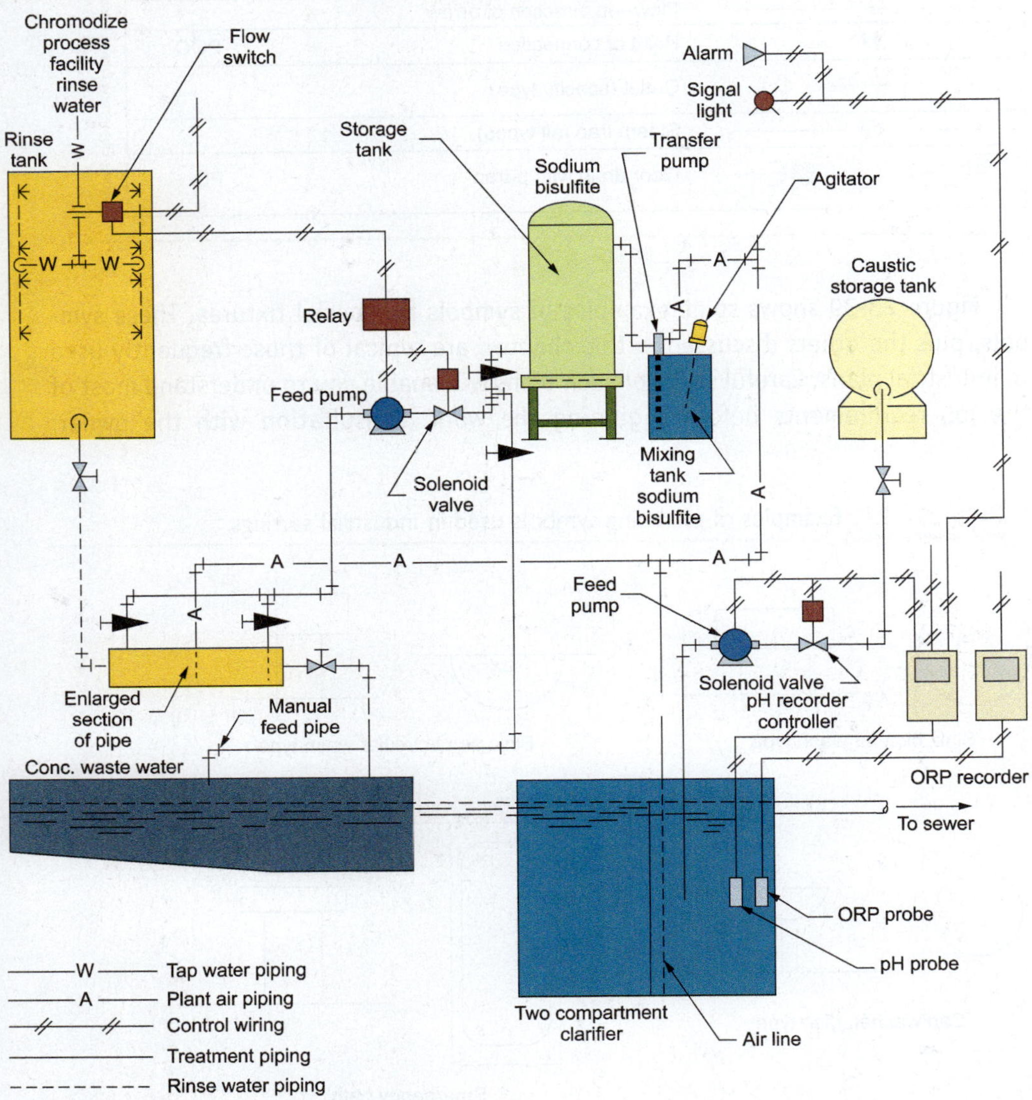

FIGURE 23-41 Plan view of piping plan.

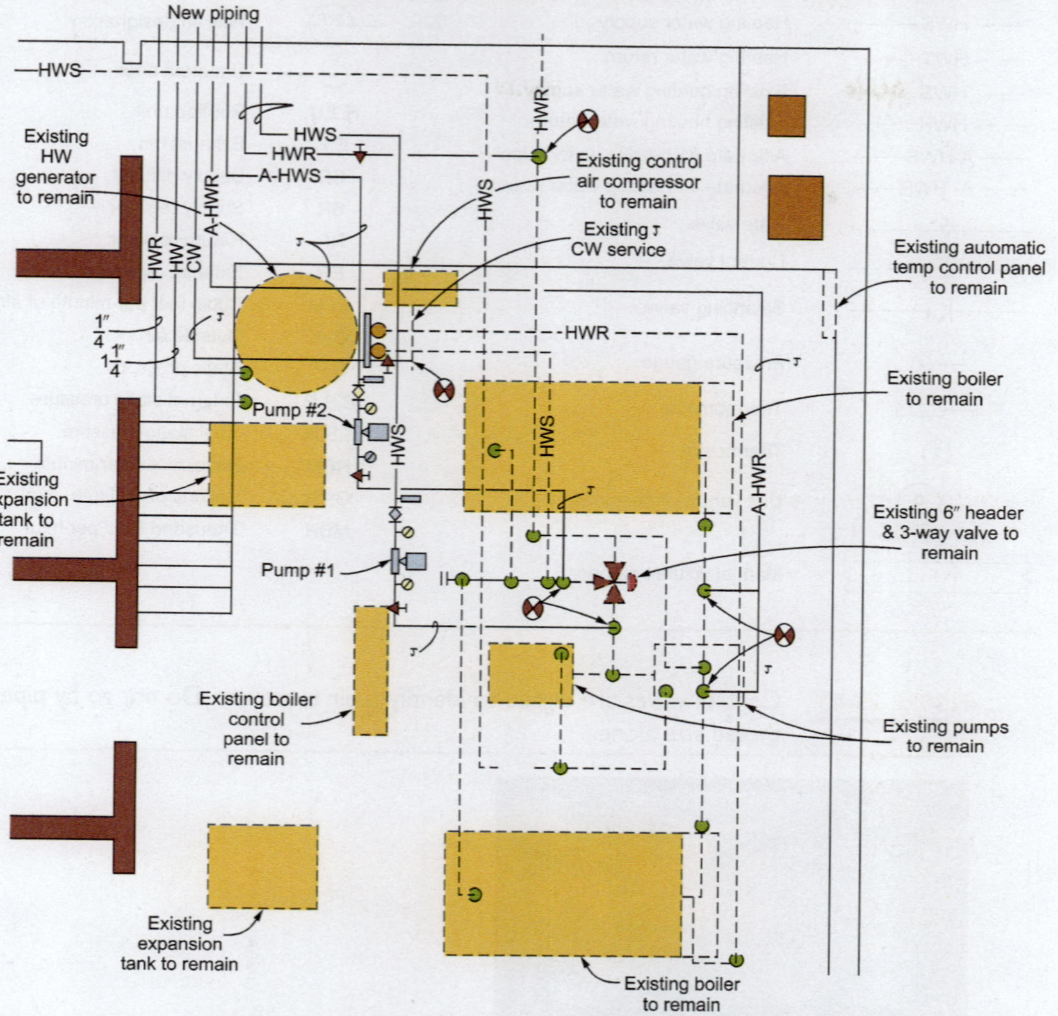

piping. Figure 23-42 shows the architect's symbols used in Figure 23-41. Notice that the piping is parallel or perpendicular to building walls and that equipment that remains unchanged is shaded to distinguish these items from new equipment. New piping is shown with solid lines and existing piping is shown with dashed lines.

When working with a hydronic system that contains control valves, it should be pointed out that not all valves are the same, even when they have the same size pipe thread and valve body. Engineers size valves so that the flow can be controlled very accurately, which means they need valves with different internal port sizes. Some manufacturers make three or four valves that all have $\frac{3}{4}$-inch pipe threads, but very different internal components. The controls engineer will supply a valve schedule with product numbers for each valve used. Figure 23-43 shows a control valve with a stamped metal tag that identifies it.

FIGURE 23-42 Mechanical symbols key for Figure 23-41.

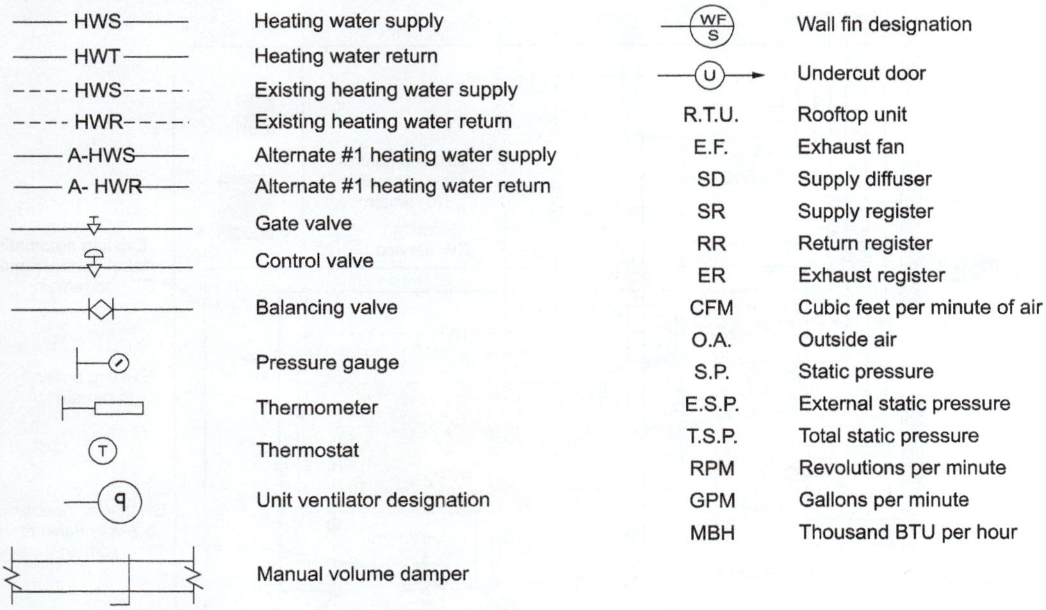

FIGURE 23-43 Control valves are tagged to identify their exact size. Do not go by pipe thread size alone.

Globe valves like the one shown in Figure 23-43 need to be installed with the flow going in the correct direction. If installed backward, they will chatter and not work properly. For this reason, control companies specify the direction of flow on the valve schedules. An incorrect installation will have to be corrected by the installer, for no additional pay.

FIGURE 23-44 Storage tank separate from main boiler.

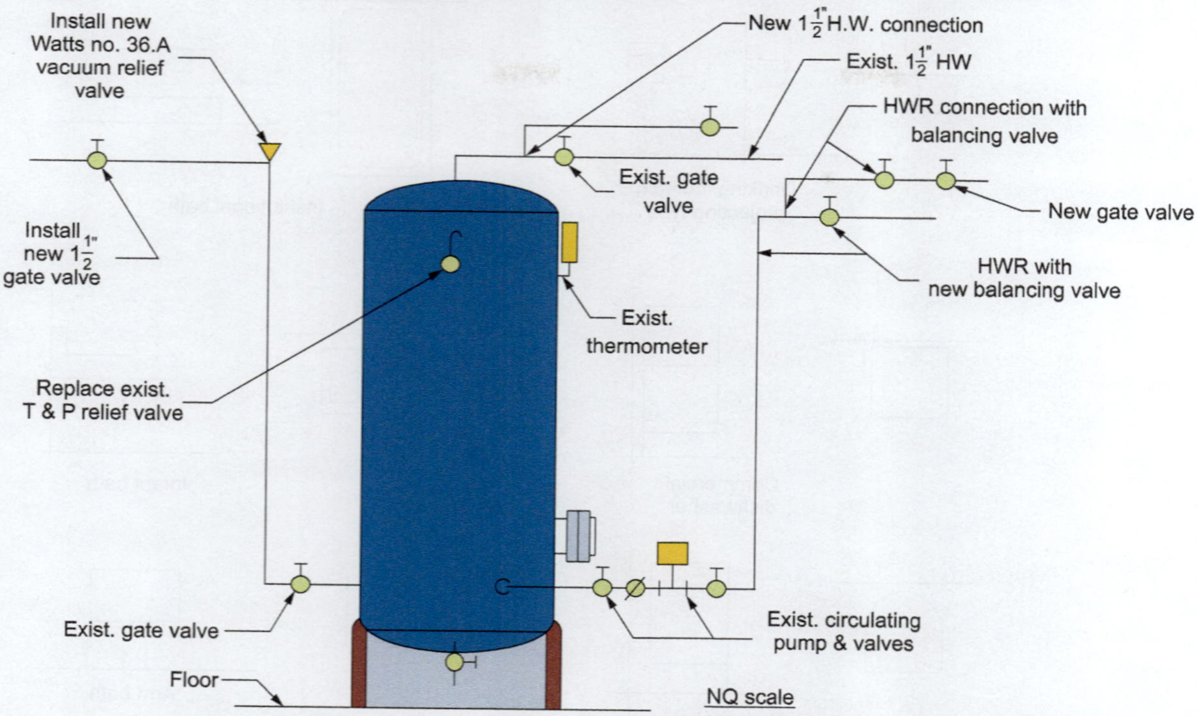

Figure 23-44 shows a small portion of the work indicated in Figure 23-41. This particular element is the piping at the hot water generator. New piping connections to existing work are shown, as well as numerous notes and details.

INSTITUTIONAL INSTALLATIONS

Institutional buildings are usually designed for a specific (or at least narrow) use by or for the public. These buildings are intended for many years of service, so the products and systems are selected for long life and ease of maintenance and resistance to vandalism.

Such specific use frequently means special purpose plumbing installations. For example, it would not be sensible to install child-size plumbing fixtures in a house, but it is often done in educational facilities for small children. Standard plumbing fixtures may be installed in a special way, for example, with unconventional spacing or mounting heights to achieve special purposes. Figure 23-45 shows symbols for these fixtures.

Figure 23-46 shows a pair of school restrooms, arranged for back-to-back installation of the fixtures. Symbols for urinals, water closets, and lavatories are used. The note VTR means *vent through roof*. The elevation view in Figure 23-47 shows the piping layout in more detail. Make certain that you can work with both figures (plan and elevation views) to get a complete picture of the job. Individual fixture venting is shown, but other venting styles

FIGURE 23-45 Plumbing fixture symbols commonly used in institutional settings.

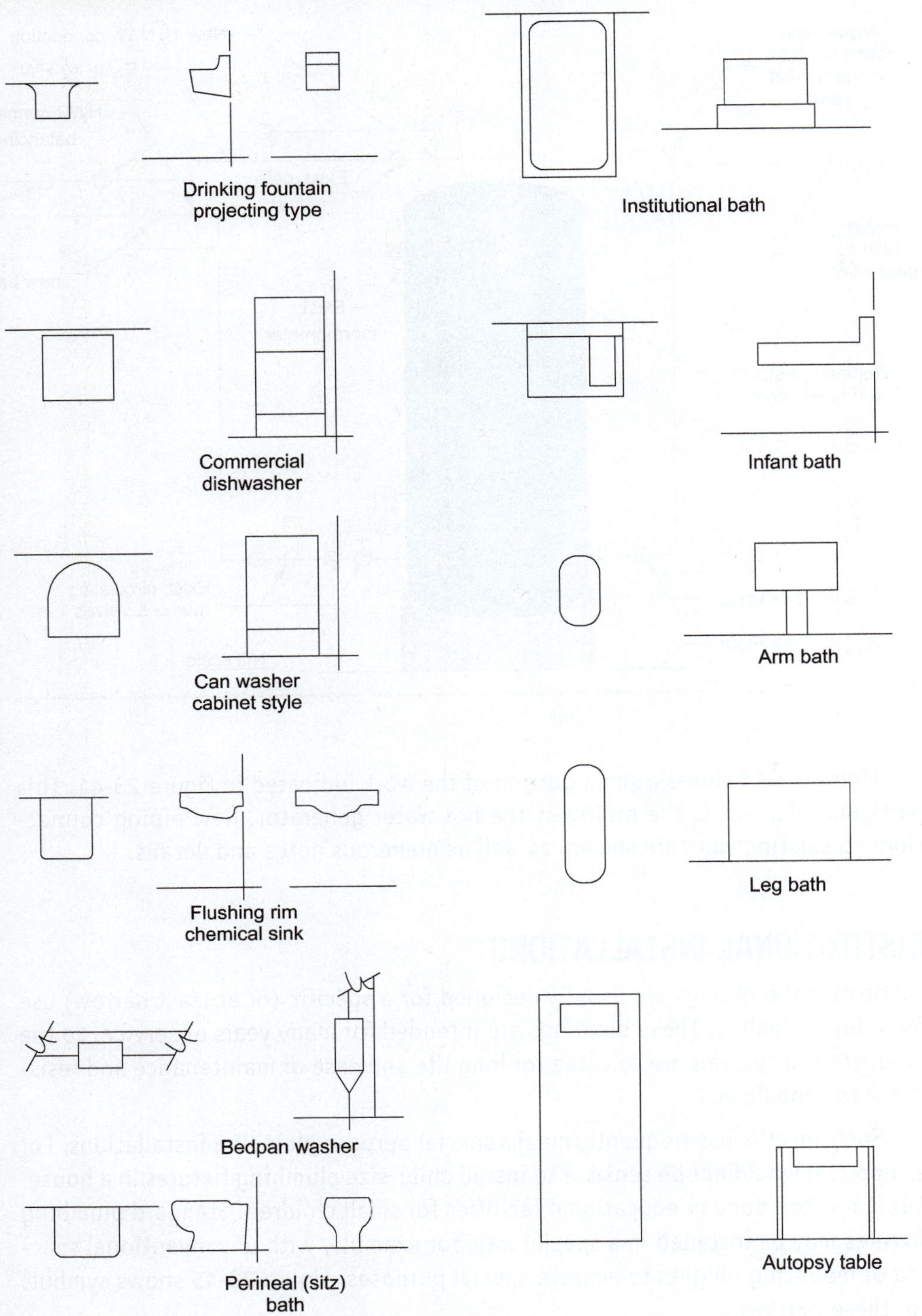

may be permitted by local code. Figure 23-48 shows the water pipe layout (appropriately sized) for these restrooms. In both cases, consult local codes for sizing methods.

FIGURE 23-46 School bathroom layout (plan view).

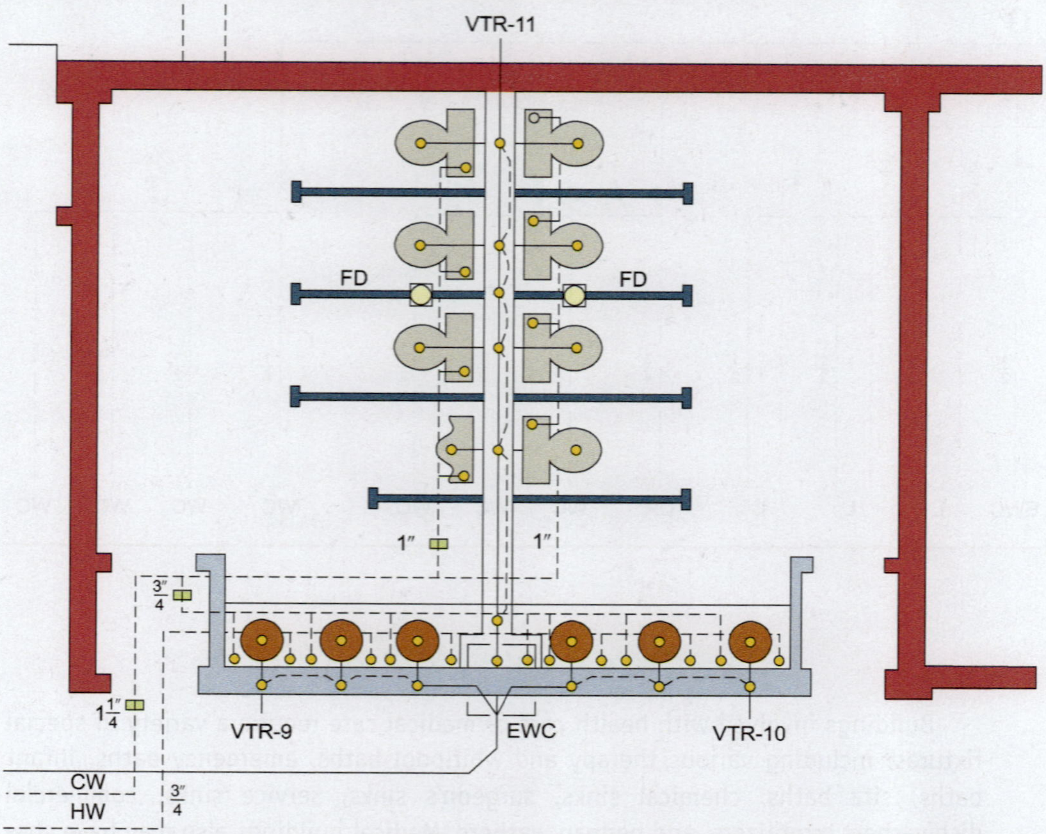

FIGURE 23-47 Single line sketch of the drainage and venting system.

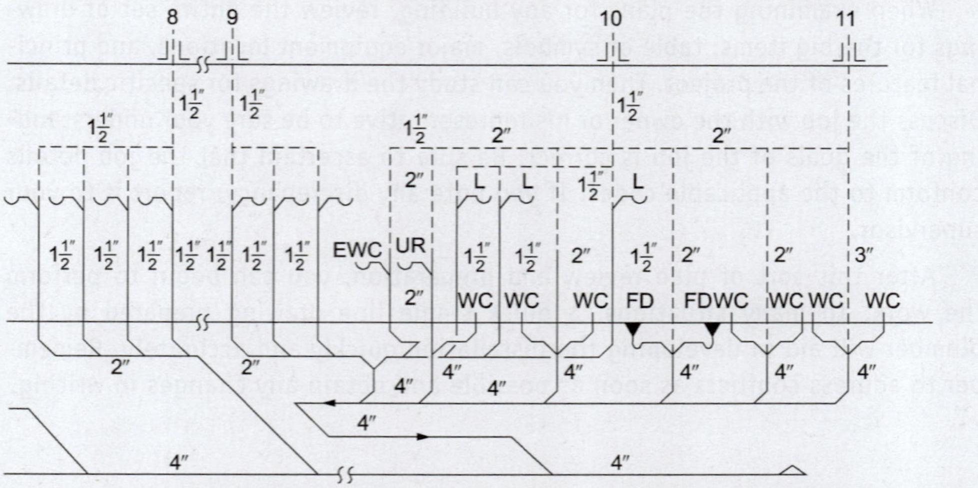

FIGURE 23-48 Water supply layout for school.

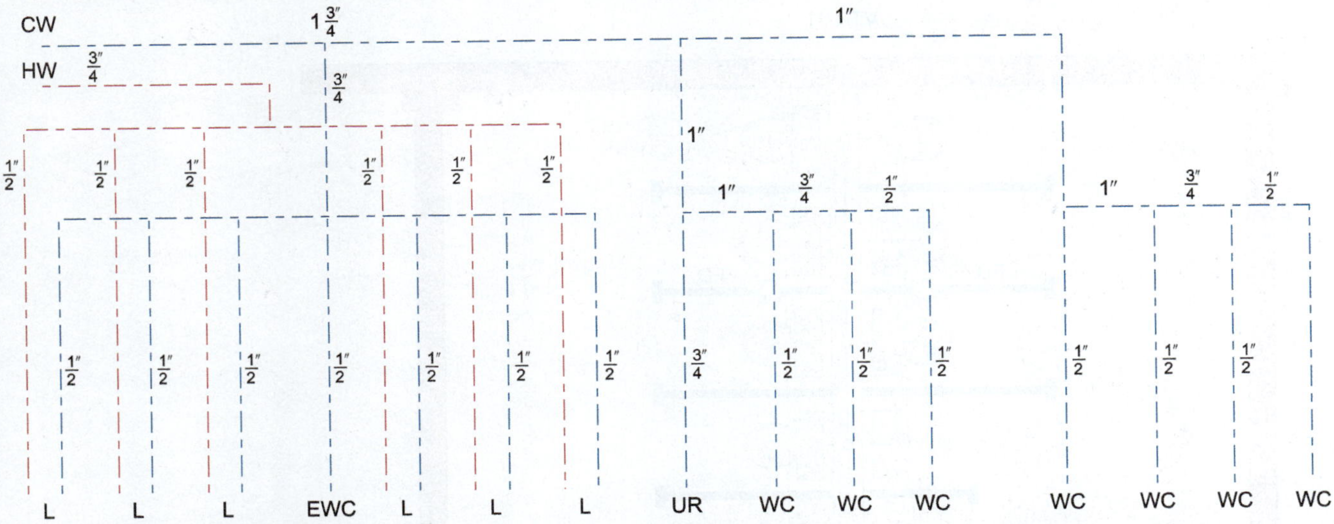

Buildings involved with health care or medical care require a variety of special fixtures, including various therapy and whirlpool baths, emergency baths, infant baths, sitz baths, chemical sinks, surgeon's sinks, service sinks, commercial dishwashers, sterilizers, and bedpan washers. Medical buildings also may have special piping systems for various gases and liquids. See Figure 23-49. Such piping should be adequately identified and precautions taken to avoid cross-connections or improper connections.

Penal institutions have special requirements for plumbing. Fixtures are usually stainless steel, solidly attached to the building. Piping is totally concealed in pipe chases so that it is inaccessible to the inmates. See Figure 23-50. Security must always be considered first in these jobs.

When examining the plans for any building, review the entire set of drawings for the big items: table of symbols, major equipment locations, and principal features of the project. Then you can study the drawings for specific details. Discuss the job with the owner or his representative to be sure your understanding of the goals of the job is correct. Be sure to ascertain that the job details conform to the applicable codes. If you note any discrepancy, report it to your supervisor.

After this sort of plan review and preparation, you can begin to perform the work. In many situations, a quick single line drawing prepared by the plumber will aid in developing the installation quickly and accurately. Remember to address conflicts as soon as possible and obtain any changes in writing.

FIGURE 23-49 Compressed gas lines must be identified to prevent accidents.

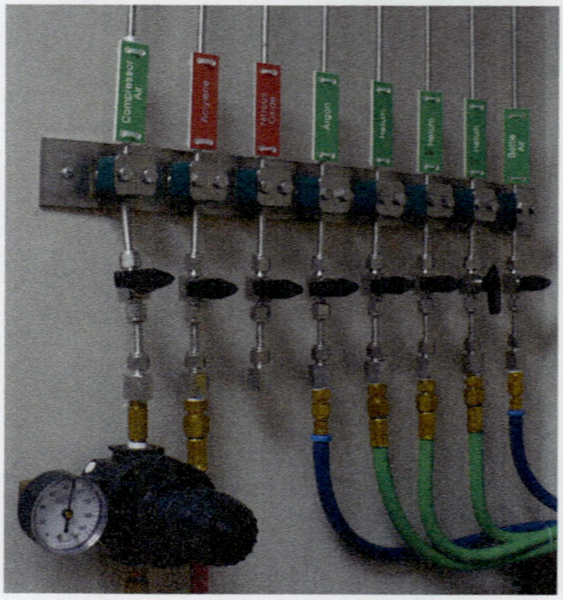

FIGURE 23-50 Combination water closet and lavatory helps conserve space and offers less chance for vandalism.

REVIEW QUESTIONS

Fill in the Blank

Determine the rough in data for the layout in Figure 23-6, using the fixtures listed below. Assume that the countertop in the diagram has been removed and use the same dimensions for placement of the pedestal sink.

1. American Standard Cadet® Elongated Toilet (Figure 23-7)
2. Kohler Memoirs™ Enameled Cast Iron Bath (Figure 23-25)
3. American Standard Town Square™ 27″ Pedestal Sink (Figure 23-26)

Review Question Rough-In Dimensions

Water Closet	
Rough-in from back wall	________
Clearance to adjacent objects (by code)	________
Height of bowl	________
Height of supply rough-in	________
Off-center dimension of supply	________
Lavatory	
Location of drain—vertical	________
Location of drain—horizontal	________
Location of water supply—vertical	________
Location of water supply—horizontal	________
Bathtub	
Location of waste—side wall	________
Location of waste—rear wall	________
Location of supply—vertical	________
Location of spout—vertical	________
Location of showerhead—vertical	________

True or False

1. ________ Single line drawings differ from rough-in sheets in that a single line drawing indicates the complete system, while a rough-in sheet shows individual items.
2. ________ Fixture symbols used in single line drawings are simple pictorial sketches of the items they represent.
3. ________ The drawing of the installation should be kept in the job file for future reference.
4. ________ The purpose of the pipe (or vent, drain, etc.) is usually represented by its line type.
5. ________ The rough-in sheets should be checked to make certain all components will match up.

Short Answer

6. What is the abbreviation for cold water?
7. What is the abbreviation for an indirect drain?
8. What is the abbreviation for fire protection water supply?
9. What is the abbreviation for a gas vent?
10. What is the abbreviation for medical compressed air?

Chapter 24

LEARNING OBJECTIVE

The student will:

- Identify the use and meaning of specialty drawings, including details, sections, and exploded views.

Details, Sections, and Exploded View Drawings

DRAWING PROJECTION REVIEW

Two types of drawings are commonly used to show building construction details:

- Orthographic
- Isometric

Orthographic projections show only one side of an object in each view, but the drawing is true to the size and shape of the object. This means that the viewer is looking perpendicularly at it. There is no distortion because of the viewing angle. Of course, one drawback is that it usually takes multiple drawings to get a complete understanding of the project.

Isometric projections represent a three-dimensional rendering of the project, allowing the viewer to get a feel for the height, width, and depth. The drawback here is that none of the sides or lengths represents their true shape and size. Each of the angles will be distorted.

Frequently, the specific arrangement of parts cannot be indicated on plans drawn to the usual scales of $\frac{1}{8}'' = 1'$ or $\frac{1}{4}'' = 1'$. When it is necessary to show such construction items, detail drawings can be used. These drawings use larger scales, such as $\frac{3}{4}'' = 1'$ or $1\frac{1}{2}'' = 1'$, to show precise features of stairways, cabinets, windows, doorways, and so on.

If a piping arrangement must be installed in a certain way, a detail drawing should be used to show the arrangement. The figures for this chapter illustrate several examples.

Figure 24-1 shows wall details of a meter room. The information presented permits decisions to be made concerning wall sleeves, hanger support points, and offsets (if needed). Note that this detail is titled *Section C.C.* A cutting plane line, labeled C.C., would have to be found on a floor plan to let you visualize exactly where the wall section is located.

If a plan includes many sections, the title will indicate the section number and the page where the section is drawn.

FIGURE 24-1 Wall section.

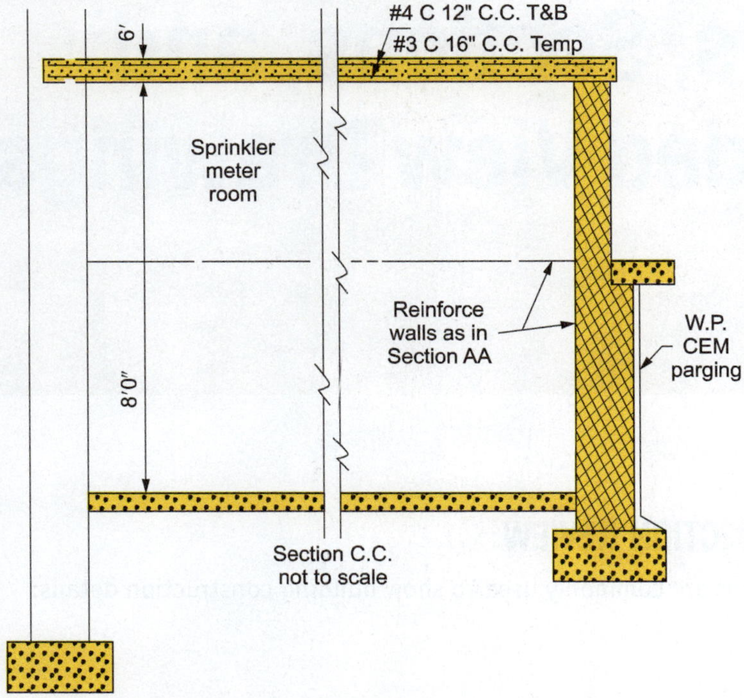

FIGURE 24-2 Sectional view of a roof drain.

Courtesy of Zurn Industries, LLC

Figure 24-2 shows a detail view of a roof drain. The drawing indicates the relationship between the roof drain, roof material, and the roof deck.

Figure 24-3 shows a detail drawing of an 80-gallon gas water heater. From this drawing, it is possible to see all of the working components and how heat is transferred. Figure 24-4 shows a freeze-proof wall hydrant and sufficient room details to locate the faucet.

Figure 24-5 shows an arrangement of lavatory, water heater, and receptor in a janitor room. The dimensions required to locate the components are shown.

Figure 24-6 shows isometric views of the items in Figures 24-4 and 24-5. This sketch shows the ability of isometric views to display the arrangement of components in three dimensions. When drawn properly, isometric views are excellent for doing material take-offs.

Table 24-1 shows a fixture table that could be used on a drawing. Each type of fixture is given a label (P-1, P-2, P-3, etc.). The label is used to identify the fixtures

FIGURE 24-3 Sectional view of a gas-fired water heater.

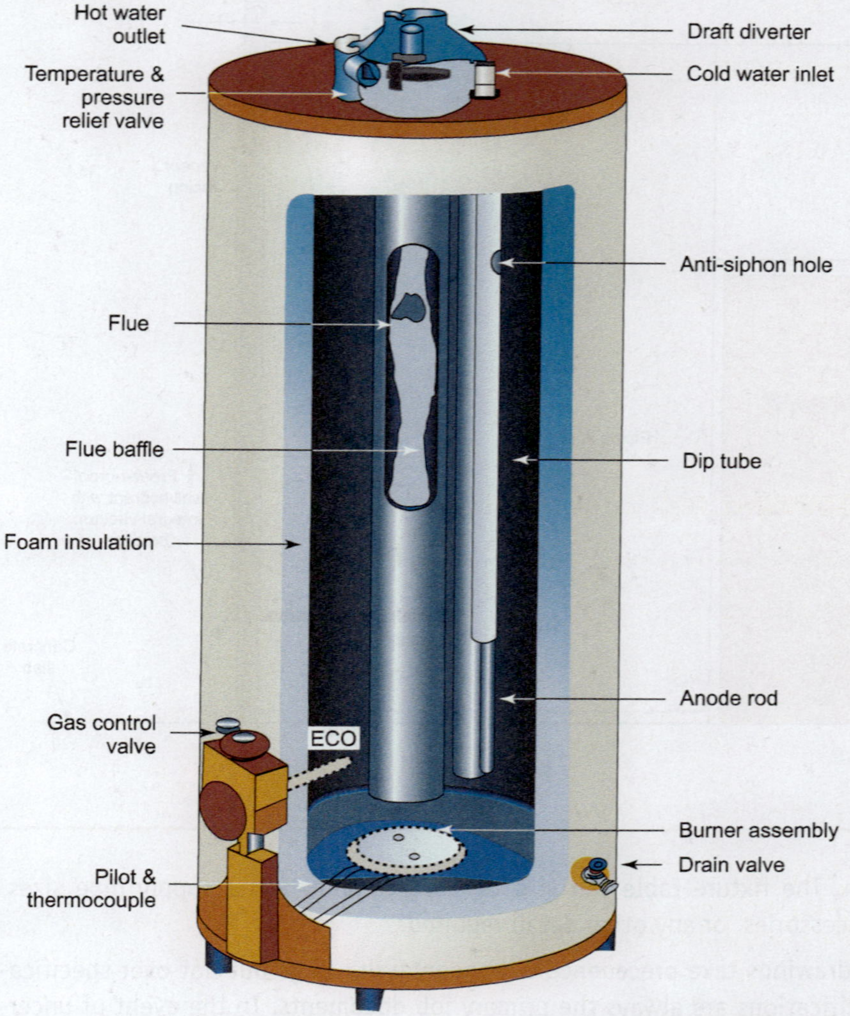

TABLE 24-1 Plumbing Fixture Schedule

		Pipe Sizes				Remarks
Fixture No.	Description	C. W.	H. W.	Vent	Waste	Catalog #
P-1	Water Closet	$\frac{1}{2}''$	—	$1\frac{1}{2}''$	3″	American Standard Cadet 2898.010 with 5311.012 Seat
P-2	Lavatory	$\frac{1}{2}''$	$\frac{1}{2}''$	$1\frac{1}{4}''$	$1\frac{1}{2}''$	American Standard Declyn 0321.026 with 2179.018 Faucet and Cast Brass Trap
P-3	Service Sink	$\frac{1}{2}''$	$\frac{1}{2}''$	$1\frac{1}{2}''$	3″	American Standard Lakewell 7692.000, 8341.075 & 7798.020
P-4	Drink Fountain	$\frac{1}{2}''$	—	$1\frac{1}{4}''$	$1\frac{1}{2}''$	Halsey Taylor OVL II Barrier Free Drinking Fountain

Note: Water closet fixtures must be elongated with open front seat.
Low consumption and water-saver-type fixtures and fixture fittings must be used as specified.

FIGURE 24-4 Water piping detail for freeze-proof wall hydrant.

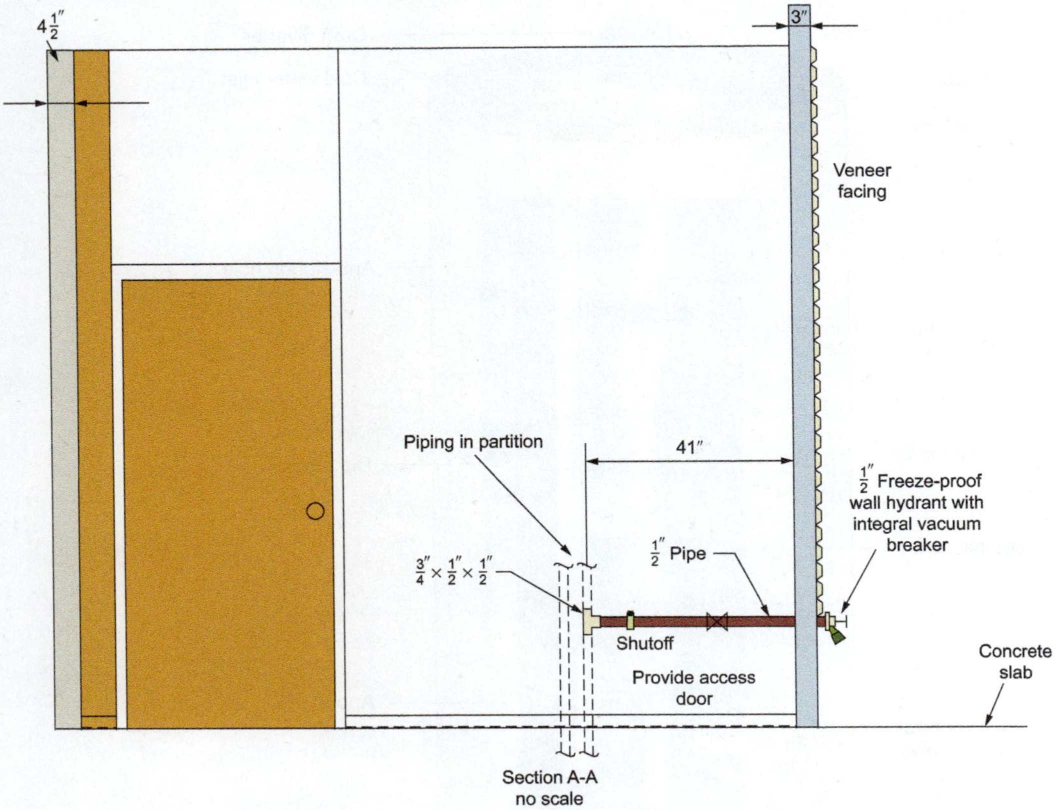

in the plan. The fixture table can also be used to show water supply pipe sizes, required accessories, or any other detail required.

Detail drawings take precedence over general drawings but not over specifications. Specifications are always the primary job documents. In the event of uncertainty regarding the location of walls, partitions, doorways, or similar details, contact the job superintendent for clarification.

SECTION VIEWS

Section views are an additional form of detail drawings. Normal drawings are orthographic projections of the top, side, and front views of an object, such as a building. Each view is an undistorted representation of the object and is usually drawn to scale. Other drawing types are needed to show the interior details of many job components.

Section views are drawn to indicate such details. A section view is drawn by imagining that the object to be shown is cut with a plane at right angles to the line of sight and showing the surface produced by the imaginary cutting plane. The resulting view will show the relationship of interior parts that make up a wall, door frame, window frame, or other assembly.

FIGURE 24-5 Fixture placement detail (not to scale).

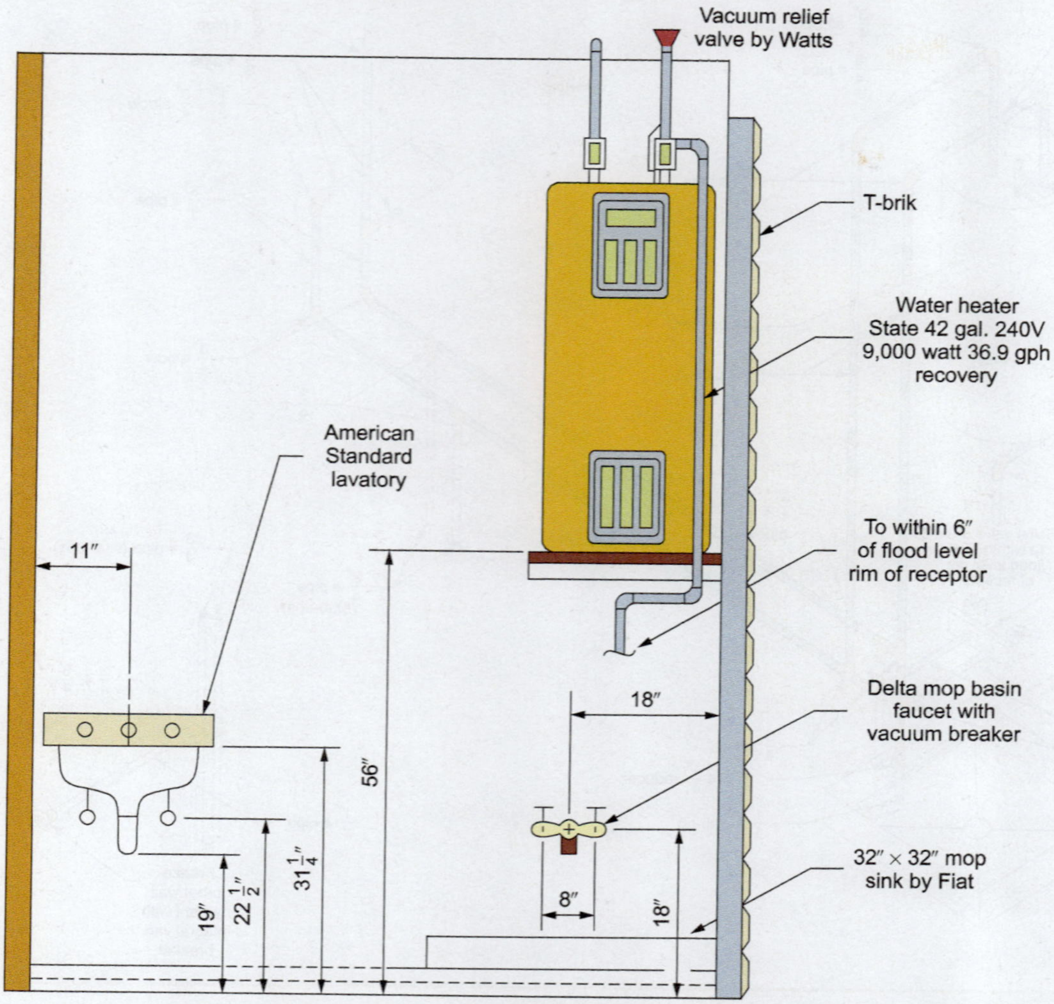

Section views are divided into types as follows, which will be further defined:

- Full section (Figure 24-7)
- Partial section
- Half section
- Offset section (Figure 24-8)
- Revolved section (Figure 24-9)
- Removed section (Figure 24-10)

The sectional views may include *cross-hatching* to enhance the drawing. A cross-hatch is a repeating geometric pattern that is used to identify the cut surface of one of the parts that make up the product being represented. When two independent parts are shown side-by-side in a sectional view, the cross-hatch patterns should be oriented differently so the viewer can distinguish between the two parts more easily.

FIGURE 24-6 Isometric views of Figures 24-4 and 24-5.

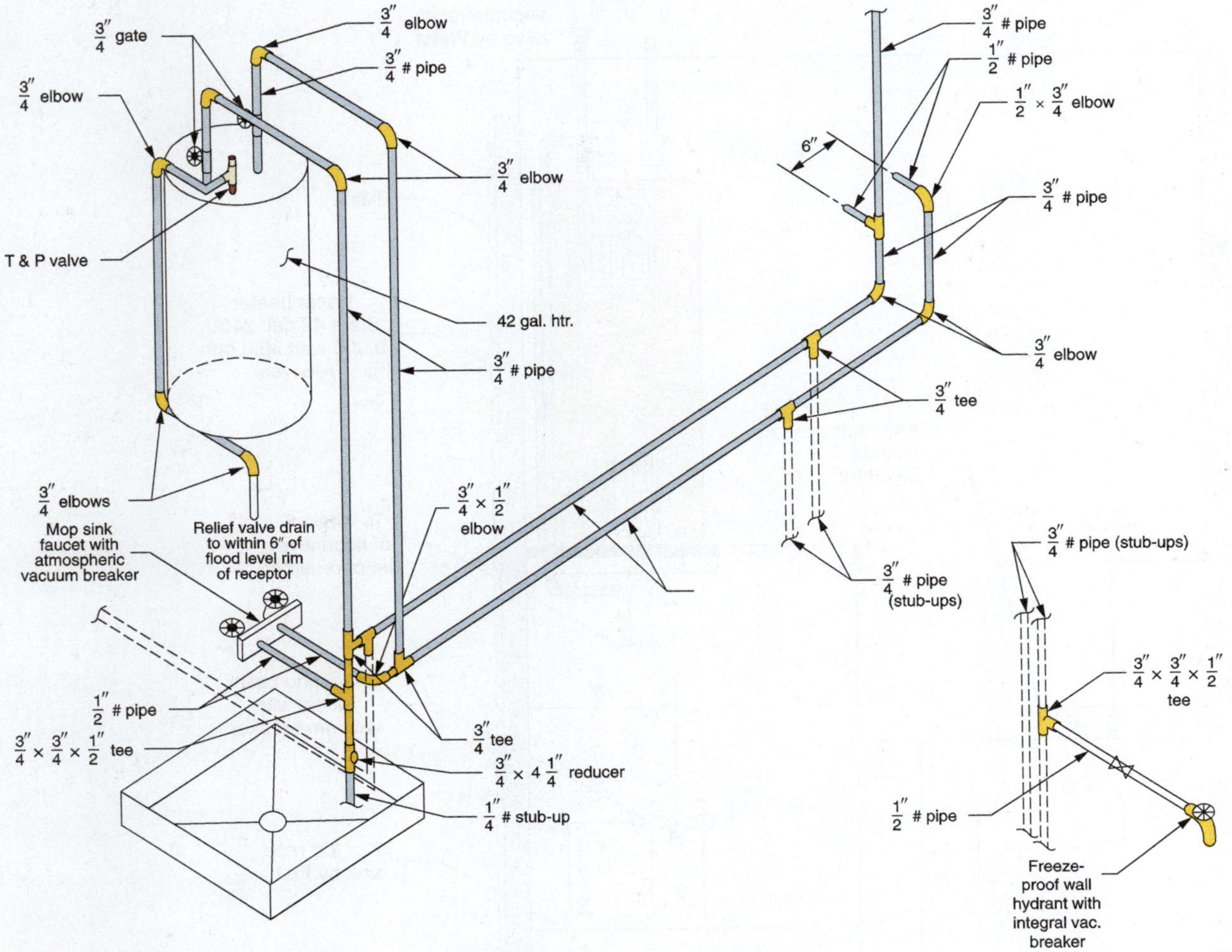

FIGURE 24-7 Full section view of water closet.

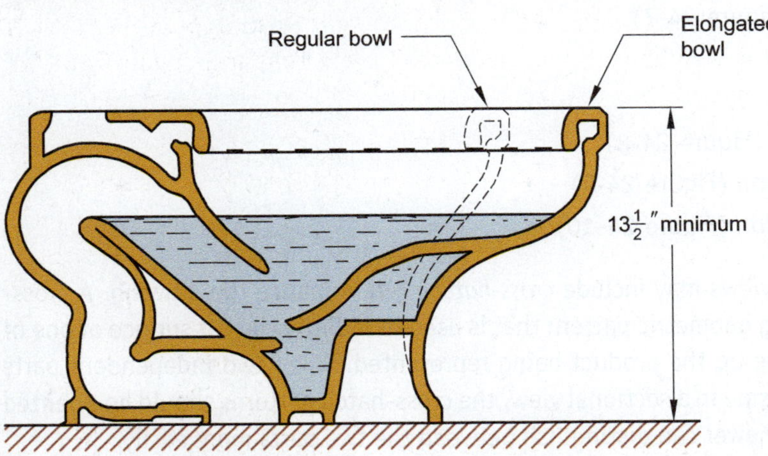

Courtesy of Plumbing–Heating–Cooling–Contractors—National Association

FIGURE 24-8 Offset sectional view. Notice that both holes are in view.

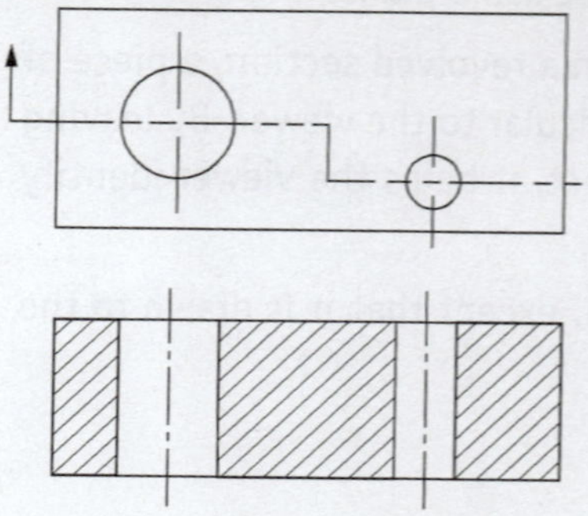

FIGURE 24-9 Revolved section of an adjustable wrench.

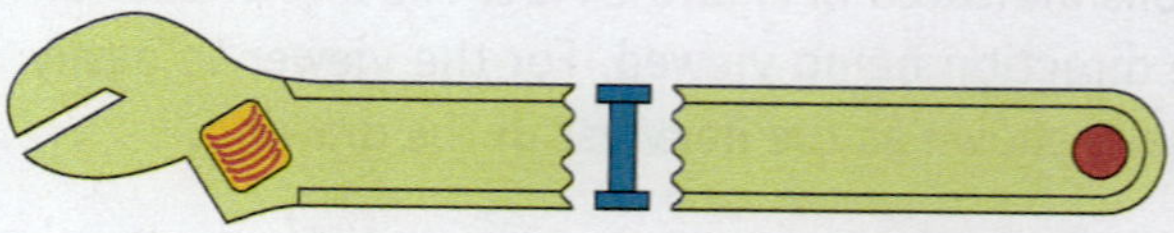

FIGURE 24-10 Removed section of a flooring chisel.

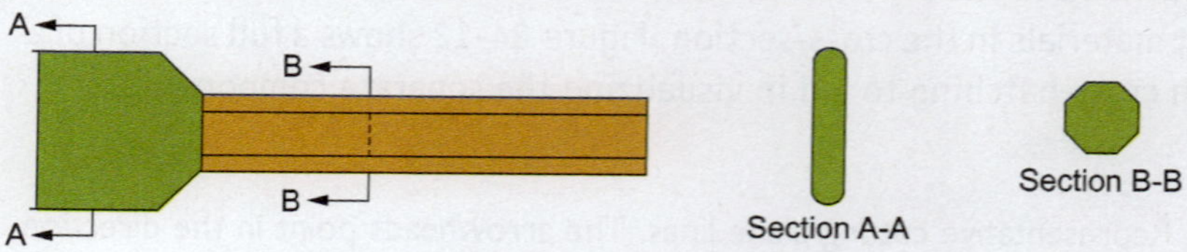

Section Types

To prepare a sectional view of an object, it is necessary to answer the following question: what should be shown and how best can it be shown? Many variations of sections are possible; the most common are illustrated in Figures 24-7 to 24-10.

The *full section* view is where the entire object is cut, usually along a centerline. Figure 24-7 shows a full section view of a common water closet type. Other design types such as wall-hung siphon jet, floor-mounted back outlet, and siphon-wash can also be represented by section views. This view allows the viewer to see details that would not normally be seen.

When the entire object does not need to be seen, a *partial section* view can be used. The partial view allows the viewer to see a portion of the hidden detail but the rest of the object in its normal view. In this case, the cutting plane to indicate the section view only passes through the portion of interest, similar to slicing off only a portion. A *half section* is used when the internal details can be clearly described by sectioning only half the object.

In some instances, the portions of the object may not lie in a straight plane. When this occurs, an *offset section* can be used. Figure 24-8 shows an example of when the cutting plane can be shifted to show different features. This is not considered a full section because the two holes are not in the same plane.

Figure 24-9 is an example of a *revolved section*. In a revolved section, a piece of the item is cut off and then turned until it is perpendicular to the viewer. By leaving the rotated part close to the regular view of the object, it helps the viewer identify where the section comes from.

A *removed section* is similar to a revolved section, except that it is drawn to the side. See Figure 24-10.

Cutting Plane

The cutting plane has standard arrangements and conventions. It is an imaginary plane that splits the object so that the section view can be developed. Cutting plane lines are shown as one of two versions indicated in Figure 24-11. The arrowheads on the cutting plane line indicate the direction being viewed. For the viewer to easily identify the cutting plane, it is usually shown as the heaviest in the drawing.

Cross-Hatching

An important part of the sectional view is cross-hatching. Cross-hatching is a pattern of fine lines drawn on the cut surface. A different pattern is used for different items and for different materials in the cross-section. Figure 24-12 shows a full section of a check valve with cross-hatching to aid in visualizing the separate components.

FIGURE 24-11 Representative cutting plane lines. The arrowheads point in the direction the viewer is looking.

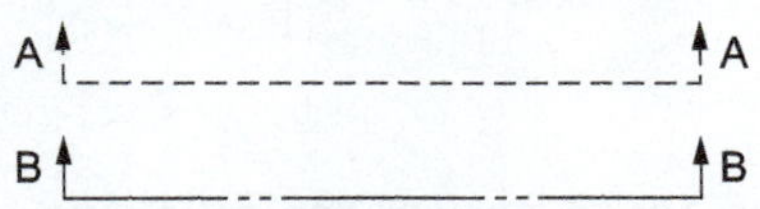

FIGURE 24-12 Full section of a lift check valve.

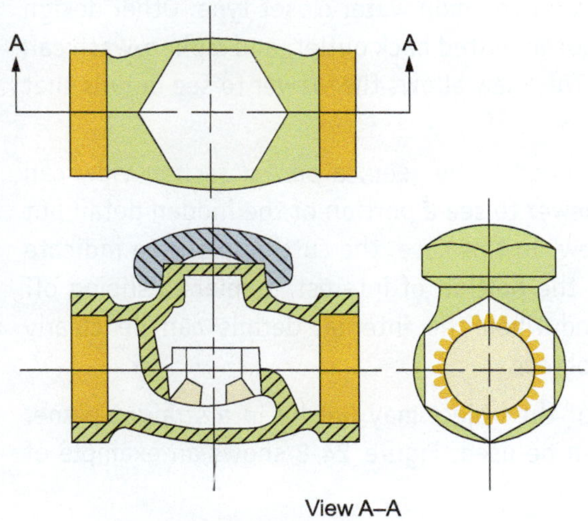

The cross-hatching for the valve cap uses lines rising to the left. The cross-hatching for the body uses lines that slope upward to the right. The surface shown with the pattern of lines is a single part of the valve, the body.

Notice that the threaded portions are not cross-hatched; this is because they were not cut by the cutting plane. Imagine the cutting plane as a saw. Only material that would be cut with the blade is cross-hatched. Figure 24-13 shows

FIGURE 24-13 Cross-hatching patterns for sectional views.

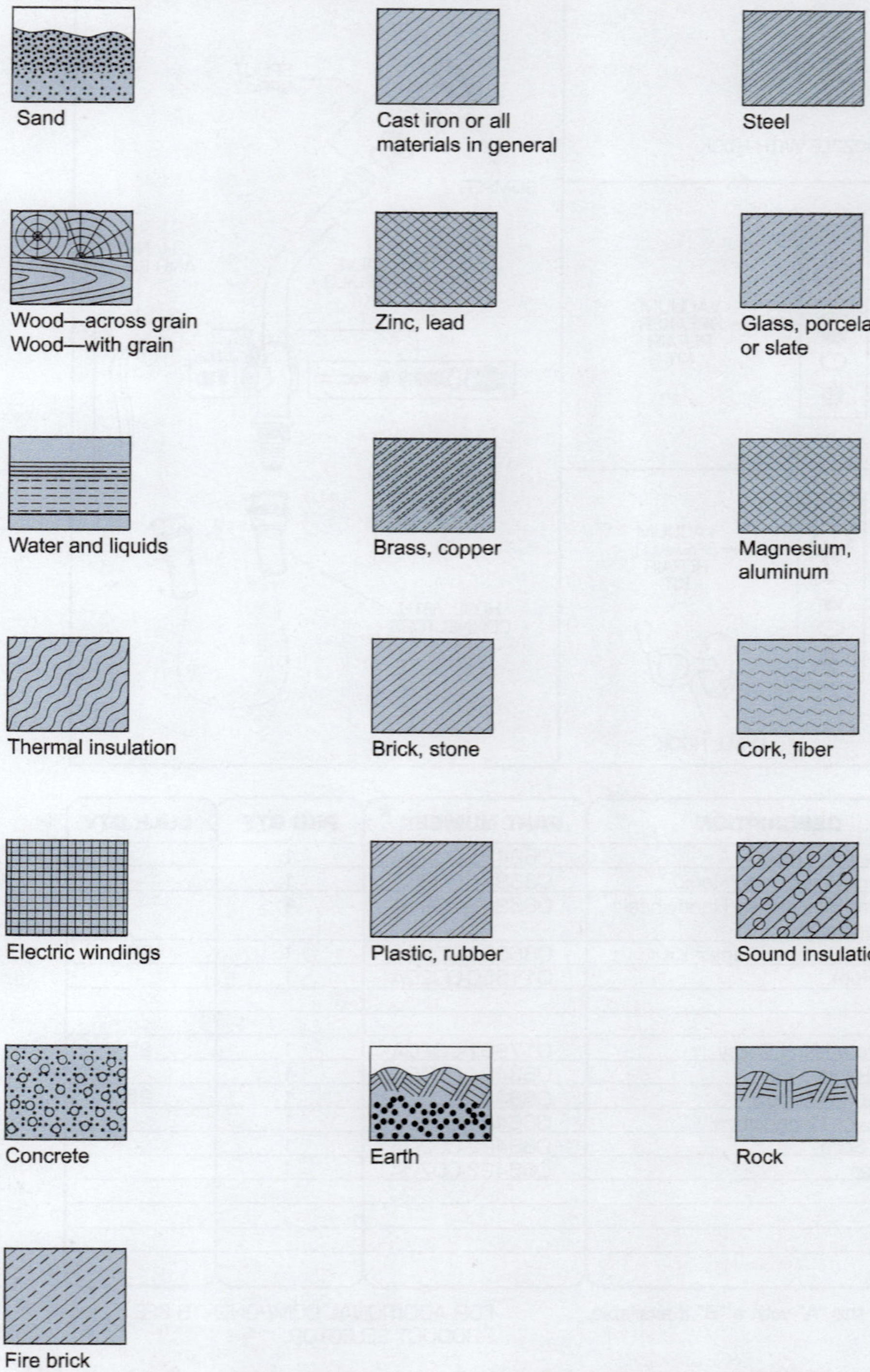

FIGURE 24-14 Exploded views can help show how parts are assembled and help identify part names and numbers.

American Standard

BED PAN CLEANSER
MODEL NUMBERS

7582.067	7836.018	7866.015
7880.024	7880.083	7880.091

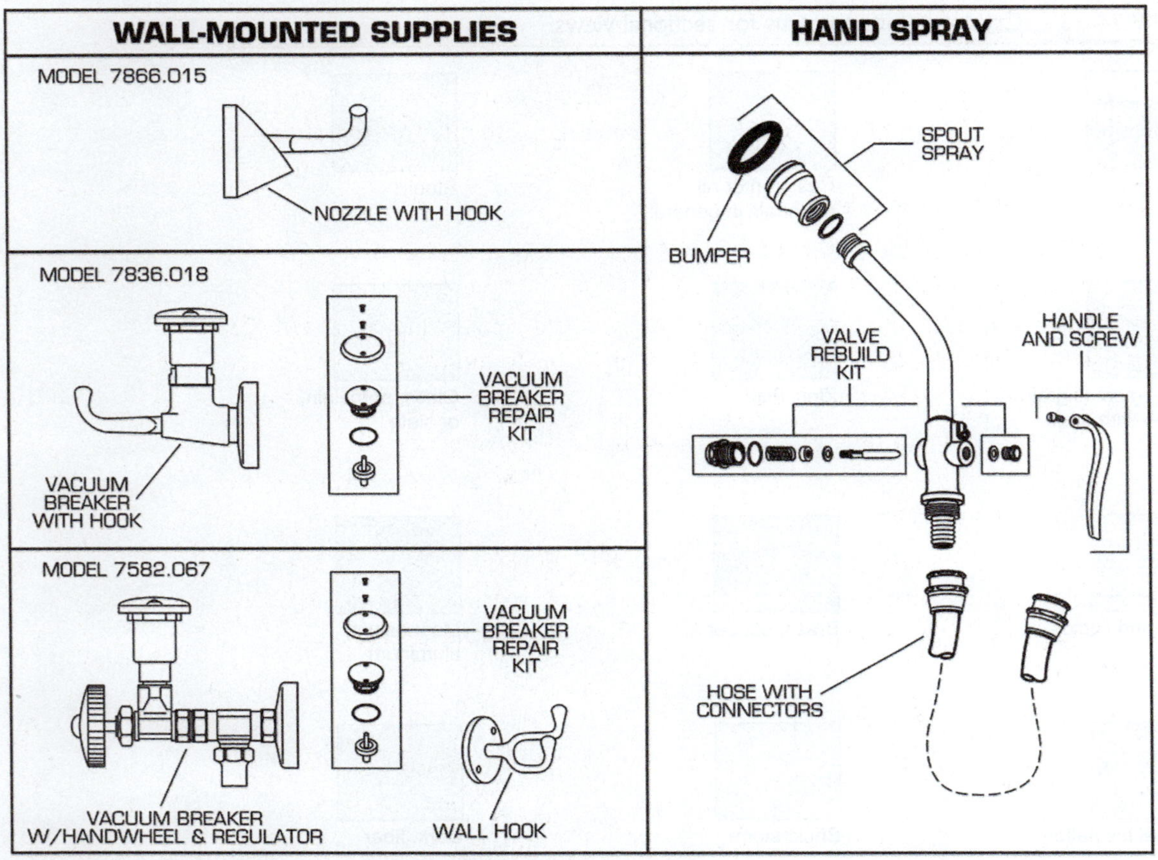

REPAIR PARTS	DESCRIPTION	PART NUMBER ★	PKG QTY	BULK QTY
Wall-Mounted Supplies	Nozzle with Hook	066425-0020A	1	-
	Vacuum Breaker w/Hook	066931-0020A	1	-
	Vacuum Breaker w/Handwheel & Regulator	066932-0020A	1	-
	Vacuum Breaker Repair Kit	066501-0020A	1	25
	Wall Hook	011350-0020A	1	-
Hand Spray Parts	Complete Hand Spray	017961-0020A	1	25
	Valve Rebuild Kit	066424-0040A	1	-
	Handle and Screw	066420-0020A	1	25
	Hose with Connectors	066421-0040A	1	-
	Spout Spray	066422-0020A	1	-
	Bumper	066423-0070A	1	-

★ For bulk packaging, replace the "A" with a "B" if available.

FOR ADDITIONAL COMPONENTS SEE PRODUCT SELECTOR.

F-11
REV. A

Courtesy of American Standard Bath & Kitchen

typical cross-hatching patterns used to indicate various material types exposed by cutting a section.

Study manufacturers' diagrams to become familiar with sectional drawings. In addition, to develop your ability to make such drawings, make sketches of items you generally use, such as a gate valve, ball valve, threaded cap, closet flange, copper pipe, and a hammer handle. Use cross-hatching variations to enhance your drawings.

EXPLODED VIEWS

Exploded views are another variation of detail drawings. Many products that we use are assemblies of a number of parts. Often, the exact order and relationship of these parts is difficult to visualize, even with sectional drawings. For such situations, exploded view drawings can be very helpful.

The exploded view drawing shows the individual parts separated from the assembly and sometimes connected to it with centerlines to show how the parts go together. Since the basic drawing is usually an isometric view, exploded drawings are not used to show the dimensions of individual parts. In many cases, exploded views are used to identify part numbers.

Figure 24-14 shows an exploded view of a medical facility faucet. Exploded views are most useful for repair and maintenance work, since each component and its function can best be seen.

You will learn how to use exploded views best if you practice making such drawings of simple products with which you are familiar. Make sketches of a few common items and include the list of materials so that you can develop the full drawing.

REVIEW QUESTIONS

True or False

1. __________ Exploded views can be used to determine dimensions of individual parts of an assembled product.
2. __________ Exploded views show the order of assembly of the parts of a complete product.
3. __________ Exploded views are useful for repair and maintenance operations.
4. __________ The arrows on a cutting plane indicate the direction a person is looking.
5. __________ Components made of different materials are cross-hatched differently.

Drawing

6. Make an isometric sketch of a kitchen cabinet and countertop that is 8 feet long, 25 inches deep, and 36 inches high. A sink that is 25 inches long and 24 inches deep should be centered in the countertop. Then make a sectional view of the countertop showing how the sink is mounted. Use cross-hatching to show the different parts.

Appendix A

Oxy-Acetylene Welding, Shielded Metal-Arc Welding, and Weld Testing

This appendix contains specific information about the characteristics of metals as they relate to welding, as well as exercises for learning a variety of welding techniques.

CHARACTERISTICS OF METALS

A metal is described physically by the following characteristics:

Crystal Structure

The molecules of a metal arrange themselves in patterns called crystals, which in turn are assembled in patterns called grains. The interactions between grains and the shape and size of crystals are influenced by the temperature history of the metal. From the time the metal was molten, the crystals and grains have been affected by the type and amount of work done on the metal to arrive at the solid shape you have in hand.

Density

Density is the weight of the metal per unit of volume. The more closely packed the grains are, the more dense the material will be.

Porosity

Porosity is the degree to which the material contains small holes or voids. Porosity in a weld is a result of improper welding equipment or techniques. For example, if the weld material is not clean (dirt or slag from previous weld), the contaminants can be dissolved into the weld bead, which will leave a void.

Specific Gravity

Specific gravity is the ratio of the weight of a volume of metal to the same volume of water at a standard temperature. It allows a person to compare the size and weight ratio of one material to another. See Table A-1.

TABLE A-1 Data about Metals

Metal or Composition	Chemical Symbol	Specific Gravity	Weight per Cubic Ft. (lbs)	Weight per Cubic In. (lbs)	Melting Point (°F)
Aluminum	Al	2.7	168.5	0.0975	1,220
Brass		8.6	536.6	0.3105	1,823
Bronze		8.78	547.9	0.3171	1,984
Copper	Cu	8.89	554.7	0.321	1,981
Gold	Au	19.3	1,204.3	0.6969	1,945
Iron, Cast	Fe	7.03–7.73	438.7–482.4	0.254–0.279	1,990–2,300
Iron, Wrought	Fe	7.8–7.9	486.7–493.0	0.282–0.285	2,750
Lead	Pb	11.342	707.7	0.4096	621
Magnesium	Mg	1.741	108.6	0.628	1,204
Mercury (68°F)	Hg	13.546	845.3	0.4892	−38
Nickel	Ni	8.8	549.1	0.3178	2,651
Silver	Ag	10.42–10.53	650.2–657.1	0.376–0.380	1,761
Steel, Carbon			489.0–490.8	0.283–0.284	2,500
Tin	Sn	7.29	454.9	0.2633	449
Zinc	Zn	7.04–7.16	439.3–446.8	0.254–0.259	778

ENERGY ABSORPTION AND TRANSMISSION CHARACTERISTICS

Energy absorption and transmission characteristics describe how the material will respond to an external energy source. In some cases, the material will simply reflect most of the energy (heat for example) and in some cases may absorb it. The characteristic responses of metals are the following:

Melting Point

The melting point is the temperature at which the metal becomes liquid. The melting point of steel is about 2,700°F. It will vary somewhat with the amount of alloying materials. For a weld to occur, the melting point must be reached to combine the base material with the filler material. See Table A-1.

Volatility

Volatility is the tendency of the material to vaporize.

Conductivity

Conductivity is the ability of a metal to conduct electricity (or heat). In cases of electricity, operations such as plasma cutting and spot welding would not be possible without the material's ability to conduct electricity. In the case of heat absorption, aluminum absorbs heat so well that it needs a larger amount of heat energy (high current) in a short time. Otherwise, the entire piece of metal would heat up.

Resistance

Resistance is the characteristic of the metal that opposes electric current (or heat) flow.

Expansion Coefficient

The expansion coefficient is the linear increase in dimension of a metal with increase in temperature, usually expressed in terms of inches change per inch of length per degree F.

Overheating Limit

The overheating limit is the highest temperature a metal can withstand before significant crystal structure change takes place. In simpler terms, it is the highest temperature to which the metal can be heated without drastically changing its properties.

Stress

Knowledge of the reaction of a metal to applied stress helps us select a metal for a given task. These stress capabilities include the ability (or lack of ability) of the metal to withstand or accommodate the following characteristics:

Plasticity

Plasticity is the ability to deform without breaking. Materials that can be bent or flattened without breaking or cracking, such as lead, are considered to be very plastic in nature.

Malleability

Malleability is the capability of metal to be worked by hammering.

Strength

Strength is the ability to withstand stress and is expressed in pounds per square inch (psi).

- *Ultimate strength* is the highest stress that can be applied before the material breaks.
- *Yield strength* is the stress at the point where the metal permanently deforms (but does not break).
- *Working strength* is a stress level safely below the yield strength.

Toughness

Toughness is the ability of a metal to resist wear from abrasive service.

Brittleness

Brittleness is the tendency to fail suddenly as stress is increased. Materials that break without any warning, such as stretching or bending, are considered brittle. They do not stand up well to impact loads. For example, a diamond is a very hard, brittle material; it will break if hit with a hammer.

Hardness

Hardness is the ability of one material to withstand penetration by another material.

Shock Resistance

Shock resistance is the ability of a metal to withstand loads that are applied suddenly.

Elasticity

Elasticity is the ability of the metal to return to its original shape after the load is removed. An example of an elastic material is rubber. A rubber band can be stretched, but as long as it is not stretched too far (past its elastic limit), it will return to its original shape.

Fatigue Failure

Fatigue failure is the mode of failure in which a metal breaks after repeated cycles of load reversal. The load may be less than the working load.

Resistance to Corrosion

Resistance to corrosion is the ability of the material to withstand chemical changes that could produce unsatisfactory structural performance.

Tables A-1 and A-2 provide additional information pertinent to metals.

HEAT TREATMENT OF LOW-CARBON STEEL

Welding of pipe may affect the characteristics of the pipe. Some of the harmful effects of welding can be reduced or eliminated by heat-treating the steel piece. Heat-treating includes annealing, hardening, case hardening, normalizing, and tempering.

The result achieved depends upon the carbon content of the steel, the temperature to which the material is heated, and the method of cooling (rapid quench hardens the material, slow cooling anneals the material).

Annealing

Annealing involves raising the temperature of the work to a level that relieves the locked-in stresses in the steel.

Hardening

Hardening makes the material very hard throughout.

Case Hardening

Case hardening produces a hard exterior surface by adding carbon to the outer surface by one of several methods.

Normalizing

Normalizing relieves internal stresses by heating the material slightly higher than with annealing and then cooling the material slightly faster than with annealing.

TABLE A-2 Physical Properties of Metals

General Classification	Specific Quality	Relating to	Measured by
Metallic Structure	Density	Compactness of a metal	Quantity/unit volume
	Porosity	Holes, inclusions, seepage susceptibility	
	Specific Gravity	Unit of measurement based upon the weight of an equal volume of water	Assigned term
Energy Absorption and Transmission	Melting Point	Temperature point at which a metal passes from a solid state to a liquid state	Degrees Fahrenheit or Celsius
	Volatility	Ease of a metal to be vaporized	
	Electrical Conductivity	Ability of the metal to allow current to flow	
	Electrical Resistance	Resistance of a material to allow current to flow	Ohms
	Weldability	Capability of a metal to flow together during a welding operation	
	Fusability	Ease of a metal to be melted	
	Thermal Conductivity	Ease with which a metal allows heat or cold to transfer across its surface	
	Expansion Coefficient	The amount of linear growth or shrinkage that occurs as temperatures differ across the surface of the metal	Linear amount/degree temperature change
	Overheating Limit	Temperature at which the internal properties are impaired	Degrees Fahrenheit or Celsius
Stress Capabilities	Plasticity	Ability to deform without breaking	
	Malleability	Capability to be shaped or forged without the loss of structural strength	
	Strength	Ability to resist load	psi units
	Toughness	Ability to withstand abrasion and wear	
	Brittleness	Tendency to fail due to suddenly applied load	
	Hardness	Ability of a material to withstand penetration from another material	Brinell hardness test or Rockwell test. File test.
	Shock Resistance	Ability to withstand maximum shocks	
	Elasticity	Ability to return to original shape after loading is removed	Modulus point-ratio of stress to strain
	Ductility	Ability to be stretched without tearing	% of stretching
	Fatigue Failure	Point at which a material breaks after cycles of alternating loads	Expressed in psi
	Resistance to Corrosion	Ability to remain chemically inert and resist metallic breakdown	

Tempering

Tempering is a process that reduces the hardness slightly after the hardening process is completed.

FORMING A PUDDLE WITH AN OXY-ACETYLENE TORCH

Under supervision of your instructor or employer, use an oxy-acetylene welding tip to practice developing a puddle and moving the puddle across a thin piece of steel plate, preferably $\frac{1}{8}$ inch thick or less. The goal is to achieve a uniform puddle that you can move in a straight line across the plate. Upon completion, the back of the plate should not be burned through, but you should be able to see the path of the puddle.

The method is as follows:

1. Put on welding goggles.
2. Select the proper-sized tip and regulator pressures.
3. Light the torch and adjust to a neutral flame.
   a. Open the acetylene gas valve first.
   b. Light the torch with a friction lighter.
   c. Open the oxygen valve until the inner cone becomes well defined (see Figure A-1).
4. Hold the torch with a comfortable grip, with the flame at about a 60° angle from the plate.
5. Start at one edge and heat until the metal puddles.
6. With a clockwise motion of the tip, the area of the puddle can be increased until it is about $\frac{3}{8}$ inch to $\frac{1}{2}$ inch in diameter. See Figure A-2.

The puddle is moved by making the clockwise flame path move about $\frac{1}{16}$ inch per rotation.

FIGURE A-1 Example of the well-defined inner cone.

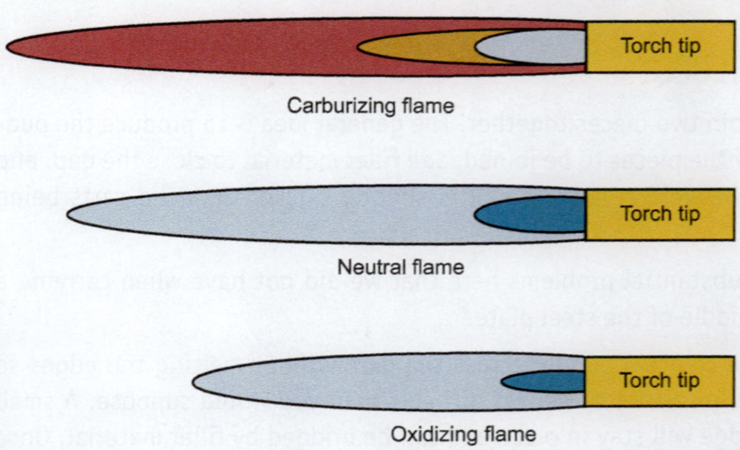

FIGURE A-2 Example of increasing the area of the puddle.

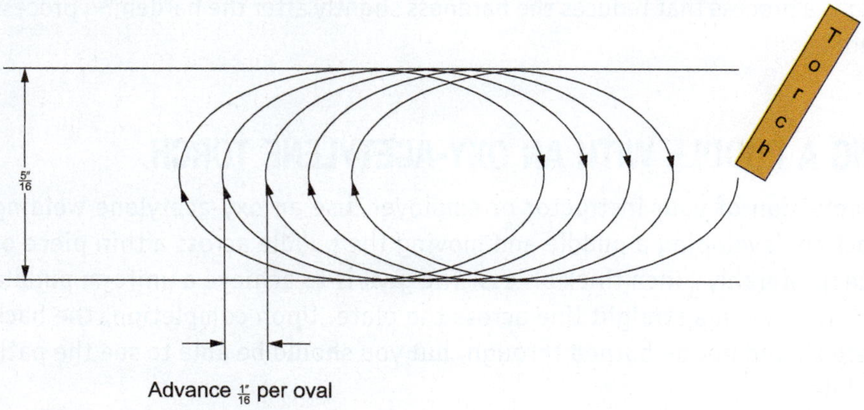

ADDING FILLER

After you are proficient at forming and controlling the puddle, practice adding filler metal to the puddle. The filler rod for most work is $\frac{1}{8}$ inch in diameter by 36 inches long, but for light work, smaller-diameter rods should be used.

Begin as described above and hold the filler rod so that the end is near the puddle (for preheating). The filler is added to the puddle, not melted by the flame. Add filler as you move the puddle from right to left. With a little practice, you will be able to add the correct amount of filler so that you achieve a uniform, slightly raised bead across the plate.

If the rod sticks, you have placed it just outside the puddle, the puddle is too cool, or the rod has too great an angle to the work so that the end of the rod is sticking at the bottom of the puddle.

Hold the rod at about a 45° angle from the work so that the filler rod is in the puddle and it will not stick to the work. The rod and torch flame must be in nearly constant motion and at approximate right angles to each other. Each motion must be done with light, easy muscular effort or you will tire very quickly.

JOINING TWO PIECES

The next task is to join two pieces together. The general idea is to produce the puddle across the gap of the pieces to be joined, add filler material to close the gap, and (usually) build up the weld area so that it is slightly thicker than the parts being joined.

There are two substantial problems here that we did not have when carrying a puddle across the middle of the steel plate.

First, how do we create a puddle across the gap without melting the edges so they disappear? This turns out to be less difficult than you would suppose. A small melt area on each edge will stay in place and can be bridged by filler material. Once joined, it is relatively easy to continue the process.

Second, how do we prevent heat distortion in our final assembly? Figure A-3 shows edge views of two plates both as they are welded and after they cool. The filler material and original pieces are molten as the weld is being placed, but much less of the joint material reaches molten temperature at the bottom of the weld than is true of the top. Thus, as the joint cools, there is more material contracting on top of the weld than on the bottom and the work will take on the V shape shown.

Another V shape is shown in Figure A-4. In this case, the work is cool as the weld is started at the left edge. As the weld progresses to the right, the joint area becomes hotter and hotter, so each increment of weld material shrinks a little less than the previous increment and the increasing temperature of the joint as a whole makes the pieces open in the direction of weld travel.

FIGURE A-3 Edge views of plates before welding and after cooling.

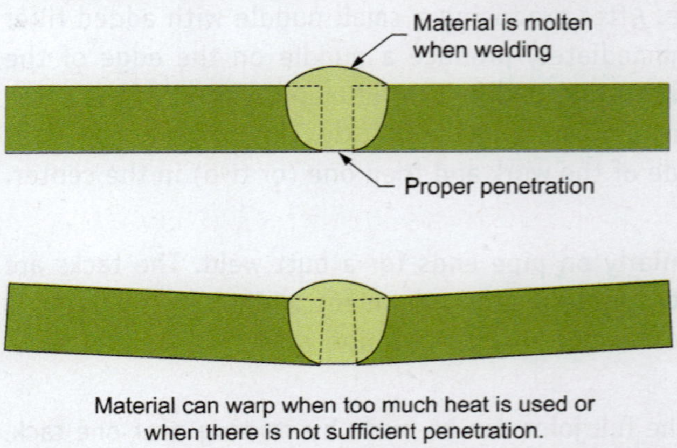

FIGURE A-4 V-shaped weld.

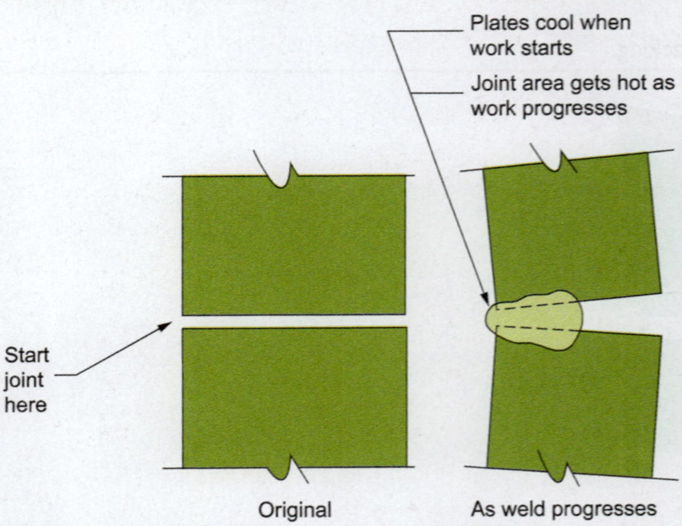

Tack Welding

The answer to both problems is to tack weld the pieces together before the continuous weld is applied. A tack weld is a short weld, $\frac{1}{4}$ inch to $\frac{1}{2}$ inch long, applied to the edges of your work. If the weld pieces are extremely thin and wide, an additional tack weld can be made in the middle. The tacks are made quickly and a minimum of material is involved, so they produce no distortion and do not melt away the edge. There is enough material, however, to control the thermal distortions described above so that the welded joint can be made.

The tack solves the first problem also. By starting the weld near a tack, we can develop a puddle that has material across the gap and thus have a surface on which to deposit filler metal. As the weld is continued, the tacks are eliminated as they are encountered.

The tack is made by clamping two pieces with edges separated by approximately the width of the filler rod. A gap too narrow will not allow for the filler material to reach the bottom of the plate. A gap too wide will be very difficult to fill without a backup plate. After producing a small puddle with added filler on the edge of one piece, immediately produce a puddle on the edge of the second piece. The puddles will merge, and by careful manipulation of the torch and filler, a short, full-depth weld can be made quickly. Immediately make another tack on the opposite side of the work and then one (or two) in the center. See Figure A-5.

Tacking is performed similarly on pipe ends for a butt weld. The tacks are placed 90° or 120° apart to maintain pipe alignment while the full joint is made. Figure A-6 shows the distortion in a partial pipe weld that was not tack-welded first.

With two pieces tacked, the full joint can be made by starting near one tack, reforming the puddle, and proceeding along the joint to complete the weld. The tack welds should be absorbed by the new welding as it proceeds.

Realize that the proper filler alloy, tip size, regulator pressures, and joint speed all combine to provide effective, economical welded joints.

FIGURE A-5 Example of tacking.

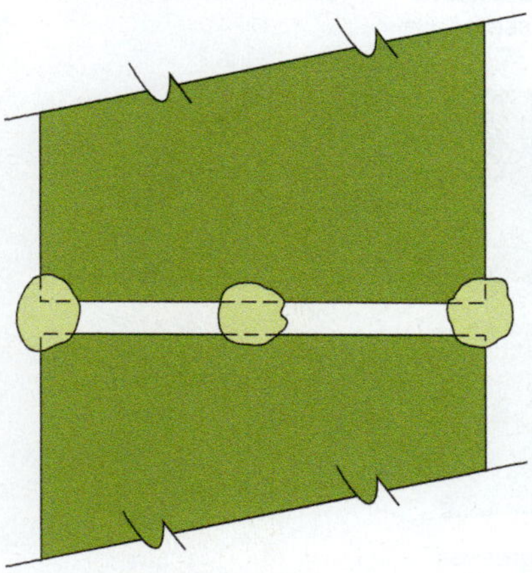

FIGURE A-6 Distortion in a partial pipe weld caused by failure to tack weld first.

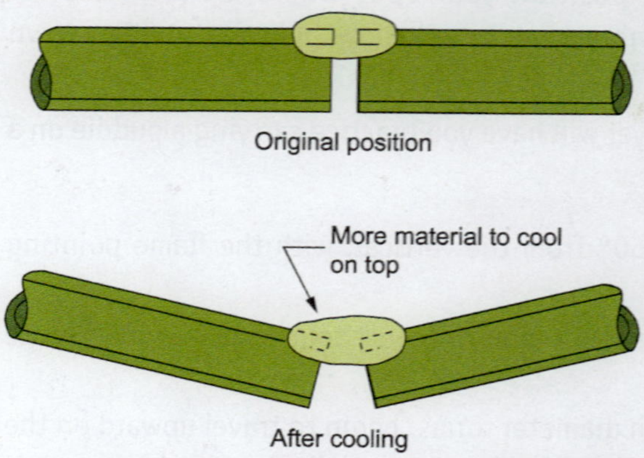

WELD CHARACTERISTICS

The characteristics to look for in a weld are these:

1. The weld material should penetrate all the way through the joint gap.
2. The weld bead should not be more than about $\frac{1}{16}$ inch above the original work surface. The weld bead should flow smoothly into the plate, with no undercut.
3. The center of the bead should be where the gap was.

Your instructor or employer will monitor your techniques as you practice these welding exercises.

OXY-ACETYLENE PIPE WELDING

Before considering the methods of oxy-acetylene pipe welding, we will consider when to use oxy-acetylene methods for pipe.

Oxy-acetylene can be used for any pipe size, but it is generally considered uneconomic for sizes over 2 inches. Sizes 2 inches and smaller involve thinner walls, so the slower weld-deposition rate is not a significant factor. Other manual methods, such as metal-arc inert gas (MIG) or tungsten inert gas (TIG) welding, require more specialized equipment and training but are much quicker. However, with separate control of heat and filler, the welder has more latitude as the weld progresses around the pipe. Furthermore, if only small pipe is to be welded, the oxy-acetylene method requires the least expensive, least bulky equipment to bring to the jobsite.

OXY-ACETYLENE VERTICAL WELDING

Vertical surface welding is accomplished by using the same techniques described above for flat welding, with only some minor alterations.

With the surfaces to be welded in a vertical position, more care is needed to keep the puddle small and as cool as possible so that molten material will not fall out of the weld area. Remember that the plastic range of molten steel gives us a viscous fluid so that we can have a puddle adequate to accept filler material and still stay in place. Always position your body where molten metal or sparks will not fall on you. Remember gravity will now pull the puddle down instead of into the plate.

Your welding instructor or employer will have you practice carrying a puddle on a vertical piece of steel.

1. Hold the torch flame at 45° to 60° from the vertical, with the flame pointing upward.
2. Hold the torch lightly so that the inner blue cone just touches the work. Rotate the flame clockwise.
3. When a puddle of $\frac{3}{16}$ inch to $\frac{3}{8}$ inch diameter forms, begin to travel upward on the plate, changing to the same semicircular motion described earlier. The extent and speed of this motion and travel speed upward are used to control the size and temperature of the puddle.

When you are able to carry the puddle and achieve a uniform ripple, you should practice adding filler material to the puddle.

1. Hold the filler rod at about 90° to the torch flame and about 45° to the work.
2. If the puddle is too hot, it will burn through the plate or the puddle will fall toward the bottom of the plate.
3. If the puddle is too cool, the rod will stick, and you will not have a uniform bead on the plate. There will also not be enough penetration of the filler material to give the weld full strength.

Tacking must be done to join two vertical plates together for the same reasons discussed above for flat welding—to control distortion and provide a start for the gap weld.

1. Clamp two plates vertically, with a $\frac{1}{16}$ inch to $\frac{1}{8}$ inch gap between them.
2. Form a small puddle at the bottom of the adjoining edge on one plate and add filler metal.
3. Immediately form a small puddle on the adjacent edge.
4. The puddles will merge and you can add filler to form a $\frac{1}{4}$ inch to $\frac{1}{2}$ inch tack.
5. Immediately repeat the process at the top of the plates, then tack one or two places in the middle region.

Once the tacks are made, the weld can be done by starting at the bottom tack and forming a puddle that includes the two edges and the top of the tack. Trying to weld from the top of the plate to the bottom leads to problems with sag and other contaminants falling into the weld zone. While this weld may look fine, it will be filled with porosity.

As filler metal is added to this puddle, the weld can be brought up the joint to the top. Remember that the torch and filler rod must be in nearly constant motion so that the puddle stays molten, but does not become so hot that it falls out of the weld region.

After the weld is complete, turn it over to check penetration. Most students will find that getting full penetration is not as big a problem as keeping the weld small.

Because the weld material wants to fall opposite the motion of travel, most find this to be a little slower and trickier to perform.

OXY-ACETYLENE OVERHEAD WELDING AND PIPE TACKING

Overhead welding is the most difficult to perform for three reasons:

- The technician performing the work is in an awkward, uncomfortable position, which makes all motions and actions difficult and very tiring.
- Control of the filler rod and torch are critical and the highest degree of coordination is essential.
- The molten material tends to fall from the work, causing great hazard to the technician, the possibility of fouling the torch and tip, and the frustration of effort lost.

This frustration mentioned is a significant job hazard. The technician is making an overhead weld because it is the only way to do the job (a more favorable method would be used if it were possible). Usually there is pressure to get the job done—many of these welds are encountered in emergency or repair situations, and haste may seem to be needed. Overhead welding cannot be hurried. You must follow these procedures to accomplish the task.

After becoming proficient at controlling the puddle and adding filler in the horizontal and vertical positions, your welding instructor or employer will have you practice controlling the puddle on a test piece clamped overhead.

Hold the flame angle at about 45° from the work, and proceed from right to left. Keep the puddle small and not too hot and you will see that the molten material will not fall from the overhead surface. Again, the plastic range of molten steel is what makes welding possible.

Next, work at adding filler material while controlling the puddle. Minimum flame size, slow travel, and careful application of the filler rod are required. You will find that only a small amount of filler can be added at any one instant.

You must form the puddle, add filler, move away to let it solidify, come back and reform the puddle, and so on. In this way, significant filler can be added, but it takes time.

The characteristic of overhead welding is that it is slow. You need minimum heat input, minimum puddle size, minimum time at any spot, and minimum filler added to a spot. These requirements conspire to make the procedure very slow.

Overhead tacking is practiced by placing two plates overhead with a $\frac{1}{16}$ inch gap between. A small puddle is formed on one plate with a small amount of filler added and a small puddle formed on the adjacent plate. The puddles will merge and the small weld can be gradually built up.

Remember:

- Puddle, filler, move away
- Puddle, filler, move away
- Puddle, filler, move away
- Then tack the other end and the center.

Once the tacking is accomplished, the continuous weld can be placed by starting near a tack. Remember, you cannot hurry this joint.

PERSONAL PROTECTION

Proper protective devices are necessary in any welding, but most especially with overhead procedures. Goggles, face shield and helmet, and gloves must be used. Consider also using leggings, foot coverings, and aprons if the exposure warrants it. The danger in being accidentally burned includes being surprised and falling off a ladder or platform, taking your equipment with you.

Never work directly below a welding task where the molten metal can fall on you. A molten bead can burn through your clothing and skin.

BASIC ELECTRICITY FOR WELDING APPLICATIONS

Electric Theory

A brief review of electrical terms and theory will help in our consideration of welding equipment and its relationship to shielded metal-arc welding.

Electrical Current

Electrical current is the rate of flow of electric charge moving on a conductor or along a conducting path. In mathematical formulas, it is abbreviated as the letter I. Electrical current is measured in amperes and its symbol is A, when expressing a current value (10 A = 10 amperes).

Electrical Potential or Voltage

Electrical potential is the force (or pressure) that causes current to flow; it is commonly referred to as voltage. Voltage is measured in units called volts and its symbol is V. In mathematical formulas, voltage can be expressed as either V or E.

Electrical Resistance

Electrical resistance is a physical property of matter that opposes the flow of an electric current. Electrical resistance to flow is measured in ohms and represented with the Greek letter omega (Ω). In mathematical formulas, resistance can be expressed as the letter R. If an object has a small resistance to flow it is called a conductor. If an object has a large resistance to flow, it is called an insulator.

Ohm's Law

Ohm's Law explains the relationship among current, voltage, and resistance.

The equation is $V = IR$.

Electrical Insulation

Materials that are very poor conductors are known as insulators. Most nonmetals are insulators. Glass, for example, is an excellent insulator.

Current Sources

Current and voltage can be either steady (DC, direct current) or varying (AC, alternating current). The voltage from a battery is an example of DC voltage. In this example, one terminal of the battery is always positive and the other is always negative. In the case of voltage from the power company or a generator, the voltage will change from positive to negative. This change cannot be seen because it occurs 60 times per second.

There are applications where direct current (DC) is preferred, but the electric utilities in the United States stopped generating direct current for sale to the public many years ago. Alternating current is much more suitable for use in large distribution systems.

Direct current and residential-type alternating current systems are single phase, meaning that they have one wire out and one wire return. In larger buildings, alternating current is distributed in three-phase wiring systems, three wires taking turns acting as the supply while the other two act as the return wire. The equipment ground conductors (wire or conduit) are used for personnel safety in case a piece of equipment was accidentally energized with electricity.

Because it is directly available from the utility, alternating current power for welding means less-expensive equipment, usually transformers with several taps for different electrode sizes. Direct current welding supplies can take their input from the building AC supply, which means a transformer and rectifier power supply, or an AC motor can be used to drive a DC generator—the motor-generator type supply.

Many times in our work, utility power is not yet available at the jobsite. In this case, engine-driven generators are used for the source of welding power, and all of these but the smallest would be DC generators.

AC shielded metal-arc welding is usually used for brackets and light equipment repair. Most welders and piping engineers do not consider AC welding satisfactory for pipe joints. AC welding delivers high deposition rates but lacks the penetration of DC welding.

DC reverse polarity welding (electrode positive and work negative) produces the highest-quality welds. This method is a little slower, but gives the best puddle and spatter control, which is important for out-of-position welds.

DC straight polarity welding produces higher deposition rates and is used for structural fabrication.

SHIELDED METAL-ARC WELDING

When arc welding is mentioned, what most people think of is technically known as shielded metal-arc welding. There are several other variations of arc welding, but this chapter will concentrate on the shielded metal-arc method of welding, sometimes called stick welding. This method is used in our industry for a variety of tasks, principally making pipe joints, supports, hangers, special tools, and jigs and fixtures.

Shielded metal-arc welding is a manual method, meaning that a technician must manipulate the welding tool and must vary the weld appearance and properties by changing or modifying the manual techniques to achieve the desired weld. The heat required to melt the steel to form the puddle is derived from an electric arc between an electrode and the work, and the filler material is delivered by the same electrode as it is consumed by the arc.

Shielded metal-arc is the most widely used of all welding processes because it is relatively easy to learn. It also has the following advantages:

- It is fast.
- It can be done with portable equipment.
- The equipment is relatively inexpensive.
- Necessary materials are widely available.
- Welds can be made in any position and under somewhat adverse weather conditions.

The disadvantages of this method compared with oxy-acetylene are:

- It is harder to work in close quarters since the heat source and filler source are the same.
- There is less flexibility for problem weld situations.

The essential components of the shielded metal-arc system are these:

- Welding machine or power source
- Cables
- Electrode holder
- Ground clamp
- Electrodes

Personal protective safety equipment includes these items:

- Helmet with a number 10 shade lens or higher
- Apron
- Gloves
- Leggings and foot protectors

Electrode

The electrode, also called the rod or stick, has a chemical coating that is consumed with the rod. The chemical coating delivers a shielding atmosphere around the molten weld area to inhibit oxidation of the molten steel. The coating also acts as an insulator so that only the ends of the stick or rod can make electric contact and start the arc.

In the late 1880s, arc welding was performed using carbon rods, an extension of the arc light, wherein an arc between two carbon rods produced a large amount of light and heat. Someone soon switched to a steel rod so that the rod would provide the filler, but it was not until the concept of shielding chemical coatings was developed that this welding method became suitable for general use.

The electric current passing through the arc releases from 2,000 to 4,000 watts, but all that power is concentrated at the ends of the arc (the electrode end and where the arc strikes the work). Because the areas are small, the temperatures become high enough to melt the steel in the workpiece and vaporize the steel in the rod.

The temperature of the workpiece and the size of the puddle are controlled by moving the arc around so that it does not remain at any one spot for more than a few milliseconds. Figure A-7 shows the average size of the weld puddle compared to the thickness of the rod.

The electric power supply is connected on one side to the work and on the other side to the electrode. The arc is started by touching the workpiece with the electrode and then pulling it away a fraction of an inch. Once started, the arc is stable because the electrode end vaporizes, providing a conducting metal atmosphere. The metal from the electrode end is deposited on the weld.

As described above, the coating on the rod produces a shielding atmosphere, much of which condenses to a coating that is deposited over the weld. After cooling, this material, called slag, must be chipped away from the weld to inspect it or before another weld pass can be placed over the previous weld. Welding over slag will cause the next weld to have porosity trapped within the weld itself.

Shielded Metal-Arc Equipment

The power supply for shielded metal-arc welding is a low-voltage, high-current device. The inherent electrical characteristic of all the sources used is constant voltage, variable current. The welding process is constant-current in nature, so the equipment designers must incorporate elements within the product to give us the constant-current characteristic that we need. These elements are usually adjustable; by connecting them to suitable hardware, the power supply (equipped with adjusting knobs or levers) is able to accommodate a wide range of welding applications.

Therefore, if you are considering buying a power supply, check the capabilities of the machine you are considering against the range of welding work you are likely to perform. There are several manufacturers who offer a range of equipment from which to choose.

Welding Power Supply

When selecting a welding power supply, consider these factors: application, power input available, safety, cost, and efficiency of operation.

FIGURE A-7 Average weld puddle compared to rod thickness.

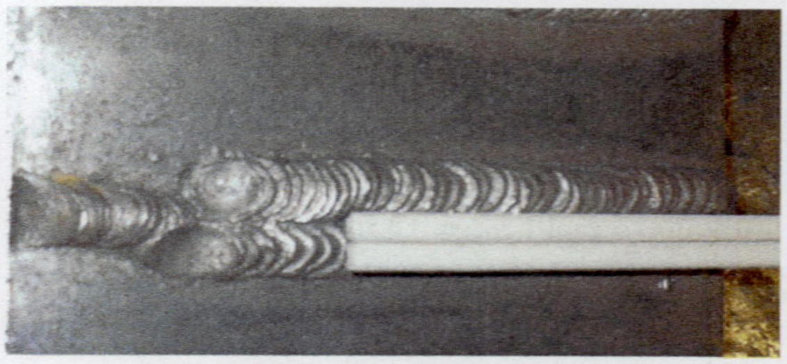

Application

- What type of welding is to be done?
- What size range is likely for the actual weld?
- What metals are to be used?
- Is the unit for shop or field use?

Power Input Available

A unit for the shop could have only one voltage and phase input, but a unit for field work should be capable of accepting at least two voltages. If you need a portable unit on the job in the early stages, you should select an engine-driven power supply.

Safety

Consider the safety implications of your choice. You may need to plan barrier fencing, provide a heavy duty trailer, pipe exhaust gas to the outdoors, and so on to minimize risk of injury and to comply with local and federal (OSHA) regulations.

Cost

For initial cost, AC transformer welders are least expensive, DC motor generators and rectifier types are in the middle price range, and engine-driven units are the most expensive. Maintenance costs follow the same ascending order, with engine-driven models significantly higher in this area.

Efficiency of Operation

The actual operating cost of any welding machine is trivial in comparison with the job being accomplished. Again, the engine-driven unit is going to be much more expensive to run—and in most cases, you also have the cost and aggravation of having to physically transport the fuel to the engine and have reserves of fuel on hand. Whenever possible, the welding power supply should derive its input from the electric utility system.

Industry Ratings

Industry ratings for power supplies may be helpful to decide on a unit or, if you already have one, to aid you in operating the supply within its capacity. Standards are developed by the National Electrical Manufacturers Association (NEMA) and include the following concepts:

Duty Cycle

The duty cycle is the percentage of time (based on a 10-minute period) that the power supply can deliver the rated current without exceeding the design temperature rise for internal parts. You are unlikely to weld more than about 70 percent of the time when you consider stick changing, fit-up, and tack welding time. Thus, 100 percent duty cycle is not needed, but less than 50 percent duty cycle may be unsatisfactory for general work.

Current Output

Current output is the amperage available for welding. A current rating of 150 to 200 amperes is adequate for nearly all piping or shop work in our field. You may find, however, that if you want an engine-driven, trailer-mounted, 2,000-watt auxiliary power, battery-starter unit, you will only be able to locate all these features with a 400-amp welding machine. The maximum current output also limits the maximum diameter of the rod that can be used.

Open-Circuit Voltage

This is the potential across the lugs before welding begins. The maximum permitted by the standards is 80 volts. Be aware that 80 volts is more than enough to give you a shock if you touch an exposed part of the electrode or holder and the work.

Power Factor

For AC input welding machines, power factor is the ratio of true power to apparent power, as measured on the input side. This consideration might be important if the welder is to be used heavily in a shop where your electric bill takes power factor into account. Check with the local utility on this issue.

Types of Power Supplies

With these considerations resolved, the types of power supply and their advantages are listed below.

AC Machines

AC machines have become more popular as the electrode manufacturers have broadened the scope and versatility of AC rods. AC welding has the following advantages:

- Faster (higher deposition rate) than DC.
- The arc is more stable.
- Arc blow is reduced. Arc blow is the swerving of the arc out of the intended direction of travel. It is caused by the magnetic fields that are set up in the metal.
- The source is highly efficient.
- It is nearly noiseless.
- There is minimal maintenance.

Disadvantages include the following:

- Difficulty in striking the arc.
- Unsuitable welds for pipe joints.

AC machines are most often transformer types with taps to select the proper current for the work to be done.

DC Machines

DC machines are popular because reverse polarity welding produces the highest-quality weld, whereas straight polarity gives high weld rates. Reverse polarity is when the electrode is connected to the positive terminal of the welding machine and the work piece is connected to the negative terminal. Advantages of DC welding are as follows:

- DC machines can be easily started.
- Weld penetration is superior.
- Welds can be performed in any position.

The principal disadvantage is that the power supply is more expensive than the AC transformers.

DC machines are either transformer-rectifier types, motor-driven generators, or engine-driven generators. The transformer-rectifier models have various internal auxiliary devices as well as load-adjusting controls so they can be used on a variety of tasks. One such device is a cooling fan. Stop welding at once if this fan quits or you may do serious damage to the rectifiers. These power supplies usually require very little maintenance.

The motor-driven or engine-driven models also have adjusting controls to vary the output for the task. These units are inherently constant-voltage sources. The auxiliary control devices permit them to serve the welding arc, which is a constant-current load. These machines, having rotating parts, will have maintenance and repair associated with bearings and shafts, and the engine models will have all the maintenance associated with engines: cooling systems, crankcase oil and oil filter, fuel problems, starting problems, ignition or injector systems, and especially exhaust gas problems. Do not operate an engine-driven welder indoors unless the exhaust gas is piped outdoors, or the area is well ventilated.

Consult manufacturers' literature or a knowledgeable local representative for assistance in exploring all the items to be considered before selecting a power supply. After selecting an appropriate power supply, we must consider four more items to form a complete shielded metal-arc welding system. These added items are the cables, electrode holder, ground clamp, and electrode.

Welding Cables

Welding cables take the current from the supply to the electrode holder and return it from the work. The cables are provided with lugs on the ends, which must be securely bolted to the power supply and the electrode holder. An improved method is to use male and female plugs on the cable ends, for versatility in developing sufficient cable length to reach the work.

Normally, cables are made of many strands of copper wire. Enough strands are assembled to achieve the total conductor capacity required for the rated welding current. The cable is then covered with paper for flexibility, nylon mesh for reinforcement, and finally covered with a heavy rubber compound for maximum electrical and mechanical protection.

The cable size to be used is based on the welding current and distance from the supply to the work. If undersized cable is used, welding stability and weld quality will be impaired and the cable will be overheated, shortening its life.

Aluminum cable is also available. It is significantly lighter than the same capacity copper cable. If aluminum is used instead of copper, the cable should be one gauge larger than the required copper size.

All joints throughout the cable loop must be in good condition or the poorly assembled joints will overheat and welding will be impaired. Keep all joints tight. Cut off and discard any cable ends that have a significant number of broken strands. Keep the insulation intact. Minimize dragging cable over the floor or across material stockpiles to avoid cable wear.

Do not use longer cables than needed and keep spare cables securely stored in tool boxes. It is unfortunate, but welding cables are often stolen if they are not locked up.

Electrode Holder

The electrode holder, sometimes called a stinger, is the device connected to the supply cable used to hold the electrode. It is good practice to use a short (about 10′) section of light (#2 AWG) welding cable, called a whip, to minimize operator fatigue. The holder attaches to the cable with an Allen screw compression connection. This connection should be checked frequently, especially if the holder becomes noticeably warm while welding. The jaws of the holder are grooved to permit holding the electrode in several different positions. These grooves should be kept clean and the spring in the holder should maintain substantial pressure on the rod.

IN THE FIELD

Always position the ground clamp as close to the welded area as possible. Electricity will take the path of least resistance, which is normally between the welding electrode and the ground clamp. But if the ground is far away or poorly connected, the weld current may travel through other parts of the machinery, which could destroy electrical components (relays, motor starters) or travel through a bearing. Welding across a bearing can cause pits on the inner mating surfaces and cause premature failure.

Be sure to use an electrode holder that has sufficient rating for the job. Do not use a 200 amp electrode holder if you are using 250 or 300 amps for the work. Figure A-8 shows how the welding cable is attached to the electrode holder.

Ground Clamp

The ground clamp is attached to the second cable from the supply. The ground clamp is attached to either the steel workbench or the work itself, if desired. Be sure that

FIGURE A-8 Proper attachment of welding cable to the electrode holder.

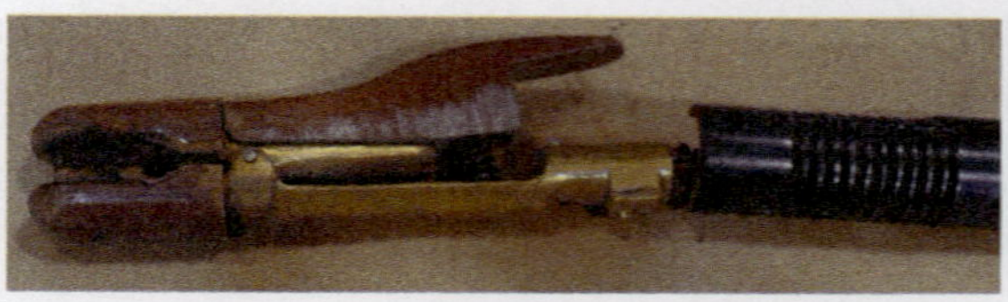

the cable connection is tight and that the ground clamp jaws make good contact with the work.

For shop work or production work in the field, it may be desirable to bolt the ground cable lug directly to the work or workbench. The circumstances of each job will dictate what is best.

When the clamp is attached, do not touch the workbench or you will be shocked.

Electrode

The welding electrode is the last component of the shielded metal-arc system. Electrodes are classified by four considerations:

- **Size:** Size is determined by the diameter of the stick itself (not including the coating).
- **Coating type:** Chemical characteristics and other properties determine whether the rod is best suited for AC or DC, direct or reverse polarity, and slag type.
- **Filler material:** The weld filler material is characterized by the alloys present, final strength, and other physical properties.
- **Operating characteristics:** The operating characteristics of the electrode determine the type of power supply, suitable position(s), deposition rate, final weld quality, and so on.

The size of the electrode influences the proper welding current. Remember, the thinner the work, the smaller the electrode must be. The rod cannot be thicker than the material being welded. The coating type, nature of filler, and operating characteristics are interrelated and are discussed below.

The American Welding Society (AWS) has developed an electrode numbering system. Here is the interpretation for the rod E-6011:

- E denotes electric arc rod.
- 60 denotes 60,000 psi tensile strength alloy.
- The next to last digit, 1, denotes position; for example:
  1. all positions
  2. horizontal and flat
  3. flat only
- The last digit, 1, refers to the type of power supply, arc, slag, penetration, or some other specific condition.

Electrodes are also identified by a color code system. The electrode has a color dot near the end of the coating and at the end that is placed in the holder. The color code is furnished by the manufacturer.

Electrodes should be stored in a cool, dry place. Many coatings will be rendered ineffective if they become moist, even if they are later dried out.

When making butt joints in piping, the first pass is made with $\frac{1}{8}$ inch or $\frac{5}{32}$ inch rod. The subsequent passes can be made with $\frac{5}{32}$ inch or $\frac{3}{16}$ inch rod. However, many technicians use $\frac{1}{8}$ inch rod for all passes, as it is easier to control and there is much less spatter. For the weld to penetrate completely through the pipe, a small gap must be present when the parts are first tacked into position. Figure A-9 shows a pipe section being joined.

Operating traits of electrodes also include the following items:

- Fast fill electrodes deposit more weld material (i.e., use a higher welding current).
- Fast freeze electrodes deposit weld fill that solidifies very quickly. This type of electrode is good for problem welds, such as overhead, or any place where a larger-than-normal gap must be filled.
- Fast follow electrodes are used when a small additional fill is needed.

In summary, electrodes should be considered and selected based on the following issues:

- Position of the weld
- Penetration required
- Code requirements, if any
- Thickness and shape of the pieces to be joined
- Use of the assembled joint
- Slag removal

FIGURE A-9 Example of a pipe section being joined.

- Joint preparation and clamping
- Current type

Personal Protection

Electric welding can be dangerous to people with pacemakers. Provide warnings and be cautious with people who might enter the work area. Safety equipment for shielded metal-arc welding and its proper use are discussed in the following paragraphs.

Protection from High-Intensity Light

The high-intensity light present during the welding process is capable of producing burns in your eyes and skin, even reflected off a wall. The welder needs a full-face-covering helmet with a welding lens of #10 to #12 shade. Fellow workers and the public should be protected from welding rays by the use of a barrier or screen.

These rays will produce severe burns on exposed flesh. Wear long-sleeved shirts and gloves, and button up shirt collars. An arc-welding hood is required over the face and to protect your eyes. These precautions also protect you from molten metal.

IN THE FIELD
Most people do not understand the danger of watching a welding process without eye protection. Make certain to shield the area from the public.

Protection from Hot Metal Globules

Molten metal spatters from the weld puddle can cause severe pain and damage if they make contact with your body. Keep your shirt buttoned and do not roll up cuffs or sleeves. Do not tuck your shirt inside your pants—if a piece of hot metal goes down your shirt and you try to shake it out, it can go down your pants. Wearing a cap inside the welding helmet will protect the top of your head from spatter.

Protection from Shock

Electrical contact with the welding circuits can produce surprise shocks. Maintain cables and connections in good condition. Do not make contact with the welding electrode and workpiece at the same time. Be especially careful when using a welding process with high-frequency AC; it is more likely to shock you.

Protection from Asphyxiation

Asphyxiation is a hazard in confined areas during any procedure that produces toxic fumes. Work only in well-ventilated areas or use an oxygen respirator. Fumes from certain metal coatings, such as galvanize, can make you very sick if inhaled in large amounts.

Protection from Falling Objects

Falling objects represent a hazard in most construction operations, not just welding. Be especially careful around temporary rigging—any dislodged item can cause serious injury.

TOOLS

Welders use a variety of different tools, but the ones most commonly used are the chipping hammer and wire brush. Welders use the chipping hammer to remove the slag to uncover the weld. After chipping, the last remnants of slag are removed with a wire brush. When running multiple passes, it is vital to remove all of the slag before running the next welding pass. Other tools include grinders, layout and clamping devices, beveling tools, and similar devices.

One note of caution: because of the large currents associated with welding, very large magnetic forces are generated, especially if the welding leads are coiled. It is best to leave your wristwatch (or any computer-driven devices) at home or you may find it magnetized and inoperable after a welding session.

CURRENT SETTINGS

The current setting for an electrode can be approximated by making it equal to the diameter of the rod in thousandths of an inch. Thus, a $\frac{1}{8}$ rod (0.125″) would use 125 amps. As a general rule, if it is difficult to start and maintain an arc, then the current can be increased. If the arc is easy to start but the puddle is large and difficult to control, then the current can be decreased. The recommended values are listed in Table A-3.

The arc length affects the arc voltage. The actual length is difficult to judge visually, so it is recommended to develop an "ear" for the sound of a proper arc. One indication that an arc is too long will be the spatter of small balls of weld material around the weld. The goal is to keep the arc short for arc stability and best filler deposition, but there is some tolerance in the arc length without significantly affecting the weld.

TABLE A-3 Welding Currents—Reverse Polarity

Size		
Fraction (in.)	Decimal (in.)	Current (amps)
$\frac{1}{16}$	0.062	60
$\frac{3}{32}$	0.093	75
$\frac{1}{8}$	0.125	100
$\frac{5}{32}$	0.156	150
$\frac{3}{16}$	0.187	200
$\frac{1}{4}$	0.250	250
$\frac{5}{16}$	0.312	300
$\frac{3}{8}$	0.375	400

As the electrode is consumed, the chemicals also vaporize and then condense in the vicinity of the arc, providing the inert, shielding atmosphere over the weld.

The rate of electrode travel controls the weld bead—too fast and the bead will not be continuous and may have an oval appearance; too slow and the bead will build up above the work surface.

The angle of the electrode to the work is also important. Many people debate whether it is better to drag the rod or push the rod. The answer, of course, is do whichever works better for you. In either case, the electrode should be about 90° to the work and about 60° to 80° from the work in the direction of travel so the rod is pointing back at the puddle. For welding in the flat position, the 90° to the work focuses most of the heat directly into the root of the weld and not across it. When welding horizontally, the rod will have to be pointed upward to counteract the tendency of the weld puddle to sag due to gravity.

STRIKING, MAINTAINING, AND CONTROLLING THE ARC

The most difficult skills for the beginner to learn are how to strike the arc, maintain the arc, and control arc blow.

The easiest way to strike the arc is to drag the electrode tip across the work (near the beginning point of the weld) with the electrode at about a 45° angle to the work, similar to striking a match. When electrical contact is made, move the tip away from the work a fraction of an inch and the arc will be established.

The arc will be maintained if the operator moves the electrode closer to the work as the electrode is consumed, thereby maintaining the arc length. This is a coordination skill learned through practice. If the electrode is moved in too quickly, the electrode will short in the puddle, extinguishing the arc. If the electrode is not moved in rapidly enough, the arc length increases until it becomes unstable and also extinguishes. The correct rate will maintain the arc until the rod burns down to minimum length (about $1\frac{1}{2}$ in.).

Arc blow is a deviation in arc path caused by magnetic forces that are more pronounced with DC arcs. The easiest way to control arc blow is to keep the arc length short. You may also relocate or split the ground to reduce arc blow. Practice is the key to developing skill at welding.

SHIELDED METAL-ARC METHODS

The basic arc-welding beads are described so that you can develop the methods used for all arc-welding processes.

Stringers

Our first task is to strike an arc and run a stringer bead across a flat plate. A stringer bead is a weld that deposits filler material in the direction of travel of the electrode. There is a minimum of filler buildup at right angles to the direction of travel. The necessary equipment is a welding power supply, protective coverings (eyes, face, hands, and skin), welding electrodes, and a piece of $\frac{1}{4}$ inch steel plate.

Set up the equipment with the plate horizontal. Be sure other people nearby know your intentions and are warned of the hazards. Start the power supply.

Strike an arc and move to the edge of the plate (if right-handed, move from left to right; the opposite if left-handed). Starting at the edge, move the electrode back and forth in the direction of the weld and gradually proceed across the plate. The electrode should be 70°–80° to the vertical in the plane at right angles to the direction of motion, and at 90° to the direction of motion. Each stringer weld should slightly overlap the previous one; this is sometimes referred to as a pad weld. This will build up the weld material in a uniform manner; otherwise, it will have valleys and resemble speed bumps in the road. See Figure A-10.

After cooling (remember, the sample will be very hot), clean off the slag with a chipping hammer and brush. Be sure to wear eye protection. Practice placing these stringers on the plate until you can produce these welds with proficiency.

IN THE FIELD

Cooling the plate in water will cool it quickly and thereby induce stresses in the weld area. While this is OK for practice, do not quick-cool welds that are made in the field.

Weave Beads

Weave beads should be practiced next. A weave bead is one in which the electrode is moved in a zigzag motion at right angles to the direction of welding, so that the bead is two to three times the diameter of the rod. The rod motion should pause slightly at each end of the weave to assure formation of the puddle and satisfactory penetration at the edge of the weave. See Figure A-11.

WELD CHARACTERISTICS

Table A-4 lists weld characteristics and adjustments or corrections necessary to produce satisfactory weld beads.

FIGURE A-10 Example of overlapping stringer welds to build up weld material uniformly.

FIGURE A-11 Example of weave beads.

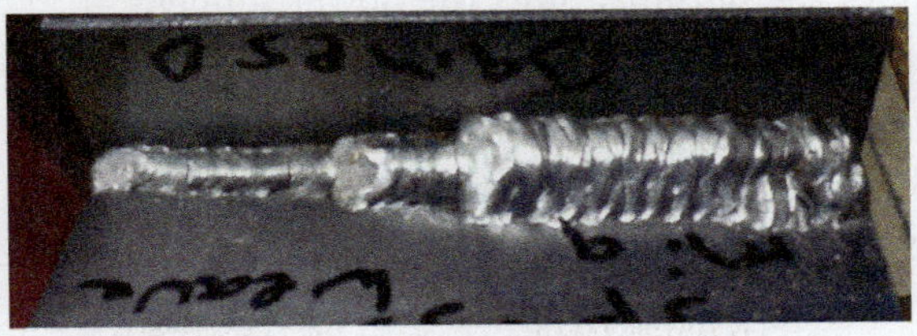

TABLE A-4 Weld Characteristics

Weld Appearance	Cause
Uniform bead width, straight, even ripple	Everything correct
Thin bead width, even ripple	Current low
Wide bead width, heavy deposits on side of bead	Current high
Standard width, uneven ripple	Voltage low
Splotchy deposit, uneven ripple	Voltage high
Wide bead, uniform wide arched ripple	Travel slow
Thin bead, pointed ripple	Travel fast

EXERCISES

The three shielded metal-arc welding exercises include the following:

1. Welding an open butt joint (there is a $\frac{1}{16}$ in. to $\frac{1}{8}$ in. gap between the edges of the plates)
2. Welding a beveled joint (the edges to be joined are cut at about 60° so a V is available for welding material and penetration)
3. Tee joint with three pass fillet weld

For each joint type, tack each setup first. Tack the plates together at the edges. Make sure to leave a gap that is approximately the width of the welding rod diameter, but no more. The arc temperature is so hot that it is difficult to start the weld at the edge of the plate without having the edge disappear. When performing shielded metal-arc welding, it is important to remember that the plate material must be no thinner than the diameter of the welding rod.

After tacking, run a stringer bead between tacks. On thin plate, $\frac{3}{16}$ inch or less, one pass should completely fill the gap. Clean the slag from the bead and inspect the opposite side for penetration. Figure A-12 shows both the top and bottom surfaces of a single pass weld.

When the plate is thicker, it is difficult to get full penetration of the plate. Figure A-13 shows how beveling the plates allows for the first pass (root pass) to completely penetrate the plate. The top two welds overlap to build up to the complete plate thickness.

FIGURE A-12 Top and bottom surfaces of a single pass weld.

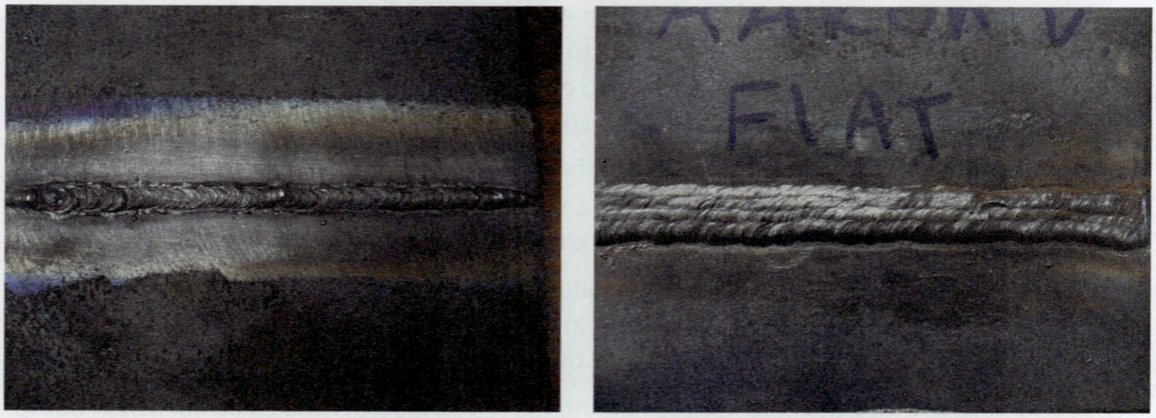

FIGURE A-13 Beveling allows the first pass to completely penetrate the plate.

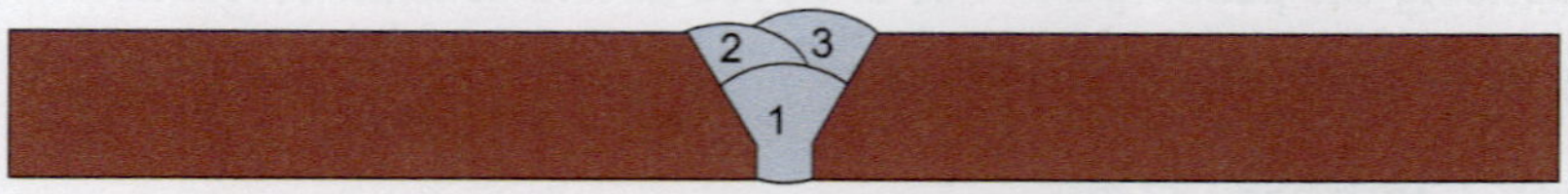

FIGURE A-14 Tee joint.

The tee joint is used when plates are joined at right angles. Because technicians sometimes need to fabricate brackets and stands, it is important for them to know how to perform this weld correctly. Figure A-14 shows a tee joint. For the vertical plate to be supported, it is important to have both legs of the weld the same length. This gives a 45 degree fillet.

After the three joint types are completed, saw through each joint so that the cross-section can be examined. If the welds have been properly made, the several weld passes will be merged together.

Practice these joint types and you will soon develop skill in this welding method.

SHIELDED METAL-ARC VERTICAL WELDING

Gravity is a problem in welding on a vertical surface. The puddle is more difficult to control and molten steel tends to fall from the plate. The rod must be held close to the work so that the arc is short, and the arc must be kept moving all the time or the puddle will become too large and too hot.

In all shielded metal-arc welding, the weld rod must be kept moving and the weld must keep advancing. When a wide or deep weld is required, multiple passes may be used to build up weld thickness to the desired depth.

This exercise requires that a plate be mounted in a vertical plane. Select the proper safety attire and weld rod and make proper power supply settings. Advise others as to your intentions and the welding hazards.

Proceed with the work by holding the rod so that it is angled up about 10° above the horizontal and close to the work. Lower your hood and strike an arc. Work at the padding exercise while holding the rod slightly uphill and pointing back at the puddle as the weld advances. If the puddle sags or significant molten metal falls out, increase the weave rate and the weld travel rate.

Continue practicing the padding exercise until you obtain consistent results.

Next, try the three joints attempted before:

- Butt joint with $\frac{1}{16}$ inch gap
- Butt joint with ends cut to form a bevel joint
- Tee joint with three fillet welds

A vertical joint between vertical plates is very difficult to master. The vertical joint is the most common joint encountered when welding piping in place, since most piping will be installed in a horizontal position, resulting in joints that are vertical.

Two options are possible for a vertical joint:

- Uphill: Start at the bottom and make the weld by moving up the joint.
- Downhill: Start at the top and make the weld by moving down the joint.

Uphill Welding

As described earlier, the uphill weld is made by holding the electrode so that it is pointing up at approximately 10° above the horizontal. The first pass should be a stringer bead to obtain penetration to the bottom of the joint without causing the edges to disappear. Work the rod up and down in small intervals, always with a net upward movement. The uphill method is the preferred method because there is less chance of the slag rolling into the weld zone.

An uphill weave bead is made by moving the weld over the gap and pausing at each end of the weave. Pausing at the ends of the motion allows the filler material to build up; otherwise, a slight undercut will occur. Each weave motion must advance the weld slightly.

The practice session should start with padding on a plate, then weld a butt joint with gap, and then weld a bevel butt joint. Uphill welding should be performed on $\frac{1}{4}$ inch or heavier material.

Downhill Welding

Downhill welding can be used on material as thin as $\frac{3}{16}$ inch. The rod position is the same as the uphill method, but the travel is slightly quicker. Most weld inspectors will not permit a weld test to be performed in the downhill motion. While the weld will appear to be alright, slag and other impurities can drop down ahead of the filler material and become trapped inside. These will lead to porosity. Again, perform the practice exercises of padding, butt weld, and bevel butt weld and tee joint.

Next, practice tacking a vertical pipe joint with three tacks 120° apart. Even though the pipe has a radius versus the flat surface of vertical plates, the required techniques for this tacking operation are the same as those needed for flat plate vertical welding.

Practice these welding methods to build your proficiency.

SHIELDED METAL-ARC OVERHEAD WELDING

Overhead welding is accomplished using the same methods as with the vertical plate joint—continuous motion, short arc, and steady weld progression.

An important aspect of overhead welding is operator comfort. As suggested earlier, use a light whip welding cable for the last few feet to the electrode holder. Position yourself as comfortably as possible. Any strain or awkwardness of position will prevent the close, easy control that is essential to successful welding.

You should not place yourself under the weld, but just to one side, so that any falling metal will not descend directly onto you or your protective gear. Fire extinguishing equipment is required. Avoid working alone, even when practicing an overhead weld.

As in the other methods, practice padding on a single plate, then work at the three butt joint types used before. Tack the plates first.

As you continue this exercise, your skill level will improve and you will be able to produce satisfactory welds.

TESTING WELDS

As welding progressed from a method to make occasional equipment repairs to the major procedure for making vital joints on major structures, it was necessary to develop methods to determine the quality of welds. If the weld can be assured to be free of defects (for critical joints) or with only certain minimum flaws (for less important members), the integrity of the building, bridge, or piping system can be assured.

The American Welding Society and other industry groups have assembled experienced people in this field to develop the requirements and methods that assure the quality of welds.

Test procedures to check these requirements will fall under one of three following broad methods:

- Destructive
- Nondestructive
- Visual

Destructive Testing

Destructive testing involves cutting, stretching, bending, or etching sample specimens of a test weld. The test itself permanently changes the weld sample so that it is not the same after the test is completed. Destructive testing is usually used to verify that a welding procedure or welding technician can produce adequate, high quality welds.

These tests generally involve cutting the example weld into sample specimens and preparing them (if necessary) for the tests to be performed. These tests include the following procedures:

Tensile Test

The sample is carefully measured so that the cross-sectional area can be computed, then the piece is pulled apart in a tensile machine. The tensile strength of the weld is computed.

Bend Test

Usually two sections of a weld are selected to be tested. One section bends a weld sample in a U shape with the root of the weld on the outside of the bend. The second section is bent with the root on the inside of the bend. Each section is then checked for cracking caused by slag inclusions, porosity, and poor penetration. Figure A-15 shows a sample that failed to pass the test.

Impact Test

The prepared specimen has a notch cut in the weld. A pendulum weight is allowed to fall on the weld to test its strength against impact. Depending on

FIGURE A-15 Example of a weld that failed a bend test.

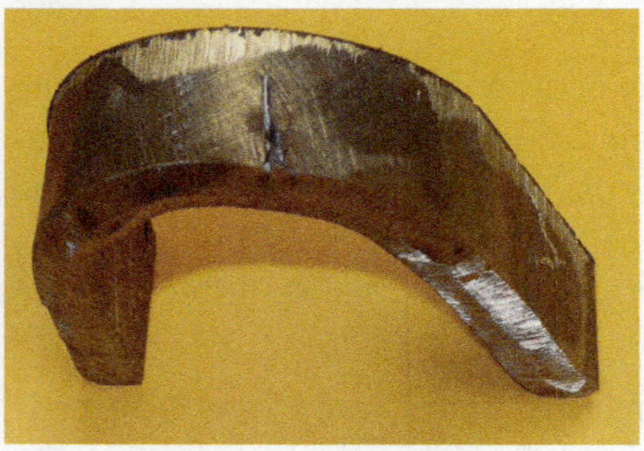

the shape of the notch cut and the instrument used, this test may be called the Charpy or Izod test.

Fatigue Test

A sample section of the weld is subjected to a specific number of reverse load cycles. Afterward, the weld is examined for cracking.

Chemical Test

A sample section of the weld is cleaned with a solution of iodine and potassium iodide, nitric acid, or hydrochloric acid. After the solutions have been removed, the boundaries of the weld and any defects will be revealed.

Nondestructive Tests

Nondestructive tests are performed on welds that are parts of completed assemblies. These tests indicate the condition or quality of the weld without changing the weld. Nondestructive methods include the following:

Radiographic Test

An X-ray or other high-energy source exposes film by passing rays through the weld. Experienced individuals can interpret the photographs for internal cracks, voids, or other weld irregularities.

Penetrant Test

Penetrant dye tests rely on capillary action of a small crack or defect to draw a fluid into the welded material. A dye is applied to the weld and then wiped off. Afterward, a special light will show any dye that enters surface flaws. In some cases, a special fluid that evaporates quickly is sprayed on the weld. This fluid leaves a white powder coating on the weld surface that will draw out any penetrated dye, thus leaving an indication of a crack. This test is especially good for nonferrous materials such as aluminum.

Magnetic Particle

In this test, an electric current is passed through the weld area while a fine iron powder is sprinkled on the weld. The electric current will magnetize the part. The idea behind this test is to take advantage of the fact that when a magnet is broken, it becomes two new magnets, each with both a north and south end. The iron powder will stick to the ends of the new magnetic sections and clearly indicate a crack. Figure A-16 shows an illustration of how this concept works.

Ultrasonic Test

In this test, a high-frequency sound wave is passed through the weld. The sound wave will pass through the part until it hits a change in the structure. In the case of a good weld, the change will be the opposite side of the weld. If there are any defects such as holes or cracks in the weld, the sound wave will bounce back sooner and the operator will observe this. Figure A-17 shows an illustration

FIGURE A-16 Example of a magnetic particle test indicating a crack.

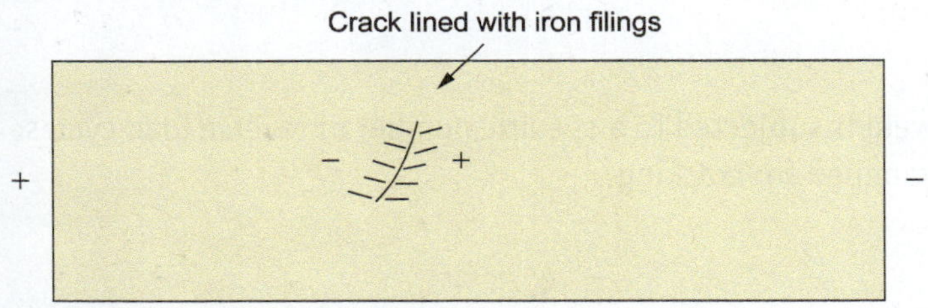

FIGURE A-17 Example of an ultrasonic weld test.

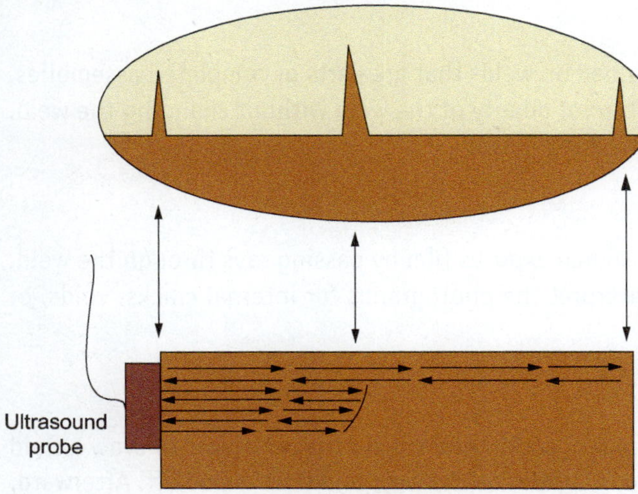

of the sound wave traveling through the part and how it is shown on the instrument.

Leak Test

If the weld is on a pipe, the pipe can be pressure tested to verify the adequacy of the weld. In many cases, water is used due to the fact that very high pressures can be created. If there is a leak, the pressure will drop very fast.

Visual Tests

Visual inspections are adequate to determine the suitability of most welds in noncritical areas. A 10-power magnifying glass is used to look for cracks, porosity, extent of penetration, inclusions (foreign materials within the weld deposit), undercutting, and degree of fusion of the base material and weld material.

The need to determine the adequacy of welds has brought about the test methods just described. Use appropriate methods to examine the welds you produce.

ADVANCED WELDING METHODS AND SYSTEMS

There are additional welding methods used in industry, of which the following three might be used by technicians in the piping trades.

TIG Welding

First is the tungsten inert gas (TIG) method. This method is extremely versatile and suitable for a variety of materials, thin gauges, or special alloys. The method is slow, but produces a very clean smooth weld. In many cases, welders will use this method as the first root pass in a pipe-welding operation.

The method is similar to oxy-acetylene, in that the heat source and filler material are independent, so each can be controlled as required by the weld condition. However, since the arc from the tungsten is hotter and focused on a smaller area, the process is quicker than oxy-acetylene welding. It also creates a much smaller heat-affected area. A tungsten electrode (which is not consumed) is used to form an arc to the work. Filler material, if required, is added in rod form to the weld puddle.

The melt area is protected from oxidation by a shielding gas that is released from an opening that surrounds the tungsten electrode mounting in the electrode holder. The shielding gas is usually argon or helium. If helium is used, the system may be referred to as heli-arc welding.

The same safety apparel is required for this method as for the shielded metal-arc system. In some cases, welders tend to use a thinner glove on the hand used to add the filler rod. This allows them better control of the rod. Figure A-18 shows a piece of stainless steel being welded.

MIG Welding

The second system used in piping is the metal-arc inert gas (MIG) process. This method is quite versatile, but somewhat less so than TIG welding. MIG welding is the highest-production manual welding method, and it is very popular on larger jobs where welding productivity is important enough to justify the somewhat greater investment in equipment. This method is sometimes called wire welding.

The filler material is purchased in wire form on a spool. The most common diameter is 0.035 inch, or about $\frac{1}{32}$ inch. When welding, the wire is fed through the center of the gun, which also releases the shielding gas. As the wire approaches the work and enters the arc region, it is consumed in the arc.

FIGURE A-18 Example of stainless steel weld using the TIG method.

The special demands on the power supply for this welding system and the wire handling and drive systems represent extra costs in the equipment. For these reasons, the equipment requires more investment than needed for other arc-welding methods.

This system as well as the TIG method requires no slag cleanup between passes. A further considerable time saver results because the continuous wire means no interruption of welding to change rods. These two considerations, plus a weld deposit rate from 30 percent to 50 percent greater than shielded metal-arc, result in 300 percent to 400 percent production rates as compared with shielded metal-arc.

The same safety apparel is required when performing MIG welding as for shielded metal-arc welding.

Resistance Welding

The third advanced system encountered in the piping field is resistance welding, also called spot welding. It is used to attach small objects to large steel surfaces. It is used in our trade to attach insulation pins to sheet metal ducts, and it could be used for other hanger devices that have to be attached to large steel areas.

The small item that is to be attached is held in a movable device, and the item and the work are connected to a charged power supply. The small item is rapidly moved to the work and held in place briefly. When they first touch, electrical energy is released and the contact area is heated enough that fusion takes place. After a second or two, the joint cools and the holder can be released.

This is a very high-speed and special-purpose welding method. This system is automatic and requires little special operator skill. The operator merely changes the amount of stored energy released at the moment of contact and activates the equipment as needed to complete the task.

OTHER WELDING SYSTEMS

Two other systems that are more likely to be used in a manufacturing facility are submerged arc welding and laser welding.

The submerged arc uses a coil of filler (much larger in diameter than MIG wire) that is fed into the arc below a pile of granular shielding material. This method produces very heavy weld deposit rates. The weld arc is not visible (hence the term *submerged*) below the granular shield material. The granular feed, wire feed, oscillation, and weld travel are all machine driven, so this is an automatic method.

Laser welding is performed by generating a high-intensity light and focusing it on the work to raise the temperature of the work material to the melting point.

Further information about welding can be obtained from the American Welding Society (http://www.aws.org), welding reference books, and welding suppliers in your area.

Practice your techniques to improve your skill in this important area of our work.

Appendix B

Answers to Review Questions

CHAPTER 1 ANSWERS

1. True
2. False
3. True
4. False
5. True

CHAPTER 2 ANSWERS

1. True
2. False
3. True
4. False
5. True
6. False
7. False
8. True
9. True
10. True

CHAPTER 3 ANSWERS

1. True
2. False
3. False
4. False
5. True

CHAPTER 4 ANSWERS

1. Acetone
2. World Wars I and II
3. Bridge building, industrial/ commercial buildings, commercial piping, etc.
4. Higher
5. All of the way

CHAPTER 5 ANSWERS

1. True
2. True
3. False
4. True
5. True

CHAPTER 6 ANSWERS

1. False
2. True
3. True
4. True

5. True
6. False
7. True
8. False
9. True
10. True

CHAPTER 7 ANSWERS

1. False
2. True
3. True
4. True
5. True
6. True
7. True
8. True
9. True
10. False
11. False
12. Hydropneumatic tank, gravity tank, booster pump
13. 5.0; 8.1; 20.6; 4 ft./sec ($1\frac{1}{2}''$ diameter pipe)
14. 5.0 ft.
15. 1.5 WSFU

CHAPTER 8 ANSWERS

1. Flow rate expected and water temperature increase needed.
2. 205,800 Btu/hr
3. 27,621 Btu/hr
4. The amount of hot water, in gallons, the water heater can supply per hour when the tank is initially full of hot water
5. 50 gallons
6. True
7. False
8. True
9. True
10. True
11. True
12. False
13. False
14. True
15. True
16. False
17. True

CHAPTER 9 ANSWERS

1. Gravity backflow, back-pressure backflow, back-siphonage backflow
2. Air gap
3. 1 inch
4. Pit or other area subject to flooding
5. Red

CHAPTER 10 ANSWERS

1. False
2. False
3. True
4. True
5. True
6. True
7. True

CHAPTER 11 ANSWERS

1. True
2. True
3. True
4. True
5. False
6.

WC branches	3″	Tub branches	$1\frac{1}{2}$″
Stack 1st to 2nd	4″	Stack drain to 1st	4″
Building drain upstream	4″	Building drain downstream	5″

CHAPTER 12 ANSWERS

1. Backwater valve, sewage pump
2. Sewer
3. 2 fps
4. True
5. True

CHAPTER 13 ANSWERS

1. Roof, two
2. 30%
3. Full-sized
4. Branch interval
5. $\frac{1}{3}$

CHAPTER 14 ANSWERS

1. 243 inches
2. 13 feet 7 inches
3. 315 inches or 26 feet 3 inches
4. 26
5. 43,560 $ft.^2$(66 ft. × 660 ft.)
6. 250 $ft.^2$ (10×5^2)
7. 17.5 ft. (approximately) (Using Table 9-1, you know the answer is between 17^2 (289) and 18^2 (324). 17.5 would give 306 $ft.^2$, which would exceed the minimum required area.) An alternate method would be to reduce 300 into 30 × 10 and then take the square roots of both the 30 and the 10. $\sqrt{300} = \sqrt{30 \times 10} = \sqrt{30} \times \sqrt{10}$; from Table 9-1, $(5.477 \times 3.162) =$ 17.32 ft.
8. 3,969; 1,764; 9,216; 961
9. 6.928, 2.449, 5, 4.243
10. 3.634, 1.817, 2.921, 2.621
11. 4 inches long (20 in. − 1.75 in. − 1.5 in. − 3 in. − 1.75 in.)/3

CHAPTER 15 ANSWERS

1.

TABLE 15-4 Offsets for $\alpha = 45°$

	1	2	3
Offset	20″	$11\frac{5}{8}$″	36″
Diagonal	$28\frac{1}{4}$″	$16\frac{1}{2}$″	$50\frac{7}{8}$″
Fitting Allowance	1″	$\frac{3}{4}$″	$\frac{3}{4}$″
Cut Length (e-e)	$26\frac{1}{4}$″	15″	$49\frac{3}{8}$″

2.

TABLE 15-5 Offsets of Various Degrees

Fitting Angle (α)	22.5°	45°	60°
Offset	20″	15″	20″
Run	$48\frac{1}{4}$″	15″	$8\frac{5}{8}$″
Diagonal	$52\frac{1}{4}$″	$21\frac{1}{4}$″	23″
Fitting Allowance	$\frac{1}{2}$″	$1\frac{1}{2}$″	2″
Cut Length (e-e)	$51\frac{1}{4}$″	$18\frac{1}{4}$″	19″

3. $2\pi(15') = 94.25'$
4. $4(100') = 400'$
5. $2(15'' + 10'') = 50''$
6. $\pi(4)^2(16)(7.48) = 6{,}016$ gallons
7. $4 \times 10 = 40\ \text{ft.}^2$

$$1{,}500 \text{ gallons} = (1{,}500)/7.48 = 200.53\ \text{ft.}^3$$
$$\text{Required width} = (200.53)/40 = 5.01'$$

8. A 6-foot-deep trench with 45° banked sides is made up of two 45° triangles, each with sides of 6 feet. Therefore, the top of the trench is 12 feet across.

$$\text{Volume} = \tfrac{1}{2}(6)(12)(300) = \tfrac{1}{2}(10{,}800\ \text{ft.}^3) = (10{,}800)/27 = 400\ \text{yd}^3$$

9.

TABLE 15-6 Lead and Oakum Takeoff for Question 9

Pipe Size	Number of Joints	Pounds Lead per Joint	Subtotal (Lead)	Pounds Oakum per Joint	Subtotal (Oakum)
2	10	2	20	0.50	5.0
3	6	3	18	0.750	4.5
4	20	4	80	1.00	20
5	0	5	0	1.25	0
6	10	6	0	1.50	0
		Total Lead =	178	Total Oakum =	44.5

CHAPTER 16 ANSWERS

1. False
2. False
3. True
4. False
5. True
6. Liquid, solid, gas
7. Heat
8. Liquid
9. Area
10. Increases

CHAPTER 17 ANSWERS

1. True
2. True
3. True
4. False (weight of pipe and its contents and supporting type)
5. True
6. True
7. True
8. True
9. False
10. False

CHAPTER 18 ANSWERS

1. False
2. True
3. True
4. True
5. True

CHAPTER 19 ANSWERS

1. True
2. True
3. True
4. True
5. True

CHAPTER 20 ANSWERS

1. True
2. False
3. False
4. False (jobsite may be different than what the manufacturer expected)
5. False

CHAPTER 21 ANSWERS

1. True
2. True
3. True
4. True
5. False

CHAPTER 22 ANSWERS

1. True
2. True
3. True
4. True
5. False
6. True

7. Safety
8. The ability to recognize, avoid, and minimize hazards at all times. A sense of priority to plan for safety.
9. Any of the following would be correct:
   - When carrying a load, drive up a ramp or grade. Never drive forward down a ramp or grade when you are carrying a load. Never make a turn while your forklift is on a ramp. Lower the forks to keep the center of gravity low.
   - Always use a proper dock board when loading a vehicle from a dock. Keep the forklift away from the edge of the loading dock.
   - Make sure the parking brake is set and wheels are chocked on the vehicle being loaded.
   - Place the forks all the way under the load. Space forks apart so they fit the load being lifted. This will help to maintain proper balance and prevent the load from falling. Never lift a load that appears to be unstable. Use belts to secure the load onto the forks.
   - Center the forks beneath the load being lifted. Lifting an uncentered load can cause the load to fall. Tilt the uprights slightly back when raising and carrying a load.
   - Do not carry any riders unless the forklift is specifically designed for them. Always keep hands and feet inside. Never speed or allow unauthorized persons to drive a forklift.
   - Never smoke when refueling or when checking the battery of a forklift. Always turn off the engine when refueling.
   - Use a properly secured safety platform when the truck is to be used as a lifting device.
   - Never carry loads that obstruct your view.
   - When the forklift is parked, fully lower the forks, put the controls in neutral, turn off the engine, set the parking brake, and remove the key.
   - When turning, reduce your speed and maneuver carefully.
   - Stay a safe distance away from other forklifts. Never drive side by side.
   - At blind corners, stop the forklift and sound the horn.
   - Know where low clearances, pipes, sprinklers, or low doorways are located.
   - Completely inspect the forklift prior to any operation of the unit. If you find anything wrong, report it to your supervisor or maintenance department.
10. Before each use
11. Training and certification

CHAPTER 23 ANSWERS

Water Closet	
Rough-in from back wall	12
Clearance to adjacent objects (by code)	Tub: 12″ dwellings, 15″ other building
Height of bowl	15″
Height of supply rough-in	8″
Off-center dimension of supply	6″
Lavatory	
Location of drain—vertical	22″
Location of drain—horizontal	centered
Location of water supply—vertical	24″
Location of water supply—horizontal	3″ off center (6″ spread)
Bathtub	
Location of waste—side wall	16″
Location of waste—rear wall	$8\frac{1}{2}''$
Location of supply—vertical	$14'' + 17\frac{7}{16}'' = 31\frac{7}{16}''$
Location of spout—vertical	$4'' + 17\frac{7}{16}'' = 21\frac{7}{16}''$
Location of showerhead—vertical	72″–78″ above finish floor (typical)

1. True
2. True
3. True
4. True
5. True
6. CW
7. D
8. F
9. GV
10. MA (medical compressed air is not the same as compressed air used in industry; it is much cleaner)

CHAPTER 24 ANSWERS

1. True
2. True
3. True
4. True
5. True

6.

FIGURE 24-15

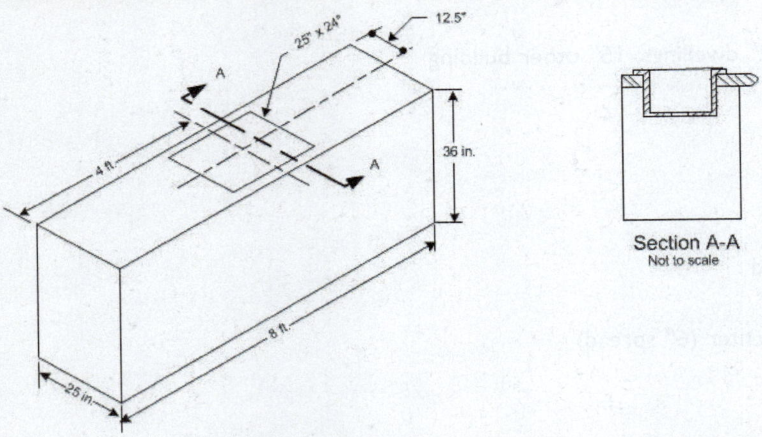

Photo by Ed Moore

Appendix C

References

The information available about our industry grows and changes daily. Rather than recommend specific books or publications, it is suggested that the student access information through the library and the Internet.

Most states provide information and code books through the state government. For state and local information, contact the governing body for your area.

Many associations and societies have a catalog of publications that are updated on a periodic basis. Some of these associations, but by no means all, are listed below.

In addition, most manufacturers publish guides and technical information about their products. Please check our list of contributors to this text.

American Gas Association (AGA)
400 N. Capitol Street NW
Suite 450
Washington, DC 20001
202-824-7000
http://www.aga.org

American Welding Society (AWS)
8669 NW 36th Street
Suite 130
Miami, FL 33166
800-443-9353
http://www.aws.org

Cast Iron Soil Pipe Institute (CISPI)
2401 Fieldcrest Drive
Mundelein, IL 60060
224-864-2910
http://www.cispi.org

Copper Development Association, Inc. (CDA)
260 Madison Avenue
New York, NY 10016
212-251-7200
http://www.copperalliance.us

National Clay Pipe Institute (NCPI)
N6369 US Highway 12
Suite A
Elkhorn, WI 53121
262-742-2904
http://www.ncpi.org

Plastic Pipe and Fittings Association (PPFA)
800 Roosevelt Road
Building C, Suite 20
Glen Ellyn, IL 60137
630-858-6540
http://www.ppfahome.org

Plastics Pipe Institute
105 Decker Court, Suite 825
Irving, TX 75062
469-499-1044
http://www.plasticpipe.org

Plumbing-Heating-Cooling Contractors—National Association
180 S. Washington Street, Suite 100
Falls Church, VA 22046-1148
800-533-7694
http://www.phccweb.org

Plumbing-Heating-Cooling Contractors—National Association Educational Foundation
180 S. Washington Street, Suite 100
Falls Church, VA 22046-1148
800-533-7694
http://foundation.phccweb.org

U.S. Department of Labor Occupational Safety and Health Administration (OSHA)
200 Constitution Avenue NW
Washington, DC 20210
800-321-6742
http://www.osha.gov

Appendix D

Contributors

A.O. Smith Water Products
500 Tennessee Waltz Parkway
Ashland City, TN 37015
800-527-1953
http://www.hotwater.com

American Society of Plumbing Engineers (ASPE)
6400 Shafer Court
Suite 250
Rosemont, IL 60018
847-296-0002
http://www.aspe.org

American Standard Kitchen & Bath
1 Centennial Plaza
Piscataway, NJ 08855
800-442-1902
http://www.americanstandard-us.com

AMTROL, Inc.
1400 Division Road
West Warwick, RI 02893
401-884-6300
http://www.amtrol.com

Bell & Gossett/Xylem Inc.
8200 N. Austin Avenue
Morton Grove, IL 60053
847-966-3700
http://www.bellgossett.com

Bradford White Corporation
725 Talamore Drive
Ambler, PA 19002
800-334-3393
http://www.bradfordwhite.com

Elkay Sales, Inc.
2222 Camden Court
Oak Brook, IL 60523
630-574-8484
http://www.elkayusa.com

John Guest USA Inc.
180 Passaic Avenue
Fairfield, NJ 07004
973-808-5600
http://www.johnguest.com

KOHLER Plumbing
444 Highland Drive
Kohler, WI 53044
800-456-4537
http://www.kohler.com

NIBCO Inc.
1516 Middlebury Street
Elkhart, IN 46516
574-295-3000
http://www.nibco.com

Plastic Pipe and Fittings Association (PPFA)
800 Roosevelt Road
Building C, Suite 312
Glen Ellyn, IL 60137
630-858-6540
http://www.ppfahome.org

Plumbing-Heating-Cooling Contractors—National Association
180 S. Washington Street
Suite 100
Falls Church, VA 22046
703-237-8100
http://www.phccweb.org

Rheem Manufacturing Company, Water Heater Division
1100 Abernathy Road
Suite 1400
Atlanta, GA 30328
800-621-5622
http://www.rheem.com

Ridge Tool Company (RIDGID)
400 Clark Street
Elyria, OH 44035
888-743-4333
http://www.ridgid.com

Sioux Chief Manufacturing Co., Inc.
P.O. Box 397
24110 S. Peculiar Drive
Peculiar, MO 64078
800-821-3944
http://www.siouxchief.com

State Water Heaters
500 Tennessee Waltz Parkway
Ashland City, TN 37015
800-365-8170
http://www.statewaterheaters.com

Uponor
5925 148th Street West
Apple Valley, MN 55124
800-321-4739
http://www.uponor-usa.com

Watts Regulator Company
815 Chestnut Street
North Andover, MA 01845
978-689-6066
http://www.watts.com

Wrightsoft Corporation
131 Hartwell Avenue
Lexington, MA 02421
800-225-8697
http://www.wrightsoft.com

Zoeller Company
3649 Cane Run Road
Louisville, KY 40211
800-928-7867
http://www.zoeller.com

Zurn Industries, LLC
1801 Pittsburgh Avenue
Erie, PA 16502
855-663-9876
http://www.zurn.com

Index

Italic page numbers refer to material in figures and tables.

A

AAVs. *See* air admittance valves (AAVs)
ABS. *See* acrylonitrile-butadiene-styrene (ABS) pipe
absorption fields. *See* leaching fields
acetone, 103
acetylene, 102–103
acetylene cylinders, 103
acetylene regulators, 105
acidity, 135
ACR copper tubing, 34
acrylonitrile-butadiene-styrene (ABS) pipe
  applications, *84*
  code status, 29
  compressed gas to be avoided, 28
  construction types, 27
  described, 26–28, *27*, *28*
  expansion and contraction, 30
  ignition temperature, *82*
  installation methods, *29*, 29–30
  installed locations, 29
  joining, 17–20, 30–32
  marking, 32–33
  nonpressure applications, 15
  physical characteristics, 20–21
  pressure ratings, *11*
  standards, *27*
  support spacing, *29*
  temperature stress rating, 24
  testing, 30
  ultraviolet radiation, 28
  uses, *27*, 28
AC shielded metal-arc welding, 535
AC welding power supply, 539
aerobic bacteria, 229
air
  components and properties, 341–343
  compressed, 343–344
  in heating lines and water lines, 343
  noises in water systems, 351–358
  pneumatics, 342
  safety issues when testing water piping systems, 345
  separators and vent, 212, *214*, 214–215, *215*
air admittance valves (AAVs), 91–93, *92*
air chambers
  described, 354, *355*
  sizing, 356–358, *358*
air compressors, 343–344, *344*
air compressor tanks, 344, *344*
air gap, 221, *222*, 223
air plenums, 46
air pressure
  Bourdon tube, 347–349, *347*
  installation tests, 349–351
  manometers, 346, *346*, *347*
  rough-in pressure test, 349–351, *350*
alkalinity, 135
alkalis, 135
alternating current (AC), 534, 535
aluminum, 523, *523*
American National Standards Institute (ANSI), 125
American Society for Testing and Materials (ASTM), 125
American Society of Mechanical Engineers (ASME), 102, 125, 215
American Society of Sanitary Engineering (ASSE), 125
American Water Works Association (AWWA), 12, 54, 125
American Welding Society (AWS), 102, 110, 112, 117, 542, 551, 556
Americans with Disabilities Act, 422, 472
anaerobic bacteria, 229
angle of repose, 451–452, *452*
annealing, 525
anode rod, *513*
ANSI/NSF standards, 14
antifreeze, 95, 96, 184
anti-siphon hole, *513*
arc, welding. *See* welding arc
arc (geometry), 317
Architect's Scale, *462*
arc welding. *See* shielded metal-arc welding
area calculations
  circles, *321*, 321–322
  rectangles, 323, *323*
  squares, *322*, 322–323
  triangles, 324, *324*
arm baths, *504*
asbestos-cement pipe, 128
aspiration, 410
atmospheric type vacuum breaker, 225, *226*
automatic compensating valves, 186
automatic gas control, 190
autopsy tables, *504*
average flow rate, 87

B

backfill, 140
backflow
  causes, 220
  degree of hazard, 221
  prevention, 221, *222*, 267, *267*
back pressure, *363*, 408, 410–411
back-pressure backflow, 220
back-siphonage backflow, 220
back vent, 381
backwater valves, 267, *267*
balanced pressure valves, 186
ball traps, 405, *406*
ball valves
  building shutoff, 144
  pressure type vacuum breaker, 224–225, *225*
  reduced pressure zone backflow preventer, 223
band clamps, 128, *129*
bases, 135
basket hitch, 433, *434*, *437*
bathrooms
  commercial, plan view, *489*
  drainage fixture units, *252–253*
  plan view, *467*

baths
drainage piping, *480*
institutional, symbol, *504*
bathtubs
rough-in sheets, *469–470, 475, 485*
symbol, *504*
See also tub and shower
beam clamps, *379*
bedpan washer, *504*, 506, *520*
belled pipe, 31
bell traps, 406, *406*
bend weld test, 552, *552*
bight, 431
blacksmiths, 101
block and tackle, 426–428
blower system, for septic tanks, 231
bonnet, 105
booster pumps, 176, *178*, 179
bored wells, 135
bottle traps, 407
bottom plates, 289
Bourdon tubes, *347*, 347–349
bowline, 431, *432*
Bradford White, 201
braided flexible connectors, *351*
branches, typical, 286
branch intervals, 286, *286*
branch loads, 286–287
branch vents, 381, *381*
brass, 132
brass manifolds, 148, *148*
brass pipe, 132
brazing, 36
defined, 117
filler materials, *111*
fluxes, 109–110
preparations, 118
technique, 119
bridle hitch, 435–436, *437*
British System of measurement, 297
brittleness, 524, *526*
bronze, 110
building distribution pipe
design goals, 151
sizing by available pressure, 157–159
building drains
combined, 268–269
dynamics of water and air in, 342–343
flow rate, 269–271, *270*
installation, 277–279
load on stacks, *274*
number of fixture units connected to, *254, 269, 376*
overview, 265
sizing plastic pipe, 88, *89*
See also building sewers
building services, 142–144
building sewers
cleanouts, 245–247, *246*
defined, 240
discharge rates and velocities, *245*
installation methods, 243–248
materials, 240
number of fixture units connected to, *254*
slopes, 256
thrust blocks, 247–248, *247*
See also building drains
building shutoff valve, 144
building subdrain systems, 393
electrical safety, 399
sewage ejectors, 399–400
sewage pumps, 398–399
sump pumps, 393–397
system installations, 400–401
building systems, flow, 154–155
building traps, 407, *407*
building water service pipes, 138
butterfly valves, 72, 353
butt fusion, 20, 58
butt joints, 543, *543*
butt welds, 530

C

can washers, *499, 504*
capillary action, trap seal loss, 411, *411*
capped-pipe air chambers, *358*
carbon, in plastics, 9–10
carbonates, 185
carbon dioxide content, 136
carbon rods, 536
carburizing flame, *527*
case hardening, 525
cast iron soil pipe
ABS joints, 32
hubbed, 41–44
hubless, 40–41, *41*
overview, 133
spacing for vent hangers and supports, 379
support intervals for stacks, *283*, 283–284
caulked-soil pipe, 327–328, *328, 329*
cave-ins, 451–454
celluloid, 9
center of gravity, 426–428, *428*
center-to-center (c-c) distance, 311, *312*
centrifugal circulator pump, 212, *213*
centrifugal sewage pumps, *394*, 395, 399
centrifugal water pumps, 154
Charpy test, 553
check valves
double check valve assembly, 223–224
gas torches, 106, *108*
pressure type vacuum breaker, 224–225, *225*
reduced pressure zone backflow preventer, 223
spring-loaded, 354–355
swing, 353
chemical sinks, *504*
chemical vent piping, 364
chemical weld, 18
cherry pickers, 449
chilled water cooling systems, 97
chipping hammer, 545
chlorinated polyvinyl chloride (CPVC) pipe
applications, *84*
availability, 46
cement for, *129*
code status, 45–46
corrosive fume handling, 97
described, 20, 44–45, 129
flange fittings, 19
ignition temperature, *82*
installation methods, 46–48, *46*
installation recommendations, 51–52
installed locations, 46
joining methods, 48–50
marking, 50–51
solvent cement joining, 17–18
standards, *44*
storage and handling, 45
temperature stress rating, 24
thermal expansion and contraction, 48
threaded joints, 20
uses, 45
chlorination, 136
chlorine, 137

choker hitch, 433, *434*, *437*
chord, 318
circles
area, *321*, 321–322
overview, *317*, 317–318
circuit vents, 381–382, *382*
circulating hot water systems, 203–204, *204*
circumference, 318
cleanouts
described, 245–247, *246*, 277, *277*
stacks, 285
Clevis hangers, 380
closed reservoirs, 140
closed tanks, 176, *177*
closet flanges, 32, 74
clove hitch, *432*, 432–433
code welding, 102
code work, 102
combination vents, 382, *382*
"come-along device," 426, *427*
commercial projects, single line drawings, 472, 475
Commercial Standards (CS), 126
common vent, *373*, 383, *383*
communication skills and professionalism
customer expectations, 3–5
organization, 5
productivity, 5
shine, 6
sort, 5–6
standardize, 6
straighten, 6
sustain, 6
community water sources, 133–134
companion flange, 72
compressed air, 343–344
compressed air piping, 25, 26
compressed gas lines, *507*
compression adapters, 50
compression clamps, 31
compression fittings, 59
computer aided drawing (CAD), 472
concrete slabs, installing plastic piping under, *171*, 172
conductivity, 523, *526*
conductors, 534
confined spaces, 458–459
contaminated supply systems, 221
continuous vents, 383, *383*
control valves
hydronic systems, 501, *502*
tagging, *502*
copper, 132
expansion/contraction, 34, 35
installation methods, 34
joining, 34
soldering and brazing, *35*, 35–40, *36*, *37*
support, 34
testing, 34
types of copper tube and use, 33–34
copper fittings, crimping, 126, *127*
Copper Tube Size (CTS), 11
copper tubing
ACR, 34
allowance in equivalent feet for friction loss, *169*
code status, 34
determining the joining method, 117
DWV, 34
outside diameter, 132
resistance to flow, 155–156
sizing based on velocity limitations, *167–168*
spacing for vent hangers and supports, 380
support intervals for stacks, 283
Type K, 33
Type L, 33
Type M, 33–34
types, 132
water flow velocity, *166*
corporation cock, 142, *142*
corrosion resistance, 13
corrosive fumes, 97
corrugated stainless steel tubing (CSST), *78*, 78–79
covered tanks, 154
CPVC. *See* chlorinated polyvinyl chloride (CPVC) pipe
CPVC cement, *129*
cracked pipes, 411
cranes, 425, 440–441
creosote, 25
crimping, 126, *127*
crimp ring fittings, 19
crimp-to-connect system, 39
cross-connection
defined, 219
protection, 221–227
See also backflow
cross-hatching, 518–519, *519*, 521
cross-linked polyethylene (PEX) pipe
applications, *84*
bending, 63, *63*
code status, 61
connection to fixtures, 63
described, 60–61
expansion and contraction, 63–64
fittings, 171
flexibility, 13
ignition temperature, *82*
individual run systems, 171, 172, *172*
installation locations, 61
installation methods, 62
joining methods, 64
manifold plumbing systems, 64
marking, 64
mechanical fitting joints, 19
overview, 130, *131*
physical characteristics, 21
pressure drop, *173*
sleeving, 63
standards, *61*
storage and handling, 61
support, 62
temperature stress rating, 24
uses, 61
crown vents, 383
crown-vent traps, 406, *406*
crystal structure, 522
CTS (Copper Tube Size) System, 21
cube roots, *305–306*, 308
cubes, *305–306*, 307–308, 326, *326*
curb cock, 142–143
curb cock box, 142–143
customer expectations, 3–5
be honest, 4
be neat and clean in appearance, 4
drive company vehicle carefully, 3
job on time, 3
listen, 4
treat customer's property better and keep work area safe/clean, 4–5
cutting plane, 518, *518*
cylinders, 324, *324*

D

D category piping, 227
DC reverse polarity welding, 535
DC shielded metal-arc welding, 539–540
DC straight polarity welding, 535
DC welding power supply, 539–540
decimals, 297
demand-type water heaters. *See* instantaneous water heaters
denominator, 295–297
density, 522, *526*
destructive weld testing, 552–553
dew point, 334
diagonal. *See* travel
dial gauges, 347, *347*, *348*
diameters, 318
Dimension Ratio (DR), 22–23
dip tube, 188, *192*, *513*
direct current (DC), 534–535
direct current welding, 535, 539–540
direct-feed system, 153–154
direct-fired instantaneous water heaters, 197, *198*
direct-fired water heaters, 182, *183*, 184
dishwasher drainage piping, *482*
dishwashers, *504*
displacement pumps, 212
disposal fields, 234–238. *See also* leaching fields
dissolved air content, 136
distribution box, 233–234, *234*
distribution layout, 138–139
distribution piping, sizing by available pressure, 157–159
distribution systems
  nonpotable water, 148
  private water, 147–148
  public water, 138–147
dosing siphon, *238*, 238–239
dosing tank, *238*, 238–239, *239*
double check valve assembly, *222*, 223–224, *224*
double wyes, 287
downhill welding, 551
DR. *See* Dimension Ratio (DR)
drafting scales, 461, *462*
drag, 122, *122*
drainage, waste and vent (DWV) systems
  ABS pipe, 28
  isometric drawing, 483
  vent terminals, 91–93
drainage fixture units (DFU)
  number connected to stacks, *254*
  offsets, 256–257, *257*
  overview, 87, 251–252, 255–256
  special circumstances, 257
  stack capacity, 281, *282*
  values for fixture types, 251–252, *252–253*, *255*, 256
  vent sizing, 366–367, *366*
drain pipe
  dynamics of water and air in, 342
  sizing, 88, *89*
  slopes, 256
  See also sewer pipe
drawings
  exploded views, *520*, 521
  fixture tables, 512, *513*, 514
  isometric projections, 511–512, *516*
  orthographic projections, 511
  overview, 511–512, 514
  sectional views, 514–515, 517–519, 521
drilled wells, 134
drinking fountains
  plumbing fixture schedule, *513*
  symbol, *504*
drip legs, 344, *344*
driven wells, 134
drum traps, 406, *406*
dry pit sump pumps, 397, *397*
dual vents, 383, *383*
dug wells, 134
duty cycle, 538
DWV copper tubing, 34
DWV pipe
  advantages, 25
  elastomeric sleeve coupling with ABS pipe, 31
DWV systems. *See* drainage, waste and vent (DWV) systems

E

elasticity, 525, *526*
elastomeric gasket couplings
  ABS pipe, 31
  PVC pipe, 72
elastomeric sealing gaskets, 18–19, *19*
elastomeric sleeve couplings, 31
electrical conductivity, 10
electrical insulation, 534
electrical resistance, 20, 534
electric current, 534
electric elements, 194–196
electric potential, 190
electric resistance, 20
electric theory, 102, 534
electric water cooler, *488*
electric water heaters
  heating elements, 195
  replacement, 187–189
  residential sizing, 201, *201*
  safety precautions, 189
  thermostats, 195, *196*
electrode holders, 540–541, *541*
electrode numbering system, 542
electrodes, 400, 536–537, 542–543
electrofusion, 20, 58–59
elevated tanks, 140, 153, *153*
emergency baths, *499*
end-to-end (e-e) distance, 311, *312*, *313*
energy cutoff (ECO), 190, 195, *196*
Energy Guide label, 201
engine-driven generators, 540
engineered systems, 388–390
equilateral triangles, 313, *314*, 320, *320*
evaporation, trap seal loss, 411
excavations, volume estimations, 327. *See also* trenches
expandable foam, 83, *83*
expansion coefficient, 24, 524, *526*
exploded views, *520*, 521
extension ladders, 443, *443*
eye protection, 113–114

F

fasteners
  compression, 416, *416*
  ease of assembly and disassembly, 416
  shear, 416, *416*
  strength requirements, 415
  tension, 416, *416*
fast fill electrodes, 543
fast follow electrodes, 543
fast freeze electrodes, 543
fatigue failure, 525, *526*

fatigue weld test, 553
faucets, 165, 356, *513*
F category piping, 227
Federal Specifications (FS), 126
feet per second (fps), 155
filler materials
  brazing, 118
  welding, 110, *111*
fillet welding, 31
  CPVC pipe socket joints, 50
  PVC pipe, 72
fire emergencies, 154
fire hydrants, 154
fire-resistive construction, 81–84
fires, with plastics, 81–84
fire stop systems, 83–84
Fire Tests of Through Penetration Fire Stops, 84
fire walls, 47
fire watch, 112–113
"first hour rating," 201
fitting allowances, *169*, 169–170, 311, *312*
fixture branch size, *161–162*, 203
fixture connections, PVC pipe, 74
fixtures
  handrails and grab bars, 422, *422*
  lavatory and service sinks, 419–420, *419*, *420*
  shower pans, 420–421, *421*
  urinal, 418–419, *419*
  water closet, 416–418
fixture tables, 512, *513*, 514
flame arrestor screen, *194*
flame cutting, 119–122
flame failure protection, 190, 197
flanges, 19
  CPVC pipe, 50
  PVC pipe, 72
flashing, 291, *292*
Flemish eye loop, *437*
floor drains, 272–273, *273*
floor mounted water closet fixtures, 417, *417*, *418*
flow, 156–157
flow control valves, 200, *200*
flow rate, 156
flue baffle, *513*
flues, *513*
fluxes, 109–110, 118
foam insulation, *513*
fold-out ladders. *See* step ladders
force, 335–336
forced-feed system, 153–154
forge welding, 101
forklifts, 441–442
45° offsets, 313
fractions, 295–297
freezer warehouses, 96
friction, 155–156
fuel-fired water heaters, replacement, 187–189. *See also* gas water heaters; oil water heaters
fuel gases, plastic pipe, 26, *27*
fuel oil, safety precautions, 189
full-port service valves, 144
full section views, 515, *516*
full-size stacks, 285–287
fusability, *526*

G

gallons per minute (gpm), 164
galvanized steel pipe
  overview, 133
  sizing based on velocity limitations, *168*
  spacing for vent hangers and supports, 377, 379
gas control valve, 190–192, *513*
gas lines, *507*
gas-metal arc welding, 101
gas piping in CSST, 78–79
gas water heaters
  components, *513*
  flame arrestor screen, *194*
  manufactured after January 1, 2004, 193–194
  manufactured prior to January 1, 2004, 190–192
  residential sizing, 201
  safety precautions, 189, 190–194
  sectional view, *513*
  thermostat, 190
  troubleshooting, *192–193*
gas welding
  equipment, 102–109
  gauges, 105, *106*
  hoses, 105–106, *107*
  pressure regulators, 104–105, *105*
  safety, 104, *104*
  torches, 106–108, *108*
gauge pressure, 16
gauges, welding, 105, *106*
geothermal ground source systems, 95–96
glass pipe, 131
globe valves, 502, *502*
gold, *523*
gravity backflow, 220
gravity-feed pressurization, 140
gravity-feed systems, 154
gravity tanks, 176, *178*
grease interceptors, 403, *404*
grease traps, 271–272, *272*
greasy waste, 271–272
  floor drains, 272–273, *273*
  grease traps, 271–272, *272*
  method of drainage, *271*
grid system, 138
grinder pumps, 398–399, *398*
grooved fittings, 77, *77*
ground clamps, 541–542
grounding, 133

H

handrails and grab bars, 422, *422*
half section view, 515, 517
hard copper tubing, 132
hardening, 525
hardness, 136, 524, *526*
HDPE pipe, 96
headers, *288*
health care buildings, 506
heat fusion joining, 20, 58–59
heating elements, 20, 195
heating lines, air inside of, 343–344
heating systems layout, *501*
heating terminals
  baseboard heating units and free standing fan coils, 208, *209*
  radiant floor heating, 208–209, *210*
heat resistance, 523
heat transfer modes, 206–207
  conduction, 207
  convection, 207
  radiation, 206–207
heli-arc welding, 555
high-density polyethylene (HDPE) pipe, 96
hitches
  grasping loads with, 433, *434*, 435–436
  load capacities, *437*
  types, 432–433, *432*, *433*
hoists, 425–426, *427*, 440–441

horizontal fixture branches
  number connected to stacks, *254*
  number of fixture units connected to, *376*
horizontal piping, connections to, *266*
horizontal storm drain capacity, *90*
hose barbs, 128, *129*
hose bibbs, 63, 147, 165
hose connection vacuum breakers, 225, *226*
hot gas welding, 72
hot water, 136, 181, *182*
hot water systems
  circulating, 185, 203–204, *204*
  general considerations, 204
  increasing capacity of, 203
  See also water heaters
hub and spigot, 72
hubbed cast iron soil pipe, 41–44
  caulked joints, 42–43, *43*
  compression joints, 43, *44*
  cutting, 41–42, *42*
hubless cast iron soil pipe, 40–41, *41*
  cutting and joining, 41
  expansion/contraction, 41
  support, 40
  testing, 40
Hunter, Roy B., 87
Hunter's curve, 87–88
Hyatt, John Wesley, 9
hydraulic jumps, 275–276, *276*, *277*
hydraulics, 340–341
hydraulic shoring, 453
hydrogen, 102
hydronic heating systems, 175, *175*, *209*
  components, 210–216
  control valves, 501, *502*
  domestic hot water (DMV) zones, *206*
  heating terminals, 208–210
  modes of heat transfer, 206–207
  overview, 205
  system sizing, 207, *208*
hydropneumatic tank, 176, *177*
Hydrostatic Design Basis, 16
hypotenuse, 313, 320

I

ice rinks, 96
ice storage systems, 97
ID controlled pipe. *See* Inside Diameter controlled pipe sizing system
impact weld test, 552–553
impellers, 395, *395*
Imperial System of measurement, 297
improper fractions, 295
indirect-fired instantaneous water heaters, 199, *199*
indirect water heaters, 184–185, *184*
individual run systems, 170–172, *172*
  cost, 174–175
  isometric sketch, *171*
  operating advantages, 174
  other uses, 175–176
  overview, 170–171
  sizing, 173–174
  system disadvantages, 174
  typical materials, 172–173
individual vents, 383, *384*
individual water sources. *See* private water sources
industrial piping layouts
  heating system, *501*
  mechanical symbols, *502*
  overview, 495–496, 499–503
  piping symbols and abbreviations, *496–499*
  plumbing symbols, *499*
  sewage treatment plant, *500*
infant baths, *504*
in-floor radiant heat systems, 94–95, *94*, *95*
inorganic materials, in sewage, 229
insert and crimping ring system, 64
insert fittings, 59, *59*
Inside Diameter controlled pipe sizing system, 10–11, 13, 23, 24
installations
  economic considerations, 126–127
  material longevity considerations, 128
installation tolerances, 465, 472
instantaneous flow rate, 87
instantaneous water heaters
  controls and accessories, 199–201, *200*
  direct-fired, 197, *198*
  gas-fired, *198*
  indirect-fired, 199, *199*
  overview, 197
  point of use, 197, 203, *203*
  sizing, 202
institutional installations, 503–504, 506
insulation
  adding before installation, 467
  electrical, 534
insulators, 534
interceptors, 403, *404*
International Association of Plumbing and Mechanical Officials (IAPMO), 125
International Plumbing Code, 134, 173, 181
IPS Schedule wall standards, 11, 21–22, *22*
iron, *523*
iron pipe size (IPS), 11
isometric drawings, 476–477
isometric projections, 511, 512, *516*
isosceles right triangle, 320, *320*
Izod test, 553

J

joining methods
  brazing, 118
  determining, 117
  lead and oakum, 328
  soldering, 118
  welding, 119
joists
  notching and drilling, 288–289, *289*, *290*
  reinforcing, *290*
jumps, hydraulic, 275–277, *276*

K

kerf, 121, *121*
kitchen sinks, *471*
knots, 431–432, *432*
Konen, Tom, 87, 251

L

ladders
  accidents, 446
  angle, 444
  care and storage, 447

inspection before use, 446
loads, 444
OSHA guidelines, 444–447
proper selection, 446
proper setup and use, 447
rungs, 444
slipping, 445
in trenches, 454
types, 442–443
large distribution lines, 138
laser welding, 556
lavatories
combination-type, *507*
elevation view, *479*, *495*
isometric drawing, *495*
plan view, *494*
plumbing fixture schedule, *513*
and service sinks, 419–420, *419*, *420*
tolerances and rough-in sheets, *464*, *476*, *490–491*
lavatory tailpiece, *343*
lawn faucets, 165
leaching fields, 231
described, 234
dosing siphon and dosing tank, 238–239
perc tests, 234
serial distribution, 239, *239*
leaching pit, 231
lead
characteristics, 523
dangers of working with, 330
estimating amounts needed, 327–330, *331*
lead and oakum joints, *328*
lead pipe, 133
lead poisoning, 329
lead wool, 74
leak weld test, 554
leg baths, 504
Legionella bacteria, 195
Legionnaires' disease, 195
life-cycle cost, 84
lift check valve, *518*
linear measure, 298–303
loads, determining, 86–87
local mains, 138
local vent, 384
loop system, 138
loop vent, 384, *384*
low-carbon steel, heat treatment, 525–527
LP gas, 189
LP gas vapor piping, 26

M

macerating toilets, *398*
magnesium, *523*
magnetic particle weld test, 553, *554*
main vent, 384, *385*
malleability, 524, *526*
malleable iron fittings, 77
manholes, 247
manifold installation, 142
manifold plumbing systems, 64. *See also* individual run systems
manifolds, brass, *147*, 148
manometers, 346, *346*, *347*
manufactured housing, 47
manufacturing industry, 5 S concept, 5–6
MAPP gas, 104
mass, 335–336
mathematics
conversions, 297–298
cube roots, *305–306*, 308
cubes, *305–306*, 307–308
decimals, 297
fractions, 295–297
squares and square roots, 304, *305–306*, 306
maximum probable demand, 88
measurement systems, 297–298
mechanical fittings
PE pipe, 59
plastic piping, 19
mechanical symbols, *502*
medical buildings, 506
medical facility faucets, 521
medical gas certification, 119
medical gas piping, 119
melting point, 523, *523*, *526*
mercury, *523*
metal-arc inert gas (MIG) welding, 555–556
metallic pipe, 132–133
metals
energy absorption and transmission, 523–525
properties and characteristics, 522, *523*, *526*
meters, public water systems, 143, *143*
meter yoke, 143–144
MIG welding, 555–556
minerals, water hardness, 136, *137*
mixed numbers, 297
mixed water temperature control, 185–187, *187*
momentum, 410
motor-driven generators, 540
municipal sewage systems. *See* public sewage systems

N

nail plates, 290
nameplates, 188, 197
National Fire Prevention Association (NFPA), 119
National Inspection, Testing and Certification Corporation (NITC), 119
National Sanitation Foundation (NSF), 125
National Standard Plumbing Code, 134, 181
natural gas, 189
natural gas piping, 26
neutral flame, 527
nickel, 523
90° offsets, 311, *314*
nitrogen, 119
testing water piping systems, 345, *345*
nitrogen tanks, *345*
nondestructive weld tests, 553–554
non-integral traps, *258–259*
nonpotable water distribution systems, 148
nonpressure plastic piping, 24
nontoxic substances, 221
normalizing, 525
numerator, 296
nut ferrule fittings, 19
nylon slings, 430, *430*

O

oakum, 32, 74, 327–330, *328*
OD controlled systems, 21
offsets
drainage fixture units, 256–257
45°, 313
90°, 311–312, *314*
with other angles, 313–317
overview, 311
travel and run chart, *316*
offset section view, 517–518, *517*

ohms, 534
Ohm's Law, 534
oil interceptor, *404*
oil water heaters, 196–197, *201*
open-circuit voltage, 539
organic materials, in sewage, 229
O-rings, 18, *19*
orthographic projections, 511
oscillation, 410
outside diameter, 132
Outside Diameter controlled pipe sizing system, 10–13, 24
overhead utilities, 456
overhead welding and tacking, 533–534, 551
overheating limit, 524, *526*
oxidizing flame, *527*
oxy-acetylene, 102–103
oxy-acetylene flat welding, 122–123
oxy-acetylene torches, *107*
  flame cutting, 119, 120, *120*
  flame types, *527*
  forming a puddle, 527, *528*
oxy-acetylene welding
  adding filler, 528
  forming a puddle, 527, *528*
  joining two pieces, 528–530, *529*
  overhead, 533–534
  personal protection, 534
  tacking, 530, *530*, *531*
  vertical, 531–533
  V-shaped weld, *529*
  weld characteristics, 531
  when to use for pipes, 531
oxygen
  dangers in testing water piping systems, 345
  dissolved in water, 136, *137*
  hydronic heating systems, 175
  as a welding gas, 103
oxygen cylinders, 103, *104*
oxygen regulators, 105

P

partial section view, 517
partition traps, 405–406, *406*
Pascal's Law, 339–340
paste thread, 20
paste thread sealants, 20, 73
PB. *See* polybutylene (PB) pipe
P category piping, 227
PE. *See* polyethylene (PE) pipe
pedestal sinks, *486*
pedestal sump pumps, 397, *397*
penetrant weld test, 553
peppermint oil test, 293
percolation test, 234
perc test. *See* percolation test
perimeter, 318–319
perineal bath, *504*
PEX. *See* cross-linked polyethylene (PEX) pipe
PEX-AL-PEX. *See* polyethylene-aluminum-polyethylene (PEX-AL-PEX) pipe
pH value, 135–136, *137*
piezo-electric igniters, 194
pigtail siphon, 348, *349*
pilot, *513*
pipe hooks, *379*
pipe/piping
  in hydronic heating systems, 210–211
  identification, 227
  standards, 125
  See also individual pipe types
pipe welding. *See* gas welding; welding
plasma arc cutting, 122, *123*
plaster traps, 403
plasticity, 524, *526*
plastic pipe
  advantages, 13–15
  applications, *84*, 85–86
  availability, 46
  chemical resistance, 98–99
  coloring for applications, 14, *15*
  corrosive fume handling systems, 97
  determination of loads in plumbing systems, 86–91
  dimensional data, *22*
  DWV and sewer service, 25
  fire-resistive construction, 81–84
  individual run systems, 170–171
  industrial applications, 98
  joining, 17–20
  nonpotable water systems, 93–97
  nonpressure applications, 15
  with potable water, 25, 86
  pressure applications, 16–17
  pressure ratings versus pressure class, *12*
  pressurized gases, 26
  in process applications, 26
  properties, 20–21
  radiant heat systems, *94*, 94–95, *95*
  resistance to flow, 155
  selection, 84–85
  sizing, 10–13, 86, *168*
  spacing for vent hangers and supports, 379–381
  standards (*see* plastic pipe standards)
  thermoplastic, 15–17
  types, 128–131
  water flow pressure drop, *89*
  See also individual types
Plastic Pipe Institute, 174
plastic pipe standards
  chemical resistance, 25
  Dimension Ratio, Standard Dimension Ratio, Standard Inside Diameter Ratio, 22–23
  expansion/contraction, 24–25
  IPS Schedules, 21–22, *22*
  nonpressure piping, 24
  pressure pipe, 23, *23*
  temperature effects, 24
plastics
  chemical components, 9–10
  combustion, 81–82
  electrical properties, 10
  flammability, 10
  history, 9
  mechanical properties, 10
  types, 9
  UV exposure, 10
plastic sewer pipe
  advantages, 25
  elastomeric sealing gaskets, 18–19, *19*
  pipe sizes, 10–13
  standards, 21
plastic waste pipe, 86, 93–97
plastic water pipe, 25, 86, 346
plenums, 46
Plumbing and Drainage Institute (PDI), 125
plumbing fixture schedule, *513*
plumbing symbols, 474
  commercial fixtures, *488*
  industrial, *499*
  institutional, *504*
plumbing traps. *See* traps
pneumatics, 342

point of use instantaneous water heaters, 197, 204–205
polluted supply systems, 221
polybutylene (PB) pipe
  applications, *84*
  freezer warehouses, 96
  ignition temperature, *82*
  overview, 52, *52*, 130, *130*
polyethylene-aluminum-polyethylene (PEX-AL-PEX) pipe
  hydronic heating systems, 175
  overview, 130–131
  radiant heat systems, 94
  uses, 61
polyethylene (PE) pipe
  applications, *84*
  availability, 55
  AWWA resources, 54
  code status, 55
  described, 53–54
  early development, 9
  expansion and contraction, 57
  freezer warehouses, 96
  fuel gas piping, 26, *27*
  gas distribution systems, 96–97
  geothermal ground source systems, 95–96
  ice rinks, 96
  ignition temperature, *82*
  installation methods, 56–57
  installed locations, 56
  joining methods, 57–60
  marking, 60
  material designation code, 54
  mechanical fitting joints, 19, *129*
  minimum wall thicknesses, *23*
  nonpressure applications, 15
  overview, 128
  physical characteristics, 20–21
  pressure ratings, *23*
  standards, 12, *53*
  temperature stress rating, 24
  uses, 55
  UV radiation, 54–55
polypropylene (PP) pipe
  applications, *84*
  nonpressure applications, 15
  physical characteristics, 20
polytetrafluoroethylene (PTFE), *82*
polyvinyl chloride (PVC) pipe
  applications, *84*
  availability, 68
  AWWA standards, 12
  chilled water cooling systems, 97
  code status, 67–68
  compressed gas use, 67
  corrosive fume handling systems, 97
  described, 64, 66
  early development, 9
  elastomeric sealing gaskets, 18–19, *19*
  expansion and contraction, 70–71
  flange fittings, 19
  ignition temperature, *82*
  installation methods, 68–70
  installation recommendations, 74–75
  installation testing, 70
  installed locations, 68
  joining methods, 71–74
  marking, 74
  nonpressure applications, 15
  overview, 130
  physical characteristics, 20–21
  pressure ratings, *23*
  Sch 40 pipe, 11
  under slab warm air ducts, 97
  solvent cement joining, 17–18
  standards, *65*
  storage and handling, 66
  supporting, *69*
  temperature limitations, 66
  temperature stress rating, 24
  threaded joints, 20
  uses, 67
  UV radiation, 66–67
polyvinylidene fluoride (PVDF), *82*
ponds, 134
porosity, 522, *526*
pounds per square inch (psi), 151, 336
pounds per square inch gauge (psig), 336
power factor, 539
power lines, 455–456
power scaffolds, 449–450
PP. *See* polypropylene (PP) pipe
press-to-connect system, 39, *39*
pressure boosting systems, 176–179
pressure class, 11–12
pressure fluctuations, 155
pressure plastic pipe, 16
  pressure rating and class, 11–12
  pressurized gases, 26, *27*
  standards, 23, *23*
pressure rating, 11–12
pressure-reducing valves, *141*, 144–145
  direct acting, *145*
  hose bibbs, 147
  sizing, 145–146
  strainers, 146, *146*
pressure regulators, 104–105, *105*
pressure relief valve, 215, *216*
pressure type vacuum breaker, *222*, 224–225, *225*
pressurization methods, 140–141
pressurized gases, 26, *27*
pressurized reservoirs, 140–141
pressurized tanks, 102, 114
Prestolite torch, 102
private sewage systems
  disposal fields, 234–235
  distribution box, 233–234, *234*
  dosing siphon and dosing tank, *238*, 238–239, *239*
  maintenance, 239
  minimum distance between components, *232*
  overview, 231
  septic tank detail, 232–233, *233*
  septic tanks, 232–233
  sizing, 235, *235–236*
private water distribution systems, 135, 147–148
private water sources, 134–135
propane, 103
PTFE. *See* polytetrafluoroethylene (PTFE)
PTFE thread tape, 20, 49, 73
P-traps
  components, *406*
  described, 407–408, *407*, *408*
public sewage systems, 230–231
public water distribution systems, 135, 138–147, 154
public water supply
  components and operation, 139, *139*
  water sources, 133
puddle, 122, 527, *528*, 537
pulleys, 426
pumping stations, 153
pumps
  booster, 176, *178*, 179
  centrifugal circulator, 212, *213*
  gravity tank systems, 176
  in hydronic heating systems, 211–212
  hydropneumatic tank system, 176
  public water systems, 141
  See also specific kinds

push-to-connect fittings, 39, *40*
putrefaction, 232
PVC. *See* polyvinyl chloride (PVC) pipe
PVDF. *See* polyvinylidene fluoride (PVDF)

Q

quick-closing valves
  water hammer, 352–355

R

radiant floor heating system, 208–209, *210*
radiant heat panels, 175
radiant heat systems, *94*, 94–95, *95*
radiographic weld test, 553
radius, 317–318
rectangles
  area, 323, *323*
  overview, 319, *319*
rectangular prisms, 325, *325*
reduced pressure zone backflow preventer, 223–224, *224*
refrigeration piping, 119
regulators, compressed gas tanks, 345, *345*
relative humidity, 342
relief valves, 185, *186*, 196
relief vent, 384, *385*
removed section view, *517*, 518
residential dual check valves, 224, *224*
resistance
  electrical, 523, *526*
  to heat, 523
  to water flow, 156
resistance heating elements, 194–195
resistance welding, 556
revolved section view, *517*, 518
Rheem, 201
rigging
  defined, 428–429
  methods, 429–430
  safety and planning, 429, 442
right triangles, isosceles, 320, *320*
riser clamps, *378*
roof drains, *512*
roof terminals, 389, *390*
rope, 431, *431*
rotation stops, *186*
rough-in pressure test, 349–351, *349*, *350*
rough-in sheets, *463–464*, *466*, *473–476*
  bathtub, *469–470*, *475*, *485*
  described, 461–462
  installation tolerances, 465, 467, 472
  kitchen sink, *487*
  lavatories, *490*, *490–491*
  purpose, 461–462
  water closets, *463*, *466*, *468*, *473–474*, *484*
run, 312–313, *314*, *315*, *316*
running traps, 407

S

saddle fusion, 20, 58
safe lighting interlock, 190
safety
  forklifts, 441–442
  hoists and cranes, 440–441
  importance, 439–440
  rigging, 442
sand interceptor, *404*
sanitary tees, 232, 287
scaffolds, 447–450
scale rulers, 461, *462*
scaling, 185
S category piping, 227
Scheme for the Identification of Piping Systems, 227
school restrooms, 503–504, *505*
scissors lifts, 449–450, *450*
screw jacks, 453, *453*
SDR. *See* Standard Dimension Ratio (SDR)
SDR-PR system, 11
sectional views
  cross-hatching, 515, 518–519, *519*, 521
  cutting plane, 518, *518*
  overview, 514
  types, 514, 517–518
seepage pits, *237*
self-siphoning, *409*, 409–410
septic systems, sizing, 235, *235–236*
septic tanks, 233
  additives, *240*
  capacity, *236*
  described, 232–233, *233*
service sinks, *513*
sewage
  defined, 229
  flow according to type of establishment, *235–236*
sewage disposal, 229–230
sewage ejectors, 387–388, 399–400
sewage pumps
  defined, 398
  described, 398–399
  duplex systems, 399
  electrical safety, 399
  installation, 393, *394*, 400
  special features, 399
sewage systems
  private, 231–239
  public, 230–231
sewage treatment, 229
sewage treatment plants, *500*
sewer pipe
  sizing plastic pipe, 90, *90*
  spacing from water piping, *139*
sewers, number of Drainage Fixture Units (DFU) units connected to, *376*. *See also* building drains; building sewers
sheet piling, 453
shielded metal-arc welding, 101
  advantages and disadvantages, 536
  components, 536
  current settings, 545–546, *545*
  electric theory, 534–535
  electrode holder, 541, *541*
  electrodes, 536–537, 542–544
  exercises, 548–549, *549*
  ground clamp, 541–542
  methods, 546–547
  overhead, 551
  overview of, 535–536
  personal protection, 536, 544
  power supply, 537–541
  striking, maintaining, and controlling the arc, 546
  testing welds, 551–554
  tools, 545
  vertical, 550–551
  weld characteristics, 547, *548*
  welding cables, 540–541
shock arrestors, 353–354, *354*, 355–356, 358
shock resistance, 525, *526*
shoring, 450–453, *453*
shower pans, 420–421, *420*, *421*

showers, *464*, *476*. *See also* tub and shower
shutoff valves, 144, 170, *172*
side vent, 387, *388*
SIDR. *See* Standard Inside Diameter Ratio (SIDR)
silver, *523*
single bath macerating systems, 398, *398*
single fixture macerating systems, 398, *398*
single leg vertical hitch, 433, *434*
single-lever faucets, 356
single line drawings
  commercial fixture symbols, *488*
  industrial piping layouts, 495–496, *496–499*
  institutional installations, 503, *504*, 506
  producing, 472, 475
  residential layout, *478*
symbols for piping, *496–499*
single pass weld, 548, *549*
single-phase wiring, 535
sinks
  elevation view, *481*
  kitchen sink, *487*
  plumbing fixture schedule, *513*
  rough-ins, *486*
  symbols, *488*
siphonage, *363*, 408–410
sitz bath, *504*, 506
sleeves, 284–285, *284*
sleeving, PEX tubing, 62
slings, 429–430, *430*
slope, building sewers, 243–245, *244*
sloping drains
  discharge rates and velocities, *245*
  general installation rules, 243–245
sludge
  private sewage systems, 231
  public sewage systems, 230–231
smoke test, 293
snatch block, 426
snow melting systems, 95
soap suds, 364, 388, *389*
socket fitting welding, 31, 50
socket fusion, 20, 58
socket joints, 50
soft copper tubing, 132
softening, 136, *137*
soil load capacities, *248*
soil pipe hub adapters, 32, 74
soil stacks, 281
soil vent, 385
solar water heating systems, *184*, *187*, 212, *213*
soldering
  apply smooth coating of flux, *36*
  defined, 117
  filler materials, 111
  flare fitting, *37*, 37–38, *38*
  fluxes, 109–110
  mechanical fittings, 37
  overview, 117
  press and push connectors, 38–40, *39*, *40*
  proper fuel pressure, 118
  proper torch tip size, 118
  using deburring tool, *35*
  using open mesh sanding cloth, *36*
solders, 118
solenoid valves, 16, 356
solids interceptor, *404*
solvent cement joining
  ABS pipe, 30–31
  CPVC pipe, 48–49
  described, 17, *17*
  PVC pipe, 71
  safety, 18
Sorensen, S. P. L., 135
specification sheets. *See* rough-in sheets
specific gravity, 522, *523*, *526*
split ring hangers, *378*
spot welding, 556
spring-loaded check valves, 354–355
squares and square roots (mathematics), 304, *305–306*, 307
squares (geometry)
  area, 322–323, *322*
  overview, 318–319, *319*
stab fittings, 19, 59
stacks
  cleanouts, 277, *277*
  defined, 281
  example layouts, *260*
  fittings and support, 285, 287, *287*
  full-size, 285–287
  limitations, 281
  loading and capacity, 282
  materials, 281
  number of fixture units connected to, *254*, *376*
  roof penetrations, 291, *292*, 293
  route of, 288–291
  running through wooden floors, 284
  sleeves and thimbles, *284*, 284–285
  supporting, 283–284, *283*
  terminal velocity, 283
  testing, 293
  types, 281
  venting, 283
stack vent, 385
stainless steel pipe, 380
Standard Dimension Ratio (SDR), 22–23, *23*
Standard Dimension Ratio-Pressure Rating (SDR-PR) system, 11
Standard Inside Diameter Ratio (SIDR), 23, *23*
standing rope, 431
static pressure tube, *349*, 349–351, *350*
steam, 334
steam control valves, *200*
steel
  characteristics, *523*
  heat treatment, 525–527
  threaded pipe, 380
steel pipe
  code status, 76
  expansion/contraction, 76
  joining, 76–77
  overview, 133
  resistance to flow, 155–156
  sizing based on velocity limitations, *168*
  stainless, 380
  support, 76
  testing, 76
  See also galvanized steel pipe
step ladders, 443, *443*, *444*
sterilization, 136–137
Stevens Institute, 87, 161, 251
stingers, 541, *541*
storage water heaters, 181–182, *183*, 184
storm drainage, *90*, 90–91
storm water stacks, 281
straight ladders, 442–443, *443*
"S" traps, 405–406, *406*
streams, 134
strength, 524, *526*

stress, 524–525
stringers, 546–547, *547*
studs
  drilling, *289*
  notching, *291*
  reinforcing, *290*
  strike protection, *290*
submerged arc welding, 556
submersible sump pumps, 395, *396*
sump pits, 394
sump pumps
  centrifugal, 395, *395*
  components, 395
  configurations, 395, 397
  connections and controls, 400–401
  defined, 393
  described, 393–395
  electrical safety, 399
  installation, 401
  purpose, 393
  submersible, 395, *396*
sumps, 394, 400–401
sump vent, 393
surge pressure, 16–17. *See also* water hammer
surge tanks, 179
suspended solids, 135, *137*
suspension brackets, *378*
swing-check valve, 353
symbols. *See* mechanical symbols; plumbing symbols
system of measurement, 297
system water pressure range, 141, *141*

T

tack welding, 530, *530*, *531*
  overhead, 533–534
"take out space," 311
tankless water heaters, 182, *183*, 197. *See also* instantaneous water heaters
tank-type water heaters, 182, *183*, 204–205
  ASME standards, 204
  general considerations, 205
tapped tees, 287
tee-wyes, 287
Teflon tape, 20, 32
temperature, plastic piping, 24. *See also* water temperature
temperature and pressure relief valve, 196, *513*
Temperature and Pressure (T&P) control valve, 185, *186*
tempered water, 181
tempering, 527
tempering valves, *187*
tensile weld test, 552
terminal length, 283
terminal velocity, 283
thermal conductivity, *526*
thermal expansion in hydronic heating systems, 211
thermal welding, 72–73
thermocouple, 190, *513*
thermoplastic pipe
  applications, *84*, 85–86
  nonpressure applications, 15
  pressure applications, 16–17
  selection, 84–85
  sizing, 86
thermoplastics, 15–17
thermosets, 9
thermostatic valves, 186
thermostats
  electric water heaters, 195, *196*
  gas water heaters, 190
thimbles, 284–285, *284*
threaded adapters
  CPVC pipe, 50
  PVC pipe, 73
threaded joints, 20
  ABS pipe joining, 32
  CPVC pipe, 49
threaded pipe, support intervals, 283
three-phase wiring, 535
thrust blocks, 247, *247*
TIG welding, 555, *555*
timber bracing, 452
timber hitch, 432, 433, *433*
tin, *523*
tip cleaners, *108*
toilets, macerating, *398*
tolerances, 465
top-mounted sump pumps, 397, *397*
top plates, *289*
torches
  oxy-acetylene, 119, *120*
  welding, 106–108, *107*
torch tips, 106–108, *108*, 118
  oxy-acetylene, 119, *120*
toughness, 524, *526*
toxic substances, 221
tracer wires, 133
transformer-rectifier power supplies, 540
transition joints
  CPVC pipe, 50
  PE pipe, 59
  PVC pipe, 73–74
trap adapters, 32, 74
trap arms, length and slope, 387, *388*
trap dip, 403
trap primer, 411, *412*
traps
  building traps, 407, *407*
  connection permitted to, 412
  defined, 403
  installation, 412
  minimum size of non-integral, *258–259*
  permitted and prohibited, 405–407, *406*
  requirements, 404–405
  trap primer, 411, *412* *See also* P-traps
trap seal
  described, 407
  dimensions, 405
  loss of, 408–411
  specialty, 403
  vent systems, 361–364, *362*, *363*
  water column pressure, 342
travel, 313, *314*, *315*, *316*
tree system, 139
trench box, 452, *453*
trenches
  building sewers, 240
  cave-ins, 451–454
  debris from working machinery, 455
  excavated debris falling into, 455
  hazards associated with, 451
  machinery falling into, 455
  materials and tools falling into, 455
  required length of, *237*
  safety, 450–451, 454, *454*
  shoring, 450–456, *452*, *453*, *454*
  sidewall failure, *454*
trench shield, 452, *453*
triangles
  area, 324, *324*
  equilateral, 313, *314*
  overview, 320, *320*, *321*
  right, 320, *324*
triangular prisms, 326–327, *327*

trunk lines, 138
tub and shower
diverter, 462, *464*, *485*
elevation view, *485*, *479*
mixed water temperature control, 185–187, *186*, *187*
rough-in, *464*, *485*
tubular "S" traps, 406, *406*, 409–410
tungsten inert gas (TIG) welding, 555, *555*
Turbo Torch, 102
turbulence, 352
Type K copper tubing, 33, 132, *168*
Type L copper tubing, 33, 132
sizing based on velocity limitations, *167*
water flow velocity, *166*
Type M copper tubing, 33–34, 132, *167*
typical branches, 286

U

ultimate strength, 524
ultrasonic weld test, 553–554, *554*
ultraviolet (UV) radiation
ABS pipe, 28
CPVC pipe, 45
PE pipe, 54–55
PEX tubing, 61
plastics, 9
PVC pipe, 66–67
underground utilities, 456
under-slab installations, 47
under slab warm air ducts, 97
uphill weave beads, 550
uphill welding, 550–551
urinals, 418–419, *419*
utilities, 456

V

vacuum relief valves, 187
valve allowances, *169*, 170
valves, symbols and abbreviations, 497–498. *See also specific kinds*
Van Stone flanges, 72
variable discharge pumps, 176
velocity
fluids in pipes, 155, 156
terminal, 283
See also water flow velocity
vent hangers, 377–381
venting
stacks, 283
trap arm slope, 387, *388*
vent pipe
hangers, 377–381
overview, 364–365
sizing plastic pipe, 88
vent stacks, 281, 385, *386*
vent systems
defined, 361
engineered, 388–390
exceptions, 371, *373*
functions, 361
hangers, 377–381
installation, 364–365, *365*
piping materials, 364–365
roof terminals, 389–390, *390*
separate vents required, 387–388
sizing, 366–367, *366*, *367*, *368*, *369–370*
soap suds, 388, *389*
tie-ins, *365*
total vent pipe area, 377
trap arm slope, 387, *388*
vent types, 381–387, *381*, *382*, *383*, *384*, *385*, 386, *387*
wet vents, 371
vent terminals, 91–93
vertical drainpipe capacity, *90*
vertical lift hitch, *437*
vertical welding, 531–533, 550–551
vibration noises, 351, *351*, *352*
visual weld tests, 554
volatility, 523, *526*
voltage, 534
volts, 534
volume calculations
cubes, 326, *326*
cylinders, 324, *324*
rectangular prisms, 325, *325*
triangular prisms, 326, *327*
V-shaped weld, 529

W

wall-hung water closet fixtures, 417–418, *418*
warm air duct, under slab, 97
waste stacks, 281
water
formula, 333
gas form, 334–335
liquid form, 334
solid form, 333–334
as a solvent, 335
water closet batteries, *465*
isometric drawing, *494*
plan and elevation views, *489*
water closet carrier, *489*
water closets
combination-type, *507*
elevation view, *480*
floor mounted, 417, *417*, *418*
full section view, *516*
plumbing fixture schedule, *513*
rough-in sheets, *463*, *466*, *468*, *473–474*, 484
tolerances, 472
wall-hung, 417–418, *418*
water compressibility, 335
water distribution piping, 128
water flow
public water distribution systems, 154
water supply fixture units, 159–165
water flow rates, *159*
water flow velocity
sizing based on, *167*
Type L copper tubing, *166*
See also velocity
water fountains, *513*
water hammer, 174, 352–355
causes, 353
corrections, 353–355
defined, 16
plastic piping, 16
shock arrestors, 353–354, *354*, 355–356
water hammer arrestors, 353–354, *354*
air chamber sizing, 356–358, *356*
See also shock arrestors
water heaters
ASME standards, 204
controls and markings, 185
direct, 182–184
general considerations, 205
indirect, *184*, 184–185
industry standards, 182
nameplate information, 197
relief valves, 196
replacement, 187–189

water heaters *(continued)*
  residential sizing, *201*, 201–202
  safety precautions, 189–193
  scaling, 185
  tankless, *183*, 197
  thermostat, 185
  types, 181–182
  vacuum relief valves, 187
  See also individual kinds
water main installations, 139–140
water meters, public water systems, *143*, 143–144
water pipe sizing
  based on velocity limitations, *167*
  overview, 88, *89*
  water supply fixture units, 159–166
water piping
  allowance in equivalent feet for friction loss, *169*, 169–170
  asbestos-cement, 128
  flushing and sterilization, 136–137
  glass, 131
  identification, 227, *227*
  metallic, 132–133
  plastic, 128–131
  residential building layout, 160
  resistance to flow, 155–156
  spacing from sewer piping, *140*
water piping systems
  installation tests, 349–351, *349*, *350*
  noises and noise correction, 351–358
  safety issues when testing, 345
water pressure
  boosting systems, 176–179
  drain lines, 342–343
  due to pumps and elevations, *339*
  versus elevation, *338*
  flow in building systems, 154–155
  flow in public water distribution systems, 154
  fluctuations, 155
  independent of shape of container, *341*, 339–340
  maintaining, 153–154
  overview, 151–152, *152*, 336–339
  Pascal's Law, 339–340
  public systems, 134, 140, 142, *142*
  velocity, friction, and flow, 155–157
water pumps, 154
water service, 128
water service connection, *142*
water source heat pumps, 96
water sources
  community or public, 134
  individual or private, 134–135
water supplies, potable, 335
water supply fixture units (WSFU), 87, 159–166
water supply layouts, *506*
water tanks
  covered, 154
  elevated, 153, *153*
water temperature
  bacterial content, 136
  limit when hot, 136
  mixed control, 185–187, *186*, *187*
  ranges for hot water, *182*
water treatment, 135–137, *137*
water vapor, 342
weave beads, 547, *548*, 550
webbed nylon slings, 430, *430*
weldability, *526*
welding
  filler materials, 110, *111*
  fittings, 77
  fluxes, 109–110
  gauges, 105
  heat treatment of low-carbon steel, 525–527
  history, 101
  laser, 556
  MIG method, 555–556
  overview, 119
  oxy-acetylene flat welding, 122–123
  personal protection, 123
  practice of, 101–102
  resistance welding, 556
  safety and safety equipment, 110, *111*, 112–114
  standards, 102
  submerged arc, 556
  testing welds, 551–554
  TIG method, 555, *555*
  See also gas welding; shielded metal-arc welding
welding arc
  blow, 546
  length, 545
  striking, maintaining, and controlling, 546
welding cables, 540–541
welding cylinders, 104, *104*, 114
welding electrode holders, 541, *541*
welding electrodes, 536–537, 542–544, 546
welding gases, 102–104
welding goggles, 113
welding puddle, 527, *527*, *528*, 537, *537*
welding rods, 110, 536–537
welding sticks, 536–537
wells, 134–135
wet vents, 371, 386, *387*
whipping, 431, *431*
whirlpool bathtub, 467, 469–470
wiping, 133
wire brushes, 545
wire rope slings, 429, *430*
wire welding, 555
working strength, 524
WSFU. *See* water supply fixture units (WSFU)

Y

yield strength, 524
yoke vents, 387, *387*

Z

zinc, *523*